|| || ||| || ||| ||| |||| || |||

W9-BFF-453

Student Solutions Manual for Stewart's

MULTIVARIABLE

CALCULUS

FOURTH EDITION

dan clegg
Palomar College

Barbara Frank
St. Andrews Presbyterian College

Brooks/Cole
Thomson Learning™

Pacific Grove • Albany • Belmont • Boston • Cincinnati • Johannesburg • London • Madrid
Melbourne • Mexico City • New York • Scottsdale • Singapore • Tokyo • Toronto

Assistant Editor: *Carol Ann Benedict*
Marketing Team: *Caroline Croley, Debra Johnston*
Production Coordinator: *Dorothy Bell*

Cover Illustration: *dan clegg*
Typesetting: *Kathi Townes, TECHarts*
Printing and Binding: *Webcom Limited*

COPYRIGHT © 1999 by Brooks/Cole
A division of Thomson Learning
The Thomson Learning logo is a trademark used herein under license.

For more information, contact:
BROOKS/COLE
511 Forest Lodge Road
Pacific Grove, CA 93950 USA
www.brookscole.com

All rights reserved. No part of this work may be reproduced, transcribed or used in any form or by any means—graphic, electronic, or mechanical, including photocopying, recording, taping, Web distribution, or information storage and/or retrieval systems—without the prior written permission of the publisher.

For permission to use material from this work, contact us by
web: www.thomsonrights.com
fax: 1-800-730-2215
phone: 1-800-730-2214

Printed in Canada

10 9 8 7 6 5 4 3 2

ISBN 0-534-35957-4

Preface

This *Student Solutions Manual* contains detailed solutions to selected exercises in the text *Multivariable Calculus, Fourth Edition* (Chapters 11–18 of *Calculus, Fourth Edition* and Chapters 10–17 of *Calculus: Early Transcendentals, Fourth Edition*) by James Stewart. Specifically, it includes solutions to the odd-numbered exercises in each chapter section, review section, True-False Quiz, and Problems Plus section. Also included are all solutions to the Concept Check questions.

The *Early Transcendentals* version of the text uses different chapter and page numbers; consequently, all section numbers and references are given in a dual format. Readers of the *Early Transcendentals* text should use the references denoted by "ET."

Each solution is presented in the context of the corresponding section of the text. In general, solutions to the initial exercises involving a new concept illustrate that concept in more detail; this knowledge is then utilized in subsequent solutions. Thus, while the intermediate steps of a solution are given, you may need to refer back to earlier exercises in the section or prior sections for additional explanation of the concepts involved. Note that, in many cases, different routes to an answer may exist which are equally valid; also, answers can be expressed in different but equivalent forms. Thus, the goal of this manual is not to give the definitive solution to each exercise, but rather to assist you as a student in understanding the concepts of the text and learning how to apply them to the challenge of solving a problem.

We would like to thank James Stewart for entrusting us with the writing of this manual and offering suggestions, Kathi Townes of TECH-arts for typesetting and producing this manual, and Brian Betsill of TECH-arts for creating the illustrations. We also thank Gary W. Ostedt and Carol Ann Benedict of Brooks/Cole Publishing Company for their trust, assistance, and patience.

dan clegg
Palomar College

Barbara Frank
St. Andrews Presbyterian College

Abbreviations and Symbols

CD   concave downward

CU   concave upward

D   the domain of f

FDT   First Derivative Test

HA   horizontal asymptote(s)

I   interval of convergence

IP   inflection point(s)

R   radius of convergence

VA   vertical asymptote(s)

$\overset{H}{=}$ indicates the use of l'Hospital's Rule.

$\overset{j}{=}$ indicates the use of Formula j in the Table of Integrals in the back endpapers.

$\overset{s}{=}$ indicates the use of the substitution $\{u = \sin x, du = \cos x\, dx\}$.

$\overset{c}{=}$ indicates the use of the substitution $\{u = \cos x, du = -\sin x\, dx\}$.

Contents

1. (a) $x = 2t + 4$, $y = t - 1$

t	-3	-2	-1	0	1	2
x	-2	0	2	4	6	8
y	-4	-3	-2	-1	0	1

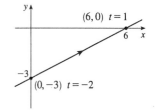

(b) $x = 2t + 4$, $y = t - 1$ $\Rightarrow$ $x = 2(y + 1) + 4 = 2y + 6$ or

$y = \frac{1}{2}x - 3$

3. (a) $x = 1 - 2t$, $y = t^2 + 4$, $0 \le t \le 3$

t	0	1	2	3
x	1	-1	-3	-5
y	4	5	8	13

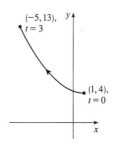

(b) $x = 1 - 2t$ $\Rightarrow$ $2t = 1 - x$ $\Rightarrow$ $t = \dfrac{1 - x}{2}$ $\Rightarrow$

$y = t^2 + 4 = \left(\dfrac{1 - x}{2}\right)^2 + 4 = \frac{1}{4}(x - 1)^2 + 4$ or

$y = \frac{1}{4}x^2 - \frac{1}{2}x + \frac{17}{4}$

5. (a) $x = \sqrt{t}$, $y = 1 - t$

t	0	1	2	3	4
x	0	1	1.414	1.732	2
y	1	0	-1	-2	-3

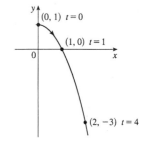

(b) $x = \sqrt{t}$ $\Rightarrow$ $t = x^2$. $y = 1 - t = 1 - x^2$. Since $t \ge 0$, $x \ge 0$.

7. (a) $x = \sin\theta$, $y = \cos\theta$, $0 \le \theta \le \pi$.

$x^2 + y^2 = \sin^2\theta + \cos^2\theta = 1$, $0 \le x \le 1$.

(b)

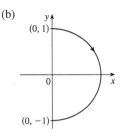

9. (a) $x = \sin^2\theta$, $y = \cos^2\theta$.

$x + y = \sin^2\theta + \cos^2\theta = 1$, $0 \le x \le 1$.

Note that the curve is at $(0, 1)$ whenever

$\theta = \pi n$ and is at $(1, 0)$ whenever $\theta = \frac{\pi}{2}n$

for every integer n.

(b)

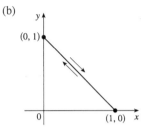

11. (a) $x = e^t$, $y = e^{-t}$, $y = 1/x$, $x > 0$

(b)

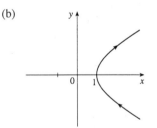

13. (a) $x = \tan\theta + \sec\theta$, $y = \tan\theta - \sec\theta$,

$-\frac{\pi}{2} < \theta < \frac{\pi}{2}$. $xy = \tan^2\theta - \sec^2\theta = -1$ $\Rightarrow$

$y = -1/x$, $x > 0$.

(b)

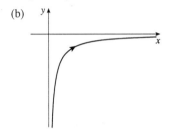

15. (a) $x = \cosh t$, $y = \sinh t$, $x^2 - y^2 = \cosh^2 t - \sinh^2 t = 1$, $x \ge 1$

(b)

17. $x^2 + y^2 = \cos^2\pi t + \sin^2\pi t = 1$, $1 \le t \le 2$, so the particle moves counterclockwise along the circle $x^2 + y^2 = 1$ from $(-1, 0)$ to $(1, 0)$, along the lower half of the circle.

19. $\left(\frac{1}{2}x\right)^2 + \left(\frac{1}{3}y\right)^2 = \sin^2 t + \cos^2 t = 1$, so the particle moves once clockwise along the ellipse $\frac{1}{4}x^2 + \frac{1}{9}y^2 = 1$, starting and ending at $(0, 3)$.

21. $x = \tan t$, $y = \cot t$, $\frac{\pi}{6} \le t \le \frac{\pi}{3}$. $y = 1/x$ for $\frac{1}{\sqrt{3}} \le x \le \sqrt{3}$. The particle moves along the first quadrant branch of the hyperbola $y = 1/x$ from $\left(\frac{1}{\sqrt{3}}, \sqrt{3}\right)$ to $\left(\sqrt{3}, \frac{1}{\sqrt{3}}\right)$.

23. From the graphs, it seems that as $t \to -\infty$, $x \to \infty$ and $y \to -\infty$. So the point $(x(t), y(t))$ will move from far out in the fourth quadrant as t increases. At $t = -\sqrt{3}$, both x and y are 0, so the graph passes through the origin. After that the graph passes through the second quadrant (x is negative, y is positive), then intersects the x-axis at $x = -9$ when $t = 0$. After this, the graph passes through the third quadrant, going through the origin again at $t = \sqrt{3}$, and then as $t \to \infty$, $x \to \infty$ and $y \to \infty$. Note that for every point $(x(t), y(t)) = (3(t^2 - 3), t^3 - 3t)$, we can substitute $-t$ to get the corresponding point $(x(-t), y(-t)) = (3[(-t)^2 - 3], (-t)^3 - 3(-t)) = (x(t), -y(t))$, and so the graph is symmetric about the x-axis. The first figure was obtained using $x_1 = t$, $y_1 = 3(t^2 - 3)$; $x_2 = t$, $y_2 = t^3 - 3t$; and $-2\pi \le t \le 2\pi$.

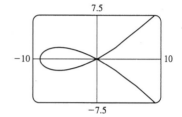

25. As $t \to -\infty$, $x \to \infty$ and $y \to -\infty$. The graph passes through the origin at $t = -1$, and then goes through the second quadrant (x negative, y positive), passing through the point $(-1, 1)$ at $t = 0$. As t increases, the graph passes through the point $(0, 2)$ at $t = 1$, and then as $t \to \infty$, both x and y approach ∞. The first figure was obtained using $x_1 = t$, $y_1 = t^4 - 1$; $x_2 = t$, $y_2 = t^3 + 1$; and $-2\pi \le t \le 2\pi$.

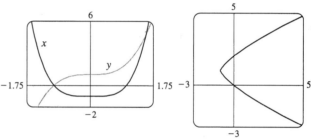

27. As in Example 4, we let $y = t$ and $x = t - 3t^3 + t^5$ and use a t-interval of $[-2\pi, 2\pi]$.

29. The circle $x^2 + y^2 = 4$ can be represented parametrically by $x = 2\cos t$, $y = 2\sin t$; $0 \le t \le 2\pi$. The circle $x^2 + (y - 1)^2 = 4$ can be represented by $x = 2\cos t$, $y = 1 + 2\sin t$; $0 \le t \le 2\pi$. This representation gives us the circle with a counterclockwise orientation starting at $(2, 1)$.

(a) To get a clockwise orientation, we could change the equations to $x = 2\cos t$, $y = 1 - 2\sin t$.

(b) To get three times around in the counterclockwise direction, we use the original equations $x = 2\cos t$, $y = 1 + 2\sin t$ with the domain expanded to $0 \le t \le 6\pi$.

(c) To start at $(0, 3)$ using the original equations, we must have $x_1 = 0$; that is, $2\cos t = 0$. Hence, $t = \frac{\pi}{2}$. So we use $x = 2\cos t$, $y = 1 + 2\sin t$; $\frac{\pi}{2} \le t \le \frac{3\pi}{2}$.

Alternatively, if we want t to start at 0, we could change the equations of the curve. For example, we could use $x = -2\sin t$, $y = 1 + 2\cos t$, $0 \le t \le \pi$.

31. (a) Let $x^2/a^2 = \sin^2 t$ and $y^2/b^2 = \cos^2 t$ to obtain

$x = a \sin t$ and $y = b \cos t$ with $0 \le t \le 2\pi$ as possible

parametric equations for the ellipse

$x^2/a^2 + y^2/b^2 = 1$.

(c) As b increases, the ellipse is stretched vertically.

(b) The equations are $x = 3 \sin t$ and

$y = b \cos t$ for $b \in \{1, 2, 4, 8\}$.

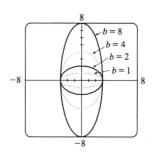

33. The case $\frac{\pi}{2} < \theta < \pi$ is illustrated. C has coordinates $(r\theta, r)$ as before, and Q

has coordinates $(r\theta, r + r \cos(\pi - \theta)) = (r\theta, r(1 - \cos\theta))$ [since

$\cos(\pi - \alpha) = \cos\pi \cos\alpha + \sin\pi \sin\alpha = -\cos\alpha$], so P has coordinates

$(r\theta - r \sin(\pi - \theta), r(1 - \cos\theta)) = (r(\theta - \sin\theta), r(1 - \cos\theta))$ [since

$\sin(\pi - \alpha) = \sin\pi \cos\alpha - \cos\pi \sin\alpha = \sin\alpha$]. Again we have the

parametric equations $x = r(\theta - \sin\theta)$, $y = r(1 - \cos\theta)$.

35. It is apparent that $x = |OQ|$ and $y = |QP| = |ST|$. From the

diagram, $x = |OQ| = a \cos\theta$ and $y = |ST| = b \sin\theta$. Thus, the

parametric equations are $x = a \cos\theta$ and $y = b \sin\theta$. To eliminate θ

we rearrange: $\sin\theta = y/b \;\Rightarrow\; \sin^2\theta = (y/b)^2$ and $\cos\theta = x/a$

$\Rightarrow\; \cos^2\theta = (x/a)^2$. Adding the two equations:

$\sin^2\theta + \cos^2\theta = 1 = x^2/a^2 + y^2/b^2$. Thus, we have an ellipse.

37. $C = (2a \cot\theta, 2a)$, so the x-coordinate of P is $x = 2a \cot\theta$. Let

$B = (0, 2a)$. Then $\angle OAB$ is a right angle and $\angle OBA = \theta$, so

$|OA| = 2a \sin\theta$ and $A = (2a \sin\theta \cos\theta, 2a \sin^2\theta)$. Thus, the

y-coordinate of P is $y = 2a \sin^2\theta$.

39. (a)

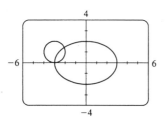

There are 2 points of intersection:

$(-3, 0)$ and approximately $(-2.1, 1.4)$.

(b) As an aid in finding collision points, set your graphing utility

to graph both curves simultaneously and closely observe the

drawing of the graphs. In this case, we have one collision

point: both particles are at $(-3, 0)$ when $t = \frac{3\pi}{2}$. [Notice that

the first curve passes through $(-2.1, 1.4)$ when $t \approx 5.5$, but

the second curve passes through $(-2.1, 1.4)$ when $t \approx 0.4$.]

(c) The circle is centered at $(3, 1)$ instead of $(-3, 1)$. There are

still 2 intersection points: $(3, 0)$ and $(2.1, 1.4)$, but there are

no collision points.

41. $x = t^2, y = t^3 - ct$. We use a graphing device to produce the graphs for various values of c with $-\pi \leq t \leq \pi$. Note that all the members of the family are symmetric about the x-axis. For $c < 0$, the graph does not cross itself, but for $c = 0$ it has a cusp at $(0, 0)$ and for $c > 0$ the graph crosses itself at $x = c$, so the loop grows larger as c increases.

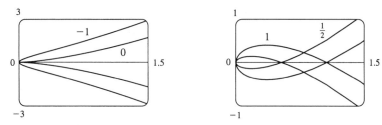

43. Note that all the Lissajous figures are symmetric about the x-axis. The parameters a and b simply stretch the graph in the x- and y-directions respectively. For $a = b = n = 1$ the graph is simply a circle with radius 1. For $n = 2$ the graph crosses itself at the origin and there are loops above and below the x-axis. In general, the figures have $n - 1$ points of intersection, all of which are on the y-axis, and a total of n closed loops.

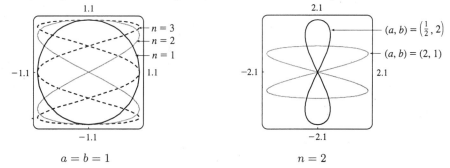

$a = b = 1$

$n = 2$

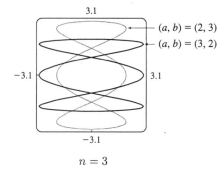

$n = 3$

11.2 Tangents and Areas

ET 10.2

1. $x = t - t^3, y = 2 - 5t \;\Rightarrow\; \dfrac{dy}{dt} = -5, \dfrac{dx}{dt} = 1 - 3t^2$, and $\dfrac{dy}{dx} = \dfrac{dy/dt}{dx/dt} = \dfrac{-5}{1 - 3t^2}$ or $\dfrac{5}{3t^2 - 1}$.

3. $x = t \ln t, y = \sin^2 t \;\Rightarrow\; \dfrac{dy}{dt} = 2 \sin t \cos t, \dfrac{dx}{dt} = t \left(\dfrac{1}{t}\right) + (\ln t) \cdot 1 = 1 + \ln t$, and

$$\frac{dy}{dx} = \frac{dy/dt}{dx/dt} = \frac{2 \sin t \cos t}{1 + \ln t}$$

5. $x = t^2 + t$, $y = t^2 - t$; $t = 0$. $\dfrac{dy}{dt} = 2t - 1$, $\dfrac{dx}{dt} = 2t + 1$, so $\dfrac{dy}{dx} = \dfrac{dy/dt}{dx/dt} = \dfrac{2t - 1}{2t + 1}$. When $t = 0$, $x = y = 0$

and $\dfrac{dy}{dx} = -1$. An equation of the tangent is $y - 0 = (-1)(x - 0)$ or $y = -x$.

7. $x = e^{\sqrt{t}}$, $y = t - \ln t^2$; $t = 1$. $\dfrac{dy}{dt} = 1 - \dfrac{2t}{t^2} = 1 - \dfrac{2}{t}$, $\dfrac{dx}{dt} = \dfrac{e^{\sqrt{t}}}{2\sqrt{t}}$, and

$\dfrac{dy}{dx} = \dfrac{dy/dt}{dx/dt} = \dfrac{1 - 2/t}{e^{\sqrt{t}}/(2\sqrt{t})} \cdot \dfrac{2t}{2t} = \dfrac{2t - 4}{\sqrt{t}e^{\sqrt{t}}}$. When $t = 1$, $(x, y) = (e, 1)$ and $\dfrac{dy}{dx} = -\dfrac{2}{e}$, so an equation of the

tangent line is $y - 1 = -\dfrac{2}{e}(x - e)$ or $y = -\dfrac{2}{e}x + 3$.

9. (a) $x = e^t$, $y = (t - 1)^2$; $(1, 1)$. $\dfrac{dy}{dt} = 2(t - 1)$, $\dfrac{dx}{dt} = e^t$, and $\dfrac{dy}{dx} = \dfrac{dy/dt}{dx/dt} = \dfrac{2(t - 1)}{e^t}$.

At $(1, 1)$, $t = 0$ and $\dfrac{dy}{dx} = -2$, so an equation of the tangent is $y - 1 = -2(x - 1)$ or $y = -2x + 3$.

(b) $x = e^t \Rightarrow t = \ln x$, so $y = (t - 1)^2 = (\ln x - 1)^2$ and $\dfrac{dy}{dx} = 2(\ln x - 1)\left(\dfrac{1}{x}\right)$. When $x = 1$,

$\dfrac{dy}{dx} = 2(-1)(1) = -2$, so an equation of the tangent is $y = -2x + 3$, as in part (a).

11. $x = 2\sin 2t$, $y = 2\sin t$; $(\sqrt{3}, 1)$.

$\dfrac{dy}{dx} = \dfrac{dy/dt}{dx/dt} = \dfrac{2\cos t}{2 \cdot 2\cos 2t} = \dfrac{\cos t}{2\cos 2t}$. The point $(\sqrt{3}, 1)$ corresponds

to $t = \dfrac{\pi}{6}$, so the slope of the tangent at that point is

$\dfrac{\cos \frac{\pi}{6}}{2\cos \frac{\pi}{3}} = \dfrac{\frac{\sqrt{3}}{2}}{2 \cdot \frac{1}{2}} = \dfrac{\sqrt{3}}{2}$. An equation of the tangent is therefore

$(y - 1) = \dfrac{\sqrt{3}}{2}(x - \sqrt{3})$ or $y = \dfrac{\sqrt{3}}{2}x - \dfrac{1}{2}$.

13. $x = t^4 - 1$, $y = t - t^2 \Rightarrow \dfrac{dy}{dt} = 1 - 2t$, $\dfrac{dx}{dt} = 4t^3$, $\dfrac{dy}{dx} = \dfrac{dy/dt}{dx/dt} = \dfrac{1 - 2t}{4t^3} = \dfrac{1}{4}t^{-3} - \dfrac{1}{2}t^{-2}$;

$\dfrac{d}{dt}\left(\dfrac{dy}{dx}\right) = -\dfrac{3}{4}t^{-4} + t^{-3}$, $\dfrac{d^2y}{dx^2} = \dfrac{d(dy/dx)/dt}{dx/dt} = \dfrac{-\frac{3}{4}t^{-4} + t^{-3}}{4t^3} \cdot \dfrac{4t^4}{4t^4} = \dfrac{-3 + 4t}{16t^7}$.

15. $x = \sin \pi t$, $y = \cos \pi t$. $\dfrac{dy}{dx} = \dfrac{dy/dt}{dx/dt} = \dfrac{-\pi \sin \pi t}{\pi \cos \pi t} = -\tan \pi t$;

$\dfrac{d^2y}{dx^2} = \dfrac{d}{dx}\left(\dfrac{dy}{dx}\right) = \dfrac{d(dy/dx)/dt}{dx/dt} = \dfrac{-\pi \sec^2 \pi t}{\pi \cos \pi t} = -\sec^3 \pi t$.

17. $x = e^{-t}$, $y = te^{2t}$. $\dfrac{dy}{dx} = \dfrac{dy/dt}{dx/dt} = \dfrac{(2t + 1)e^{2t}}{-e^{-t}} = -(2t + 1)e^{3t}$;

$\dfrac{d}{dt}\left(\dfrac{dy}{dx}\right) = -3(2t + 1)e^{3t} - 2e^{3t} = -(6t + 5)e^{3t}$;

$\dfrac{d^2y}{dx^2} = \dfrac{d}{dx}\left(\dfrac{dy}{dx}\right) = \dfrac{d(dy/dx)/dt}{dx/dt} = \dfrac{-(6t + 5)e^{3t}}{-e^{-t}} = (6t + 5)e^{4t}$.

19. $x = t\left(t^2 - 3\right) = t^3 - 3t$, $y = 3\left(t^2 - 3\right)$. $\dfrac{dx}{dt} = 3t^2 - 3 = 3\left(t - 1\right)\left(t + 1\right)$; $\dfrac{dy}{dt} = 6t$. $\dfrac{dy}{dt} = 0 \iff t = 0$

$\iff (x, y) = (0, -9)$. $\dfrac{dx}{dt} = 0 \iff t = \pm 1 \iff (x, y) = (-2, -6)$ or $(2, -6)$. So there is a horizontal

tangent at $(0, -9)$ and there are vertical tangents at $(-2, -6)$ and $(2, -6)$.

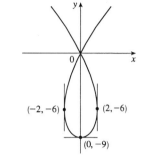

	$t < -1$	$-1 < t < 0$	$0 < t < 1$	$t > 1$
dx/dt	$+$	$-$	$-$	$+$
dy/dt	$-$	$-$	$+$	$+$
x	$\rightarrow$	$\leftarrow$	$\leftarrow$	$\rightarrow$
y	$\downarrow$	$\downarrow$	$\uparrow$	$\uparrow$
curve	$\searrow$	$\swarrow$	$\nwarrow$	$\nearrow$

21. $x = \dfrac{3t}{1 + t^3}$, $y = \dfrac{3t^2}{1 + t^3}$. $\dfrac{dx}{dt} = \dfrac{\left(1 + t^3\right)3 - 3t\left(3t^2\right)}{\left(1 + t^3\right)^2} = \dfrac{3 - 6t^3}{\left(1 + t^3\right)^2}$,

$\dfrac{dy}{dt} = \dfrac{\left(1 + t^3\right)\left(6t\right) - 3t^2\left(3t^2\right)}{\left(1 + t^3\right)^2} = \dfrac{6t - 3t^4}{\left(1 + t^3\right)^2} = \dfrac{3t\left(2 - t^3\right)}{\left(1 + t^3\right)^2}$. $\dfrac{dy}{dt} = 0 \iff t = 0$ or $\sqrt[3]{2} \iff$

$(x, y) = (0, 0)$ or $\left(\sqrt[3]{2}, \sqrt[3]{4}\right)$. $\dfrac{dx}{dt} = 0 \iff t^3 = \tfrac{1}{2} \iff t = 2^{-1/3}$

$\iff (x, y) = \left(\sqrt[3]{4}, \sqrt[3]{2}\right)$. There are horizontal tangents at $(0, 0)$ and

$\left(\sqrt[3]{2}, \sqrt[3]{4}\right)$, and there are vertical tangents at $\left(\sqrt[3]{4}, \sqrt[3]{2}\right)$ and $(0, 0)$. [The

vertical tangent at $(0, 0)$ is undetectable by the methods of this section

because that tangent corresponds to the limiting position of the point (x, y)

as $t \to \pm\infty$.] In the following table, $\alpha = \sqrt[3]{2}$.

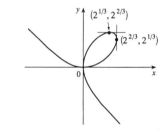

	$t < -1$	$-1 < t < 0$	$0 < t < 1/\alpha$	$1/\alpha < t < \alpha$	$t > \alpha$
dx/dt	$+$	$+$	$+$	$-$	$-$
dy/dt	$-$	$-$	$+$	$+$	$-$
x	$\rightarrow$	$\rightarrow$	$\rightarrow$	$\leftarrow$	$\leftarrow$
y	$\downarrow$	$\downarrow$	$\uparrow$	$\uparrow$	$\downarrow$
curve	$\searrow$	$\searrow$	$\nearrow$	$\nwarrow$	$\swarrow$

23. From the graph, it appears that the leftmost point on the curve $x = t^4 - t^2$,

$y = t + \ln t$ is about $(-0.25, 0.36)$. To find the exact coordinates, we find

the value of t for which the graph has a vertical tangent, that is,

$0 = dx/dt = 4t^3 - 2t \iff 2t\left(2t^2 - 1\right) = 0 \iff$

$2t\left(\sqrt{2}t + 1\right)\left(\sqrt{2}t - 1\right) = 0 \iff t = 0$ or $\pm\dfrac{1}{\sqrt{2}}$. The negative and

0 roots are inadmissible since $y(t)$ is only defined for $t > 0$, so the leftmost point must be

$$\left(x\left(\tfrac{1}{\sqrt{2}}\right), y\left(\tfrac{1}{\sqrt{2}}\right)\right) = \left(\left(\tfrac{1}{\sqrt{2}}\right)^4 - \left(\tfrac{1}{\sqrt{2}}\right)^2, \tfrac{1}{\sqrt{2}} + \ln\tfrac{1}{\sqrt{2}}\right) = \left(-\tfrac{1}{4}, \tfrac{1}{\sqrt{2}} - \tfrac{1}{2}\ln 2\right)$$

25. We graph the curve $x = t^4 - 2t^3 - 2t^2$,

$y = t^3 - t$ in the viewing rectangle
$[-2, 1.1]$ by $[-0.5, 0.5]$. This rectangle
corresponds approximately to
$t \in [-1, 0.8]$. We estimate that the curve
has horizontal tangents at about

$(-1, -0.4)$ and $(-0.17, 0.39)$ and vertical tangents at about $(0, 0)$ and $(-0.19, 0.37)$. We calculate

$\dfrac{dy}{dx} = \dfrac{dy/dt}{dx/dt} = \dfrac{3t^2 - 1}{4t^3 - 6t^2 - 4t}$. The horizontal tangents occur when $dy/dt = 3t^2 - 1 = 0 \Leftrightarrow t = \pm\frac{1}{\sqrt{3}}$, so

both horizontal tangents are shown in our graph. The vertical tangents occur when $dx/dt = 2t\left(2t^2 - 3t - 2\right) = 0$

$\Leftrightarrow 2t\left(2t + 1\right)\left(t - 2\right) = 0 \Leftrightarrow t = 0, -\frac{1}{2}$ or 2. It seems that we have missed one vertical tangent, and indeed
if we plot the curve on the t-interval $[-1.2, 2.2]$ we see that there is another vertical tangent at $(-8, 6)$.

27. $x = \cos t$, $y = \sin t \cos t$. $\frac{dx}{dt} = -\sin t$,

$\frac{dy}{dt} = -\sin^2 t + \cos^2 t = \cos 2t$. $(x, y) = (0, 0) \Leftrightarrow \cos t = 0 \Leftrightarrow$

t is an odd multiple of $\frac{\pi}{2}$. When $t = \frac{\pi}{2}$, $\frac{dx}{dt} = -1$ and $\frac{dy}{dt} = -1$, so

$\frac{dy}{dx} = 1$. When $t = \frac{3\pi}{2}$, $\frac{dx}{dt} = 1$ and $\frac{dy}{dt} = -1$. So $\frac{dy}{dx} = -1$. Thus,

$y = x$ and $y = -x$ are both tangent to the curve at $(0, 0)$.

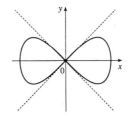

29. (a) $x = r\theta - d\sin\theta$, $y = r - d\cos\theta$; $\dfrac{dx}{d\theta} = r - d\cos\theta$, $\dfrac{dy}{d\theta} = d\sin\theta$. So $\dfrac{dy}{dx} = \dfrac{d\sin\theta}{r - d\cos\theta}$.

(b) If $0 < d < r$, then $|d\cos\theta| \le d < r$, so $r - d\cos\theta \ge r - d > 0$. This shows that $dx/d\theta$ never vanishes, so
the trochoid can have no vertical tangent if $d < r$.

31. The line with parametric equations $x = -7t$, $y = 12t - 5$ is $y = 12\left(-\frac{1}{7}x\right) - 5$, which has slope $-\frac{12}{7}$. The curve

$x = t^3 + 4t$, $y = 6t^2$ has slope $\dfrac{dy}{dx} = \dfrac{dy/dt}{dx/dt} = \dfrac{12t}{3t^2 + 4}$. This equals $-\frac{12}{7} \Leftrightarrow 3t^2 + 4 = -7t \Leftrightarrow$

$(3t + 4)(t + 1) = 0 \Leftrightarrow t = -1$ or $t = -\frac{4}{3} \Leftrightarrow (x, y) = (-5, 6)$ or $\left(-\frac{208}{27}, \frac{32}{3}\right)$.

33. $A = \int_0^1 (y - 1)\, dx = \int_{\pi/2}^0 \left(e^t - 1\right)(-\sin t)\, dt = \int_0^{\pi/2}\left(e^t \sin t - \sin t\right) dt \overset{98}{=} \left[\frac{1}{2}e^t\left(\sin t - \cos t\right) + \cos t\right]_0^{\pi/2}$

$= \frac{1}{2}\left(e^{\pi/2} - 1\right)$

35. By symmetry of the ellipse about the x- and y-axes,

$$A = 4\int_0^a y\, dx = 4\int_{\pi/2}^0 b\sin\theta\left(-a\sin\theta\right) d\theta = 4ab\int_0^{\pi/2}\sin^2\theta\, d\theta = 4ab\int_0^{\pi/2}\frac{1}{2}\left(1 - \cos 2\theta\right) d\theta$$

$$= 2ab\left[\theta - \frac{1}{2}\sin 2\theta\right]_0^{\pi/2} = 2ab\left(\frac{\pi}{2}\right) = \pi ab$$

37. $A = \int_0^{2\pi r} y\, dx = \int_0^{2\pi}\left(r - d\cos\theta\right)\left(r - d\cos\theta\right) d\theta = \int_0^{2\pi}\left(r^2 - 2dr\cos\theta + d^2\cos^2\theta\right) d\theta$

$= \left[r^2\theta - 2dr\sin\theta + \frac{1}{2}d^2\left(\theta + \frac{1}{2}\sin 2\theta\right)\right]_0^{2\pi} = 2\pi r^2 + \pi d^2$

39. The graph of $x = \sin t - 2\cos t$, $y = 1 + \sin t \cos t$ is symmetric
about the y-axis. The graph intersects the y-axis when $x = 0$ $\Rightarrow$
$\sin t - 2\cos t = 0$ $\Rightarrow$ $\sin t = 2\cos t$ $\Rightarrow$ $\tan t = 2$ $\Rightarrow$
$t = \tan^{-1} 2 + n\pi$. The left loop is traced in a clockwise direction
from $t = \tan^{-1} 2 - \pi$ to $t = \tan^{-1} 2$, so the area of the loop is given
(as in Example 4) by

$$A = \int_{\tan^{-1} 2 - \pi}^{\tan^{-1} 2} y \, dx \approx \int_{-2.0344}^{1.1071} (1 + \sin t \cos t)(\cos t + 2\sin t) \, dt \approx 0.8944$$

This integral can be evaluated exactly; its value is $\frac{2}{5}\sqrt{5}$.

41. The coordinates of T are $(r\cos\theta, r\sin\theta)$. Since TP was unwound from
arc TA, TP has length $r\theta$. Also $\angle PTQ = \angle PTR - \angle QTR = \frac{1}{2}\pi - \theta$,
so P has coordinates

$x = r\cos\theta + r\theta\cos\left(\frac{1}{2}\pi - \theta\right) = r(\cos\theta + \theta\sin\theta)$,

$y = r\sin\theta - r\theta\sin\left(\frac{1}{2}\pi - \theta\right) = r(\sin\theta - \theta\cos\theta)$.

11.3 Arc Length and Surface Area   ET 10.3

1. $x = t - t^2$, $y = \frac{4}{3}t^{3/2}$, $1 \le t \le 2$. $dx/dt = 1 - 2t$ and $dy/dt = 2t^{1/2}$, so

$(dx/dt)^2 + (dy/dt)^2 = (1 - 2t)^2 + \left(2t^{1/2}\right)^2 = 1 - 4t + 4t^2 + 4t = 1 + 4t^2$ and

$L = \int_\alpha^\beta \sqrt{(dx/dt)^2 + (dy/dt)^2} \, dt = \int_1^2 \sqrt{1 + 4t^2} \, dt$.

3. $x = t\sin t$, $y = t\cos t$, $0 \le t \le \frac{\pi}{2}$. $dx/dt = t\cos t + \sin t$ and $dy/dt = t(-\sin t) + \cos t$, so

$$(dx/dt)^2 + (dy/dt)^2 = (t\cos t + \sin t)^2 + (\cos t - t\sin t)^2$$
$$= t^2\cos^2 t + 2t\sin t\cos t + \sin^2 t + \cos^2 t - 2t\sin t\cos t + t^2\sin^2 t$$
$$= t^2\left(\cos^2 t + \sin^2 t\right) + \sin^2 t + \cos^2 t = t^2 + 1$$

and $L = \int_0^{\pi/2} \sqrt{t^2 + 1} \, dt$.

5. $x = t^3$, $y = t^2$, $0 \le t \le 4$. $(dx/dt)^2 + (dy/dt)^2 = (3t^2)^2 + (2t)^2 = 9t^4 + 4t^2$.

$L = \int_0^4 \sqrt{(dx/dt)^2 + (dy/dt)^2} \, dt = \int_0^4 \sqrt{9t^4 + 4t^2} \, dt = \int_0^4 t\sqrt{9t^2 + 4} \, dt = \frac{1}{18} \int_4^{148} \sqrt{u} \, du$

(where $u = 9t^2 + 4$). So $L = \frac{1}{18} \left(\frac{2}{3}\right) \left[u^{3/2}\right]_4^{148} = \frac{1}{27} \left(148^{3/2} - 4^{3/2}\right) = \frac{8}{27} \left(37^{3/2} - 1\right)$.

7. $x = \dfrac{t}{1+t}$, $y = \ln(1+t)$, $0 \le t \le 2$. $\dfrac{dx}{dt} = \dfrac{(1+t)\cdot 1 - t\cdot 1}{(1+t)^2} = \dfrac{1}{(1+t)^2}$ and $\dfrac{dy}{dt} = \dfrac{1}{1+t}$, so

$\left(\dfrac{dx}{dt}\right)^2 + \left(\dfrac{dy}{dt}\right)^2 = \dfrac{1}{(1+t)^4} + \dfrac{1}{(1+t)^2} = \dfrac{1}{(1+t)^4}\left[1 + (1+t)^2\right] = \dfrac{t^2 + 2t + 2}{(1+t)^4}$ and

$L = \displaystyle\int_0^2 \dfrac{\sqrt{t^2 + 2t + 2}}{(1+t)^2}\,dt = \int_1^3 \dfrac{\sqrt{u^2 + 1}}{u^2}\,du \;[u = t+1,\; du = dt] \;\overset{24}{=}\; \left[-\dfrac{\sqrt{u^2+1}}{u} + \ln\left(u + \sqrt{u^2+1}\right)\right]_1^3$

$= -\dfrac{\sqrt{10}}{3} + \ln\left(3 + \sqrt{10}\right) + \sqrt{2} - \ln\left(1 + \sqrt{2}\right)$

9. $x = e^t \cos t$, $y = e^t \sin t$, $0 \le t \le \pi$.

$\left(\dfrac{dx}{dt}\right)^2 + \left(\dfrac{dy}{dt}\right)^2 = \left[e^t(\cos t - \sin t)\right]^2 + \left[e^t(\sin t + \cos t)\right]^2$

$= e^{2t}\left(2\cos^2 t + 2\sin^2 t\right) = 2e^{2t}$

$\Rightarrow\quad L = \displaystyle\int_0^\pi \sqrt{2}\,e^t\,dt = \sqrt{2}\,(e^\pi - 1)$

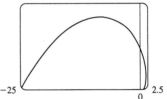

11. $x = e^t - t$, $y = 4e^{t/2}$, $-8 \le t \le 3$.

$(dx/dt)^2 + (dy/dt)^2 = (e^t - 1)^2 + \left(2e^{t/2}\right)^2 = e^{2t} - 2e^t + 1 + 4e^t$

$= e^{2t} + 2e^t + 1 = (e^t + 1)^2$

$L = \displaystyle\int_{-8}^3 \sqrt{(e^t + 1)^2}\,dt = \int_{-8}^3 (e^t + 1)\,dt = \left[e^t + t\right]_{-8}^3$

$= (e^3 + 3) - (e^{-8} - 8) = e^3 - e^{-8} + 11$

13. $x = \ln t$ and $y = e^{-t}$ $\Rightarrow$ $\dfrac{dx}{dt} = \dfrac{1}{t}$ and $\dfrac{dy}{dt} = -e^{-t}$ $\Rightarrow$ $L = \displaystyle\int_1^2 \sqrt{t^{-2} + e^{-2t}}\,dt$. Using Simpson's

Rule with $n = 10$, $\Delta x = (2-1)/10 = 0.1$ and $f(t) = \sqrt{t^{-2} + e^{-2t}}$ we get

$L \approx \dfrac{0.1}{3}\left[f(1.0) + 4f(1.1) + 2f(1.2) + \cdots + 2f(1.8) + 4f(1.9) + f(2.0)\right] \approx 0.7314$.

15. $x = \sin^2 \theta$, $y = \cos^2 \theta$, $0 \le \theta \le 3\pi$.

$(dx/d\theta)^2 + (dy/d\theta)^2 = (2\sin\theta\cos\theta)^2 + (-2\cos\theta\sin\theta)^2 = 8\sin^2\theta\cos^2\theta = 2\sin^2 2\theta$ $\Rightarrow$

$\text{Distance} = \displaystyle\int_0^{3\pi} \sqrt{2}\,|\sin 2\theta|\,d\theta = 6\sqrt{2}\int_0^{\pi/2} \sin 2\theta\,d\theta \text{ (by symmetry)} = \left[-3\sqrt{2}\cos 2\theta\right]_0^{\pi/2}$

$= -3\sqrt{2}\,(-1 - 1) = 6\sqrt{2}$

The full curve is traversed as θ goes from 0 to $\frac{\pi}{2}$, because the curve is the segment of $x + y = 1$ that lies in the first quadrant (since $x, y \ge 0$), and this segment is completely traversed as θ goes from 0 to $\frac{\pi}{2}$.

Thus $L = \displaystyle\int_0^{\pi/2} \sin 2\theta\,d\theta = \sqrt{2}$, as above.

17. $x = a \sin \theta$, $y = b \cos \theta$, $0 \le \theta \le 2\pi$.

$$\left(\tfrac{dx}{d\theta}\right)^2 + \left(\tfrac{dy}{d\theta}\right)^2 = (a \cos \theta)^2 + (-b \sin \theta)^2 = a^2 \cos^2 \theta + b^2 \sin^2 \theta = a^2 \left(1 - \sin^2 \theta\right) + b^2 \sin^2 \theta$$

$$= a^2 - (a^2 - b^2) \sin^2 \theta = a^2 - c^2 \sin^2 \theta = a^2 \left(1 - \frac{c^2}{a^2} \sin^2 \theta\right) = a^2 \left(1 - e^2 \sin^2 \theta\right)$$

So $L = 4 \int_0^{\pi/2} \sqrt{a^2 \left(1 - e^2 \sin^2 \theta\right)} \, d\theta$ (by symmetry) $= 4a \int_0^{\pi/2} \sqrt{1 - e^2 \sin^2 \theta} \, d\theta$

19. (a) Notice that $0 \le t \le 2\pi$ does not give the complete curve
because $x(0) \ne x(2\pi)$. In fact, we must take $t \in [0, 4\pi]$ in
order to obtain the complete curve, since the first term in each
of the parametric equations has period 2π and the second has
period $\frac{4\pi}{11}$, and the least common integer multiple of these
two numbers is 4π.

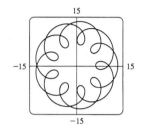

(b) We use the CAS to find the derivatives dx/dt and dy/dt, and then use Theorem 4 to find the arc length. Recent
versions of Maple express the integral $\int_0^{4\pi} \sqrt{(dx/dt)^2 + (dy/dt)^2} \, dt$ as $88E\left(2\sqrt{2}i\right)$, where $E(x)$ is the

elliptic integral $\int_0^1 \frac{\sqrt{1 - x^2 t^2}}{\sqrt{1 - t^2}} \, dt$ and i is the imaginary number $\sqrt{-1}$. Some earlier versions of Maple (as
well as Mathematica) cannot do the integral exactly, so we use the command
`evalf(Int(sqrt(diff(x,t)^2+diff(y,t)^2),t=0..4*Pi));` to estimate the length, and find
that the arc length is approximately 294.03.

21. $x = t^3$ and $y = t^4$ $\Rightarrow$ $dx/dt = 3t^2$ and $dy/dt = 4t^3$. So
$S = \int_0^1 2\pi t^4 \sqrt{9t^4 + 16t^6} \, dt = \int_0^1 2\pi t^6 \sqrt{9 + 16t^2} \, dt$.

23. $x = t^3$, $y = t^2$, $0 \le t \le 1$. $\left(\dfrac{dx}{dt}\right)^2 + \left(\dfrac{dy}{dt}\right)^2 = \left(3t^2\right)^2 + (2t)^2 = 9t^4 + 4t^2$.

$$S = \int_0^1 2\pi y \sqrt{\left(\frac{dx}{dt}\right)^2 + \left(\frac{dy}{dt}\right)^2} \, dt = \int_0^1 2\pi t^2 \sqrt{9t^4 + 4t^2} \, dt$$

$$= 2\pi \int_4^{13} \frac{u - 4}{9} \sqrt{u} \left(\tfrac{1}{18} du\right) \text{ (where } u = 9t^2 + 4) = \tfrac{\pi}{81} \left[\tfrac{2}{5} u^{5/2} - \tfrac{8}{3} u^{3/2}\right]_4^{13} = \tfrac{2\pi}{1215}\left(247\sqrt{13} + 64\right)$$

25. $x = a \cos^3 \theta$, $y = a \sin^3 \theta$, $0 \le \theta \le \tfrac{\pi}{2}$.
$$\left(\frac{dx}{d\theta}\right)^2 + \left(\frac{dy}{d\theta}\right)^2 = \left(-3a \cos^2 \theta \sin \theta\right)^2 + \left(3a \sin^2 \theta \cos \theta\right)^2 = 9a^2 \sin^2 \theta \cos^2 \theta.$$
$S = \int_0^{\pi/2} 2\pi a \sin^3 \theta \, 3a \sin \theta \cos \theta \, d\theta = 6\pi a^2 \int_0^{\pi/2} \sin^4 \theta \cos \theta \, d\theta = \tfrac{6}{5}\pi a^2 \left[\sin^5 \theta\right]_0^{\pi/2} = \tfrac{6}{5}\pi a^2$

27. $x = t + t^3$, $y = t - \dfrac{1}{t^2}$, $1 \le t \le 2$. $\dfrac{dx}{dt} = 1 + 3t^2$ and $\dfrac{dy}{dt} = 1 + \dfrac{2}{t^3}$, so
$$\left(\frac{dx}{dt}\right)^2 + \left(\frac{dy}{dt}\right)^2 = (1 + 3t^2)^2 + \left(1 + \frac{2}{t^3}\right)^2 \text{ and}$$

$$S = \int 2\pi y \, ds = \int_1^2 2\pi \left(t - \frac{1}{t^2}\right) \sqrt{(1 + 3t^2)^2 + \left(1 + \frac{2}{t^3}\right)^2} \, dt \approx 59.101.$$

29. $\left(\dfrac{dx}{dt}\right)^2 + \left(\dfrac{dy}{dt}\right)^2 = (6t)^2 + (6t^2)^2 = 36t^2 \left(1 + t^2\right) \;\Rightarrow$

$S = \int_0^5 2\pi x \sqrt{(dx/dt)^2 + (dy/dt)^2}\, dt = \int_0^5 2\pi \left(3t^2\right) 6t\sqrt{1+t^2}\, dt = 18\pi \int_0^5 t^2 \sqrt{1+t^2}\, 2t\, dt$

$= 18\pi \int_1^{26} (u-1)\sqrt{u}\, du \;\text{(where } u = 1 + t^2) \; = 18\pi \int_1^{26} \left(u^{3/2} - u^{1/2}\right) du = 18\pi \left[\frac{2}{5}u^{5/2} - \frac{2}{3}u^{3/2}\right]_1^{26}$

$= 18\pi \left[\left(\frac{2}{5} \cdot 676\sqrt{26} - \frac{2}{3} \cdot 26\sqrt{26}\right) - \left(\frac{2}{5} - \frac{2}{3}\right)\right] = \frac{24}{5}\pi \left(949\sqrt{26} + 1\right)$

31. $x = a\cos\theta,\; y = b\sin\theta,\; 0 \le \theta \le 2\pi.$

$(dx/d\theta)^2 + (dy/d\theta)^2 = (-a\sin\theta)^2 + (b\cos\theta)^2 = a^2\sin^2\theta + b^2\cos^2\theta = a^2\left(1 - \cos^2\theta\right) + b^2\cos^2\theta$

$$= a^2 - \left(a^2 - b^2\right)\cos^2\theta = a^2 - c^2\cos^2\theta = a^2\left(1 - \frac{c^2}{a^2}\cos^2\theta\right) = a^2\left(1 - e^2\cos^2\theta\right)$$

(a) $S = \int_0^\pi 2\pi b\sin\theta a\sqrt{1 - e^2\cos^2\theta}\, d\theta = 2\pi ab \int_{-e}^{e} \sqrt{1 - u^2}\left(\frac{1}{e}\right) du \;\text{(where } u = -e\cos\theta,\; du = e\sin\theta\, d\theta)$

$= \frac{4\pi ab}{e} \int_0^e \left(1 - u^2\right)^{1/2} du = \frac{4\pi ab}{e} \int_0^{\sin^{-1}e} \cos^2 v\, dv \;\text{(where } u = \sin v) = \frac{2\pi ab}{e} \int_0^{\sin^{-1}e} (1 + \cos 2v)\, dv$

$= \frac{2\pi ab}{e}\left[v + \frac{1}{2}\sin 2v\right]_0^{\sin^{-1}e} = \frac{2\pi ab}{e}\left[v + \sin v \cos v\right]_0^{\sin^{-1}e} = \frac{2\pi ab}{e}\left(\sin^{-1} e + e\sqrt{1 - e^2}\right)$

But $\sqrt{1 - e^2} = \sqrt{1 - \dfrac{c^2}{a^2}} = \sqrt{\dfrac{a^2 - c^2}{a^2}} = \sqrt{\dfrac{b^2}{a^2}} = \dfrac{b}{a}$, so $S = \dfrac{2\pi ab}{e}\sin^{-1} e + 2\pi b^2.$

(b) $S = \int_{-\pi/2}^{\pi/2} 2\pi a\cos\theta a\sqrt{1 - e^2\cos^2\theta}\, d\theta = 4\pi a^2 \int_0^{\pi/2} \cos\theta \sqrt{(1 - e^2) + e^2\sin^2\theta}\, d\theta$

$= \dfrac{4\pi a^2\left(1 - e^2\right)}{e} \int_0^{\pi/2} \dfrac{e}{\sqrt{1 - e^2}}\cos\theta\sqrt{1 + \left(\dfrac{e\sin\theta}{\sqrt{1 - e^2}}\right)^2}\, d\theta$

$= \dfrac{4\pi a^2\left(1 - e^2\right)}{e} \int_0^{e/\sqrt{1 - e^2}} \sqrt{1 + u^2}\, du \;\left(\text{where } u = \dfrac{e\sin\theta}{\sqrt{1 - e^2}}\right)$

$= \dfrac{4\pi a^2\left(1 - e^2\right)}{e} \int_0^{\sin^{-1}e} \sec^3 v\, dv \;\;\text{(where } u = \tan v,\; du = \sec^2 v\, dv)$

$= \dfrac{2\pi a^2\left(1 - e^2\right)}{e}\left[\sec v \tan v + \ln|\sec v + \tan v|\right]_0^{\sin^{-1}e}$

$= \dfrac{2\pi a^2\left(1 - e^2\right)}{e}\left[\dfrac{1}{\sqrt{1 - e^2}}\dfrac{e}{\sqrt{1 - e^2}} + \ln\left|\dfrac{1}{\sqrt{1 - e^2}} + \dfrac{e}{\sqrt{1 - e^2}}\right|\right]$

$= 2\pi a^2 + \dfrac{2\pi a^2\left(1 - e^2\right)}{e} \ln\sqrt{\dfrac{1 + e}{1 - e}} = 2\pi a^2 + \dfrac{2\pi b^2}{e}\dfrac{1}{2}\ln\left(\dfrac{1 + e}{1 - e}\right) \;\left(\text{since } 1 - e^2 = \dfrac{b^2}{a^2}\right)$

$= 2\pi\left[a^2 + \dfrac{b^2}{2e}\ln\dfrac{1 + e}{1 - e}\right]$

33. (a) $\phi = \tan^{-1}\left(\dfrac{dy}{dx}\right)$ $\Rightarrow$ $\dfrac{d\phi}{dt} = \dfrac{d}{dt}\tan^{-1}\left(\dfrac{dy}{dx}\right) = \dfrac{1}{1+(dy/dx)^2}\left[\dfrac{d}{dt}\left(\dfrac{dy}{dx}\right)\right]$. But $\dfrac{dy}{dx} = \dfrac{dy/dt}{dx/dt} = \dfrac{\dot{y}}{\dot{x}}$

$\Rightarrow$ $\dfrac{d}{dt}\left(\dfrac{dy}{dx}\right) = \dfrac{d}{dt}\left(\dfrac{\dot{y}}{\dot{x}}\right) = \dfrac{\ddot{y}\dot{x} - \ddot{x}\dot{y}}{\dot{x}^2}$ $\Rightarrow$ $\dfrac{d\phi}{dt} = \dfrac{1}{1+(\dot{y}/\dot{x})^2}\left(\dfrac{\ddot{y}\dot{x} - \ddot{x}\dot{y}}{\dot{x}^2}\right) = \dfrac{\dot{x}\ddot{y} - \ddot{x}\dot{y}}{\dot{x}^2 + \dot{y}^2}.$

Using the Chain Rule, and the fact that $s = \displaystyle\int_0^t \sqrt{\left(\dfrac{dx}{dt}\right)^2 + \left(\dfrac{dy}{dt}\right)^2}\,dt$ $\Rightarrow$

$\dfrac{ds}{dt} = \sqrt{\left(\dfrac{dx}{dt}\right)^2 + \left(\dfrac{dy}{dt}\right)^2} = (\dot{x}^2 + \dot{y}^2)^{1/2}$, we have that

$\dfrac{d\phi}{ds} = \dfrac{d\phi/dt}{ds/dt} = \left(\dfrac{\dot{x}\ddot{y} - \ddot{x}\dot{y}}{\dot{x}^2 + \dot{y}^2}\right)\dfrac{1}{(\dot{x}^2 + \dot{y}^2)^{1/2}} = \dfrac{\dot{x}\ddot{y} - \ddot{x}\dot{y}}{(\dot{x}^2 + \dot{y}^2)^{3/2}}.$ So

$\kappa = \left|\dfrac{d\phi}{ds}\right| = \left|\dfrac{\dot{x}\ddot{y} - \ddot{x}\dot{y}}{(\dot{x}^2 + \dot{y}^2)^{3/2}}\right| = \dfrac{|\dot{x}\ddot{y} - \ddot{x}\dot{y}|}{(\dot{x}^2 + \dot{y}^2)^{3/2}}.$

(b) $x = x$ and $y = f(x)$ $\Rightarrow$ $\dot{x} = 1$, $\ddot{x} = 0$ and $\dot{y} = \dfrac{dy}{dx}$, $\ddot{y} = \dfrac{d^2y}{dx^2}$.

So $\kappa = \dfrac{|1\cdot(d^2y/dx^2) - 0\cdot(dy/dx)|}{[1+(dy/dx)^2]^{3/2}} = \dfrac{|d^2y/dx^2|}{[1+(dy/dx)^2]^{3/2}}.$

35. $x = \theta - \sin\theta$ $\Rightarrow$ $\dot{x} = 1 - \cos\theta$ $\Rightarrow$ $\ddot{x} = \sin\theta$, and $y = 1 - \cos\theta$ $\Rightarrow$ $\dot{y} = \sin\theta$ $\Rightarrow$ $\ddot{y} = \cos\theta$.

Therefore, $\kappa = \dfrac{|\cos\theta - \cos^2\theta - \sin^2\theta|}{[(1-\cos\theta)^2 + \sin^2\theta]^{3/2}} = \dfrac{|\cos\theta - (\cos^2\theta + \sin^2\theta)|}{(1 - 2\cos\theta + \cos^2\theta + \sin^2\theta)^{3/2}} = \dfrac{|\cos\theta - 1|}{(2 - 2\cos\theta)^{3/2}}.$ The top

of the arch is characterized by a horizontal tangent, and from Example 1 in Section 11.2 [ET 10.2], the tangent is

horizontal when $\theta = (2n-1)\pi$, so take $n = 1$ and substitute $\theta = \pi$ into the expression for κ:

$\kappa = \dfrac{|\cos\pi - 1|}{(2 - 2\cos\pi)^{3/2}} = \dfrac{|-1 - 1|}{[2 - 2(-1)]^{3/2}} = \dfrac{1}{4}.$

11.4   Polar Coordinates   ET 10.4

1. (a) By adding 2π to $\frac{\pi}{2}$, we obtain the   (b) $\left(-2, \frac{\pi}{4}\right)$ (c) $(3, 2)$

point $\left(1, \frac{5\pi}{2}\right)$. The direction

opposite $\frac{\pi}{2}$ is $\frac{3\pi}{2}$, so $\left(-1, \frac{3\pi}{2}\right)$ is a

point that satisfies the $r < 0$

requirement.

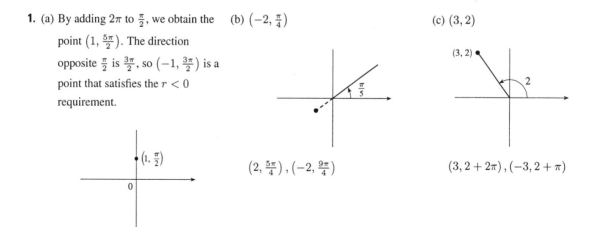

$\left(2, \frac{5\pi}{4}\right), \left(-2, \frac{9\pi}{4}\right)$ $(3, 2 + 2\pi), (-3, 2 + \pi)$

3. (a)

(b)

(c)

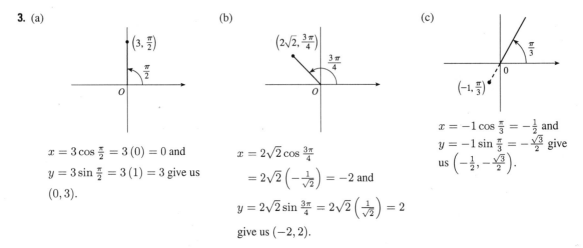

$x = 3\cos\frac{\pi}{2} = 3(0) = 0$ and
$y = 3\sin\frac{\pi}{2} = 3(1) = 3$ give us
$(0, 3)$.

$x = 2\sqrt{2}\cos\frac{3\pi}{4}$
$= 2\sqrt{2}\left(-\frac{1}{\sqrt{2}}\right) = -2$ and
$y = 2\sqrt{2}\sin\frac{3\pi}{4} = 2\sqrt{2}\left(\frac{1}{\sqrt{2}}\right) = 2$
give us $(-2, 2)$.

$x = -1\cos\frac{\pi}{3} = -\frac{1}{2}$ and
$y = -1\sin\frac{\pi}{3} = -\frac{\sqrt{3}}{2}$ give
us $\left(-\frac{1}{2}, -\frac{\sqrt{3}}{2}\right)$.

5. (a) $x = 1$ and $y = 1 \Rightarrow r = \sqrt{1^2 + 1^2} = \sqrt{2}$ and $\theta = \tan^{-1}\left(\frac{1}{1}\right) = \frac{\pi}{4}$. Since $(1, 1)$ is in the first quadrant, the polar coordinates are (i) $\left(\sqrt{2}, \frac{\pi}{4}\right)$ and (ii) $\left(-\sqrt{2}, \frac{5\pi}{4}\right)$.

(b) $x = 2\sqrt{3}$ and $y = -2 \Rightarrow r = \sqrt{\left(2\sqrt{3}\right)^2 + (-2)^2} = 4$ and $\theta = \tan^{-1}\left(-\frac{2}{2\sqrt{3}}\right) = -\frac{\pi}{6}$. Since $\left(2\sqrt{3}, -2\right)$ is in the fourth quadrant and $0 \le \theta \le 2\pi$, the polar coordinates are (i) $\left(4, \frac{11\pi}{6}\right)$ and (ii) $\left(-4, \frac{5\pi}{6}\right)$.

7. $r > 1$ **9.** $0 \le r \le 2,\ \frac{\pi}{2} \le \theta \le \pi$ **11.** $2 < r < 3,\ \frac{5\pi}{3} \le \theta \le \frac{7\pi}{3}$

13. $\left(1, \frac{\pi}{6}\right)$ is $\left(\frac{\sqrt{3}}{2}, \frac{1}{2}\right)$ Cartesian and $\left(3, \frac{3\pi}{4}\right)$ is $\left(-\frac{3}{\sqrt{2}}, \frac{3}{\sqrt{2}}\right)$ Cartesian. The square of the distance between them is $\left(\frac{\sqrt{3}}{2} + \frac{3}{\sqrt{2}}\right)^2 + \left(\frac{1}{2} - \frac{3}{\sqrt{2}}\right)^2 = \frac{1}{4}\left(40 + 6\sqrt{6} - 6\sqrt{2}\right)$, so the distance is $\frac{1}{2}\sqrt{40 + 6\sqrt{6} - 6\sqrt{2}}$.

15. $r = 2 \Leftrightarrow \sqrt{x^2 + y^2} = 2 \Leftrightarrow x^2 + y^2 = 4$, a circle of radius 2 centered at the origin.

17. $r = 3\sin\theta \Rightarrow r^2 = 3r\sin\theta \Leftrightarrow x^2 + y^2 = 3y \Leftrightarrow x^2 + \left(y - \frac{3}{2}\right)^2 = \left(\frac{3}{2}\right)^2$, a circle of radius $\frac{3}{2}$ centered at $\left(0, \frac{3}{2}\right)$. The first two equations are actually equivalent since $r^2 = 3r\sin\theta \Rightarrow r(r - 3\sin\theta) = 0 \Rightarrow r = 0$ or $r = 3\sin\theta$. But $r = 3\sin\theta$ gives the point $r = 0$ (the pole) when $\theta = 0$. Thus, the single equation $r = 3\sin\theta$ is equivalent to the compound condition ($r = 0$ or $r = 3\sin\theta$).

19. $r^2 = \sin 2\theta = 2\sin\theta\cos\theta \Leftrightarrow r^4 = 2r\sin\theta\, r\cos\theta \Leftrightarrow \left(x^2 + y^2\right)^2 = 2yx$

21. $y = 5 \Leftrightarrow r\sin\theta = 5$

23. $x^2 + y^2 = 25 \Leftrightarrow r^2 = 25 \Rightarrow r = 5$

25. $2xy = 1$ $\Leftrightarrow$ $2r\cos\theta\, r\sin\theta = 1$ $\Leftrightarrow$ $r^2\sin 2\theta = 1$ $\Leftrightarrow$ $r^2 = \csc 2\theta$

27. (a) The description leads immediately to the polar equation $\theta = \frac{\pi}{6}$, and the Cartesian equation
$y = \tan\left(\frac{\pi}{6}\right) x = \frac{1}{\sqrt{3}}x$ is slightly more difficult to derive.

(b) The easier description here is the Cartesian equation $x = 3$.

29. $r = -2\sin\theta$ $\Leftrightarrow$ $r^2 = -2r\sin\theta$ (since the possibility $r = 0$ is covered by the equation
$r = -2\sin\theta$) $\Leftrightarrow$ $x^2 + y^2 = -2y$ $\Leftrightarrow$
$x^2 + y^2 + 2y + 1 = 1$ $\Leftrightarrow$ $x^2 + (y+1)^2 = 1$.

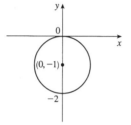

31. $r = \csc\theta = \dfrac{1}{\sin\theta}$ $\Leftrightarrow$ $r\sin\theta = 1$. (The right-hand equation implies that $\sin\theta \neq 0$, so we can divide by $\sin\theta$ to get the left-hand equation)
$\Leftrightarrow$ $y = 1$.

33. As in Example 4, $r = 5$ represents the circle with center O and radius 5.

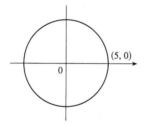

35. $r = \sin\theta$ $\Leftrightarrow$ $r^2 = r\sin\theta$ $\Leftrightarrow$ $x^2 + y^2 = y$
$\Leftrightarrow$ $x^2 + \left(y - \frac{1}{2}\right)^2 = \left(\frac{1}{2}\right)^2$. The reasoning here is the same as in Exercise 29. This is a circle of radius $\frac{1}{2}$ centered at $\left(0, \frac{1}{2}\right)$.

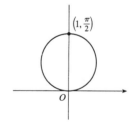

37. $r = 2\left(1 - \sin\theta\right)$. This curve is a cardioid.

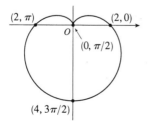

39. $r = \theta, \theta \geq 0$

41. $r = 1/\theta$

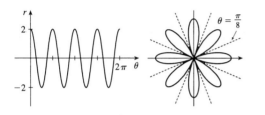

43. $r = \sin 2\theta$

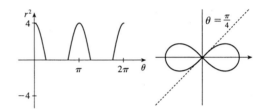

45. $r = 2\cos 4\theta$

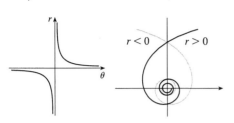

47. $r^2 = 4\cos 2\theta$

49. $r = 2\cos\left(\frac{3}{2}\theta\right)$

51. $x = (r)\cos\theta = (4 + 2\sec\theta)\cos\theta = 4\cos\theta + 2$. Now, $r \to \infty$ ⇒

$(4 + 2\sec\theta) \to \infty$ ⇒ $\theta \to \left(\frac{\pi}{2}\right)^-$ or $\theta \to \left(\frac{3\pi}{2}\right)^+$ (since we need only

consider $0 \le \theta < 2\pi$), so $\lim\limits_{r\to\infty} x = \lim\limits_{\theta\to\pi/2^-} (4\cos\theta + 2) = 2$. Also,

$r \to -\infty$ ⇒ $(4 + 2\sec\theta) \to -\infty$ ⇒ $\theta \to \left(\frac{\pi}{2}\right)^+$ or $\theta \to \left(\frac{3\pi}{2}\right)^-$, so

$\lim\limits_{r\to-\infty} x = \lim\limits_{\theta\to\pi/2^+} (4\cos\theta + 2) = 2$. Therefore, $\lim\limits_{r\to\pm\infty} x = 2$ ⇒ $x = 2$

is a vertical asymptote.

53. To show that $x = 1$ is an asymptote we must prove $\lim\limits_{r\to\pm\infty} x = 1$.

$x = (r)\cos\theta = (\sin\theta\tan\theta)\cos\theta = \sin^2\theta$. Now, $r \to \infty$ ⇒

$\sin\theta\tan\theta \to \infty$ ⇒ $\theta \to \left(\frac{\pi}{2}\right)^-$, so $\lim\limits_{r\to\infty} x = \lim\limits_{\theta\to\pi/2^-} \sin^2\theta = 1$. Also,

$r \to -\infty$ ⇒ $\sin\theta\tan\theta \to -\infty$ ⇒ $\theta \to \left(\frac{\pi}{2}\right)^+$, so

$\lim\limits_{r\to-\infty} x = \lim\limits_{\theta\to\pi/2^+} \sin^2\theta = 1$.

Therefore, $\lim\limits_{r\to\pm\infty} x = 1$ ⇒ $x = 1$ is a vertical asymptote. Also notice that $x = \sin^2\theta \ge 0$ for all θ, and

$x = \sin^2\theta \le 1$ for all θ. And $x \ne 1$, since the curve is not defined at odd multiples of $\frac{\pi}{2}$. Therefore, the curve lies

entirely within the vertical strip $0 \le x < 1$.

55. (a) We see that the curve crosses itself at the origin, where $r = 0$ (in fact the inner loop corresponds to negative r-values,) so we solve the equation of the limaçon for $r = 0$ $\Leftrightarrow$ $c\sin\theta = -1$ $\Leftrightarrow$ $\sin\theta = -1/c$. Now if $|c| < 1$, then this equation has no solution and hence there is no inner loop. But if $c < -1$, then on the interval $(0, 2\pi)$ the equation has the two solutions $\theta = \sin^{-1}(-1/c)$ and $\theta = \pi - \sin^{-1}(-1/c)$, and if $c > 1$, the solutions are $\theta = \pi + \sin^{-1}(1/c)$ and $\theta = 2\pi - \sin^{-1}(1/c)$. In each case, $r < 0$ for θ between the two solutions, indicating a loop.

(b) For $0 < c < 1$, the dimple (if it exists) is characterized by the fact that y has a local maximum at $\theta = \frac{3\pi}{2}$. So we determine for what c-values $\dfrac{d^2 y}{d\theta^2}$ is negative at $\theta = \frac{3\pi}{2}$, since by the Second Derivative Test this indicates a maximum: $y = r\sin\theta = \sin\theta + c\sin^2\theta$ $\Rightarrow$ $\dfrac{dy}{d\theta} = \cos\theta + 2c\sin\theta\cos\theta = \cos\theta + c\sin 2\theta$ $\Rightarrow$ $\dfrac{d^2 y}{d\theta^2} = -\sin\theta + 2c\cos 2\theta$. At $\theta = \frac{3\pi}{2}$, this is equal to $-(-1) + 2c(-1) = 1 - 2c$, which is negative only for $c > \frac{1}{2}$. A similar argument shows that for $-1 < c < 0$, y only has a local minimum at $\theta = \frac{\pi}{2}$ (indicating a dimple) for $c < -\frac{1}{2}$.

57. Using Equation 3 with $r = 3\cos\theta$, we have

$$\frac{dy}{dx} = \frac{dy/d\theta}{dx/d\theta} = \frac{(dr/d\theta)(\sin\theta) + r\cos\theta}{(dr/d\theta)(\cos\theta) - r\sin\theta} = \frac{-3\sin\theta\sin\theta + 3\cos\theta\cos\theta}{-3\sin\theta\cos\theta - 3\cos\theta\sin\theta} = \frac{3(\cos^2\theta - \sin^2\theta)}{-3(2\sin\theta\cos\theta)}$$

$$= -\frac{\cos 2\theta}{\sin 2\theta} = -\cot 2\theta = \frac{1}{\sqrt{3}} \text{ when } \theta = \frac{\pi}{3}$$

Another Solution: $r = 3\cos\theta$ $\Rightarrow$ $x = r\cos\theta = 3\cos^2\theta$, $y = r\sin\theta = 3\sin\theta\cos\theta$ $\Rightarrow$

$$\frac{dy}{dx} = \frac{dy/d\theta}{dx/d\theta} = \frac{-3\sin^2\theta + 3\cos^2\theta}{-6\cos\theta\sin\theta} = \frac{\cos 2\theta}{-\sin 2\theta} = -\cot 2\theta = \frac{1}{\sqrt{3}} \text{ when } \theta = \frac{\pi}{3}$$

59. $r = 1/\theta$ $\Rightarrow$ $x = r\cos\theta = (\cos\theta)/\theta$, $y = r\sin\theta = (\sin\theta)/\theta$ $\Rightarrow$

$$\frac{dy}{dx} = \frac{dy/d\theta}{dx/d\theta} = \frac{\sin\theta(-1/\theta^2) + (1/\theta)\cos\theta}{\cos\theta(-1/\theta^2) - (1/\theta)\sin\theta} \cdot \frac{\theta^2}{\theta^2} = \frac{-\sin\theta + \theta\cos\theta}{-\cos\theta - \theta\sin\theta} = -\pi \text{ when } \theta = \pi$$

61. $r = 1 + \cos\theta$ $\Rightarrow$ $x = r\cos\theta = \cos\theta + \cos^2\theta$, $y = r\sin\theta = \sin\theta + \sin\theta\cos\theta$ $\Rightarrow$

$$\frac{dy}{dx} = \frac{dy/d\theta}{dx/d\theta} = \frac{\cos\theta + \cos^2\theta - \sin^2\theta}{-\sin\theta - 2\cos\theta\sin\theta} = \frac{\cos\theta + \cos 2\theta}{-\sin\theta - \sin 2\theta} = -1 \text{ when } \theta = \frac{\pi}{6}$$

63. $r = 3\cos\theta$ $\Rightarrow$ $x = r\cos\theta = 3\cos\theta\cos\theta$, $y = r\sin\theta = 3\cos\theta\sin\theta$ $\Rightarrow$ $dy/d\theta = -3\sin^2\theta + 3\cos^2\theta = 3\cos 2\theta = 0$ $\Rightarrow$ $2\theta = \frac{\pi}{2}$ or $\frac{3\pi}{2}$ $\Leftrightarrow$ $\theta = \frac{\pi}{4}$ or $\frac{3\pi}{4}$. So the tangent is horizontal at $\left(\frac{3}{\sqrt{2}}, \frac{\pi}{4}\right)$ and $\left(-\frac{3}{\sqrt{2}}, \frac{3\pi}{4}\right)$ $\left[\text{same as } \left(\frac{3}{\sqrt{2}}, -\frac{\pi}{4}\right)\right]$. $dx/d\theta = -6\sin\theta\cos\theta = -3\sin 2\theta = 0$ $\Rightarrow$ $2\theta = 0$ or π $\Leftrightarrow$ $\theta = 0$ or $\frac{\pi}{2}$. So the tangent is vertical at $(3, 0)$ and $\left(0, \frac{\pi}{2}\right)$.

65. $r = 1 + \cos\theta \Rightarrow x = r\cos\theta = \cos\theta\,(1 + \cos\theta),\; y = r\sin\theta = \sin\theta\,(1 + \cos\theta) \Rightarrow$

$dy/d\theta = (1 + \cos\theta)\cos\theta - \sin^2\theta = 2\cos^2\theta + \cos\theta - 1 = (2\cos\theta - 1)(\cos\theta + 1) = 0 \Rightarrow \cos\theta = \frac{1}{2}$ or

$-1 \Rightarrow \theta = \frac{\pi}{3}, \pi,$ or $\frac{5\pi}{3} \Rightarrow$ horizontal tangent at $\left(\frac{3}{2}, \frac{\pi}{3}\right),\, (0, \pi),$ and $\left(\frac{3}{2}, \frac{5\pi}{3}\right).$

$dx/d\theta = -(1 + \cos\theta)\sin\theta - \cos\theta\sin\theta = -\sin\theta\,(1 + 2\cos\theta) = 0 \Rightarrow \sin\theta = 0$ or $\cos\theta = -\frac{1}{2} \Rightarrow$

$\theta = 0, \pi, \frac{2\pi}{3},$ or $\frac{4\pi}{3} \Rightarrow$ vertical tangent at $(2, 0),\, \left(\frac{1}{2}, \frac{2\pi}{3}\right),$ and $\left(\frac{1}{2}, \frac{4\pi}{3}\right).$ Note that the tangent is horizontal, not

vertical when $\theta = \pi,$ since $\displaystyle\lim_{\theta \to \pi} \frac{dy/d\theta}{dx/d\theta} = 0.$

67. $r = \cos 2\theta \Rightarrow x = r\cos\theta = \cos 2\theta \cos\theta,\; y = r\sin\theta = \cos 2\theta \sin\theta \Rightarrow$

$$dy/d\theta = -2\sin 2\theta \sin\theta + \cos 2\theta \cos\theta = -4\sin^2\theta \cos\theta + \left(\cos^3\theta - \sin^2\theta \cos\theta\right)$$

$$= \cos\theta\,(\cos^2\theta - 5\sin^2\theta) = \cos\theta\,(1 - 6\sin^2\theta) = 0 \Rightarrow$$

$\cos\theta = 0$ or $\sin\theta = \pm\frac{1}{\sqrt{6}} \Rightarrow \theta = \frac{\pi}{2}, \frac{3\pi}{2}, \alpha, \pi - \alpha, \pi + \alpha,$ or $2\pi - \alpha$ (where $\alpha = \sin^{-1}\frac{1}{\sqrt{6}}$).

So the tangent is horizontal at $\left(-1, \frac{\pi}{2}\right), \left(-1, \frac{3\pi}{2}\right), \left(\frac{2}{3}, \alpha\right), \left(\frac{2}{3}, \pi - \alpha\right), \left(\frac{2}{3}, \pi + \alpha\right),$ and $\left(\frac{2}{3}, 2\pi - \alpha\right).$

$$dx/d\theta = -2\sin 2\theta \cos\theta - \cos 2\theta \sin\theta = -4\sin\theta \cos^2\theta - (2\cos^2\theta - 1)\sin\theta$$

$$= \sin\theta\,(1 - 6\cos^2\theta) = 0 \Rightarrow$$

$\sin\theta = 0$ or $\cos\theta = \pm\frac{1}{\sqrt{6}} \Rightarrow \theta = 0, \pi, \frac{\pi}{2} - \alpha, \frac{\pi}{2} + \alpha, \frac{3\pi}{2} - \alpha,$ or $\frac{3\pi}{2} + \alpha$ (where $\alpha = \cos^{-1}\frac{1}{\sqrt{6}}$).

So the tangent is vertical at $(1, 0),\, (1, \pi),\, \left(\frac{2}{3}, \frac{3\pi}{2} - \alpha\right), \left(\frac{2}{3}, \frac{3\pi}{2} + \alpha\right),\left(\frac{2}{3}, \frac{\pi}{2} - \alpha\right),$ and $\left(\frac{2}{3}, \frac{\pi}{2} + \alpha\right).$

69. $r = a\sin\theta + b\cos\theta \Rightarrow r^2 = ar\sin\theta + br\cos\theta \Rightarrow x^2 + y^2 = ay + bx \Rightarrow$

$\left(x - \frac{1}{2}b\right)^2 + \left(y - \frac{1}{2}a\right)^2 = \frac{1}{4}\left(a^2 + b^2\right),$ and this is a circle with center $\left(\frac{1}{2}b, \frac{1}{2}a\right)$ and radius $\frac{1}{2}\sqrt{a^2 + b^2}.$

Note for Exercises 71–75: Maple is able to plot polar curves using the `polarplot` command, or using the `coords=polar` option in a regular `plot` command. In Mathematica, use `PolarPlot`. In Derive, change to `Polar` under `Options State`. If your graphing device cannot plot polar equations, you must convert to parametric equations. For example, in Exercise 71,

$$x = r\cos\theta = [1 + 2\sin(\theta/2)]\cos\theta \qquad y = r\sin\theta = [1 + 2\sin(\theta/2)]\sin\theta$$

71. $r = 1 + 2\sin(\theta/2).$ The parameter interval is $[0, 4\pi].$

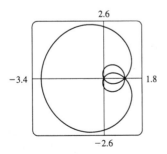

73. $r = e^{\sin\theta} - 2\cos(4\theta).$ The parameter interval is $[0, 2\pi].$

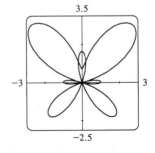

75. $r = \sin(9\theta/4)$. The parameter interval is $[0, 8\pi]$.

77.

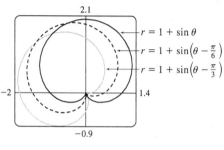

It appears that the graph of $r = 1 + \sin\left(\theta - \frac{\pi}{6}\right)$ is the same shape as the graph of $r = 1 + \sin\theta$, but rotated counterclockwise about the origin by $\frac{\pi}{6}$. Similarly, the graph of $r = 1 + \sin\left(\theta - \frac{\pi}{3}\right)$ is rotated by $\frac{\pi}{3}$. In general, the graph of $r = f(\theta - \alpha)$ is the same shape as that of $r = f(\theta)$, but rotated counterclockwise through α about the origin. That is, for any point (r_0, θ_0) on the curve $r = f(\theta)$, the point $(r_0, \theta_0 + \alpha)$ is on the curve $r = f(\theta - \alpha)$, since $r_0 = f(\theta_0) = f((\theta_0 + \alpha) - \alpha)$.

79. (a) $r = \sin n\theta$. From the graphs, it seems that when n is even, the number of loops in the curve (called a rose) is $2n$, and when n is odd, the number of loops is simply n.

This is because in the case of n odd, every point on the graph is traversed twice, due to the fact that

$$r(\theta + \pi) = \sin[n(\theta + \pi)] = \sin n\theta \cos n\pi + \cos n\theta \sin n\pi = \begin{cases} \sin n\theta & \text{if } n \text{ is even} \\ -\sin n\theta & \text{if } n \text{ is odd} \end{cases}$$

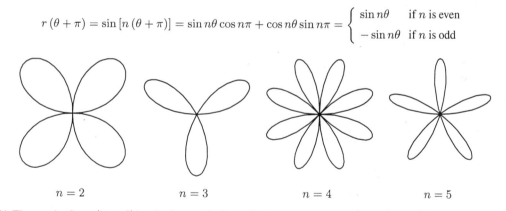

(b) The graph of $r = |\sin n\theta|$ has $2n$ loops whether n is odd or even, since $r(\theta + \pi) = r(\theta)$.

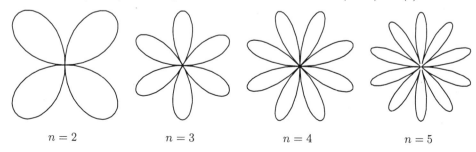

81. $r = \dfrac{1 - a\cos\theta}{1 + a\cos\theta}$. We start with $a = 0$, since in this case the curve is simply the circle $r = 1$.

As a increases, the graph moves to the left, and its right side becomes flattened. As a increases through about 0.4, the right side seems to grow a dimple, which upon closer investigation (with narrower θ-ranges) seems to appear at $a \approx 0.42$ (the actual value is $\sqrt{2} - 1$). As $a \to 1$, this dimple becomes more pronounced, and the curve begins to stretch out horizontally, until at $a = 1$ the denominator vanishes at $\theta = \pi$, and the dimple becomes an actual cusp. For $a > 1$ we must choose our parameter interval carefully, since $r \to \infty$ as $1 + a\cos\theta \to 0 \iff$ $\theta \to \pm\cos^{-1}(-1/a)$. As a increases from 1, the curve splits into two parts. The left part has a loop, which grows larger as a increases, and the right part grows broader vertically, and its left tip develops a dimple when $a \approx 2.42$ (actually, $\sqrt{2} + 1$). As a increases, the dimple grows more and more pronounced.

If $a < 0$, we get the same graph as we do for the corresponding positive a-value, but with a rotation through π about the pole, as happened when c was replaced with $-c$ in Exercise 80.

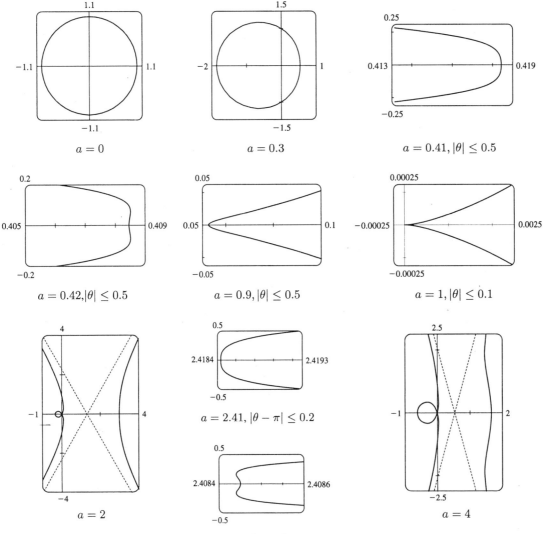

83. $\tan \psi = \tan(\phi - \theta) = \dfrac{\tan \phi - \tan \theta}{1 + \tan \phi \tan \theta} = \dfrac{\dfrac{dy}{dx} - \tan \theta}{1 + \dfrac{dy}{dx}\tan \theta} = \dfrac{\dfrac{dy/d\theta}{dx/d\theta} - \tan \theta}{1 + \dfrac{dy/d\theta}{dx/d\theta}\tan \theta}$

$= \dfrac{\dfrac{dy}{d\theta} - \dfrac{dx}{d\theta}\tan \theta}{\dfrac{dx}{d\theta} + \dfrac{dy}{d\theta}\tan \theta} = \dfrac{\left(\dfrac{dr}{d\theta}\sin \theta + r\cos \theta\right) - \tan \theta\left(\dfrac{dr}{d\theta}\cos \theta - r\sin \theta\right)}{\left(\dfrac{dr}{d\theta}\cos \theta - r\sin \theta\right) + \tan \theta\left(\dfrac{dr}{d\theta}\sin \theta + r\cos \theta\right)}$

$= \dfrac{r\cos \theta + r \cdot \dfrac{\sin^2 \theta}{\cos \theta}}{\dfrac{dr}{d\theta}\cos \theta + \dfrac{dr}{d\theta}\cdot\dfrac{\sin^2 \theta}{\cos \theta}} = \dfrac{r\cos^2 \theta + r\sin^2 \theta}{\dfrac{dr}{d\theta}\cos^2 \theta + \dfrac{dr}{d\theta}\sin^2 \theta} = \dfrac{r}{dr/d\theta}$

11.5 Areas and Lengths in Polar Coordinates   ET 10.5

1. $r = \sqrt{\theta},\, 0 \le \theta \le \frac{\pi}{4}.$ $A = \int_0^{\pi/4} \frac{1}{2}r^2\,d\theta = \int_0^{\pi/4} \frac{1}{2}\left(\sqrt{\theta}\right)^2 d\theta = \int_0^{\pi/4} \frac{1}{2}\theta\,d\theta = \left[\frac{1}{4}\theta^2\right]_0^{\pi/4} = \frac{1}{64}\pi^2$

3. $r = \sin \theta,\, \frac{\pi}{3} \le \theta \le \frac{2\pi}{3}.$

$A = \int_{\pi/3}^{2\pi/3} \frac{1}{2}\sin^2 \theta\,d\theta = \frac{1}{4}\int_{\pi/3}^{2\pi/3}\left(1 - \cos 2\theta\right) d\theta = \frac{1}{4}\left[\theta - \frac{1}{2}\sin 2\theta\right]_{\pi/3}^{2\pi/3}$

$= \frac{1}{4}\left[\frac{2\pi}{3} - \frac{1}{2}\sin\frac{4\pi}{3} - \frac{\pi}{3} + \frac{1}{2}\sin\frac{2\pi}{3}\right] = \frac{1}{4}\left[\frac{2\pi}{3} - \frac{1}{2}\left(-\frac{\sqrt{3}}{2}\right) - \frac{\pi}{3} + \frac{1}{2}\left(\frac{\sqrt{3}}{2}\right)\right] = \frac{1}{4}\left(\frac{\pi}{3} + \frac{\sqrt{3}}{2}\right) = \frac{\pi}{12} + \frac{\sqrt{3}}{8}$

5. $r = \theta,\, 0 \le \theta \le \pi.$ $A = \int_0^{\pi} \frac{1}{2}\theta^2\,d\theta = \left[\frac{1}{6}\theta^3\right]_0^{\pi} = \frac{1}{6}\pi^3$

7. $r = 4 + 3\sin \theta,\, -\frac{\pi}{2} \le \theta \le \frac{\pi}{2}.$

$A = \int_{-\pi/2}^{\pi/2} \frac{1}{2}(4 + 3\sin \theta)^2\,d\theta = \frac{1}{2}\int_{-\pi/2}^{\pi/2}\left(16 + 24\sin \theta + 9\sin^2 \theta\right) d\theta$

$= \frac{1}{2}\int_{-\pi/2}^{\pi/2}\left(16 + 9\sin^2 \theta\right) d\theta \quad \left(\text{by Theorem 5.5.6(b) [ET 5.5.7(b)]}\right)$

$= \frac{1}{2}\cdot 2\int_0^{\pi/2}\left[16 + 9\cdot\frac{1}{2}(1 - \cos 2\theta)\right] d\theta \quad \left(\text{by Theorem 5.5.6(a) [ET 5.5.7(a)]}\right)$

$= \int_0^{\pi/2}\left(\frac{41}{2} - \frac{9}{2}\cos 2\theta\right) d\theta = \left[\frac{41}{2}\theta - \frac{9}{4}\sin 2\theta\right]_0^{\pi/2} = \left(\frac{41\pi}{4} - 0\right) - (0 - 0) = \frac{41\pi}{4}$

9. $A = \int_0^{\pi} \frac{1}{2}(5\sin \theta)^2\,d\theta$

$= \frac{25}{4}\int_0^{\pi}(1 - \cos 2\theta)\,d\theta$

$= \frac{25}{4}\left[\theta - \frac{1}{2}\sin 2\theta\right]_0^{\pi} = \frac{25}{4}\pi$

11. $A = 4\int_0^{\pi/4} \frac{1}{2}r^2\,d\theta = 2\int_0^{\pi/4}(4\cos 2\theta)\,d\theta$

$= 8\int_0^{\pi/4}\cos 2\theta\,d\theta = 4\left[\sin 2\theta\right]_0^{\pi/4} = 4$

13. $A = 2 \int_{-\pi/2}^{\pi/2} \frac{1}{2} (4 - \sin \theta)^2 \, d\theta = \int_{-\pi/2}^{\pi/2} (16 - 8 \sin \theta + \sin^2 \theta) \, d\theta$

$\quad = \int_{-\pi/2}^{\pi/2} (16 + \sin^2 \theta) \, d\theta \quad \left(\text{by Theorem 5.5.6(b) [ET 5.5.7(b)]} \right)$

$\quad = 2 \int_0^{\pi/2} (16 + \sin^2 \theta) \, d\theta \quad \left(\text{by Theorem 5.5.6(a) [ET 5.5.7(a)]} \right)$

$\quad = 2 \int_0^{\pi/2} \left[16 + \frac{1}{2} (1 - \cos 2\theta) \right] d\theta = 2 \left[\frac{33}{2} \theta - \frac{1}{4} \sin 2\theta \right]_0^{\pi/2}$

$\quad = \frac{33\pi}{2}$

15. By symmetry, the total area is twice the area enclosed above the polar axis, so

$\quad A = 2 \int_0^\pi \frac{1}{2} r^2 d\theta = \int_0^\pi [2 + \cos 6\theta]^2 \, d\theta = \int_0^\pi (4 + 4 \cos 6\theta + \cos^2 6\theta) \, d\theta$

$\quad = \left[4\theta + 4 \left(\frac{1}{6} \sin 6\theta \right) + \left(\frac{1}{24} \sin 12\theta + \frac{1}{2} \theta \right) \right]_0^\pi = 4\pi + \frac{\pi}{2} = \frac{9\pi}{2}$

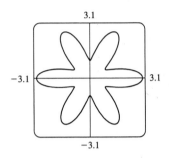

3.1

-3.1   3.1

-3.1

17. $A = \int_0^{\pi/2} \frac{1}{2} \sin^2 2\theta \, d\theta = \int_0^{\pi/2} \frac{1}{4} (1 - \cos 4\theta) \, d\theta = \frac{1}{4} \left[\theta - \frac{1}{4} \sin 4\theta \right]_0^{\pi/2} = \frac{\pi}{8}$

19. $r = 0 \;\Rightarrow\; 3 \cos 5\theta = 0 \;\Rightarrow\; 5\theta = \frac{\pi}{2} \;\Rightarrow\; \theta = \frac{\pi}{10}$.

$\quad A = \int_{-\pi/10}^{\pi/10} \frac{1}{2} (3 \cos 5\theta)^2 \, d\theta = \int_0^{\pi/10} 9 \cos^2 5\theta \, d\theta = \frac{9}{2} \int_0^{\pi/10} (1 + \cos 10\theta) \, d\theta = \frac{9}{2} \left[\theta + \frac{1}{10} \sin 10\theta \right]_0^{\pi/10} = \frac{9\pi}{20}$

21.

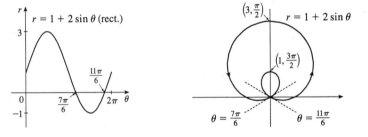

This is a limaçon, with inner loop traced out between $\theta = \frac{7\pi}{6}$ and $\frac{11\pi}{6}$ [found by solving $r = 0$].

$\quad A = 2 \int_{7\pi/6}^{3\pi/2} \frac{1}{2} (1 + 2 \sin \theta)^2 \, d\theta = \int_{7\pi/6}^{3\pi/2} (1 + 4 \sin \theta + 4 \sin^2 \theta) \, d\theta = [\theta - 4 \cos \theta + 2\theta - \sin 2\theta]_{7\pi/6}^{3\pi/2}$

$\quad = \left(\frac{9\pi}{2} \right) - \left(\frac{7\pi}{2} + 2\sqrt{3} - \frac{\sqrt{3}}{2} \right) = \pi - \frac{3\sqrt{3}}{2}$

23. $1 - \cos \theta = \frac{3}{2} \;\Rightarrow\; \cos \theta = -\frac{1}{2} \;\Rightarrow\; \theta = \frac{2\pi}{3} \text{ or } \frac{4\pi}{3} \;\Rightarrow\;$

$\quad A = 2 \int_{2\pi/3}^{\pi} \frac{1}{2} \left[(1 - \cos \theta)^2 - \left(\frac{3}{2} \right)^2 \right] d\theta = \int_{2\pi/3}^{\pi} \left(-\frac{5}{4} - 2 \cos \theta + \cos^2 \theta \right) d\theta$

$\quad = \left[-\frac{5}{4} \theta - 2 \sin \theta \right]_{2\pi/3}^{\pi} + \frac{1}{2} \int_{2\pi/3}^{\pi} (1 + \cos 2\theta) \, d\theta$

$\quad = -\frac{5}{12} \pi + \sqrt{3} + \frac{1}{2} \left[\theta + \frac{1}{2} \sin 2\theta \right]_{2\pi/3}^{\pi}$

$\quad = -\frac{5}{12} \pi + \sqrt{3} + \frac{1}{6} \pi + \frac{\sqrt{3}}{8} = \frac{9\sqrt{3}}{8} - \frac{1}{4} \pi$

$\theta = \frac{2\pi}{3}$

$r = \frac{3}{2}$

$r = 1 - \cos \theta$

25. $4 \sin \theta = 2 \iff \sin \theta = \frac{1}{2} \iff \theta = \frac{\pi}{6}$ or $\frac{5\pi}{6} \iff$

$$A = 2 \int_{\pi/6}^{\pi/2} \frac{1}{2} \left[(4 \sin \theta)^2 - 2^2 \right] d\theta = \int_{\pi/6}^{\pi/2} \left(16 \sin^2 \theta - 4 \right) d\theta$$

$$= \int_{\pi/6}^{\pi/2} \left[8 \left(1 - \cos 2\theta \right) - 4 \right] d\theta = \left[4\theta - 4 \sin 2\theta \right]_{\pi/6}^{\pi/2}$$

$$= \frac{4}{3}\pi + 2\sqrt{3}$$

$r = 4 \sin \theta$

$\theta = \frac{\pi}{6}$

$r = 2$

27. $3 \cos \theta = 1 + \cos \theta \iff \cos \theta = \frac{1}{2} \implies \theta = \frac{\pi}{3}$ or $-\frac{\pi}{3}$.

$$A = 2 \int_0^{\pi/3} \frac{1}{2} \left[(3 \cos \theta)^2 - (1 + \cos \theta)^2 \right] d\theta$$

$$= \int_0^{\pi/3} \left(8 \cos^2 \theta - 2 \cos \theta - 1 \right) d\theta = \int_0^{\pi/3} \left[4 \left(1 + \cos 2\theta \right) - 2 \cos \theta - 1 \right] d\theta$$

$$= \left[3\theta + 2 \sin 2\theta - 2 \sin \theta \right]_0^{\pi/3} = \pi + \sqrt{3} - \sqrt{3} = \pi$$

29. $A = 2 \int_0^{\pi/4} \frac{1}{2} \sin^2 \theta \, d\theta = \int_0^{\pi/4} \frac{1}{2} \left(1 - \cos 2\theta \right) d\theta$

$$= \left[\frac{1}{2}\theta - \frac{1}{4} \sin 2\theta \right]_0^{\pi/4}$$

$$= \frac{1}{8}\pi - \frac{1}{4}$$

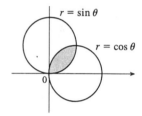

$r = \sin \theta$

$r = \cos \theta$

0

31. $\sin 2\theta = \cos 2\theta \implies \tan 2\theta = 1 \implies 2\theta = \frac{\pi}{4} \implies \theta = \frac{\pi}{8} \implies$

$$A = 16 \int_0^{\pi/8} \frac{1}{2} \sin^2 2\theta \, d\theta = 4 \int_0^{\pi/8} \left(1 - \cos 4\theta \right) d\theta$$

$$= 4 \left[\theta - \frac{1}{4} \sin 4\theta \right]_0^{\pi/8} = \frac{1}{2}\pi - 1$$

$\theta = \frac{\pi}{4}$

$r = \sin 2\theta$

$r = \cos 2\theta$

33. $A = 2 \left[\int_{-\pi/2}^{-\pi/6} \frac{1}{2} \left(3 + 2 \sin \theta \right)^2 d\theta + \int_{-\pi/6}^{\pi/2} \frac{1}{2} 2^2 \, d\theta \right]$

$$= \int_{-\pi/2}^{-\pi/6} \left(9 + 12 \sin \theta + 4 \sin^2 \theta \right) d\theta + \left[4\theta \right]_{-\pi/6}^{\pi/2}$$

$$= \left[9\theta - 12 \cos \theta + 2\theta - \sin 2\theta \right]_{-\pi/2}^{-\pi/6} + \frac{8\pi}{3} = \frac{19\pi}{3} - \frac{11\sqrt{3}}{2}$$

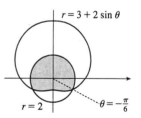

$r = 3 + 2 \sin \theta$

$r = 2$

$\theta = -\frac{\pi}{6}$

35. $A = 2 \left[\int_0^{2\pi/3} \frac{1}{2} \left(\frac{1}{2} + \cos \theta \right)^2 d\theta - \int_{2\pi/3}^{\pi} \frac{1}{2} \left(\frac{1}{2} + \cos \theta \right)^2 d\theta \right]$

$$= \int_0^{2\pi/3} \left(\frac{1}{4} + \cos \theta + \cos^2 \theta \right) d\theta - \int_{2\pi/3}^{\pi} \left(\frac{1}{4} + \cos \theta + \cos^2 \theta \right) d\theta$$

$$= \left[\frac{\theta}{4} + \sin \theta + \frac{\theta}{2} + \frac{\sin 2\theta}{4} \right]_0^{2\pi/3} - \left[\frac{\theta}{4} + \sin \theta + \frac{\theta}{2} + \frac{\sin 2\theta}{4} \right]_{2\pi/3}^{\pi}$$

$$= \left(\frac{\pi}{2} + \frac{3\sqrt{3}}{8} \right) - \left(\frac{3\pi}{4} \right) + \left(\frac{\pi}{2} + \frac{3\sqrt{3}}{8} \right) = \frac{1}{4} \left(\pi + 3\sqrt{3} \right)$$

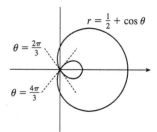

$r = \frac{1}{2} + \cos \theta$

$\theta = \frac{2\pi}{3}$

$\theta = \frac{4\pi}{3}$

37. The two circles intersect at the pole since $(0, 0)$ satisfies the first equation

and $\left(0, \frac{\pi}{2}\right)$ the second. The other intersection point $\left(\frac{1}{\sqrt{2}}, \frac{\pi}{4}\right)$ occurs

where $\sin \theta = \cos \theta$.

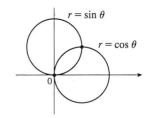

39. The curves intersect at the pole since $\left(0, \frac{\pi}{2}\right)$ satisfies $r = \cos \theta$ and

$(0, 0)$ satisfies $r = 1 - \cos \theta$. $\cos \theta = 1 - \cos \theta \Longrightarrow \cos \theta = \frac{1}{2} \Longrightarrow$

$\theta = \frac{\pi}{3}$ or $\frac{5\pi}{3} \Longrightarrow$ the other intersection points are $\left(\frac{1}{2}, \frac{\pi}{3}\right)$ and

$\left(\frac{1}{2}, \frac{5\pi}{3}\right)$.

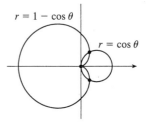

41. The pole is a point of intersection. $\sin \theta = \sin 2\theta = 2 \sin \theta \cos \theta \Leftrightarrow$

$\sin \theta \left(1 - 2 \cos \theta\right) = 0 \Leftrightarrow \sin \theta = 0$ or $\cos \theta = \frac{1}{2} \Rightarrow \theta = 0$,

$\pi, \frac{\pi}{3}, -\frac{\pi}{3} \Rightarrow \left(\frac{\sqrt{3}}{2}, \frac{\pi}{3}\right)$ and $\left(\frac{\sqrt{3}}{2}, \frac{2\pi}{3}\right)$ (by symmetry) are the

other intersection points.

43.

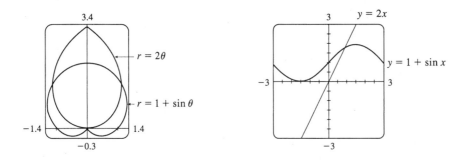

From the first graph, we see that the pole is one point of intersection. By zooming in or using the cursor, we
estimate the θ-values of the intersection points to be about 0.89 and $\pi - 0.89 \approx 2.25$. (The first of these values
may be more easily estimated by plotting $y = 1 + \sin x$ and $y = 2x$ in rectangular coordinates; see the second
graph.)

By symmetry, the total area contained is twice the area contained in the first quadrant, that is,

$$A \approx 2 \int_0^{0.89} \tfrac{1}{2} (2\theta)^2 \, d\theta + 2 \int_{0.89}^{\pi/2} \tfrac{1}{2} (1 + \sin \theta)^2 \, d\theta$$

$$= \left[\tfrac{4}{3}\theta^3\right]_0^{0.89} + \left[\theta - 2\cos \theta + \left(\tfrac{1}{2}\theta - \tfrac{1}{4}\sin 2\theta\right)\right]_{0.89}^{\pi/2} \approx 3.46$$

45. $L = \int_a^b \sqrt{r^2 + (dr/d\theta)^2} \, d\theta = \int_0^{3\pi/4} \sqrt{(5 \cos \theta)^2 + (-5 \sin \theta)^2} \, d\theta = 5 \int_0^{3\pi/4} \sqrt{\cos^2 \theta + \sin^2 \theta} \, d\theta$

$= 5 \int_0^{3\pi/4} d\theta = \tfrac{15}{4} \pi$

47. $L = \int_a^b \sqrt{r^2 + (dr/d\theta)^2}\, d\theta = \int_0^{2\pi} \sqrt{(2^\theta)^2 + [(\ln 2)\, 2^\theta]^2}\, d\theta = \int_0^{2\pi} 2^\theta \sqrt{1 + \ln^2 2}\, d\theta$

$\quad = \left[\sqrt{1 + \ln^2 2} \left(\dfrac{2^\theta}{\ln 2} \right) \right]_0^{2\pi} = \dfrac{\sqrt{1 + \ln^2 2}\, (2^{2\pi} - 1)}{\ln 2}$

49. $L = \int_0^{2\pi} \sqrt{\left(\theta^2\right)^2 + (2\theta)^2}\, d\theta = \int_0^{2\pi} \theta \sqrt{\theta^2 + 4}\, d\theta = \frac{1}{2} \cdot \frac{2}{3} \left[\left(\theta^2 + 4\right)^{3/2} \right]_0^{2\pi} = \frac{8}{3} \left[\left(\pi^2 + 1\right)^{3/2} - 1 \right]$

51. From Figure 4 in Example 1,

$$L = \int_{-\pi/4}^{\pi/4} \sqrt{r^2 + (r')^2}\, d\theta = 2 \int_0^{\pi/4} \sqrt{\cos^2 2\theta + 4\sin^2 2\theta}\, d\theta \approx 2\,(1.211056) \approx 2.4221$$

53. $L = 2 \int_0^{2\pi} \sqrt{\cos^8 (\theta/4) + \cos^6 (\theta/4)\sin^2 (\theta/4)}\, d\theta$

$\quad = 2 \int_0^{2\pi} \left|\cos^3 (\theta/4)\right| \sqrt{\cos^2 (\theta/4) + \sin^2 (\theta/4)}\, d\theta$

$\quad = 2 \int_0^{2\pi} \left|\cos^3 (\theta/4)\right| d\theta = 8 \int_0^{\pi/2} \cos^3 u\, du \quad$ (where $u = \frac{1}{4}\theta$)

$\quad = 8 \left[\sin u - \frac{1}{3}\sin^3 u \right]_0^{\pi/2} = \frac{16}{3}$

Note that the curve is retraced after every interval of length 4π.

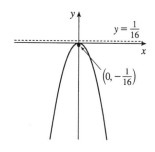

55. (a) From (11.3.5 [ET 10.3.5]),

$\quad S = \int_a^b 2\pi y \sqrt{(dx/d\theta)^2 + (dy/d\theta)^2}\, d\theta$

$\qquad = \int_a^b 2\pi y \sqrt{r^2 + (dr/d\theta)^2}\, d\theta$ (see the derivation of Equation 5) $= \int_a^b 2\pi r \sin\theta \sqrt{r^2 + (dr/d\theta)^2}\, d\theta$

(b) $r^2 = \cos 2\theta \;\Rightarrow\; 2r\dfrac{dr}{d\theta} = -2\sin 2\theta \;\Rightarrow\; \left(\dfrac{dr}{d\theta}\right)^2 = \dfrac{\sin^2 2\theta}{r^2} = \dfrac{\sin^2 2\theta}{\cos 2\theta}.$

$\quad S = 2 \int_0^{\pi/4} 2\pi \sqrt{\cos 2\theta} \sin\theta \sqrt{\cos 2\theta + \left(\sin^2 2\theta\right)/\cos 2\theta}\, d\theta = 4\pi \int_0^{\pi/4} \sin\theta\, d\theta$

$\quad = [-4\pi \cos\theta]_0^{\pi/4} = -4\pi \left(\frac{1}{\sqrt{2}} - 1 \right) = 2\pi \left(2 - \sqrt{2} \right)$

11.6 Conic Sections   ET 10.6

1. $x = 2y^2 \;\Rightarrow\; y^2 = \frac{1}{2}x.$ $4p = \frac{1}{2},$ so $p = \frac{1}{8}.$ The vertex is $(0, 0)$, the focus is $\left(\frac{1}{8}, 0\right)$, and the directrix is $x = -\frac{1}{8}$.

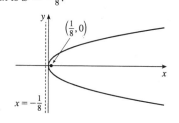

3. $4x^2 = -y \;\Rightarrow\; x^2 = -\frac{1}{4}y.$ $4p = -\frac{1}{4},$ so $p = -\frac{1}{16}.$ The vertex is $(0, 0)$, the focus is $\left(0, -\frac{1}{16}\right)$, and the directrix is $y = \frac{1}{16}.$

5. $(x + 2)^2 = 8(y - 3)$. $4p = 8$, so $p = 2$. The vertex is $(-2, 3)$, the focus is $(-2, 5)$, and the directrix is $y = 1$.

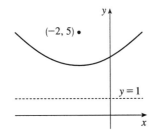

7. $2x + y^2 - 8y + 12 = 0 \Rightarrow$
$$(y - 4)^2 = -2(x - 2) \Rightarrow p = -\tfrac{1}{2} \Rightarrow$$
vertex $(2, 4)$, focus $\left(\tfrac{3}{2}, 4\right)$, directrix $x = \tfrac{5}{2}$

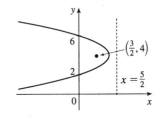

9. The equation has the form $y^2 = 4px$, where $p < 0$. Since the parabola passes through $(-1, 1)$, we have $1^2 = 4p(-1)$, so $4p = -1$ and an equation is $y^2 = -x$ or $x = -y^2$. $4p = -1$, so $p = -\tfrac{1}{4}$ and the focus is $\left(-\tfrac{1}{4}, 0\right)$ while the directrix is $x = \tfrac{1}{4}$.

11. $x^2/16 + y^2/4 = 1 \Rightarrow a = 4, b = 2$,
$c = \sqrt{16 - 4} = 2\sqrt{3} \Rightarrow$ center $(0, 0)$, vertices $(\pm 4, 0)$, foci $\left(\pm 2\sqrt{3}, 0\right)$

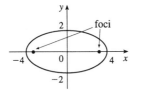

13. $25x^2 + 9y^2 = 225 \Leftrightarrow \tfrac{1}{9}x^2 + \tfrac{1}{25}y^2 = 1 \Rightarrow$
$a = 5, b = 3, c = 4 \Rightarrow$ center $(0, 0)$, vertices $(0, \pm 5)$, foci $(0, \pm 4)$

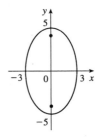

15. $9x^2 - 18x + 4y^2 = 27 \Leftrightarrow$
$$\frac{(x - 1)^2}{4} + \frac{y^2}{9} = 1 \Rightarrow a = 3, b = 2, c = \sqrt{5}$$
$\Rightarrow$ center $(1, 0)$, vertices $(1, \pm 3)$, foci $\left(1, \pm\sqrt{5}\right)$

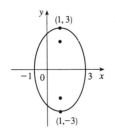

17. The center is $(0, 0)$, $a = 3$, and $b = 2$, so an equation is $\dfrac{x^2}{4} + \dfrac{y^2}{9} = 1$. $c = \sqrt{a^2 - b^2} = \sqrt{5}$, so the foci are $\left(0, \pm\sqrt{5}\right)$.

19. $\dfrac{x^2}{144} - \dfrac{y^2}{25} = 1 \;\Rightarrow\; a = 12, b = 5,$

$c = \sqrt{144 + 25} = 13 \;\Rightarrow\;$ center $(0,0)$, vertices

$(\pm 12, 0)$, foci $(\pm 13, 0)$, asymptotes $y = \pm \frac{5}{12} x$

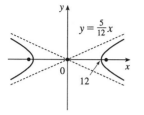

21. $9y^2 - x^2 = 9 \;\Rightarrow\; y^2 - \frac{1}{9} x^2 = 1 \;\Rightarrow\; a = 1,$

$b = 3, c = \sqrt{10} \;\Rightarrow\;$ center $(0,0)$, vertices

$(0, \pm 1)$, foci $\left(0, \pm \sqrt{10}\right)$, asymptotes $y = \pm \frac{1}{3} x$

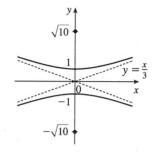

23. $2y^2 - 4y - 3x^2 + 12x = -8 \;\Leftrightarrow\; \dfrac{(x-2)^2}{6} - \dfrac{(y-1)^2}{9} = 1 \;\Rightarrow\; a = \sqrt{6}, b = 3, c = \sqrt{15} \;\Rightarrow\;$ center

$(2, 1)$, vertices $\left(2 \pm \sqrt{6}, 1\right)$, foci $\left(2 \pm \sqrt{15}, 1\right)$, asymptotes $y - 1 = \pm \frac{3}{\sqrt{6}} (x - 2)$

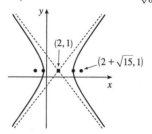

25. The parabola with vertex $(0,0)$ and focus $(0, -2)$ opens downward and has $p = -2$, so its equation is
$x^2 = 4py = -8y.$

27. Vertex at $(2, 0)$, $p = 1$, opens to right $\;\Rightarrow\; y^2 = 4p(x - 2) = 4(x - 2)$

29. The parabola must have equation $y^2 = 4px$, so $(-4)^2 = 4p(1) \;\Rightarrow\; p = 4 \;\Rightarrow\; y^2 = 16x.$

31. The ellipse with foci $(\pm 2, 0)$ and vertices $(\pm 5, 0)$ has center $(0,0)$ and a horizontal major axis, with $a = 5$ and

$c = 2$, so $b = \sqrt{a^2 - c^2} = \sqrt{21}$. An equation is $\dfrac{x^2}{25} + \dfrac{y^2}{21} = 1.$

33. Center $(3, 0)$, $c = 1$, $a = 3 \;\Rightarrow\; b = \sqrt{8} = 2\sqrt{2} \;\Rightarrow\; \frac{1}{8}(x - 3)^2 + \frac{1}{9} y^2 = 1$

35. Center $(2, 2)$, $c = 2$, $a = 3 \;\Rightarrow\; b = \sqrt{5} \;\Rightarrow\; \frac{1}{9}(x - 2)^2 + \frac{1}{5}(y - 2)^2 = 1$

37. Center $(0, 0)$, vertical axis, $c = 3$, $a = 1 \;\Rightarrow\; b = \sqrt{8} = 2\sqrt{2} \;\Rightarrow\; y^2 - \frac{1}{8} x^2 = 1$

39. Center $(4, 3)$, horizontal axis, $c = 3$, $a = 2 \;\Rightarrow\; b = \sqrt{5} \;\Rightarrow\; \frac{1}{4}(x - 4)^2 - \frac{1}{5}(y - 3)^2 = 1$

41. Center $(0, 0)$, horizontal axis, $a = 3$, $\frac{b}{a} = 2 \;\Rightarrow\; b = 6 \;\Rightarrow\; \frac{1}{9} x^2 - \frac{1}{36} y^2 = 1$

43. In Figure 8, we see that the point on the ellipse closest to a focus is the closer vertex (which is a distance $a - c$ from
it) while the farthest point is the other vertex (at a distance of $a + c$). So for this lunar orbit,
$(a - c) + (a + c) = 2a = (1728 + 110) + (1728 + 314)$, or $a = 1940$; and
$(a + c) - (a - c) = 2c = 314 - 110$, or $c = 102$. Thus, $b^2 = a^2 - c^2 = 3{,}753{,}196$, and the equation is

$\dfrac{x^2}{3{,}763{,}600} + \dfrac{y^2}{3{,}753{,}196} = 1.$

45. (a) Set up the coordinate system so that A is $(-200, 0)$ and B is $(200, 0)$.

$|PA| - |PB| = (1200)(980) = 1{,}176{,}000 \text{ ft} = \frac{2450}{11} \text{ mi} = 2a \;\Rightarrow\; a = \frac{1225}{11}$, and $c = 200$ so

$b^2 = c^2 - a^2 = \frac{3{,}339{,}375}{121} \;\Rightarrow\; \dfrac{121x^2}{1{,}500{,}625} - \dfrac{121y^2}{3{,}339{,}375} = 1.$

(b) Due north of $B \;\Rightarrow\; x = 200 \;\Rightarrow\; \dfrac{(121)(200)^2}{1{,}500{,}625} - \dfrac{121y^2}{3{,}339{,}375} = 1 \;\Rightarrow\; y = \dfrac{133{,}575}{539} \approx 248 \text{ mi}$

47. The function whose graph is the upper branch of this hyperbola is concave upward. The

function is $y = f(x) = a\sqrt{1 + \dfrac{x^2}{b^2}} = \dfrac{a}{b}\sqrt{b^2 + x^2}$, so $y' = \dfrac{a}{b}x\left(b^2 + x^2\right)^{-1/2}$ and

$y'' = \dfrac{a}{b}\left[\left(b^2 + x^2\right)^{-1/2} - x^2\left(b^2 + x^2\right)^{-3/2}\right] = ab\left(b^2 + x^2\right)^{-3/2} > 0$ for all x, and so f is concave upward.

49. (a) ellipse

(b) hyperbola

(c) empty graph (no curve)

(d) In case (a), $a^2 = k$, $b^2 = k - 16$, and $c^2 = a^2 - b^2 = 16$, so the foci are at $(\pm 4, 0)$. In case (b), $k - 16 < 0$, so $a^2 = k$, $b^2 = 16 - k$, and $c^2 = a^2 + b^2 = 16$, and so again the foci are at $(\pm 4, 0)$.

51. Use the parametrization $x = 2\cos t$, $y = \sin t$, $0 \le t \le 2\pi$ to get

$$L = 4\int_0^{\pi/2}\sqrt{(dx/dt)^2 + (dy/dt)^2}\,dt = 4\int_0^{\pi/2}\sqrt{4\sin^2 t + \cos^2 t}\,dt = 4\int_0^{\pi/2}\sqrt{3\sin^2 t + 1}\,dt$$

Using Simpson's Rule with $n = 10$,

$$L \approx \tfrac{4}{3}\left(\tfrac{\pi}{20}\right)\left[f(0) + 4f\left(\tfrac{\pi}{20}\right) + 2f\left(\tfrac{\pi}{10}\right) + \cdots + 2f\left(\tfrac{2\pi}{5}\right) + 4f\left(\tfrac{9\pi}{20}\right) + f\left(\tfrac{\pi}{2}\right)\right]$$

with $f(t) = \sqrt{3\sin^2 t + 1}$, so $L \approx 9.69$.

53. $\dfrac{x^2}{a^2} + \dfrac{y^2}{b^2} = 1 \;\Rightarrow\; \dfrac{2x}{a^2} + \dfrac{2yy'}{b^2} = 0 \;\Rightarrow\; y' = -\dfrac{b^2 x}{a^2 y}$ $(y \ne 0)$. Thus, the slope of the tangent line at P is

$-\dfrac{b^2 x_1}{a^2 y_1}$. The slope of $F_1 P$ is $\dfrac{y_1}{x_1 + c}$ and of $F_2 P$ is $\dfrac{y_1}{x_1 - c}$. By the formula from Problems Plus, we have

$$\tan\alpha = \dfrac{\dfrac{y_1}{x_1 + c} + \dfrac{b^2 x_1}{a^2 y_1}}{1 - \dfrac{b^2 x_1 y_1}{a^2 y_1(x_1 + c)}} = \dfrac{a^2 y_1^2 + b^2 x_1(x_1 + c)}{a^2 y_1(x_1 + c) - b^2 x_1 y_1} = \dfrac{a^2 b^2 + b^2 c x_1}{c^2 x_1 y_1 + a^2 c y_1} \quad \left(\begin{array}{l}\text{using } b^2 x_1^2 + a^2 y_1^2 = a^2 b^2 \\ \text{and } a^2 - b^2 = c^2\end{array}\right)$$

$$= \dfrac{b^2\left(cx_1 + a^2\right)}{cy_1\left(cx_1 + a^2\right)} = \dfrac{b^2}{cy_1}$$

and

$$\tan\beta = \dfrac{-\dfrac{y_1}{x_1 - c} - \dfrac{b^2 x_1}{a^2 y_1}}{1 - \dfrac{b^2 x_1 y_1}{a^2 y_1(x_1 - c)}} = \dfrac{-a^2 y_1^2 - b^2 x_1(x_1 - c)}{a^2 y_1(x_1 - c) - b^2 x_1 y_1} = \dfrac{-a^2 b^2 + b^2 c x_1}{c^2 x_1 y_1 - a^2 c y_1} = \dfrac{b^2\left(cx_1 - a^2\right)}{cy_1\left(cx_1 - a^2\right)} = \dfrac{b^2}{cy_1}$$

So $\alpha = \beta$.

11.7 Conic Sections in Polar Coordinates ET 10.7

1. The directrix $y = 6$ is above the focus at the origin, so we use the form with "$+ e \sin \theta$" in the denominator.

$$r = \frac{ed}{1 + e \sin \theta} = \frac{\frac{7}{4} \cdot 6}{1 + \frac{7}{4} \sin \theta} = \frac{42}{4 + 7 \sin \theta}$$

3. The directrix $x = -5$ is to the left of the focus at the origin, so we use the form with "$- e \cos \theta$" in the

denominator. $r = \dfrac{ed}{1 - e \cos \theta} = \dfrac{\frac{3}{4} \cdot 5}{1 - \frac{3}{4} \cos \theta} = \dfrac{15}{4 - 3 \cos \theta}$

5. $r = 5 \sec \theta \iff x = r \cos \theta = 5$, so $r = \dfrac{ed}{1 + e \cos \theta} = \dfrac{4 \cdot 5}{1 + 4 \cos \theta} = \dfrac{20}{1 + 4 \cos \theta}$

7. Focus $(0,0)$, vertex $\left(5, \frac{\pi}{2}\right)$ $\Rightarrow$ directrix $y = 10$ $\Rightarrow$ $r = \dfrac{ed}{1 + e \sin \theta} = \dfrac{10}{1 + \sin \theta}$

9. $r = \dfrac{4}{1 + 3 \cos \theta}$

(a) $e = 3$

(b) Since $e = 3 > 1$, the conic is a hyperbola.

(c) $ed = 4 \Rightarrow d = \frac{4}{3} \Rightarrow$ directrix $x = \frac{4}{3}$

(d) The vertices are $(1, 0)$ and $(-2, \pi) = (2, 0)$; the center is $\left(\frac{3}{2}, 0\right)$; the

asymptotes are parallel to $\theta = \pm \cos^{-1}\left(-\frac{1}{3}\right)$

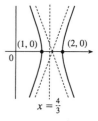

11. $r = \dfrac{2}{1 - \cos \theta}$

(a) $e = 1$

(b) Parabola

(c) $ed = 2 \Rightarrow d = 2 \Rightarrow$ directrix $x = -2$

(d) Vertex $(-1, 0) = (1, \pi)$

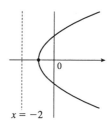

13. $r = \dfrac{3}{1 + \frac{1}{2} \sin \theta}$

(a) $e = \frac{1}{2}$

(b) Ellipse

(c) $ed = 3 \Rightarrow d = 6 \Rightarrow$ directrix $y = 6$

(d) Vertices $\left(2, \frac{\pi}{2}\right)$ and $\left(6, \frac{3\pi}{2}\right)$; center $\left(2, \frac{3\pi}{2}\right)$

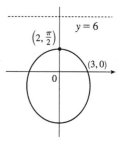

15. $r = \dfrac{7/2}{1 - \frac{5}{2}\sin\theta}$

   (a) $e = \frac{5}{2}$

   (b) Hyperbola

   (c) $ed = \frac{7}{2}$ $\Rightarrow$ $d = \frac{7}{5}$ $\Rightarrow$ directrix $y = -\frac{7}{5}$

   (d) Center $\left(\frac{5}{3}, \frac{3\pi}{2}\right)$; vertices $\left(-\frac{7}{3}, \frac{\pi}{2}\right) = \left(\frac{7}{3}, \frac{3\pi}{2}\right)$ and $\left(1, \frac{3\pi}{2}\right)$

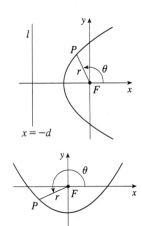

17. (a) The equation is $r = \dfrac{1}{4 - 3\cos\theta} = \dfrac{1/4}{1 - \frac{3}{4}\cos\theta}$, so $e = \frac{3}{4}$ and

   $ed = \frac{1}{4}$ $\Rightarrow$ $d = \frac{1}{3}$. The conic is an ellipse, and the equation of its

   directrix is $x = r\cos\theta = -\frac{1}{3}$ $\Rightarrow$ $r = -\dfrac{1}{3\cos\theta}$. We must be

   careful in our choice of parameter values in this equation

   $(-1 \le \theta \le 1$ works well).

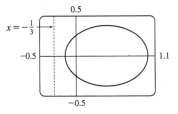

(b) The equation is obtained by
replacing θ with $\theta - \frac{\pi}{3}$ in the
equation of the original conic
(see Example 4), so

$$r = \dfrac{1}{4 - 3\cos\left(\theta - \frac{\pi}{3}\right)}.$$

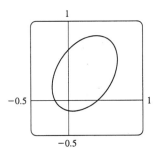

19. For $e < 1$ the curve is an ellipse. It is nearly circular when e is close to 0.
As e increases, the graph is stretched out to the right, and grows larger
(that is, its right-hand focus moves to the right while its left-hand focus
remains at the origin.) At $e = 1$, the curve becomes a parabola with focus
at the origin.

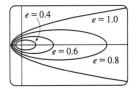

21. $|PF| = e\,|Pl|$ $\Rightarrow$ $r = e\,[d - r\cos(\pi - \theta)] = e\,(d + r\cos\theta)$ $\Rightarrow$

   $r\,(1 - e\cos\theta) = ed$ $\Rightarrow$ $r = \dfrac{ed}{1 - e\cos\theta}$

23. $|PF| = e\,|Pl|$ $\Rightarrow$ $r = e\,[d - r\sin(\theta - \pi)] = e\,(d + r\sin\theta)$ $\Rightarrow$

   $r\,(1 - e\sin\theta) = ed$ $\Rightarrow$ $r = \dfrac{ed}{1 - e\sin\theta}$

25. (a) If the directrix is $x = -d$, then $r = \dfrac{ed}{1 - e \cos \theta}$ [see Figure 2(b)], and, from (4), $a^2 = \dfrac{e^2 d^2}{(1 - e^2)^2}$ $\Rightarrow$

$ed = a \left(1 - e^2\right)$. Therefore, $r = \dfrac{a \left(1 - e^2\right)}{1 - e \cos \theta}$.

(b) $e = 0.017$ and the major axis $= 2a = 2.99 \times 10^8$ $\Rightarrow$ $a = 1.495 \times 10^8$.

Therefore $r = \dfrac{1.495 \times 10^8 \left[1 - (0.017)^2\right]}{1 - 0.017 \cos \theta} \approx \dfrac{1.49 \times 10^8}{1 - 0.017 \cos \theta}$.

27. Here $2a = $ length of major axis $= 36.18$ AU $\Rightarrow$ $a = 18.09$ AU and $e = 0.97$. By Exercise 25(a), the equation

of the orbit is $r = \dfrac{18.09 \left[1 - (0.97)^2\right]}{1 - 0.97 \cos \theta} \approx \dfrac{1.07}{1 - 0.97 \cos \theta}$. By Exercise 26(a), the maximum distance from the

comet to the sun is $18.09 \left(1 + 0.97\right) \approx 35.64$ AU or about 3.314 billion miles.

29. The minimum distance is at perihelion where $4.6 \times 10^7 = r = a \left(1 - e\right) = a \left(1 - 0.206\right) = a \left(0.794\right)$

$\Rightarrow$ $a = 4.6 \times 10^7 / 0.794$. So the maximum distance, which is at aphelion, is

$r = a \left(1 + e\right) = \left(4.6 \times 10^7 / 0.794\right) \times 10^7 \left(1.206\right) \approx 7.0 \times 10^7$ km.

31. From Exercise 29, we have $e = 0.206$ and $a \left(1 - e\right) = 4.6 \times 10^7$ km. Thus, $a = 4.6 \times 10^7 / 0.794$. From

Exercise 25, we can write the equation of Mercury's orbit as $r = a \dfrac{1 - e^2}{1 - e \cos \theta}$. So since

$\dfrac{dr}{d\theta} = \dfrac{-a \left(1 - e^2\right) e \sin \theta}{(1 - e \cos \theta)^2}$ $\Rightarrow$

$r^2 + \left(\dfrac{dr}{d\theta}\right)^2 = \dfrac{a^2 \left(1 - e^2\right)^2}{(1 - e \cos \theta)^2} + \dfrac{a^2 \left(1 - e^2\right)^2 e^2 \sin^2 \theta}{(1 - e \cos \theta)^4} = \dfrac{a^2 \left(1 - e^2\right)^2}{(1 - e \cos \theta)^4} \left(1 - 2e \cos \theta + e^2\right)$

the length of the orbit is

$$L = \int_0^{2\pi} \sqrt{r^2 + (dr/d\theta)^2} \, d\theta = a \left(1 - e^2\right) \int_0^{2\pi} \dfrac{\sqrt{1 + e^2 - 2e \cos \theta}}{(1 - e \cos \theta)^2} \, d\theta \approx 3.6 \times 10^8 \text{ km}$$

This seems reasonable, since Mercury's orbit is nearly circular, and the circumference of a circle of radius a is

$2\pi a \approx 3.6 \times 10^8$ km.

11   Review

ET 10

CONCEPT CHECK

1. (a) A parametric curve is a set of points of the form $(x, y) = (f(t), g(t))$, where f and g are continuous functions
of a variable t.

(b) Sketching a parametric curve, like sketching the graph of a function, is difficult to do in general. We can plot
points on the curve by finding $f(t)$ and $g(t)$ for various values of t, either by hand or with a calculator or
computer. Sometimes, when f and g are given by formulas, we can eliminate t from the equations $x = f(t)$ and
$y = g(t)$ to get a Cartesian equation relating x and y. It may be easier to graph that equation than to work with
the original formulas for x and y in terms of t.

2. (a) You can find $\dfrac{dy}{dx}$ as a function of t by calculating $\dfrac{dy}{dx} = \dfrac{dy/dt}{dx/dt}$ (if $dx/dt \neq 0$).

(b) Calculate the area as $\int_a^b y \, dx = \int_\alpha^\beta g(t) \, f'(t) \, dt$ [or $\int_\beta^\alpha g(t) \, f'(t) \, dt$ if the leftmost point is $(f(\beta), g(\beta))$
rather than $(f(\alpha), g(\alpha))$].

3. (a) $L = \int_\alpha^\beta \sqrt{(dx/dt)^2 + (dy/dt)^2}\, dt = \int_\alpha^\beta \sqrt{[f'(t)]^2 + [g'(t)]^2}\, dt$

(b) $S = \int_\alpha^\beta 2\pi y \sqrt{(dx/dt)^2 + (dy/dt)^2}\, dt = \int_\alpha^\beta 2\pi g(t) \sqrt{[f'(t)]^2 + [g'(t)]^2}\, dt$

4. (a) See Figure 5 in Section 11.4 [ET 10.4].

(b) $x = r\cos\theta,\ y = r\sin\theta$

(c) To find a polar representation (r, θ) with $r \geq 0$ and $0 \leq \theta < 2\pi$, first calculate $r = \sqrt{x^2 + y^2}$. Then θ is specified by $\cos\theta = x/r$ and $\sin\theta = y/r$.

5. (a) Calculate $\dfrac{dy}{dx} = \dfrac{dy/d\theta}{dx/d\theta} = \dfrac{(d/d\theta)(r\sin\theta)}{(d/d\theta)(r\cos\theta)} = \dfrac{(dr/d\theta)\sin\theta + r\cos\theta}{(dr/d\theta)\cos\theta - r\sin\theta}$, where $r = f(\theta)$.

(b) Calculate $A = \int_a^b \frac{1}{2}r^2\, d\theta = \int_a^b \frac{1}{2}[f(\theta)]^2\, d\theta$

(c) $L = \int_a^b \sqrt{(dx/d\theta)^2 + (dy/d\theta)^2}\, d\theta = \int_a^b \sqrt{r^2 + (dr/d\theta)^2}\, d\theta = \int_a^b \sqrt{[f(\theta)]^2 + [f'(\theta)]^2}\, d\theta$

6. (a) A parabola is a set of points in a plane whose distances from a fixed point F (the focus) and a fixed line ℓ (the directrix) are equal.

(b) $x^2 = 4py;\ y^2 = 4px$

7. (a) An ellipse is a set of points in a plane the sum of whose distances from two fixed points (the foci) is a constant.

(b) $\dfrac{x^2}{a^2} + \dfrac{y^2}{a^2 - c^2} = 1.$

8. (a) A hyperbola is a set of points in a plane the difference of whose distances from two fixed points (the foci) is a constant.

(b) $\dfrac{x^2}{a^2} - \dfrac{y^2}{c^2 - a^2} = 1$

(c) $y = \pm\dfrac{\sqrt{c^2 - a^2}}{a}x$

9. (a) If a conic section has focus F and corresponding directrix ℓ, then the eccentricity e is the fixed ratio $|PF|\,/\,|P\ell|$ for points P of the conic section.

(b) $e < 1$ for an ellipse; $e > 1$ for a hyperbola; $e = 1$ for a parabola.

(c) $x = d:\ r = \dfrac{ed}{1 + e\cos\theta}.\quad x = -d:\ r = \dfrac{ed}{1 - e\cos\theta}.\quad y = d:\ r = \dfrac{ed}{1 + e\sin\theta}.\quad y = -d:\ r = \dfrac{ed}{1 - e\sin\theta}.$

EXERCISES

1. $x = t^2 + 4t,\ y = 2 - t,\ -4 \leq t \leq 1.\ t = 2 - y$, so

$x = (2 - y)^2 + 4(2 - y) = 4 - 4y + y^2 + 8 - 4y = y^2 - 8y + 12.$ This is part of a parabola with vertex $(-4, 4)$, opening to the right.

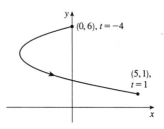

3. $x = \tan\theta$, $y = \cot\theta$. $y = 1/\tan\theta = 1/x$. The whole curve is traced out
as θ ranges over the open interval $\left(-\frac{\pi}{2}, \frac{\pi}{2}\right)$ [or any open interval of the
form $\left(-\frac{\pi}{2} + n\pi, \frac{\pi}{2} + n\pi\right)$, where n is an integer].

5. $r = 1 + 3\cos\theta$

7. $r^2 = \sec 2\theta \implies$
$r^2 \cos 2\theta = 1 \implies$
$r^2 \left(\cos^2\theta - \sin^2\theta\right) = 1 \implies$
$r^2 \cos^2\theta - r^2 \sin^2\theta = 1 \implies$
$x^2 - y^2 = 1$, a hyperbola

9. $r = 2\cos^2\left(\theta/2\right) = 1 + \cos\theta$

11. $r = \dfrac{1}{1 + \cos\theta} \implies e = 1 \implies$ parabola; $d = 1 \implies$ directrix $x = 1$
and vertex $\left(\frac{1}{2}, 0\right)$; y-intercepts are $\left(1, \frac{\pi}{2}\right)$ and $\left(1, \frac{3\pi}{2}\right)$.

13. $x + y = 2 \iff r\cos\theta + r\sin\theta = 2 \iff r\left(\cos\theta + \sin\theta\right) = 2 \iff r = \dfrac{2}{\cos\theta + \sin\theta}$

15. $r = (\sin \theta)/\theta$. As $\theta \to \pm\infty$,

$r \to 0$.

17. $x = \ln t$, $y = 1 + t^2$; $t = 1$. $\dfrac{dy}{dt} = 2t$ and $\dfrac{dx}{dt} = \dfrac{1}{t}$, so $\dfrac{dy}{dx} = \dfrac{dy/dt}{dx/dt} = \dfrac{2t}{1/t} = 2t^2$. When $t = 1$, $\dfrac{dy}{dx} = 2$.

19. $\dfrac{dy}{dx} = \dfrac{(dr/d\theta)\sin\theta + r\cos\theta}{(dr/d\theta)\cos\theta - r\sin\theta} = \dfrac{\sin\theta + \theta\cos\theta}{\cos\theta - \theta\sin\theta} = \dfrac{\frac{1}{\sqrt{2}} + \frac{\pi}{4} \cdot \frac{1}{\sqrt{2}}}{\frac{1}{\sqrt{2}} - \frac{\pi}{4} \cdot \frac{1}{\sqrt{2}}} = \dfrac{4 + \pi}{4 - \pi}$ when $\theta = \frac{\pi}{4}$.

21. $\dfrac{dy}{dx} = \dfrac{dy/dt}{dx/dt} = \dfrac{t\cos t + \sin t}{-t\sin t + \cos t}$. $\dfrac{d^2 y}{dx^2} = \dfrac{\frac{d}{dt}\left(\frac{dy}{dx}\right)}{dx/dt}$, where

$\dfrac{d}{dt}\left(\dfrac{dy}{dx}\right) = \dfrac{(-t\sin t + \cos t)(-t\sin t + 2\cos t) - (t\cos t + \sin t)(-t\cos t - 2\sin t)}{(-t\sin t + \cos t)^2} = \dfrac{t^2 + 2}{(-t\sin t + \cos t)^2}$

$\Rightarrow \dfrac{d^2 y}{dx^2} = \dfrac{t^2 + 2}{(-t\sin t + \cos t)^3}$.

23. We graph the curve for $-2.2 \le t \le 1.2$. By zooming in or using a cursor,

we find that the lowest point is about $(1.4, 0.75)$. To find the exact values,

we find the t-value at which $dy/dt = 2t + 1 = 0 \;\Leftrightarrow\; t = -\frac{1}{2} \;\Leftrightarrow\;$

$(x, y) = \left(\frac{11}{8}, \frac{3}{4}\right)$.

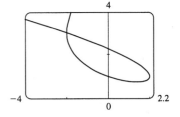

25. $\dfrac{dx}{dt} = -2a\sin t + 2a\sin 2t = 2a\sin t(2\cos t - 1) = 0 \;\Leftrightarrow\; \sin t = 0$ or $\cos t = \frac{1}{2} \;\Rightarrow\; t = 0, \frac{\pi}{3}, \pi,$ or $\frac{5\pi}{3}$.

$\dfrac{dy}{dt} = 2a\cos t - 2a\cos 2t = 2a\left(1 + \cos t - 2\cos^2 t\right) = 2a\left(1 - \cos t\right)(1 + 2\cos t) = 0 \;\Rightarrow\; t = 0, \frac{2\pi}{3},$

or $\frac{4\pi}{3}$.

Thus the graph has vertical tangents where

$t = \frac{\pi}{3}, \pi$ and $\frac{5\pi}{3}$, and horizontal tangents where

$t = \frac{2\pi}{3}$ and $\frac{4\pi}{3}$. To determine what the slope is

where $t = 0$, we use l'Hospital's Rule to evaluate

$\lim\limits_{t \to 0} \dfrac{dy/dt}{dx/dt} = 0$, so there is a horizontal tangent

there.

t	x	y
0	a	0
$\frac{\pi}{3}$	$\frac{3}{2}a$	$\frac{\sqrt{3}}{2}a$
$\frac{2\pi}{3}$	$-\frac{1}{2}a$	$\frac{3\sqrt{3}}{2}a$
π	$-3a$	0
$\frac{4\pi}{3}$	$-\frac{1}{2}a$	$-\frac{3\sqrt{3}}{2}a$
$\frac{5\pi}{3}$	$\frac{3}{2}a$	$-\frac{\sqrt{3}}{2}a$

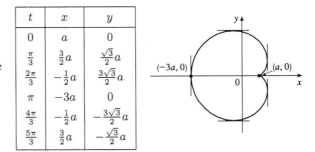

27. This curve has 10 "petals". For instance, for $-\frac{\pi}{10} \le \theta \le \frac{\pi}{10}$, there are two petals, one with $r > 0$ and one with

$r < 0$.

$$A = 10 \int_{-\pi/10}^{\pi/10} \tfrac{1}{2} r^2 \, d\theta = 5 \int_{-\pi/10}^{\pi/10} 9\cos 5\theta \, d\theta = 90 \int_{0}^{\pi/10} \cos 5\theta \, d\theta = [18\sin 5\theta]_{0}^{\pi/10} = 18$$

29. The curves intersect where $4 \cos \theta = 2$; that is, at $\left(2, \frac{\pi}{3}\right)$ and $\left(2, -\frac{\pi}{3}\right)$.

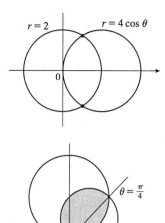

31. The curves intersect where $2 \sin \theta = \sin \theta + \cos \theta \Rightarrow$
$\sin \theta = \cos \theta \Rightarrow \theta = \frac{\pi}{4}$, and also at the origin (at which $\theta = \frac{3\pi}{4}$ on the
second curve).

$$A = \int_0^{\pi/4} \tfrac{1}{2} (2 \sin \theta)^2 \, d\theta + \int_{\pi/4}^{3\pi/4} \tfrac{1}{2} (\sin \theta + \cos \theta)^2 \, d\theta$$

$$= \int_0^{\pi/4} (1 - \cos 2\theta) \, d\theta + \tfrac{1}{2} \int_{\pi/4}^{3\pi/4} (1 + \sin 2\theta) \, d\theta$$

$$= \left[\theta - \tfrac{1}{2} \sin 2\theta\right]_0^{\pi/4} + \left[\tfrac{1}{2}\theta - \tfrac{1}{4} \cos 2\theta\right]_{\pi/4}^{3\pi/4} = \tfrac{1}{2}(\pi - 1)$$

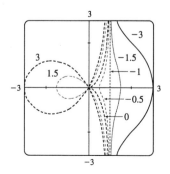

33. $x = 3t^2$, $y = 2t^3$.

$$L = \int_0^2 \sqrt{(dx/dt)^2 + (dy/dt)^2} \, dt = \int_0^2 \sqrt{(6t)^2 + (6t^2)^2} \, dt = 6 \int_0^2 t\sqrt{1 + t^2} \, dt$$

$$= \left[2\left(1 + t^2\right)^{3/2}\right]_0^2 = 2\left(5\sqrt{5} - 1\right)$$

35. $L = \int_\pi^{2\pi} \sqrt{r^2 + (dr/d\theta)^2} \, d\theta = \int_\pi^{2\pi} \sqrt{(1/\theta)^2 + \left(-1/\theta^2\right)^2} \, d\theta$

$$\overset{24}{=} \int_\pi^{2\pi} \frac{\sqrt{\theta^2 + 1}}{\theta^2} \, d\theta = \left[-\frac{\sqrt{\theta^2 + 1}}{\theta} + \ln\left|\theta + \sqrt{\theta^2 + 1}\right|\right]_\pi^{2\pi}$$

$$= \frac{\sqrt{\pi^2 + 1}}{\pi} - \frac{\sqrt{4\pi^2 + 1}}{2\pi} + \ln\left|\frac{2\pi + \sqrt{4\pi^2 + 1}}{\pi + \sqrt{\pi^2 + 1}}\right|$$

37.

$$S = \int_1^4 2\pi y \sqrt{(dx/dt)^2 + (dy/dt)^2} \, dt = \int_1^4 2\pi \left(\tfrac{1}{3}t^3 + \tfrac{1}{2}t^{-2}\right) \sqrt{\left(2/\sqrt{t}\right)^2 + (t^2 - t^{-3})^2} \, dt$$

$$= 2\pi \int_1^4 \left(\tfrac{1}{3}t^3 + \tfrac{1}{2}t^{-2}\right) \sqrt{(t^2 + t^{-3})^2} \, dt = 2\pi \int_1^4 \left(\tfrac{1}{3}t^5 + \tfrac{5}{6} + \tfrac{1}{2}t^{-5}\right) dt = 2\pi \left[\tfrac{1}{18}t^6 + \tfrac{5}{6}t - \tfrac{1}{8}t^{-4}\right]_1^4 = \frac{471{,}295}{1024}\pi$$

39. For all c except -1, the curve is asymptotic to the line $x = 1$. For
$c < -1$, the curve bulges to the right near $y = 0$. As c increases, the
bulge becomes smaller, until at $c = -1$ the curve is the straight line $x = 1$.
As c continues to increase, the curve bulges to the left, until at $c = 0$ there
is a cusp at the origin. For $c > 0$, there is a loop to the left of the origin,
whose size and roundness increase as c increases. Note that the x-intercept
of the curve is always $-c$.

41. Ellipse, center $(0, 0)$, $a = 3$, $b = 2\sqrt{2}$, $c = 1$ $\Rightarrow$
foci $(\pm 1, 0)$, vertices $(\pm 3, 0)$.

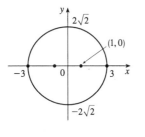

43. $6\left(y^2 - 6y + 9\right) = -(x + 1)$ $\Leftrightarrow$
$(y - 3)^2 = -\frac{1}{6}(x + 1)$, a parabola with vertex
$(-1, 3)$, opening to the left, $p = -\frac{1}{24}$ $\Rightarrow$ focus
$\left(-\frac{25}{24}, 3\right)$ and directrix $x = -\frac{23}{24}$.

45. The parabola opens upward with vertex $(0, 4)$ and $p = 2$, so its equation is $(x - 0)^2 = 4 \cdot 2 (y - 4)$ $\Leftrightarrow$
$x^2 = 8(y - 4)$.

47. The hyperbola has center $(0, 0)$ and foci on the x-axis. $c = 3$ and $b/a = \frac{1}{2}$ (from the asymptotes)
$\Rightarrow$ $9 = c^2 = a^2 + b^2 = (2b)^2 + b^2 = 5b^2$ $\Rightarrow$ $b = \frac{3}{\sqrt{5}}$ $\Rightarrow$ $a = \frac{6}{\sqrt{5}}$ $\Rightarrow$ the equation is
$\dfrac{x^2}{36/5} - \dfrac{y^2}{9/5} = 1$ $\Leftrightarrow$ $5x^2 - 20y^2 = 36$.

49. $x^2 = -y + 100$ has its vertex at $(0, 100)$, so one of the vertices of the ellipse is $(0, 100)$. Another form of the
equation of a parabola is $x^2 = 4p(y - 100)$ so $4p(y - 100) = -y + 100$ $\Rightarrow$ $4py - 4p(100) = 100 - y$ $\Rightarrow$
$4p = \dfrac{100 - y}{y - 100}$ $\Rightarrow$ $p = -\frac{1}{4}$. Therefore the shared focus is found at $\left(0, \frac{399}{4}\right)$ so $2c = \frac{399}{4} - 0$ $\Rightarrow$ $c = \frac{399}{8}$
and the center of the ellipse is $\left(0, \frac{399}{8}\right)$. So $a = 100 - \frac{399}{8} = \frac{401}{8}$ and $b^2 = a^2 - c^2 = \frac{401^2 - 399^2}{8^2} = 25$. So the
equation of the ellipse is $\dfrac{x^2}{b^2} + \dfrac{\left(y - \frac{399}{8}\right)^2}{a^2} = 1$ $\Rightarrow$ $\dfrac{x^2}{25} + \dfrac{\left(y - \frac{399}{8}\right)^2}{\left(\frac{401}{8}\right)^2} = 1$ or $\dfrac{x^2}{25} + \dfrac{(8y - 399)^2}{160{,}801} = 1$.

51. Directrix $x = 4$ $\Rightarrow$ $d = 4$, so $e = \frac{1}{3}$ $\Rightarrow$ $r = \dfrac{ed}{1 + e\cos\theta} = \dfrac{4}{3 + \cos\theta}$.

53. In polar coordinates, an equation for the circle is $r = 2a\sin\theta$. Thus, the coordinates of Q are
$x = r\cos\theta = 2a\sin\theta\cos\theta$ and $y = r\sin\theta = 2a\sin^2\theta$. The coordinates of R are $x = 2a\cot\theta$ and $y = 2a$.
Since P is the midpoint of QR, we use the midpoint formula to get $x = a(\sin\theta\cos\theta + \cot\theta)$ and
$y = a\left(1 + \sin^2\theta\right)$.

Problems Plus

1. $x = \int_1^t \frac{\cos u}{u}\, du$, $y = \int_1^t \frac{\sin u}{u}\, du$, so by FTC1, we have $\frac{dx}{dt} = \frac{\cos t}{t}$ and $\frac{dy}{dt} = \frac{\sin t}{t}$. Vertical tangent lines

occur when $\frac{dx}{dt} = 0 \iff \cos t = 0$. The parameter value corresponding to $(x, y) = (0, 0)$ is $t = 1$, so the

nearest vertical tangent occurs when $t = \frac{\pi}{2}$. Therefore, the arc length between these points is

$$L = \int_1^{\pi/2} \sqrt{\left(\frac{dx}{dt}\right)^2 + \left(\frac{dy}{dt}\right)^2}\, dt = \int_1^{\pi/2} \sqrt{\frac{\cos^2 t}{t^2} + \frac{\sin^2 t}{t^2}}\, dt = \int_1^{\pi/2} \frac{dt}{t} = [\ln t]_1^{\pi/2} = \ln \frac{\pi}{2}$$

3. (a) If $\tan \theta = \sqrt{\frac{y}{C - y}}$, then $\tan^2 \theta = \frac{y}{C - y}$, so $C \tan^2 \theta - y \tan^2 \theta = y$ and

$$y = \frac{C \tan^2 \theta}{1 + \tan^2 \theta} = \frac{C \tan^2 \theta}{\sec^2 \theta} = C \tan^2 \theta \cos^2 \theta = C \sin^2 \theta = \frac{C}{2}\,(1 - \cos 2\theta). \text{ Now}$$

$$dx = \sqrt{\frac{y}{C - y}}\, dy = \tan \theta \cdot \frac{C}{2} \cdot 2 \sin 2\theta\, d\theta = C \tan \theta \cdot 2 \sin \theta \cos \theta\, d\theta = 2C \sin^2 \theta\, d\theta = C\,(1 - \cos 2\theta)\, d\theta$$

Thus, $x = C\left(\theta - \frac{1}{2} \sin 2\theta\right) + K$ for some constant K. When $\theta = 0$, we have $y = 0$. We require that $x = 0$
when $\theta = 0$ so that the curve passes through the origin when $\theta = 0$. This yields $K = 0$. We now have
$x = \frac{1}{2} C\,(2\theta - \sin 2\theta)$, $y = \frac{1}{2} C\,(1 - \cos 2\theta)$.

(b) Setting $\phi = 2\theta$ and $r = \frac{1}{2} C$, we get $x = r\,(\phi - \sin \phi)$, $y = r\,(1 - \cos \phi)$. Comparison with Equations 11.1.1
[ET 10.1.1] shows that the curve is a cycloid.

5. (a) If (a, b) lies on the curve, then there is some parameter value t_1 such that $\frac{3t_1}{1 + t_1^3} = a$ and $\frac{3t_1^2}{1 + t_1^3} = b$. If

$t_1 = 0$, the point is $(0, 0)$, which lies on the line $y = x$. If $t_1 \neq 0$, then the point corresponding to $t = \frac{1}{t_1}$ is

given by $x = \frac{3\,(1/t_1)}{1 + (1/t_1)^3} = \frac{3t_1^2}{t_1^3 + 1} = b$, $y = \frac{3\,(1/t_1)^2}{1 + (1/t_1)^3} = \frac{3t_1}{t_1^3 + 1} = a$. So (b, a) also lies on the curve.

[Another way to see this is to do part (e) first; the result is immediate.] The curve intersects the line $y = x$ when

$\frac{3t}{1 + t^3} = \frac{3t^2}{1 + t^3} \implies t = t^2 \implies t = 0$ or 1, so the points are $(0, 0)$ and $\left(\frac{3}{2}, \frac{3}{2}\right)$.

(b) $\frac{dy}{dt} = \frac{(1 + t^3)\,(6t) - 3t^2\,(3t^2)}{(1 + t^3)^2} = \frac{6t - 3t^4}{(1 + t^3)^2} = 0$ when $6t - 3t^4 = 3t\,(2 - t^3) = 0 \implies t = 0$ or $t = \sqrt[3]{2}$,

so there are horizontal tangents at $(0, 0)$ and $\left(\sqrt[3]{2}, \sqrt[3]{4}\right)$. Using the symmetry from part (a), we see that there
are vertical tangents at $(0, 0)$ and $\left(\sqrt[3]{4}, \sqrt[3]{2}\right)$.

(c) Notice that as $t \to -1^+$, we have $x \to -\infty$ and $y \to \infty$. As $t \to -1^-$, we have $x \to \infty$ and $y \to -\infty$. Also

$$y - (-x - 1) = y + x + 1 = \frac{3t + 3t^2 + (1 + t^3)}{1 + t^3} = \frac{(t + 1)^3}{1 + t^3} = \frac{(t + 1)^2}{t^2 - t + 1} \to 0 \text{ as } t \to -1. \text{ So}$$

$y = -x - 1$ is a slant asymptote.

(d) $\dfrac{dx}{dt} = \dfrac{(1+t^3)(3) - 3t(3t^2)}{(1+t^3)^2} = \dfrac{3 - 6t^3}{(1+t^3)^2}$ and from part (b) we have $\dfrac{dy}{dt} = \dfrac{6t - 3t^4}{(1+t^3)^2}$. So

$\dfrac{dy}{dx} = \dfrac{dy/dt}{dx/dt} = \dfrac{t(2 - t^3)}{1 - 2t^3}$. Also $\dfrac{d^2y}{dx^2} = \dfrac{\dfrac{d}{dt}\left(\dfrac{dy}{dx}\right)}{dx/dt} = \dfrac{2(1+t^3)^4}{3(1 - 2t^3)^3} > 0 \iff t < \frac{1}{\sqrt[3]{2}}$. So the curve is

concave upward there and has a minimum point at $(0,0)$ and a maximum point at $\left(\sqrt[3]{2}, \sqrt[3]{4}\right)$. Using this together with the information from parts (a), (b), and (c), we sketch the curve.

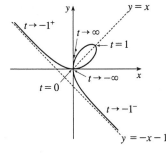

(e) $x^3 + y^3 = \left(\dfrac{3t}{1+t^3}\right)^3 + \left(\dfrac{3t^2}{1+t^3}\right)^3 = \dfrac{27t^3 + 27t^6}{(1+t^3)^3} = \dfrac{27t^3(1+t^3)}{(1+t^3)^3} = \dfrac{27t^3}{(1+t^3)^2}$ and

$3xy = 3\left(\dfrac{3t}{1+t^3}\right)\left(\dfrac{3t^2}{1+t^3}\right) = \dfrac{27t^3}{(1+t^3)^2}$, so $x^3 + y^3 = 3xy$.

(f) We start with the equation from part (e) and substitute $x = r\cos\theta$, $y = r\sin\theta$. Then $x^3 + y^3 = 3xy \Rightarrow$

$r^3\cos^3\theta + r^3\sin^3\theta = 3r^2\cos\theta\sin\theta$. For $r \ne 0$, this gives $r = \dfrac{3\cos\theta\sin\theta}{\cos^3\theta + \sin^3\theta}$. Dividing numerator and

denominator by $\cos^3\theta$, we obtain $r = \dfrac{3\left(\dfrac{1}{\cos\theta}\right)\dfrac{\sin\theta}{\cos\theta}}{1 + \dfrac{\sin^3\theta}{\cos^3\theta}} = \dfrac{3\sec\theta\tan\theta}{1 + \tan^3\theta}$.

(g) The loop corresponds to $\theta \in \left(0, \frac{\pi}{2}\right)$, so its area is

$$A = \int_0^{\pi/2} \dfrac{r^2}{2}\,d\theta = \dfrac{1}{2}\int_0^{\pi/2}\left(\dfrac{3\sec\theta\tan\theta}{1 + \tan^3\theta}\right)^2 d\theta = \dfrac{9}{2}\int_0^{\pi/2}\dfrac{\sec^2\theta\tan^2\theta}{(1 + \tan^3\theta)^2}\,d\theta$$

$$= \dfrac{9}{2}\int_0^{\infty}\dfrac{u^2\,du}{(1 + u^3)^2} \quad (\text{put } u = \tan\theta) \quad = \lim_{b\to\infty}\dfrac{9}{2}\left[-\dfrac{1}{3}(1 + u^3)^{-1}\right]_0^b = \dfrac{3}{2}$$

(h) By symmetry, the area between the folium and the line $y = -x - 1$ is equal to the enclosed area in the third quadrant, plus twice the enclosed area in the fourth quadrant. The area in the third quadrant is $\frac{1}{2}$, and since

$y = -x - 1 \Rightarrow r\sin\theta = -r\cos\theta - 1 \Rightarrow r = -\dfrac{1}{\sin\theta + \cos\theta}$, the area in the fourth quadrant is

$\dfrac{1}{2}\int_{-\pi/2}^{-\pi/4}\left[\left(-\dfrac{1}{\sin\theta + \cos\theta}\right)^2 - \left(\dfrac{3\sec\theta\tan\theta}{1 + \tan^3\theta}\right)^2\right] d\theta = \dfrac{1}{2}$ (using a CAS). Therefore, the total area is

$\frac{1}{2} + 2\left(\frac{1}{2}\right) = \frac{3}{2}$.

1. (a) A sequence is an ordered list of numbers. It can also be defined as a function whose domain is the set of positive integers.

(b) The terms a_n approach 8 as n becomes large. In fact, we can make a_n as close to 8 as we like by taking n sufficiently large.

(c) The terms a_n become large as n becomes large.

3. $a_n = 1 - (0.2)^n$, so the sequence is $\{0.8, 0.96, 0.992, 0.9984, 0.99968, \ldots\}$.

5. $a_n = \dfrac{3(-1)^n}{n!}$, so the sequence is $\left\{ \dfrac{-3}{1}, \dfrac{3}{2}, \dfrac{-3}{6}, \dfrac{3}{24}, \dfrac{-3}{120}, \ldots \right\} = \left\{ -3, \dfrac{3}{2}, -\dfrac{1}{2}, \dfrac{1}{8}, -\dfrac{1}{40}, \ldots \right\}$.

7. $a_n = \sin \dfrac{n\pi}{2}$, so the sequence is $\{1, 0, -1, 0, 1, \ldots\}$.

9. The numerators are all 1 and the denominators are powers of 2, so $a_n = \dfrac{1}{2^n}$.

11. $\{2, 7, 12, 17, \ldots\}$. Each term is larger than the preceding one by 5, so
$a_n = a_1 + d(n-1) = 2 + 5(n-1) = 5n - 3$.

13. $\left\{1, -\dfrac{2}{3}, \dfrac{4}{9}, -\dfrac{8}{27}, \ldots \right\}$. Each term is $-\dfrac{2}{3}$ times the preceding one, so $a_n = \left(-\dfrac{2}{3}\right)^{n-1}$.

15. $a_n = n(n-1)$. $a_n \to \infty$ as $n \to \infty$, so the sequence diverges.

17. $a_n = \dfrac{3 + 5n^2}{n + n^2} = \dfrac{5 + 3/n^2}{1 + 1/n}$, so $a_n \to \dfrac{5+0}{1+0} = 5$ as $n \to \infty$. Converges

19. $a_n = \dfrac{2^n}{3^{n+1}} = \dfrac{1}{3}\left(\dfrac{2}{3}\right)^n$, so $\lim\limits_{n\to\infty} a_n = \dfrac{1}{3} \lim\limits_{n\to\infty} \left(\dfrac{2}{3}\right)^n = \dfrac{1}{3} \cdot 0 = 0$ by (7) with $r = \dfrac{2}{3}$. Converges

21. $a_n = \dfrac{(-1)^{n-1} n}{n^2 + 1} = \dfrac{(-1)^{n-1}}{n + 1/n}$, so $0 \le |a_n| = \dfrac{1}{n + 1/n} \le \dfrac{1}{n} \to 0$ as $n \to \infty$, so $a_n \to 0$ by the Squeeze Theorem and Theorem 5. Converges

23. $a_n = 2 + \cos n\pi$, so
$\{a_n\} = \{2 + \cos\pi, 2 + \cos 2\pi, 2 + \cos 3\pi, 2 + \cos 4\pi, \ldots\} = \{2 - 1, 2 + 1, 2 - 1, 2 + 1, \ldots\}$
$= \{1, 3, 1, 3, \ldots\}$
This sequence oscillates between 1 and 3, so it diverges.

25. $0 < \dfrac{3 + (-1)^n}{n^2} \le \dfrac{4}{n^2}$ and $\lim\limits_{n\to\infty} \dfrac{4}{n^2} = 0$, so $\left\{ \dfrac{3 + (-1)^n}{n^2} \right\}$ converges to 0 by the Squeeze Theorem.

27. $\lim\limits_{x\to\infty} \dfrac{\ln(x^2)}{x} = \lim\limits_{x\to\infty} \dfrac{2\ln x}{x} \overset{H}{=} \lim\limits_{x\to\infty} \dfrac{2/x}{1} = 0$, so by Theorem 2, $\left\{ \dfrac{\ln(n^2)}{n} \right\}$ converges to 0.

29. $b_n = \sqrt{n+2} - \sqrt{n} = (\sqrt{n+2} - \sqrt{n}) \dfrac{\sqrt{n+2} + \sqrt{n}}{\sqrt{n+2} + \sqrt{n}} = \dfrac{2}{\sqrt{n+2} + \sqrt{n}} < \dfrac{2}{2\sqrt{n}} = \dfrac{1}{\sqrt{n}} \to 0$ as $n \to \infty$. So

by the Squeeze Theorem with $a_n = 0$ and $c_n = 1/\sqrt{n}$, $\left\{\sqrt{n+2} - \sqrt{n}\right\}$ converges to 0.

31. $\lim\limits_{x \to \infty} \dfrac{x}{2^x} \overset{\text{H}}{=} \lim\limits_{x \to \infty} \dfrac{1}{(\ln 2)\, 2^x} = 0$, so by Theorem 2, $\left\{n2^{-n}\right\}$ converges to 0.

33. $0 \le \dfrac{\cos^2 n}{2^n} \le \dfrac{1}{2^n}$ [since $0 \le \cos^2 n \le 1$], so since $\lim\limits_{n \to \infty} \dfrac{1}{2^n} = 0$, $\left\{\dfrac{\cos^2 n}{2^n}\right\}$ converges to 0 by the Squeeze

Theorem.

35. The series converges, since

$$a_n = \dfrac{1 + 2 + 3 + \cdots + n}{n^2} = \dfrac{n(n+1)/2}{n^2} \quad \text{[sum of the first } n \text{ positive integers]} \quad = \dfrac{n+1}{2n} = \dfrac{1 + 1/n}{2} \to \dfrac{1}{2}$$

as $n \to \infty$.

37. $a_n = \dfrac{1}{2} \cdot \dfrac{2}{2} \cdot \dfrac{3}{2} \cdot \ldots \cdot \dfrac{(n-1)}{2} \cdot \dfrac{n}{2} \ge \dfrac{1}{2} \cdot \dfrac{n}{2} = \dfrac{n}{4} \to \infty$ as $n \to \infty$, so $\{a_n\}$ diverges.

39.

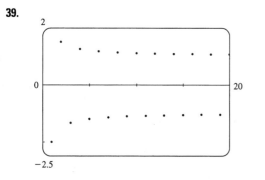

From the graph, we see that the sequence

$\left\{(-1)^n \dfrac{n+1}{n}\right\}$ is divergent, since it oscillates

between 1 and -1 (approximately).

41.

From the graph, it appears that the sequence

converges to about 0.78.

$$\lim_{n \to \infty} \dfrac{2n}{2n+1} = \lim_{n \to \infty} \dfrac{2}{2 + 1/n} = 1, \text{ so}$$

$$\lim_{n \to \infty} \arctan\left(\dfrac{2n}{2n+1}\right) = \arctan 1 = \dfrac{\pi}{4}.$$

43.

From the graph, it appears that the sequence converges to 0.

$$0 < a_n = \dfrac{n^3}{n!} = \dfrac{n}{n} \cdot \dfrac{n}{(n-1)} \cdot \dfrac{n}{(n-2)} \cdot \dfrac{n}{(n-3)} \cdot \dfrac{1}{(n-3)} \ldots \cdot \dfrac{1}{3} \cdot \dfrac{1}{2} \cdot \dfrac{1}{1}$$

$$\le \dfrac{n^2}{(n-1)(n-2)(n-3)} \quad \text{(for } n \ge 4\text{)}$$

$$= \dfrac{1/n}{(1 - 1/n)(1 - 2/n)(1 - 3/n)} \to 0 \text{ as } n \to \infty$$

So by the Squeeze Theorem, $\{n^3/n!\}$ converges to 0.

45.

From the graph, it appears that the sequence approaches 0.

$$0 < a_n = \frac{1 \cdot 3 \cdot 5 \cdots \cdots (2n-1)}{(2n)^n} = \frac{1}{2n} \cdot \frac{3}{2n} \cdot \frac{5}{2n} \cdots \cdots \frac{2n-1}{2n}$$

$$\leq \frac{1}{2n} \cdot (1) \cdot (1) \cdots \cdots (1) = \frac{1}{2n} \to 0 \text{ as } n \to \infty$$

So by the Squeeze Theorem, $\left\{ \dfrac{1 \cdot 3 \cdot 5 \cdots \cdots (2n-1)}{(2n)^n} \right\}$ converges

to 0.

47. (a) $a_n = 1000\,(1.06)^n$ $\Rightarrow$ $a_1 = 1060$, $a_2 = 1123.60$, $a_3 = 1191.02$, $a_4 = 1262.48$, and $a_5 = 1338.23$.

(b) $\lim\limits_{n \to \infty} a_n = 1000 \lim\limits_{n \to \infty} (1.06)^n$, so the sequence diverges by (7) with $r = 1.06 > 1$.

49. If $|r| \geq 1$, then $\{r^n\}$ diverges by (7), so $\{nr^n\}$ diverges also, since $|nr^n| = n\,|r^n| \geq |r^n|$. If $|r| < 1$ then

$$\lim_{x \to \infty} xr^x = \lim_{x \to \infty} \frac{x}{r^{-x}} \overset{\text{H}}{=} \lim_{x \to \infty} \frac{1}{(-\ln r)\,r^{-x}} = \lim_{x \to \infty} \frac{r^x}{-\ln r} = 0, \text{ so } \lim_{n \to \infty} nr^n = 0, \text{ and hence } \{nr^n\} \text{ converges}$$

whenever $|r| < 1$.

51. Since $\{a_n\}$ is a decreasing sequence, $a_n > a_{n+1}$ for all $n \geq 1$. Because all of its terms lie between 5 and 8, $\{a_n\}$ is a bounded sequence. By the Monotonic Sequence Theorem, $\{a_n\}$ is convergent, that is, $\{a_n\}$ has a limit L. L must be less than 8 since $\{a_n\}$ is decreasing, so $5 \leq L < 8$.

53. $a_n = \dfrac{1}{2n+3}$ is decreasing since $a_{n+1} = \dfrac{1}{2(n+1)+3} = \dfrac{1}{2n+5} < \dfrac{1}{2n+3} = a_n$ for each $n \geq 1$. The

sequence is bounded since $0 < a_n \leq \frac{1}{5}$ for all $n \geq 1$.

55. $a_n = \cos(n\pi/2)$ is not monotonic. The first few terms are $0, -1, 0, 1, 0, -1, 0, 1, \ldots$. In fact, the sequence consists of the terms $0, -1, 0, 1$ repeated over and over again in that order. The sequence is bounded since $|a_n| \leq 1$ for all $n \geq 1$.

57. $a_n = \dfrac{n}{n^2+1}$ defines a decreasing sequence since for $f(x) = \dfrac{x}{x^2+1}$,

$$f'(x) = \frac{(x^2+1)(1) - x(2x)}{(x^2+1)^2} = \frac{1-x^2}{(x^2+1)^2} \leq 0 \text{ for } x \geq 1. \text{ The sequence is bounded since } 0 < a_n \leq \tfrac{1}{2} \text{ for all}$$

$n \geq 1$.

59. $a_1 = 2^{1/2}$, $a_2 = 2^{3/4}$, $a_3 = 2^{7/8}$, $\ldots$, so $a_n = 2^{(2^n-1)/2^n} = 2^{1-(1/2^n)}$. $\lim\limits_{n \to \infty} a_n = \lim\limits_{n \to \infty} 2^{1-(1/2^n)} = 2^1 = 2$.

Alternate Solution: Let $L = \lim\limits_{n \to \infty} a_n$. (We could show the limit exists by showing that $\{a_n\}$ is bounded and

increasing.) So L must satisfy $L = \sqrt{2 \cdot L}$ $\Rightarrow$ $L^2 = 2L$ $\Rightarrow$ $L(L-2) = 0$ ($L \neq 0$ since the sequence

increases), so $L = 2$.

61. We show by induction that $\{a_n\}$ is increasing and bounded above by 3.

Let P_n be the proposition that $a_{n+1} > a_n$ and $0 < a_n < 3$. Clearly P_1 is true. Assume that P_n is true. Then

$$a_{n+1} > a_n \quad \Rightarrow \quad \frac{1}{a_{n+1}} < \frac{1}{a_n} \quad \Rightarrow \quad -\frac{1}{a_{n+1}} > -\frac{1}{a_n}.$$

Now $a_{n+2} = 3 - \dfrac{1}{a_{n+1}} > 3 - \dfrac{1}{a_n} = a_{n+1} \quad \Leftrightarrow \quad P_{n+1}$. This proves that $\{a_n\}$ is increasing and bounded above

by 3, so $1 = a_1 < a_n < 3$, that is, $\{a_n\}$ is bounded, and hence convergent by the Monotonic Sequence Theorem. If

$L = \lim\limits_{n\to\infty} a_n$, then $\lim\limits_{n\to\infty} a_{n+1} = L$ also, so L must satisfy $L = 3 - 1/L \quad \Rightarrow \quad L^2 - 3L + 1 = 0 \quad \Rightarrow$

$L = \frac{3 \pm \sqrt{5}}{2}$. But $L > 1$, so $L = \frac{3 + \sqrt{5}}{2}$.

63. (a) Let a_n be the number of rabbit pairs in the nth month. Clearly $a_1 = 1 = a_2$. In the nth month, each pair that is

2 or more months old (that is, a_{n-2} pairs) will produce a new pair to add to the a_{n-1} pairs already present.

Thus, $a_n = a_{n-1} + a_{n-2}$, so that $\{a_n\} = \{f_n\}$, the Fibonacci sequence.

(b) $a_n = \dfrac{f_{n+1}}{f_n} \quad \Rightarrow \quad a_{n-1} = \dfrac{f_n}{f_{n-1}} = \dfrac{f_{n-1} + f_{n-2}}{f_{n-1}} = 1 + \dfrac{f_{n-2}}{f_{n-1}} = 1 + \dfrac{1}{f_{n-1}/f_{n-2}} = 1 + \dfrac{1}{a_{n-2}}$. If

$L = \lim\limits_{n\to\infty} a_n$, then $L = \lim\limits_{n\to\infty} a_{n-1}$ and $L = \lim\limits_{n\to\infty} a_{n-2}$, so L must satisfy $L = 1 + \dfrac{1}{L} \quad \Rightarrow$

$L^2 - L - 1 = 0 \quad \Rightarrow \quad L = \frac{1 + \sqrt{5}}{2}$ (since L must be positive).

65. (a)

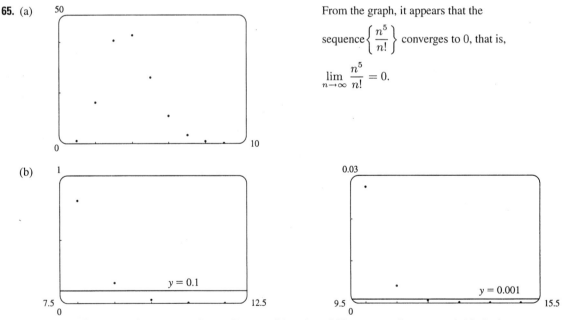

From the graph, it appears that the

sequence $\left\{\dfrac{n^5}{n!}\right\}$ converges to 0, that is,

$$\lim_{n\to\infty} \frac{n^5}{n!} = 0.$$

(b)

From the first graph, it seems that the smallest possible value of N corresponding to $\varepsilon = 0.1$ is 9, since

$n^5/n! < 0.1$ whenever $n \geq 10$, but $9^5/9! > 0.1$. From the second graph, it seems that for $\varepsilon = 0.001$, the

smallest possible value for N is 11.

67. If $\lim\limits_{n\to\infty} |a_n| = 0$ then $\lim\limits_{n\to\infty} -|a_n| = 0$, and since $-|a_n| \leq a_n \leq |a_n|$, we have that $\lim\limits_{n\to\infty} a_n = 0$ by the Squeeze

Theorem.

69. (a) First we show that $a > a_1 > b_1 > b$.

$$a_1 - b_1 = \frac{a+b}{2} - \sqrt{ab} = \frac{1}{2}\left(a - 2\sqrt{ab} + b\right) = \frac{1}{2}\left(\sqrt{a} - \sqrt{b}\right)^2 > 0 \quad \text{(since } a > b) \quad \Rightarrow \quad a_1 > b_1. \text{ Also}$$

$$a - a_1 = a - \frac{1}{2}(a+b) = \frac{1}{2}(a-b) > 0 \text{ and } b - b_1 = b - \sqrt{ab} = \sqrt{b}\left(\sqrt{b} - \sqrt{a}\right) < 0, \text{ so } a > a_1 > b_1 > b.$$

In the same way we can show that $a_1 > a_2 > b_2 > b_1$ and so the given assertion is true for $n = 1$. Suppose it is
true for $n = k$, that is, $a_k > a_{k+1} > b_{k+1} > b_k$. Then

$$a_{k+2} - b_{k+2} = \frac{1}{2}(a_{k+1} + b_{k+1}) - \sqrt{a_{k+1}b_{k+1}} = \frac{1}{2}\left(a_{k+1} - 2\sqrt{a_{k+1}b_{k+1}} + b_{k+1}\right)$$

$$= \frac{1}{2}\left(\sqrt{a_{k+1}} - \sqrt{b_{k+1}}\right)^2 > 0$$

$$a_{k+1} - a_{k+2} = a_{k+1} - \frac{1}{2}(a_{k+1} + b_{k+1}) = \frac{1}{2}(a_{k+1} - b_{k+1}) > 0$$

and $b_{k+1} - b_{k+2} = b_{k+1} - \sqrt{a_{k+1}b_{k+1}} = \sqrt{b_{k+1}}\left(\sqrt{b_{k+1}} - \sqrt{a_{k+1}}\right) < 0 \quad \Rightarrow$

$a_{k+1} > a_{k+2} > b_{k+2} > b_{k+1}$, so the assertion is true for $n = k + 1$. Thus, it is true for all n by mathematical
induction.

(b) From part (a) we have $a > a_n > a_{n+1} > b_{n+1} > b_n > b$, which shows that both sequences, $\{a_n\}$ and $\{b_n\}$,
are monotonic and bounded. So they are both convergent by the Monotonic Sequence Theorem.

(c) Let $\lim\limits_{n \to \infty} a_n = \alpha$ and $\lim\limits_{n \to \infty} b_n = \beta$. Then $\lim\limits_{n \to \infty} a_{n+1} = \lim\limits_{n \to \infty} \dfrac{a_n + b_n}{2} \quad \Rightarrow \quad \alpha = \dfrac{\alpha + \beta}{2} \quad \Rightarrow$

$2\alpha = \alpha + \beta \quad \Rightarrow \quad \alpha = \beta$.

12.2 Series   ET 11.2

1. (a) A sequence is an ordered list of numbers whereas a series is the *sum* of a list of numbers.

(b) A series is convergent if the sequence of partial sums is a convergent sequence. A series is divergent if it is not
convergent.

3.

n	s_n
1	3.33333
2	4.44444
3	4.81481
4	4.93827
5	4.97942
6	4.99314
7	4.99771
8	4.99924
9	4.99975
10	4.99992
11	4.99997
12	4.99999

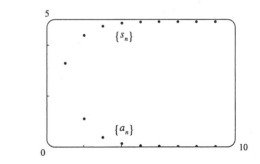

From the graph, it seems that the series converges. In fact, it is a geometric

series with $a = \frac{10}{3}$ and $r = \frac{1}{3}$, so its sum is $\sum\limits_{n=1}^{\infty} \dfrac{10}{3^n} = \dfrac{10/3}{1 - 1/3} = 5$. Note that

the dot corresponding to $n = 1$ is part of both $\{a_n\}$ and $\{s_n\}$.

5.

n	s_n
1	0.50000
2	1.16667
3	1.91667
4	2.71667
5	3.55000
6	4.40714
7	5.28214
8	6.17103
9	7.07103
10	7.98012

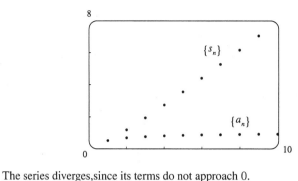

The series diverges, since its terms do not approach 0.

7.

n	s_n
1	0.64645
2	0.80755
3	0.87500
4	0.91056
5	0.93196
6	0.94601
7	0.95581
8	0.96296
9	0.96838
10	0.97259

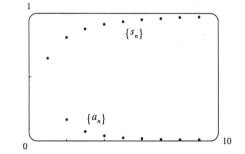

From the graph, it seems that the series converges to 1. To find the sum, we write

$$s_n = \sum_{i=1}^{n} \left(\frac{1}{i^{1.5}} - \frac{1}{(i+1)^{1.5}} \right) = \left(1 - \frac{1}{2^{1.5}} \right) + \left(\frac{1}{2^{1.5}} - \frac{1}{3^{1.5}} \right)$$

$$+ \left(\frac{1}{3^{1.5}} - \frac{1}{4^{1.5}} \right) + \cdots + \left(\frac{1}{n^{1.5}} - \frac{1}{(n+1)^{1.5}} \right) = 1 - \frac{1}{(n+1)^{1.5}}$$

So the sum is $\lim_{n \to \infty} s_n = 1$.

9. (a) $\lim_{n \to \infty} a_n = \lim_{n \to \infty} \dfrac{2n}{3n+1} = \dfrac{2}{3}$, so the *sequence* $\{a_n\}$ is convergent by (12.1.1 [ET 11.1.1]).

(b) Since $\lim_{n \to \infty} a_n = \frac{2}{3} \neq 0$, the *series* $\sum_{n=1}^{\infty} a_n$ is divergent by the Test for Divergence (7).

11. $4 + \frac{8}{5} + \frac{16}{25} + \frac{32}{125} + \cdots$ is a geometric series with $a = 4$ and $r = \frac{2}{5}$. Since $|r| = \frac{2}{5} < 1$, the series converges to $\frac{4}{1 - 2/5} = \frac{4}{3/5} = \frac{20}{3}$.

13. $-2 + \frac{5}{2} - \frac{25}{8} + \frac{125}{32} - \cdots$ is a geometric series with $a = -2$ and $r = \frac{5/2}{-2} = -\frac{5}{4}$. Since $|r| = \frac{5}{4} > 1$, the series diverges by (4).

15. $\sum_{n=1}^{\infty} 5 \left(\frac{2}{3}\right)^{n-1}$ is a geometric series with $a = 5$ and $r = \frac{2}{3}$. Since $|r| = \frac{2}{3} < 1$, the series converges to

$\frac{a}{1-r} = \frac{5}{1-2/3} = \frac{5}{1/3} = 15$.

17. $\sum_{n=1}^{\infty} \frac{(-3)^{n-1}}{4^n} = \frac{1}{4} \sum_{n=1}^{\infty} \left(-\frac{3}{4}\right)^{n-1}$. The latter series is geometric with $a = 1$ and $r = -\frac{3}{4}$. Since $|r| = \frac{3}{4} < 1$, it

converges to $\frac{1}{1-(-3/4)} = \frac{4}{7}$. Thus, the given series converges to $\left(\frac{1}{4}\right)\left(\frac{4}{7}\right) = \frac{1}{7}$.

19. For $\sum_{n=1}^{\infty} 3^{-n} 8^{n+1} = \sum_{n=1}^{\infty} 8 \left(\frac{8}{3}\right)^n$, $a = \frac{64}{3}$ and $r = \frac{8}{3} > 1$, so the series diverges.

21. $\sum_{n=1}^{\infty} \frac{n}{n+5}$ diverges since $\lim_{n\to\infty} a_n = \lim_{n\to\infty} \frac{n}{n+5} = 1 \neq 0$. [Use (7), the Test for Divergence.]

23. Converges. $s_n = \sum_{i=1}^{n} \frac{1}{i(i+2)} = \sum_{i=1}^{n} \left(\frac{1/2}{i} - \frac{1/2}{i+2}\right)$ (using partial fractions) $= \frac{1}{2} \sum_{i=1}^{n} \left(\frac{1}{i} - \frac{1}{i+2}\right)$. The

latter sum is a telescoping series:

$\left(1 - \frac{1}{3}\right) + \left(\frac{1}{2} - \frac{1}{4}\right) + \left(\frac{1}{3} - \frac{1}{5}\right) + \cdots + \left(\frac{1}{n-1} - \frac{1}{n+1}\right) + \left(\frac{1}{n} - \frac{1}{n+2}\right) = 1 + \frac{1}{2} - \frac{1}{n+1} - \frac{1}{n+2}$

Thus, $\sum_{n=1}^{\infty} \frac{1}{n(n+2)} = \frac{1}{2} \lim_{n\to\infty} \left(1 + \frac{1}{2} - \frac{1}{n+1} - \frac{1}{n+2}\right) = \frac{1}{2}\left(1 + \frac{1}{2}\right) = \frac{3}{4}$.

25. $\sum_{n=1}^{\infty} [2(0.1)^n + (0.2)^n] = 2 \sum_{n=1}^{\infty} (0.1)^n + \sum_{n=1}^{\infty} (0.2)^n$. These are convergent geometric series and so by

Theorem 8, their sum is also convergent. $2\left(\frac{0.1}{1-0.1}\right) + \frac{0.2}{1-0.2} = \frac{2}{9} + \frac{1}{4} = \frac{17}{36}$

27. $\lim_{n\to\infty} a_n = \lim_{n\to\infty} \frac{n}{\sqrt{1+n^2}} = \lim_{n\to\infty} \frac{1}{\sqrt{1+1/n^2}} = 1 \neq 0$, so the series diverges by the Test for Divergence.

29. Converges. $\sum_{n=1}^{\infty} \frac{3^n + 2^n}{6^n} = \sum_{n=1}^{\infty} \left(\frac{3^n}{6^n} + \frac{2^n}{6^n}\right) = \sum_{n=1}^{\infty} \left[\left(\frac{1}{2}\right)^n + \left(\frac{1}{3}\right)^n\right] = \frac{1/2}{1-1/2} + \frac{1/3}{1-1/3} = 1 + \frac{1}{2} = \frac{3}{2}$

31. $\lim_{n\to\infty} a_n = \lim_{n\to\infty} \arctan n = \frac{\pi}{2} \neq 0$, so the series diverges by the Test for Divergence.

33. $s_n = (\ln 1 - \ln 2) + (\ln 2 - \ln 3) + (\ln 3 - \ln 4) + \cdots + [\ln n - \ln(n+1)] = \ln 1 - \ln(n+1) = -\ln(n+1)$

(telescoping series). Thus, $\lim_{n\to\infty} s_n = -\infty$, so the series is divergent.

35. $0.\overline{2} = \frac{2}{10} + \frac{2}{10^2} + \cdots = \frac{2/10}{1-1/10} = \frac{2}{9}$

37. $3.\overline{417} = 3 + \frac{417}{10^3} + \frac{417}{10^6} + \cdots = 3 + \frac{417/10^3}{1-1/10^3} = 3 + \frac{417}{999} = \frac{3414}{999} = \frac{1138}{333}$

39. $0.12\overline{3456} = \frac{123}{1000} + \frac{0.000456}{1-0.001} = \frac{123}{1000} + \frac{456}{999,000} = \frac{123,333}{999,000} = \frac{41,111}{333,000}$

41. $\sum_{n=1}^{\infty} \frac{x^n}{3^n}$ is a geometric series with $r = \frac{x}{3}$, so the series converges $\Leftrightarrow |r| < 1 \Leftrightarrow \frac{|x|}{3} < 1 \Leftrightarrow |x| < 3$. In

that case, the sum of the series is $\frac{x/3}{1-x/3} = \frac{x}{3-x}$.

43. $\sum_{n=0}^{\infty} 4^n x^n = \sum_{n=0}^{\infty} (4x)^n$ is a geometric series with $r = 4x$, so the series converges $\Leftrightarrow |r| < 1 \Leftrightarrow$

$4|x| < 1 \Leftrightarrow |x| < \frac{1}{4}$. In that case, the sum of the series is $\frac{1}{1-4x}$.

45. $\sum_{n=0}^{\infty} \left(\dfrac{1}{x}\right)^n$ is geometric with $r = \dfrac{1}{x}$, so it converges whenever $\left|\dfrac{1}{x}\right| < 1 \Leftrightarrow |x| > 1 \Leftrightarrow x > 1$ or $x < -1$,

and the sum is $\dfrac{1}{1 - 1/x} = \dfrac{x}{x - 1}$.

47. After defining f, We use `convert(f,parfrac);` in Maple, `Apart` in Mathematica, or `Expand Rational`

and `Simplify` in Derive to find that the general term is $\dfrac{1}{(4n + 1)(4n - 3)} = -\dfrac{1/4}{4n + 1} + \dfrac{1/4}{4n - 3}$. So the nth

partial sum is

$$s_n = \sum_{k=1}^{n} \left(-\frac{1/4}{4k + 1} + \frac{1/4}{4k - 3}\right) = \frac{1}{4}\left(\frac{1}{4k - 3} - \frac{1}{4k + 1}\right)$$

$$= \frac{1}{4}\left[\left(1 - \frac{1}{5}\right) + \left(\frac{1}{5} - \frac{1}{9}\right) + \left(\frac{1}{9} - \frac{1}{13}\right) + \cdots + \left(\frac{1}{4n - 3} - \frac{1}{4n + 1}\right)\right] = \frac{1}{4}\left(1 - \frac{1}{4n + 1}\right)$$

The series converges to $\lim_{n\to\infty} s_n = \frac{1}{4}$. This can be confirmed by directly computing the sum using

`sum(f,1..infinity);` (in Maple), `Sum[f, {n,1,Infinity}]` (in Mathematica), or `Calculus Sum`

(from 1 to ∞) and `Simplify` (in Derive).

49. For $n = 1$, $a_1 = 0$ since $s_1 = 0$. For $n > 1$,

$$a_n = s_n - s_{n-1} = \frac{n - 1}{n + 1} - \frac{(n - 1) - 1}{(n - 1) + 1} = \frac{(n - 1)n - (n + 1)(n - 2)}{(n + 1)n} = \frac{2}{n(n + 1)}$$

Also, $\sum_{n=1}^{\infty} a_n = \lim_{n\to\infty} s_n = \lim_{n\to\infty} \dfrac{1 - 1/n}{1 + 1/n} = 1$.

51. (a) The first step in the chain occurs when the local government spends D dollars. The people who receive it spend

a fraction c of those D dollars, that is, Dc dollars. Those who receive the Dc dollars spend a fraction c of it,

that is, Dc^2 dollars. Continuing in this way, we see that the total spending after n transactions is

$$S_n = D + Dc + Dc^2 + \cdots + Dc^{n-1} = \frac{D(1 - c^n)}{1 - c} \text{ by (3).}$$

(b) $\lim_{n\to\infty} S_n = \lim_{n\to\infty} \dfrac{D(1 - c^n)}{1 - c} = \dfrac{D}{1 - c} \lim_{n\to\infty} (1 - c^n) = \dfrac{D}{1 - c}$ (since $0 < c < 1 \Rightarrow \lim_{n\to\infty} c^n = 0$)

$= \dfrac{D}{s}$ (since $c + s = 1$) $= kD$ (since $k = 1/s$)

If $c = 0.8$, then $s = 1 - c = 0.2$ and the multiplier is $k = 1/s = 5$.

53. $\sum_{n=2}^{\infty} (1 + c)^{-n}$ is a geometric series with $a = (1 + c)^{-2}$ and $r = (1 + c)^{-1}$, so the series converges when

$\left|(1 + c)^{-1}\right| < 1 \Leftrightarrow |1 + c| > 1 \Leftrightarrow 1 + c > 1$ or $1 + c < -1 \Leftrightarrow c > 0$ or $c < -2$. We calculate the

sum of the series and set it equal to 2: $\dfrac{(1 + c)^{-2}}{1 - (1 + c)^{-1}} = 2 \Leftrightarrow \left(\dfrac{1}{1 + c}\right)^2 = 2 - 2\left(\dfrac{1}{1 + c}\right) \Leftrightarrow$

$1 = 2(1 + c)^2 - 2(1 + c) = 0 \Leftrightarrow 2c^2 + 2c - 1 = 0 \Leftrightarrow c = \dfrac{-2 \pm \sqrt{12}}{4} = \dfrac{\pm\sqrt{3} - 1}{2}$. However, the negative

root is inadmissible because $-2 < \dfrac{-\sqrt{3} - 1}{2} < 0$. So $c = \dfrac{\sqrt{3} - 1}{2}$.

55. Let d_n be the diameter of C_n. We draw lines from the centers of the C_i to the center of D (or C), and using the Pythagorean Theorem, we can write $1^2 + \left(1 - \frac{1}{2}d_1\right)^2 = \left(1 + \frac{1}{2}d_1\right)^2$ $\Leftrightarrow$

$1 = \left(1 + \frac{1}{2}d_1\right)^2 - \left(1 - \frac{1}{2}d_1\right)^2 = 2d_1$ (difference of squares)

$\Rightarrow \ d_1 = \frac{1}{2}$. Similarly,

$1 = \left(1 + \frac{1}{2}d_2\right)^2 - \left(1 - d_1 - \frac{1}{2}d_2\right)^2 = 2d_2 + 2d_1 - d_1^2 - d_1 d_2$

$= (2 - d_1)(d_1 + d_2)$ $\Leftrightarrow$

$d_2 = \dfrac{1}{2 - d_1} - d_1 = \dfrac{(1 - d_1)^2}{2 - d_1}$, $1 = \left(1 + \frac{1}{2}d_3\right)^2 - \left(1 - d_1 - d_2 - \frac{1}{2}d_3\right)^2$ $\Leftrightarrow$ $d_3 = \dfrac{[1 - (d_1 + d_2)]^2}{2 - (d_1 + d_2)}$, and in

general, $d_{n+1} = \dfrac{\left(1 - \sum_{i=1}^{n} d_i\right)^2}{2 - \sum_{i=1}^{n} d_i}$. If we actually calculate d_2 and d_3 from the formulas above, we find that they

are $\dfrac{1}{6} = \dfrac{1}{2 \cdot 3}$ and $\dfrac{1}{12} = \dfrac{1}{3 \cdot 4}$ respectively, so we suspect that in general, $d_n = \dfrac{1}{n(n+1)}$. To prove this, we use

induction: assume that for all $k \leq n$, $d_k = \dfrac{1}{k(k+1)} = \dfrac{1}{k} - \dfrac{1}{k+1}$. Then

$\displaystyle\sum_{i=1}^{n} d_i = 1 - \dfrac{1}{n+1} = \dfrac{n}{n+1}$ (telescoping sum). Substituting this into our formula for d_{n+1}, we get

$d_{n+1} = \dfrac{\left[1 - \dfrac{n}{n+1}\right]^2}{2 - \left(\dfrac{n}{n+1}\right)} = \dfrac{\dfrac{1}{(n+1)^2}}{\dfrac{n+2}{n+1}} = \dfrac{1}{(n+1)(n+2)}$, and the induction is complete.

Now, we observe that the partial sums $\sum_{i=1}^{n} d_i$ of the diameters of the circles approach 1 as $n \to \infty$; that is,

$\displaystyle\sum_{n=1}^{\infty} a_n = \sum_{n=1}^{\infty} \dfrac{1}{n(n+1)} = 1$, which is what we wanted to prove.

57. The series $1 - 1 + 1 - 1 + 1 - 1 + \cdots$ diverges (geometric series with $r = -1$) so we cannot say that $0 = 1 - 1 + 1 - 1 + 1 - 1 + \cdots$.

59. $\sum_{n=1}^{\infty} c a_n = \lim_{n \to \infty} \sum_{i=1}^{n} c a_i = \lim_{n \to \infty} c \sum_{i=1}^{n} a_i = c \lim_{n \to \infty} \sum_{i=1}^{n} a_i = c \sum_{n=1}^{\infty} a_n$, which exists by hypothesis.

61. Suppose on the contrary that $\sum (a_n + b_n)$ converges. Then by Theorem 8(iii), so would $\sum [(a_n + b_n) - a_n] = \sum b_n$, a contradiction.

63. The partial sums $\{s_n\}$ form an increasing sequence, since $s_n - s_{n-1} = a_n > 0$ for all n. Also, the sequence $\{s_n\}$ is bounded since $s_n \leq 1000$ for all n. So by Theorem 12.1.10 [ET 11.1.10], the sequence of partial sums converges, that is, the series $\sum a_n$ is convergent.

65. (a) At the first step, only the interval $\left(\frac{1}{3}, \frac{2}{3}\right)$ (length $\frac{1}{3}$) is removed. At the second step, we remove the intervals

$\left(\frac{1}{9}, \frac{2}{9}\right)$ and $\left(\frac{7}{9}, \frac{8}{9}\right)$, which have a total length of $2 \cdot \left(\frac{1}{3}\right)^2$. At the third step, we remove 2^2 intervals, each of

length $\left(\frac{1}{3}\right)^3$. In general, at the nth step we remove 2^{n-1} intervals, each of length $\left(\frac{1}{3}\right)^n$, for a length of

$2^{n-1} \cdot \left(\frac{1}{3}\right)^n = \frac{1}{3}\left(\frac{2}{3}\right)^{n-1}$. Thus, the total length of all removed intervals is $\sum_{n=1}^{\infty} \frac{1}{3}\left(\frac{2}{3}\right)^{n-1} = \frac{1/3}{1-2/3} = 1$

(geometric series with $a = \frac{1}{3}$ and $r = \frac{2}{3}$). Notice that at the nth step, the leftmost interval that is removed is

$\left(\left(\frac{1}{3}\right)^n, \left(\frac{2}{3}\right)^n\right)$, so we never remove 0, and 0 is in the Cantor set. Also, the rightmost interval removed is

$\left(1 - \left(\frac{2}{3}\right)^n, 1 - \left(\frac{1}{3}\right)^n\right)$, so 1 is never removed. Some other numbers in the Cantor set are $\frac{1}{3}, \frac{2}{3}, \frac{1}{9}, \frac{2}{9}, \frac{7}{9}$, and $\frac{8}{9}$.

(b) The area removed at the first step is $\frac{1}{9}$; at the second step, $8 \cdot \left(\frac{1}{9}\right)^2$; at the third step, $(8)^2 \cdot \left(\frac{1}{9}\right)^3$. In general, the

area removed at the nth step is $(8)^{n-1}\left(\frac{1}{9}\right)^n = \frac{1}{9}\left(\frac{8}{9}\right)^{n-1}$, so the total area of all removed squares is

$\sum_{n=1}^{\infty} \frac{1}{9}\left(\frac{8}{9}\right)^{n-1} = \frac{1/9}{1-8/9} = 1.$

67. (a) $\sum_{n=1}^{\infty} \frac{n}{(n+1)!} \Rightarrow s_1 = \frac{1}{1\cdot 2} = \frac{1}{2}, s_2 = \frac{1}{2} + \frac{2}{1\cdot 2\cdot 3} = \frac{5}{6}, s_3 = \frac{5}{6} + \frac{3}{1\cdot 2\cdot 3\cdot 4} = \frac{23}{24},$

$s_4 = \frac{23}{24} + \frac{4}{1\cdot 2\cdot 3\cdot 4\cdot 5} = \frac{119}{120}.$ The denominators are $(n+1)!$, so a guess would be $s_n = \frac{(n+1)!-1}{(n+1)!}.$

(b) For $n = 1$, $s_1 = \frac{1}{2} = \frac{2!-1}{2!}$, so the formula holds for $n = 1$. Assume $s_k = \frac{(k+1)!-1}{(k+1)!}$. Then

$$s_{k+1} = \frac{(k+1)!-1}{(k+1)!} + \frac{k+1}{(k+2)!} = \frac{(k+1)!-1}{(k+1)!} + \frac{k+1}{(k+1)!\,(k+2)}$$

$$= \frac{(k+2)! - (k+2) + k + 1}{(k+2)!} = \frac{(k+2)!-1}{(k+2)!}$$

Thus, the formula is true for $n = k+1$. So by induction, the guess is correct.

(c) $\lim_{n\to\infty} s_n = \lim_{n\to\infty} \frac{(n+1)!-1}{(n+1)!} = \lim_{n\to\infty}\left[1 - \frac{1}{(n+1)!}\right] = 1$ and so $\sum_{n=0}^{\infty} \frac{n}{(n+1)!} = 1.$

12.3   The Integral Test and Estimates of Sums   ET 11.3

1. The picture shows that $a_2 = \dfrac{1}{2^{1.3}} < \displaystyle\int_1^2 \dfrac{1}{x^{1.3}}\,dx,$

$a_3 = \dfrac{1}{3^{1.3}} < \displaystyle\int_2^3 \dfrac{1}{x^{1.3}}\,dx,$ and so on, so $\displaystyle\sum_{n=2}^{\infty} \dfrac{1}{n^{1.3}} < \displaystyle\int_1^{\infty} \dfrac{1}{x^{1.3}}\,dx.$ The

integral converges by (8.8.2 [ET 7.8.2]) with $p = 1.3 > 1$, so the series
converges.

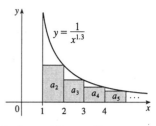

3. The function $f(x) = 1/x^4$ is continuous, positive, and decreasing on $[1, \infty)$, so the Integral Test applies.

$$\int_1^{\infty} \frac{1}{x^4}\,dx = \lim_{b\to\infty} \int_1^b x^{-4}\,dx = \lim_{b\to\infty} \left[\frac{x^{-3}}{-3}\right]_1^b = \lim_{b\to\infty} \left(-\frac{1}{3b^3} + \frac{1}{3}\right) = \frac{1}{3}, \text{ so } \sum_{n=1}^{\infty} \frac{1}{n^4} \text{ converges.}$$

5. The function $f(x) = 1/(3x+1)$ is continuous, positive, and decreasing on $[1, \infty)$, so the Integral Test applies.

$$\int_1^{\infty} \frac{dx}{3x+1} = \lim_{b\to\infty} \int_1^b \frac{dx}{3x+1} = \lim_{b\to\infty} \left[\tfrac{1}{3}\ln(3x+1)\right]_1^b = \lim_{b\to\infty} \left[\tfrac{1}{3}\ln(3b+1) - \tfrac{1}{3}\ln 4\right] = \infty$$

so the improper integral diverges, and so does the series $\sum_{n=1}^{\infty} 1/(3n+1)$.

7. $f(x) = xe^{-x}$ is continuous and positive on $[1, \infty)$. $f'(x) = -xe^{-x} + e^{-x} = e^{-x}(1-x) < 0$ for $x > 1$, so f is decreasing on $[1, \infty)$. Thus, the Integral Test applies.

$\int_1^{\infty} xe^{-x}\,dx = \lim_{b\to\infty} \int_1^b xe^{-x}\,dx = \lim_{b\to\infty} \left[-xe^{-x} - e^{-x}\right]_1^b$ (by parts) $= \lim_{b\to\infty} \left[-be^{-b} - e^{-b} + e^{-1} + e^{-1}\right] = 2/e$

since $\lim_{b\to\infty} be^{-b} = \lim_{b\to\infty} (b/e^b) \overset{\text{H}}{=} \lim_{b\to\infty} (1/e^b) = 0$ and $\lim_{b\to\infty} e^{-b} = 0$. Thus, $\sum_{n=1}^{\infty} ne^{-n}$ converges.

9. $\sum_{n=5}^{\infty} (1/n^{1.0001})$ is a p-series, $p = 1.0001 > 1$, so it converges.

11. $1 + \frac{1}{8} + \frac{1}{27} + \frac{1}{64} + \frac{1}{125} + \cdots = \sum_{n=1}^{\infty} (1/n^3)$. This is a p-series with $p = 3 > 1$, so it converges by (1).

13. $\displaystyle\sum_{n=1}^{\infty} \frac{5 - 2\sqrt{n}}{n^3} = 5 \sum_{n=1}^{\infty} \frac{1}{n^3} - 2 \sum_{n=1}^{\infty} \frac{1}{n^{5/2}}$ by Theorem 12.2.8 [ET 11.2.8], since $\sum_{n=1}^{\infty} \frac{1}{n^3}$ and $\sum_{n=1}^{\infty} \frac{1}{n^{5/2}}$ both

converge by (1) (with $p = 3$ and $p = \frac{5}{2}$). Thus, $\displaystyle\sum_{n=1}^{\infty} \frac{5 - 2\sqrt{n}}{n^3}$ converges.

15. $f(x) = xe^{-x^2}$ is continuous and positive on $[1, \infty)$, and since $f'(x) = e^{-x^2}(1 - 2x^2) < 0$ for $x > 1$, f is decreasing as well. Thus, we can use the Integral Test.

$\int_1^{\infty} xe^{-x^2}\,dx = \lim_{t\to\infty} \left[-\tfrac{1}{2}e^{-x^2}\right]_1^t = 0 - \left(-\tfrac{1}{2}e^{-1}\right) = 1/(2e)$. Since the integral converges, the series converges.

17. $f(x) = \dfrac{x}{x^2+1}$ is continuous and positive on $[1, \infty)$, and since $f'(x) = \dfrac{1 - x^2}{(x^2+1)^2} < 0$ for $x > 1$, f is also

decreasing. Using the Integral Test, $\displaystyle\int_1^{\infty} \frac{x}{x^2+1}\,dx = \lim_{t\to\infty} \left[\frac{\ln(x^2+1)}{2}\right]_1^t = \infty$, so the series diverges.

19. $f(x) = \dfrac{1}{x \ln x}$ is continuous and positive on $[2, \infty)$, and also decreasing since $f'(x) = -\dfrac{1 + \ln x}{x^2(\ln x)^2} < 0$ for

$x > 2$, so we can use the Integral Test. $\displaystyle\int_2^{\infty} \frac{1}{x \ln x}\,dx = \lim_{t\to\infty} \left[\ln(\ln x)\right]_2^t = \lim_{t\to\infty} \left[\ln(\ln t) - \ln(\ln 2)\right] = \infty$, so
the series diverges.

21. $f(x) = \dfrac{\arctan x}{1 + x^2}$ is continuous and positive on $[1, \infty)$. $f'(x) = \dfrac{1 - 2x \arctan x}{(1 + x^2)^2} < 0$ for $x > 1$, since

$2x \arctan x \geq \frac{\pi}{2} > 1$ for $x \geq 1$. So f is decreasing and we can use the Integral Test.

$\displaystyle\int_1^\infty \dfrac{\arctan x}{1 + x^2}\, dx = \lim_{t\to\infty} \left[\tfrac{1}{2} \arctan^2 x\right]_1^t = \dfrac{(\pi/2)^2}{2} - \dfrac{(\pi/4)^2}{2} = \dfrac{3\pi^2}{32}$, so the series converges.

23. $f(x) = \dfrac{1}{x^2 + 2x + 2}$ is continuous and positive on $[1, \infty)$, and $f'(x) = -\dfrac{2x + 2}{(x^2 + 2x + 2)^2} < 0$

for $x \geq 1$, so f is decreasing and we can use the Integral Test.

$\displaystyle\int_1^\infty \dfrac{1}{x^2 + 2x + 2}\, dx = \int_1^\infty \dfrac{1}{(x+1)^2 + 1}\, dx = \lim_{t\to\infty} \left[\arctan(x+1)\right]_1^t = \dfrac{\pi}{2} - \arctan 2$, so the series converges

as well.

25. We have already shown (in Exercise 19) that when $p = 1$ the series $\displaystyle\sum_{n=2}^\infty \dfrac{1}{n(\ln n)^p}$ diverges, so assume that $p \neq 1$.

$f(x) = \dfrac{1}{x(\ln x)^p}$ is continuous and positive on $[2, \infty)$, and $f'(x) = -\dfrac{p + \ln x}{x^2(\ln x)^{p+1}} < 0$ if $x > e^{-p}$, so that f is

eventually decreasing and we can use the Integral Test.

$\displaystyle\int_2^\infty \dfrac{1}{x(\ln x)^p}\, dx = \lim_{t\to\infty} \left[\dfrac{(\ln x)^{1-p}}{1-p}\right]_2^t \text{ (for } p \neq 1) = \lim_{t\to\infty} \left[\dfrac{(\ln t)^{1-p}}{1-p}\right] - \dfrac{(\ln 2)^{1-p}}{1-p}$

This limit exists whenever $1 - p < 0 \iff p > 1$, so the series converges for $p > 1$.

27. Clearly the series cannot converge if $p \geq -\frac{1}{2}$, because then $\displaystyle\lim_{n\to\infty} n(1 + n^2)^p \neq 0$. Also, if $p = -1$ the series

diverges (see Exercise 17). So assume $p < -\frac{1}{2}, p \neq -1$. Then $f(x) = x(1 + x^2)^p$ is continuous,

positive, and eventually decreasing on $[1, \infty)$, and we can use the Integral Test.

$\displaystyle\int_1^\infty x(1 + x^2)^p\, dx = \lim_{t\to\infty} \left[\dfrac{1}{2} \cdot \dfrac{(1 + x^2)^{p+1}}{p+1}\right]_1^t = \lim_{t\to\infty} \dfrac{1}{2} \cdot \dfrac{(1 + t^2)^{p+1}}{p+1} - \dfrac{2^p}{p+1}$. This limit exists and is

finite $\iff p + 1 < 0 \iff p < -1$, so the series converges whenever $p < -1$.

29. Since this is a p-series with $p = x$, $\zeta(x)$ is defined when $x > 1$. Unless specified otherwise, the domain of a

function f is the set of numbers x such that the expression for $f(x)$ makes sense and defines a real number. So, in

the case of a series, it's the set of numbers x such that the series is convergent.

31. (a) $f(x) = \dfrac{1}{x^2}$ is positive and continuous and $f'(x) = -\dfrac{2}{x^3}$ is negative for $x > 1$, and so the Integral

Test applies. $\displaystyle\sum_{n=1}^\infty \dfrac{1}{n^2} \approx s_{10} = \dfrac{1}{1^2} + \dfrac{1}{2^2} + \dfrac{1}{3^2} + \cdots + \dfrac{1}{10^2} \approx 1.549768$.

$R_{10} \leq \displaystyle\int_{10}^\infty \dfrac{1}{x^2}\, dx = \lim_{t\to\infty} \left[\dfrac{-1}{x}\right]_{10}^t = \lim_{t\to\infty}\left(-\dfrac{1}{t} + \dfrac{1}{10}\right) = \dfrac{1}{10}$, so the error is at most 0.1.

(b) $s_{10} + \displaystyle\int_{11}^\infty \dfrac{1}{x^2}\, dx \leq s \leq s_{10} + \int_{10}^\infty \dfrac{1}{x^2}\, dx \implies s_{10} + \tfrac{1}{11} \leq s \leq s_{10} + \tfrac{1}{10} \implies$

$1.549768 + 0.090909 = 1.640677 \leq s \leq 1.549768 + 0.1 = 1.649768$, so we get $s \approx 1.64522$ (the average of

1.640677 and 1.649768) with error ≤ 0.005 (the maximum of $1.649768 - 1.64522$ and $1.64522 - 1.640677$,

rounded up).

(c) $R_n \leq \displaystyle\int_n^\infty \dfrac{1}{x^2}\, dx = \dfrac{1}{n}$. So $R_n < 0.001$ if $\dfrac{1}{n} < \dfrac{1}{1000} \iff n > 1000$.

33. $f(x) = x^{-3/2}$ is positive and continuous and $f'(x) = -\frac{3}{2}x^{-5/2}$ is negative for $x > 1$, so the Integral Test applies.

From the end of Example 6, we see that the error is at most half the length of the interval. From (3), the interval is

$\left(s_n + \int_{n+1}^{\infty} f(x)\,dx,\ s_n + \int_n^{\infty} f(x)\,dx\right)$, so its length is $\int_n^{\infty} f(x)\,dx - \int_{n+1}^{\infty} f(x)\,dx$. Thus, we need n such

that

$$0.01 > \frac{1}{2}\left(\int_n^{\infty} x^{-3/2}\,dx - \int_{n+1}^{\infty} x^{-3/2}\,dx\right) = \frac{1}{2}\left(\lim_{t\to\infty}\left[\frac{-2}{\sqrt{x}}\right]_n^t - \lim_{t\to\infty}\left[\frac{-2}{\sqrt{x}}\right]_{n+1}^t\right) = \frac{1}{\sqrt{n}} - \frac{1}{\sqrt{n+1}}$$

$\Leftrightarrow\ n > 13.08$. Again from the end of Example 6, we approximate s by the midpoint of this interval. In general,

the midpoint is $\frac{1}{2}\left[\left(s_n + \int_{n+1}^{\infty} f(x)\,dx\right) + \left(s_n + \int_n^{\infty} f(x)\,dx\right)\right] = s_n + \frac{1}{2}\left(\int_{n+1}^{\infty} f(x)\,dx + \int_n^{\infty} f(x)\,dx\right)$. So

using $n = 14$, we have $s \approx s_{14} + \frac{1}{2}\left(\int_{14}^{\infty} x^{-3/2}\,dx + \int_{15}^{\infty} x^{-3/2}\,dx\right) = 2.0872 + \frac{1}{\sqrt{14}} + \frac{1}{\sqrt{15}} \approx 2.6127$. Any

larger value of n will also work. For instance, $s \approx s_{30} + \frac{1}{\sqrt{30}} + \frac{1}{\sqrt{31}} \approx 2.6124$.

35. (a) From the figure, $a_2 + a_3 + \cdots + a_n \le \int_1^n f(x)\,dx$, so with

$$f(x) = \frac{1}{x},$$

$$\frac{1}{2} + \frac{1}{3} + \frac{1}{4} + \cdots + \frac{1}{n} \le \int_1^n \frac{1}{x}\,dx = \ln n$$

Thus, $s_n = 1 + \frac{1}{2} + \frac{1}{3} + \frac{1}{4} + \cdots + \frac{1}{n} \le 1 + \ln n$.

(b) By part (a), $s_{10^6} \le 1 + \ln 10^6 \approx 14.82 < 15$ and $s_{10^9} \le 1 + \ln 10^9 \approx 21.72 < 22$.

37. $b^{\ln n} = \left(e^{\ln b}\right)^{\ln n} = \left(e^{\ln n}\right)^{\ln b} = n^{\ln b} = \dfrac{1}{n^{-\ln b}}$. This is a p-series, which converges for all b such that

$-\ln b > 1\ \Leftrightarrow\ \ln b < -1\ \Leftrightarrow\ b < e^{-1}\ \Leftrightarrow\ b < 1/e$.

12.4 The Comparison Tests   ET 11.4

1. (a) We cannot say anything about $\sum a_n$. If $a_n > b_n$ for all n and $\sum b_n$ is convergent, then $\sum a_n$ could be

convergent or divergent. (See Note 2 on page 757 [ET 723].)

(b) If $a_n < b_n$ for all n, then $\sum a_n$ is convergent. [This is part (i) of the Comparison Test.]

3. $\dfrac{1}{n^2 + n + 1} < \dfrac{1}{n^2}$ for all $n \ge 1$, so $\displaystyle\sum_{n=1}^{\infty} \dfrac{1}{n^2 + n + 1}$ converges by comparison with $\displaystyle\sum_{n=1}^{\infty} \dfrac{1}{n^2}$, which converges

because it is a p-series with $p = 2 > 1$.

5. $\dfrac{5}{2 + 3^n} < \dfrac{5}{3^n}$ for all $n \ge 1$, so $\displaystyle\sum_{n=1}^{\infty} \dfrac{5}{2 + 3^n}$ converges by comparison with $\displaystyle\sum_{n=1}^{\infty} \dfrac{5}{3^n} = 5\displaystyle\sum_{n=1}^{\infty} \dfrac{1}{3^n}$, which converges

because $\displaystyle\sum_{n=1}^{\infty} \dfrac{1}{3^n}$ is a convergent geometric series with $r = \frac{1}{3}$.

7. $\dfrac{n+1}{n^2} > \dfrac{n}{n^2} = \dfrac{1}{n}$ for all $n \ge 1$, so $\displaystyle\sum_{n=1}^{\infty} \dfrac{n+1}{n^2}$ diverges by comparison with the harmonic series $\displaystyle\sum_{n=1}^{\infty} \dfrac{1}{n}$.

9. $\dfrac{3}{n2^n} \le \dfrac{3}{2^n}$. $\displaystyle\sum_{n=1}^{\infty} \dfrac{3}{2^n}$ is a geometric series with $|r| = \frac{1}{2} < 1$, and hence converges, so $\displaystyle\sum_{n=1}^{\infty} \dfrac{3}{n2^n}$ converges also, by

the Comparison Test.

11. $\dfrac{1}{\sqrt{n\,(n+1)\,(n+2)}} < \dfrac{1}{\sqrt{n\cdot n\cdot n}} = \dfrac{1}{n^{3/2}}$ and since $\displaystyle\sum_{n=1}^{\infty}\dfrac{1}{n^{3/2}}$ converges ($p = \frac{3}{2} > 1$), so does

$\displaystyle\sum_{n=1}^{\infty}\dfrac{1}{\sqrt{n\,(n+1)\,(n+2)}}$ by the Comparison Test.

13. If $a_n = \dfrac{n^2+1}{n^3-1}$ and $b_n = \dfrac{1}{n}$, then $\displaystyle\lim_{n\to\infty}\dfrac{a_n}{b_n} = \lim_{n\to\infty}\dfrac{n^3+n}{n^3-1} = \lim_{n\to\infty}\dfrac{1+1/n^2}{1-1/n^3} = 1$, so $\displaystyle\sum_{n=2}^{\infty}\dfrac{n^2+1}{n^3-1}$ diverges by

the Limit Comparison Test with the divergent (partial) harmonic series $\displaystyle\sum_{n=2}^{\infty}\dfrac{1}{n}$.

15. $\dfrac{3+\cos n}{3^n} \le \dfrac{4}{3^n}$ since $\cos n \le 1$. $\displaystyle\sum_{n=1}^{\infty}\dfrac{4}{3^n}$ is a geometric series with $|r| = \frac{1}{3} < 1$ so it converges, and so

$\displaystyle\sum_{n=1}^{\infty}\dfrac{3+\cos n}{3^n}$ converges by the Comparison Test.

17. $\dfrac{n}{\sqrt{n^5+4}} < \dfrac{n}{\sqrt{n^5}} = \dfrac{1}{n^{3/2}}$. $\displaystyle\sum_{n=1}^{\infty}\dfrac{1}{n^{3/2}}$ is a convergent p-series ($p = \frac{3}{2} > 1$) so $\displaystyle\sum_{n=1}^{\infty}\dfrac{n}{\sqrt{n^5+4}}$ converges by the Comparison Test.

19. $\dfrac{2^n}{1+3^n} < \dfrac{2^n}{3^n} = \left(\dfrac{2}{3}\right)^n$. $\displaystyle\sum_{n=1}^{\infty}\left(\dfrac{2}{3}\right)^n$ is a convergent geometric series ($|r| = \frac{2}{3} < 1$), so $\displaystyle\sum_{n=1}^{\infty}\dfrac{2^n}{1+3^n}$ converges by the Comparison Test.

21. Use the Limit Comparison Test with $a_n = \dfrac{1}{1+\sqrt{n}}$ and $b_n = \dfrac{1}{\sqrt{n}}$: $\displaystyle\lim_{n\to\infty}\dfrac{a_n}{b_n} = \lim_{n\to\infty}\dfrac{\sqrt{n}}{1+\sqrt{n}} = 1 > 0$. Since

$\displaystyle\sum_{n=1}^{\infty}\dfrac{1}{\sqrt{n}}$ is a divergent p-series ($p = \frac{1}{2} \le 1$), $\displaystyle\sum_{n=1}^{\infty}\dfrac{1}{1+\sqrt{n}}$ also diverges.

23. Let $a_n = \dfrac{n^2+1}{n^4+1}$ and $b_n = \dfrac{1}{n^2}$. Then $\displaystyle\lim_{n\to\infty}\dfrac{a_n}{b_n} = \lim_{n\to\infty}\dfrac{n^4+n^2}{n^4+1} = 1$. Since $\displaystyle\sum_{n=1}^{\infty}\dfrac{1}{n^2}$ is a convergent p-series

($p = 2 > 1$), so is $\displaystyle\sum_{n=1}^{\infty}\dfrac{n^2+1}{n^4+1}$ by the Limit Comparison Test.

25. If $a_n = \dfrac{1+n+n^2}{\sqrt{1+n^2+n^6}}$ and $b_n = \dfrac{1}{n}$, then $\displaystyle\lim_{n\to\infty}\dfrac{a_n}{b_n} = \lim_{n\to\infty}\dfrac{n+n^2+n^3}{\sqrt{1+n^2+n^6}} = \lim_{n\to\infty}\dfrac{1/n^2+1/n+1}{\sqrt{1/n^6+1/n^4+1}} = 1$,

so $\displaystyle\sum_{n=1}^{\infty}\dfrac{1+n+n^2}{\sqrt{1+n^2+n^6}}$ diverges by the Limit Comparison Test with the divergent harmonic series $\displaystyle\sum_{n=1}^{\infty}\dfrac{1}{n}$.

27. Let $a_n = \dfrac{n+1}{n2^n}$ and $b_n = \dfrac{1}{2^n}$. Then $\displaystyle\lim_{n\to\infty}\dfrac{a_n}{b_n} = \lim_{n\to\infty}\dfrac{n+1}{n} = 1$. Since $\displaystyle\sum_{n=1}^{\infty}\dfrac{1}{2^n}$ is a convergent geometric

series ($|r| = \frac{1}{2} < 1$), $\displaystyle\sum_{n=1}^{\infty}\dfrac{n+1}{n2^n}$ converges by the Limit Comparison Test.

29. Clearly $n! = n\,(n-1)\,(n-2)\cdots(3)\,(2) \ge 2\cdot2\cdot2\cdot\cdots\cdot2\cdot2 = 2^{n-1}$, so $\dfrac{1}{n!} \le \dfrac{1}{2^{n-1}}$. $\displaystyle\sum_{n=1}^{\infty}\dfrac{1}{2^{n-1}}$ is a

convergent geometric series ($|r| = \frac{1}{2} < 1$) so $\displaystyle\sum_{n=1}^{\infty}\dfrac{1}{n!}$ converges by the Comparison Test.

31. Use the Limit Comparison Test with $a_n = \sin\left(\dfrac{1}{n}\right)$ and $b_n = \dfrac{1}{n}$:

$\displaystyle\lim_{n\to\infty}\dfrac{a_n}{b_n} = \lim_{n\to\infty}\dfrac{\sin(1/n)}{1/n} = \lim_{\theta\to0}\dfrac{\sin\theta}{\theta} = 1 > 0$. Since $\sum_{n=1}^{\infty}b_n$ is the divergent harmonic series,

$\sum_{n=1}^{\infty}\sin(1/n)$ also diverges.

33. $\sum\limits_{n=1}^{10} \dfrac{1}{n^4+n^2} = \dfrac{1}{2} + \dfrac{1}{20} + \dfrac{1}{90} + \cdots + \dfrac{1}{10,100} \approx 0.567975.$ Now $\dfrac{1}{n^4+n^2} < \dfrac{1}{n^4}$, so using the reasoning and

notation of Example 5, the error is $R_{10} \le T_{10} = \sum\limits_{n=11}^{\infty} \dfrac{1}{n^4} \le \int_{10}^{\infty} \dfrac{dx}{x^4} = \lim\limits_{t\to\infty} \left[-\dfrac{x^{-3}}{3} \right]_{10}^{t} = \dfrac{1}{3000} = 0.000\overline{3}.$

35. $\sum\limits_{n=1}^{10} \dfrac{1}{1+2^n} = \dfrac{1}{3} + \dfrac{1}{5} + \dfrac{1}{9} + \cdots + \dfrac{1}{1025} \approx 0.76352.$ Now $\dfrac{1}{1+2^n} < \dfrac{1}{2^n}$, so the error is

$R_{10} \le T_{10} = \sum\limits_{n=11}^{\infty} \dfrac{1}{2^n} = \dfrac{1/2^{11}}{1-1/2}$ (geometric series) $\approx 0.00098.$

37. Since $\dfrac{d_n}{10^n} \le \dfrac{9}{10^n}$ for each n, and since $\sum\limits_{n=1}^{\infty} \dfrac{9}{10^n}$ is a convergent geometric series $(|r| = \tfrac{1}{10} < 1)$,

$0.d_1 d_2 d_3 \ldots = \sum\limits_{n=1}^{\infty} \dfrac{d_n}{10^n}$ will always converge by the Comparison Test.

39. Since $\sum a_n$ converges, $\lim\limits_{n\to\infty} a_n = 0$, so there exists N such that $|a_n - 0| < 1$ for all $n > N \;\Rightarrow\; 0 \le a_n < 1$

for all $n > N \;\Rightarrow\; 0 \le a_n^2 \le a_n.$ Since $\sum a_n$ converges, so does $\sum a_n^2$ by the Comparison Test.

41. (a) We wish to prove that if $\lim\limits_{n\to\infty} \dfrac{a_n}{b_n} = \infty$ and $\sum b_n$ diverges, then so does $\sum a_n.$ So suppose on the contrary that

$\sum a_n$ converges. Since $\lim\limits_{n\to\infty} \dfrac{a_n}{b_n} = \infty$, we have that $\lim\limits_{n\to\infty} \dfrac{b_n}{a_n} = 0$, so by the extension of the Limit

Comparison Test proved in Exercise 40(a), if $\sum a_n$ converges, so must $\sum b_n.$ But this contradicts our

hypothesis, so $\sum a_n$ must diverge.

(b) If $a_n = \dfrac{1}{\ln n}$ and $b_n = \dfrac{1}{n}$ for $n \ge 2$, then $\lim\limits_{n\to\infty} \dfrac{a_n}{b_n} = \lim\limits_{n\to\infty} \dfrac{n}{\ln n} = \lim\limits_{x\to\infty} \dfrac{x}{\ln x} \overset{\text{H}}{=} \lim\limits_{x\to\infty} \dfrac{1}{1/x} = \lim\limits_{x\to\infty} x = \infty,$

so by part (a), $\sum\limits_{n=2}^{\infty} \dfrac{1}{\ln n}$ is divergent.

43. $\lim\limits_{n\to\infty} na_n = \lim\limits_{n\to\infty} \dfrac{a_n}{1/n}$, so we apply the Limit Comparison Test with $b_n = \dfrac{1}{n}.$ Since $\lim\limits_{n\to\infty} na_n > 0$ we know that

either both series converge or both series diverge, and we also know that $\sum\limits_{n=0}^{\infty} \dfrac{1}{n}$ diverges (p-series with $p = 1$).

Therefore, $\sum a_n$ must be divergent.

45. Yes. Since $\sum a_n$ converges, its terms approach 0 as $n \to \infty$, so $\lim\limits_{n\to\infty} \dfrac{\sin a_n}{a_n} = 1$ by Theorem 3.5.2 [ET 3.4.2].

Thus, $\sum \sin a_n$ converges by the Limit Comparison Test.

12.5 Alternating Series   ET 11.5

1. (a) An alternating series is a series whose terms are alternately positive and negative.

(b) An alternating series $\sum_{n=1}^{\infty} (-1)^{n-1} b_n$ converges if $0 < b_{n+1} \le b_n$ for all n and $\lim\limits_{n\to\infty} b_n = 0.$ (This is the

Alternating Series Test.)

(c) The error involved in using the partial sum s_n as an approximation to the total sum s is the remainder

$R_n = s - s_n$ and the size of the error is smaller than b_{n+1}, that is, $|R_n| \le b_{n+1}.$ (This is the Alternating Series

Estimation Theorem.)

3. $\dfrac{4}{7} - \dfrac{4}{8} + \dfrac{4}{9} - \dfrac{4}{10} + \dfrac{4}{11} - \cdots = \sum_{n=1}^{\infty} (-1)^{n-1} \dfrac{4}{n+6}$. $b_n = \dfrac{4}{n+6} > 0$, $\{b_n\}$ is decreasing, and $\lim\limits_{n\to\infty} b_n = 0$, so the series converges by the Alternating Series Test.

5. $b_n = \dfrac{1}{\sqrt{n}} > 0$, $\{b_n\}$ is decreasing, and $\lim\limits_{n\to\infty} b_n = 0$, so $\sum_{n=1}^{\infty} \dfrac{(-1)^{n-1}}{\sqrt{n}}$ converges by the Alternating Series Test.

7. $a_n = (-1)^n \dfrac{2n}{4n+1}$, so $|a_n| = \dfrac{2n}{4n+1} \to \dfrac{1}{2}$ as $n \to \infty$. Therefore, $\lim\limits_{n\to\infty} a_n \neq 0$ (in fact the limit does not exist)

and the series $\sum_{n=1}^{\infty} (-1)^n \dfrac{2n}{4n+1}$ diverges by the Test for Divergence.

9. $b_n = \dfrac{1}{4n^2+1} > 0$, $\{b_n\}$ is decreasing, and $\lim\limits_{n\to\infty} b_n = 0$, so the series $\sum_{n=1}^{\infty} \dfrac{(-1)^{n+1}}{4n^2+1}$ converges by the Alternating

Series Test.

11. $\sum_{n=1}^{\infty} (-1)^{n-1} \dfrac{\sqrt{n}}{n+4}$. $b_n = \dfrac{\sqrt{n}}{n+4} > 0$ for all n. Let $f(x) = \dfrac{\sqrt{x}}{x+4}$. Then $f'(x) = \dfrac{4-x}{2\sqrt{x}\,(x+4)^2} < 0$ if $x > 4$,

so $\{b_n\}$ is decreasing after $n = 4$. $\lim\limits_{n\to\infty} \dfrac{\sqrt{n}}{n+4} = \lim\limits_{n\to\infty} \dfrac{1}{\sqrt{n}+4/\sqrt{n}} = 0$. So the series converges by the

Alternating Series Test.

13. $\sum_{n=2}^{\infty} (-1)^n \dfrac{n}{\ln n}$. $\lim\limits_{n\to\infty} \dfrac{n}{\ln n} \overset{\text{H}}{=} \lim\limits_{n\to\infty} \dfrac{1}{1/n} = \infty$, so the series diverges by the Test for Divergence.

15. $\sum_{n=1}^{\infty} \dfrac{\cos n\pi}{n^{3/4}} = \sum_{n=1}^{\infty} \dfrac{(-1)^n}{n^{3/4}}$. $b_n = \dfrac{1}{n^{3/4}}$ is decreasing and positive and $\lim\limits_{n\to\infty} \dfrac{1}{n^{3/4}} = 0$, so the series converges by

the Alternating Series Test.

17. $\sum_{n=1}^{\infty} (-1)^n \sin \dfrac{\pi}{n}$. $b_n = \sin \dfrac{\pi}{n} > 0$ for $n \geq 2$ and $\sin \dfrac{\pi}{n} \geq \sin \dfrac{\pi}{n+1}$, and $\lim\limits_{n\to\infty} \sin \dfrac{\pi}{n} = \sin 0 = 0$, so the series

converges by the Alternating Series Test.

19. $\dfrac{n^n}{n!} = \dfrac{n \cdot n \cdot \cdots \cdot n}{1 \cdot 2 \cdot \cdots \cdot n} \geq n \;\Rightarrow\; \lim\limits_{n\to\infty} \dfrac{n^n}{n!} = \infty \;\Rightarrow\; \lim\limits_{n\to\infty} \dfrac{(-1)^n n^n}{n!}$ does not exist. So the series diverges by

the Test for Divergence.

21.

n	a_n	s_n
1	1	1
2	-0.35355	0.64645
3	0.19245	0.83890
4	-0.125	0.71390
5	0.08944	0.80334
6	-0.06804	0.73530
7	0.05399	0.78929
8	-0.04419	0.74510
9	0.03704	0.78214
10	-0.03162	0.75051

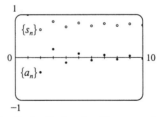

By the Alternating Series Estimation Theorem, the error in the

approximation $\sum_{n=1}^{\infty} \dfrac{(-1)^{n-1}}{n^{3/2}} \approx 0.75051$ is

$|s - s_{10}| \leq b_{11} = 1/(11)^{3/2} \approx 0.0275$ (to four decimal places, rounded up).

23. With $b_n = 1/n^2$, $b_{10} = 1/10^2 = 0.01$ and $b_{11} = 1/11^2 = 1/121 \approx 0.008 < 0.01$, so by the Alternating Series Estimation Theorem, $n = 10$.

25. $b_7 = 2^7/7! \approx 0.025 > 0.01$ and $b_8 = 2^8/8! \approx 0.006 < 0.01$, so by the Alternating Series Estimation Theorem, $n = 7$. (That is, since the 8th term is less than the desired error, we need to add the first 7 terms to get the sum to the desired accuracy.)

27. $\displaystyle\sum_{n=1}^{\infty} \frac{(-1)^{n-1}}{(2n-1)!}$. $b_5 = \dfrac{1}{(2\cdot 5 - 1)!} = \dfrac{1}{362{,}880} < 0.00001$, so $\displaystyle\sum_{n=1}^{\infty} \frac{(-1)^{n-1}}{(2n-1)!} \approx \sum_{n=1}^{4} \frac{(-1)^{n-1}}{(2n-1)!} \approx 0.8415$.

29. $b_6 = \dfrac{1}{2^6 6!} = \dfrac{1}{46{,}080} \approx 0.000022 < 0.00001$, so $\displaystyle\sum_{n=0}^{\infty} \frac{(-1)^n}{2^n n!} \approx \sum_{n=0}^{5} \frac{(-1)^n}{2^n n!} \approx 0.6065$.

31. $\displaystyle\sum_{n=1}^{\infty} \frac{(-1)^{n-1}}{n} = 1 - \frac{1}{2} + \frac{1}{3} - \frac{1}{4} + \cdots + \frac{1}{49} - \frac{1}{50} + \frac{1}{51} - \frac{1}{52} + \cdots$. The 50th partial sum of this series is an

underestimate, since $\displaystyle\sum_{n=1}^{\infty} \frac{(-1)^{n-1}}{n} = s_{50} + \left(\frac{1}{51} - \frac{1}{52}\right) + \left(\frac{1}{53} - \frac{1}{54}\right) + \cdots$, and the terms in parentheses are all positive. The result can be seen geometrically in Figure 1.

33. Clearly $b_n = \dfrac{1}{n+p}$ is decreasing and eventually positive and $\lim\limits_{n\to\infty} b_n = 0$ for any p. So the series converges (by the Alternating Series Test) for any p for which every b_n is defined, that is, $n + p \neq 0$ for $n \geq 1$, or p is not a negative integer.

35. $\sum b_{2n} = \sum 1/(2n)^2$ clearly converges (by comparison with the p-series for $p = 2$). So suppose that $\sum (-1)^{n-1} b_n$ converges. Then by Theorem 12.2.8(ii) [ET 11.2.8(ii)], so does

$\displaystyle\sum \left[(-1)^{n-1} b_n + b_n\right] = 2\left(1 + \frac{1}{3} + \frac{1}{5} + \cdots\right) = 2\sum \frac{1}{2n-1}$. But this diverges by comparison with the

harmonic series, a contradiction. Therefore, $\sum (-1)^{n-1} b_n$ must diverge. The Alternating Series Test does not apply since $\{b_n\}$ is not decreasing.

12.6 Absolute Convergence and the Ratio and Root Tests ET 11.6

1. (a) Since $\lim\limits_{n\to\infty} \left|\dfrac{a_{n+1}}{a_n}\right| = 8 > 1$, part (ii) of the Ratio Test tells us that $\sum a_n$ is divergent.

(b) Since $\lim\limits_{n\to\infty} \left|\dfrac{a_{n+1}}{a_n}\right| = 0.8 < 1$, part (i) of the Ratio Test tells us that $\sum a_n$ is convergent.

(c) Since $\lim\limits_{n\to\infty} \left|\dfrac{a_{n+1}}{a_n}\right| = 1$, the Ratio Test fails and $\sum a_n$ might converge or it might diverge.

3. $\displaystyle\sum_{n=1}^{\infty} \frac{1}{n\sqrt{n}} = \sum_{n=1}^{\infty} \frac{1}{n^{3/2}}$ is a convergent p-series ($p = \frac{3}{2} > 1$), so the given series is absolutely convergent.

5. Using the Ratio Test, $\lim\limits_{n\to\infty} \left|\dfrac{a_{n+1}}{a_n}\right| = \lim\limits_{n\to\infty} \left|\dfrac{(-3)^{n+1}/(n+1)^3}{(-3)^n/n^3}\right| = 3\lim\limits_{n\to\infty} \left(\dfrac{n}{n+1}\right)^3 = 3 > 1$, so the series diverges.

7. $\displaystyle\sum_{n=1}^{\infty} \frac{(-1)^n}{5+n}$ converges by the Alternating Series Test, but $\displaystyle\sum_{n=1}^{\infty} \frac{1}{5+n}$ diverges by the Limit Comparison Test with

the harmonic series $\displaystyle\sum_{n=1}^{\infty} \frac{1}{n}$, so the given series is conditionally convergent.

9. $\lim\limits_{n\to\infty} |a_n| = \lim\limits_{n\to\infty} \dfrac{n}{5+n} = \lim\limits_{n\to\infty} \dfrac{1}{5/n+1} = 1$, so $\lim\limits_{n\to\infty} a_n \neq 0$. Thus, the given series is divergent by the Test

for Divergence.

11. $\lim\limits_{n\to\infty} \left| \dfrac{a_{n+1}}{a_n} \right| = \lim\limits_{n\to\infty} \dfrac{1/(2n+2)!}{1/(2n)!} = \lim\limits_{n\to\infty} \dfrac{(2n)!}{(2n+2)!} = \lim\limits_{n\to\infty} \dfrac{(2n)!}{(2n+2)(2n+1)(2n)!} =$

$\lim\limits_{n\to\infty} \dfrac{1}{(2n+2)(2n+1)} = 0 < 1$, so the series is absolutely convergent by the Ratio Test. Of course, absolute

convergence is the same as convergence for this series, since all of its terms are positive.

13. $\left| \dfrac{\sin 2n}{n^2} \right| \leq \dfrac{1}{n^2}$ and $\sum\limits_{n=1}^{\infty} \dfrac{1}{n^2}$ converges (p-series, $p = 2 > 1$), so $\sum\limits_{n=1}^{\infty} \dfrac{\sin 2n}{n^2}$ converges absolutely by the

Comparison Test.

15. $\lim\limits_{n\to\infty} \left| \dfrac{a_{n+1}}{a_n} \right| = \lim\limits_{n\to\infty} \left[\dfrac{(n+1)\,3^{n+1}}{4^n} \cdot \dfrac{4^{n-1}}{n \cdot 3^n} \right] = \lim\limits_{n\to\infty} \left(\dfrac{3}{4} \cdot \dfrac{n+1}{n} \right) = \dfrac{3}{4} < 1$, so the series is absolutely

convergent by the Ratio Test.

17. $\lim\limits_{n\to\infty} \left| \dfrac{a_{n+1}}{a_n} \right| = \lim\limits_{n\to\infty} \left[\dfrac{10^{n+1}}{(n+2)\,4^{2n+3}} \cdot \dfrac{(n+1)\,4^{2n+1}}{10^n} \right] = \lim\limits_{n\to\infty} \left(\dfrac{10}{4^2} \cdot \dfrac{n+1}{n+2} \right) = \dfrac{5}{8} < 1$, so the series is

absolutely convergent by the Ratio Test. Since the terms of this series are positive, absolute convergence is the same

as convergence.

19. $\lim\limits_{n\to\infty} \left| \dfrac{a_{n+1}}{a_n} \right| = \lim\limits_{n\to\infty} \dfrac{(n+1)!/10^{n+1}}{n!/10^n} = \lim\limits_{n\to\infty} \dfrac{n+1}{10} = \infty$, so the series diverges by the Ratio Test.

21. $\dfrac{|\cos(n\pi/3)|}{n!} \leq \dfrac{1}{n!}$ and $\sum\limits_{n=1}^{\infty} \dfrac{1}{n!}$ converges (Exercise 12.4.29 [ET 11.4.29]), so the given series converges absolutely

by the Comparison Test.

23. $\lim\limits_{n\to\infty} \sqrt[n]{|a_n|} = \lim\limits_{n\to\infty} \left(\dfrac{n^n}{3^{1+3n}} \right)^{1/n} = \lim\limits_{n\to\infty} \dfrac{n}{\sqrt[n]{3} \cdot 3^3} = \infty$, so the series is divergent by the Root Test.

Or: $\lim\limits_{n\to\infty} \left| \dfrac{a_{n+1}}{a_n} \right| = \lim\limits_{n\to\infty} \left[\dfrac{(n+1)^{n+1}}{3^{4+3n}} \cdot \dfrac{3^{1+3n}}{n^n} \right] = \lim\limits_{n\to\infty} \left[\dfrac{1}{3^3} \cdot \left(\dfrac{n+1}{n} \right)^n (n+1) \right]$

$= \dfrac{1}{27} \lim\limits_{n\to\infty} \left(1 + \dfrac{1}{n} \right)^n \lim\limits_{n\to\infty} (n+1) = \tfrac{1}{27} e \lim\limits_{n\to\infty} (n+1) = \infty$

so the series is divergent by the Ratio Test.

25. $\lim\limits_{n\to\infty} \sqrt[n]{|a_n|} = \lim\limits_{n\to\infty} \dfrac{n^2+1}{2n^2+1} = \lim\limits_{n\to\infty} \dfrac{1+1/n^2}{2+1/n^2} = \dfrac{1}{2} < 1$, so the series is absolutely convergent by the

Root Test.

27. $\lim\limits_{n\to\infty} \left| \dfrac{a_{n+1}}{a_n} \right| = \lim\limits_{n\to\infty} \dfrac{(n+1)!/[1 \cdot 3 \cdot 5 \cdots (2n+1)]}{n!/[1 \cdot 3 \cdot 5 \cdots (2n-1)]} = \lim\limits_{n\to\infty} \dfrac{n+1}{2n+1} = \dfrac{1}{2} < 1$, so the series converges

absolutely by the Ratio Test.

29. $\sum\limits_{n=1}^{\infty} \dfrac{2 \cdot 4 \cdot 6 \cdots (2n)}{n!} = \sum\limits_{n=1}^{\infty} \dfrac{2^n n!}{n!} = \sum\limits_{n=1}^{\infty} 2^n$ which diverges by the Test for Divergence since $\lim\limits_{n\to\infty} 2^n = \infty$.

31. By the recursive definition, $\lim\limits_{n\to\infty} \left| \dfrac{a_{n+1}}{a_n} \right| = \lim\limits_{n\to\infty} \left| \dfrac{5n+1}{4n+3} \right| = \dfrac{5}{4} > 1$, so the series diverges by the Ratio Test.

33. (a) $\lim\limits_{n\to\infty} \left| \dfrac{1/(n+1)^3}{1/n^3} \right| = \lim\limits_{n\to\infty} \dfrac{n^3}{(n+1)^3} = \lim\limits_{n\to\infty} \dfrac{1}{(1+1/n)^3} = 1$. Inconclusive.

(b) $\lim\limits_{n\to\infty} \left| \dfrac{(n+1)}{2^{n+1}} \cdot \dfrac{2^n}{n} \right| = \lim\limits_{n\to\infty} \dfrac{n+1}{2n} = \lim\limits_{n\to\infty} \left(\dfrac{1}{2} + \dfrac{1}{2n} \right) = \dfrac{1}{2}$. Conclusive (convergent).

(c) $\lim\limits_{n\to\infty} \left| \dfrac{(-3)^n}{\sqrt{n+1}} \cdot \dfrac{\sqrt{n}}{(-3)^{n-1}} \right| = 3 \lim\limits_{n\to\infty} \sqrt{\dfrac{n}{n+1}} = 3 \lim\limits_{n\to\infty} \sqrt{\dfrac{1}{1+1/n}} = 3$. Conclusive (divergent).

(d) $\lim\limits_{n\to\infty} \left| \dfrac{\sqrt{n+1}}{1+(n+1)^2} \cdot \dfrac{1+n^2}{\sqrt{n}} \right| = \lim\limits_{n\to\infty} \left[\sqrt{1+\dfrac{1}{n}} \cdot \dfrac{1/n^2+1}{1/n^2+(1+1/n)^2} \right] = 1$. Inconclusive.

35. (a) $\lim\limits_{n\to\infty} \left| \dfrac{a_{n+1}}{a_n} \right| = \lim\limits_{n\to\infty} \dfrac{|x|^{n+1}/(n+1)!}{|x|^n/n!} = |x| \lim\limits_{n\to\infty} \dfrac{1}{n+1} = 0$, so by the Ratio Test the series converges for all x.

(b) Since the series of part (a) always converges, we must have $\lim\limits_{n\to\infty} \dfrac{x^n}{n!} = 0$ by Theorem 12.2.6 [ET 11.2.6].

37. (a) $s_5 = \sum\limits_{n=1}^{5} \dfrac{1}{n2^n} = \dfrac{1}{2} + \dfrac{1}{8} + \dfrac{1}{24} + \dfrac{1}{64} + \dfrac{1}{160} = \dfrac{661}{960} \approx 0.68854$. Now the ratios

$r_n = \dfrac{a_{n+1}}{a_n} = \dfrac{n2^n}{(n+1)\,2^{n+1}} = \dfrac{n}{2(n+1)}$ form an increasing sequence, since

$r_{n+1} - r_n = \dfrac{n+1}{2(n+2)} - \dfrac{n}{2(n+1)} = \dfrac{(n+1)^2 - n(n+2)}{2(n+1)(n+2)} = \dfrac{1}{2(n+1)(n+2)} > 0$. So by

Exercise 36(b), the error is less than $\dfrac{a_6}{1 - \lim\limits_{n\to\infty} r_n} = \dfrac{1/(6 \cdot 2^6)}{1 - 1/2} = \dfrac{1}{192} \approx 0.00521$.

(b) The error in using s_n as an approximation to the sum is $R_n = \dfrac{a_{n+1}}{1 - \frac{1}{2}} = \dfrac{2}{(n+1)\,2^{n+1}}$. We want

$R_n < 0.00005 \quad \Leftrightarrow \quad \dfrac{1}{(n+1)\,2^n} < 0.00005 \quad \Leftrightarrow \quad (n+1)\,2^n > 20{,}000$. To find such an n we can use trial

and error or a graph. We calculate $(11+1)\,2^{11} = 24{,}576$, so $s_{11} = \sum\limits_{n=1}^{11} \dfrac{1}{n2^n} \approx 0.693109$ is within 0.00005

of the actual sum.

39. Summing the inequalities $-|a_i| \le a_i \le |a_i|$ for $i = 1, 2, \ldots, n$, we get $-\sum_{i=1}^{n} |a_i| \le \sum_{i=1}^{n} a_i \le \sum_{i=1}^{n} |a_i|$
$\Rightarrow -\lim\limits_{n\to\infty} \sum_{i=1}^{n} |a_i| \le \lim\limits_{n\to\infty} \sum_{i=1}^{n} a_i \le \lim\limits_{n\to\infty} \sum_{i=1}^{n} |a_i| \Rightarrow -\sum_{n=1}^{\infty} |a_n| \le \sum_{n=1}^{\infty} a_n \le \sum_{n=1}^{\infty} |a_n| \Rightarrow$
$\left| \sum_{n=1}^{\infty} a_n \right| \le \sum_{n=1}^{\infty} |a_n|$.

41. (a) Since $\sum a_n$ is absolutely convergent, and since $|a_n^+| \le |a_n|$ and $|a_n^-| \le |a_n|$ (because a_n^+ and a_n^- each equal either a_n or 0), we conclude by the Comparison Test that both $\sum a_n^+$ and $\sum a_n^-$ must be absolutely convergent. (Or use Theorem 12.2.8 [ET 11.2.8].)

(b) We will show by contradiction that both $\sum a_n^+$ and $\sum a_n^-$ must diverge. For suppose that $\sum a_n^+$ converged. Then so would $\sum \left(a_n^+ - \frac{1}{2} a_n \right)$ by Theorem 12.2.8 [ET 11.2.8]. But

$\sum \left(a_n^+ - \frac{1}{2} a_n \right) = \sum \left[\frac{1}{2} \left(a_n + |a_n| \right) - \frac{1}{2} a_n \right] = \frac{1}{2} \sum |a_n|$, which diverges because $\sum a_n$ is only conditionally convergent. Hence, $\sum a_n^+$ can't converge. Similarly, neither can $\sum a_n^-$.

12.7 Strategy for Testing Series

1. $\lim\limits_{n\to\infty} a_n = \lim\limits_{n\to\infty} \dfrac{n^2-1}{n^2+1} = \lim\limits_{n\to\infty} \dfrac{1-1/n^2}{1+1/n} = 1 \neq 0$, so the series diverges by the Test for Divergence.

3. $\dfrac{1}{n^2+n} < \dfrac{1}{n^2}$ for all $n \geq 1$, so $\sum\limits_{n=1}^{\infty} \dfrac{1}{n^2+n}$ converges by the Comparison Test with $\sum\limits_{n=1}^{\infty} \dfrac{1}{n^2}$, a p-series that converges because $p = 2 > 1$.

5. $\lim\limits_{n\to\infty} \left| \dfrac{a_{n+1}}{a_n} \right| = \lim\limits_{n\to\infty} \left(\dfrac{3^{n+2}}{2^{3n+3}} \cdot \dfrac{2^{3n}}{3^{n+1}} \right) = \lim\limits_{n\to\infty} \dfrac{3}{2^3} = \dfrac{3}{8} < 1$, so the series is absolutely convergent by the Ratio Test.

7. $\sum\limits_{k=1}^{\infty} k^{-1.7} = \sum\limits_{k=1}^{\infty} \dfrac{1}{k^{1.7}}$ is a convergent p-series ($p = 1.7 > 1$).

9. $\lim\limits_{n\to\infty} \left| \dfrac{a_{n+1}}{a_n} \right| = \lim\limits_{n\to\infty} \dfrac{(n+1)/e^{n+1}}{n/e^n} = \dfrac{1}{e} \lim\limits_{n\to\infty} \dfrac{n+1}{n} = \dfrac{1}{e} < 1$, so the series converges by the Ratio Test.

11. $b_n = \dfrac{1}{n \ln n} > 0$ for $n \geq 2$, $\{b_n\}$ is decreasing, and $\lim\limits_{n\to\infty} b_n = 0$, so $\sum\limits_{n=2}^{\infty} \dfrac{(-1)^{n+1}}{n \ln n}$ converges by the Alternating

Series Test. The series is conditionally convergent since $\sum\limits_{n=2}^{\infty} \dfrac{1}{n \ln n}$ diverges by Exercise 12.3.19 [ET 11.3.19].

13. Let $f(x) = \dfrac{2}{x (\ln x)^3}$. $f(x)$ is clearly positive and decreasing for $x \geq 2$, so we apply the Integral Test.

$\displaystyle\int_2^{\infty} \dfrac{2}{x (\ln x)^3} dx = \lim\limits_{t\to\infty} \left[\dfrac{-1}{(\ln x)^2} \right]_2^t = 0 - \dfrac{-1}{(\ln 2)^2}$, which is finite, so $\sum\limits_{n=2}^{\infty} \dfrac{2}{n (\ln n)^3}$ converges.

15. $\lim\limits_{n\to\infty} \left| \dfrac{a_{n+1}}{a_n} \right| = \lim\limits_{n\to\infty} \dfrac{3^{n+1} (n+1)^2 / (n+1)!}{3^n n^2 / n!} = 3 \lim\limits_{n\to\infty} \dfrac{n+1}{n^2} = 0$, so the series converges by the Ratio Test.

17. $\dfrac{3^n}{5^n+n} \leq \dfrac{3^n}{5^n} = \left(\dfrac{3}{5} \right)^n$. Since $\sum\limits_{n=1}^{\infty} \left(\dfrac{3}{5} \right)^n$ is a convergent geometric series ($|r| = \dfrac{3}{5} < 1$), $\sum\limits_{n=1}^{\infty} \dfrac{3^n}{5^n+n}$ converges by the Comparison Test.

19. $\lim\limits_{n\to\infty} \left| \dfrac{a_{n+1}}{a_n} \right| = \lim\limits_{n\to\infty} \dfrac{\dfrac{(n+1)!}{2\cdot5\cdot8\cdots\cdots(3n+5)}}{\dfrac{n!}{2\cdot5\cdot8\cdots\cdots(3n+2)}} = \lim\limits_{n\to\infty} \dfrac{n+1}{3n+5} = \dfrac{1}{3} < 1$, so the series converges by the Ratio Test.

21. Use the Limit Comparison Test with $a_i = \dfrac{1}{\sqrt{i(i+1)}}$ and $b_i = \dfrac{1}{i}$.

$\lim\limits_{i\to\infty} \dfrac{a_i}{b_i} = \lim\limits_{i\to\infty} \dfrac{i}{\sqrt{i(i+1)}} = \lim\limits_{i\to\infty} \dfrac{1}{\sqrt{1+1/i}} = 1$. Since $\sum\limits_{i=1}^{\infty} b_i$ diverges (harmonic series) so does

$\sum\limits_{i=1}^{\infty} \dfrac{1}{\sqrt{i(i+1)}}$.

23. $\lim\limits_{n\to\infty} 2^{1/n} = 2^0 = 1$, so $\lim\limits_{n\to\infty} (-1)^n 2^{1/n}$ does not exist and the series diverges by the Test for Divergence.

25. Let $f(x) = \dfrac{\ln x}{\sqrt{x}}$. Then $f'(x) = \dfrac{2 - \ln x}{2x^{3/2}} < 0$ when $\ln x > 2$ or $x > e^2$, so $\dfrac{\ln n}{\sqrt{n}}$ is decreasing for $n > e^2$.

By l'Hospital's Rule, $\displaystyle\lim_{n\to\infty} \frac{\ln n}{\sqrt{n}} = \lim_{n\to\infty} \frac{1/n}{1/(2\sqrt{n})} = \lim_{n\to\infty} \frac{2}{\sqrt{n}} = 0$, so the series converges by the Alternating

Series Test.

27. $\displaystyle\sum_{n=1}^{\infty} \frac{(-2)^{2n}}{n^n} = \sum_{n=1}^{\infty} \left(\frac{4}{n}\right)^n$. $\displaystyle\lim_{n\to\infty} \sqrt[n]{|a_n|} = \lim_{n\to\infty} \frac{4}{n} = 0$, so the series converges by the Root Test.

29. $\displaystyle\int_2^{\infty} \frac{\ln x}{x^2}\,dx = \lim_{t\to\infty} \left[-\frac{\ln x}{x} - \frac{1}{x}\right]_1^t$ (using integration by parts) $\overset{\text{H}}{=} 1$. So $\displaystyle\sum_{n=1}^{\infty} \frac{\ln n}{n^2}$ converges by the Integral Test,

and since $\dfrac{k \ln k}{(k+1)^3} < \dfrac{k \ln k}{k^3} = \dfrac{\ln k}{k^2}$, the given series converges by the Comparison Test.

31. $\displaystyle\lim_{n\to\infty} \left|\frac{a_{n+1}}{a_n}\right| = \lim_{n\to\infty} \frac{2^{n+1}/(2n+3)!}{2^n/(2n+1)!} = 2\lim_{n\to\infty} \frac{1}{(2n+3)(2n+2)} = 0$, so the series converges by the Ratio

Test.

33. $0 < \dfrac{\tan^{-1} n}{n^{3/2}} < \dfrac{\pi/2}{n^{3/2}}$. $\displaystyle\sum_{n=1}^{\infty} \frac{\pi/2}{n^{3/2}} = \frac{\pi}{2} \sum_{n=1}^{\infty} \frac{1}{n^{3/2}}$ which is a convergent p-series ($p = \frac{3}{2} > 1$), so

$\displaystyle\sum_{n=1}^{\infty} \frac{\tan^{-1} n}{n^{3/2}}$ converges by the Comparison Test.

35. $\displaystyle\lim_{n\to\infty} \sqrt[n]{|a_n|} = \lim_{n\to\infty} \left(\frac{n}{n+1}\right)^{n^2/n} = \lim_{n\to\infty} \frac{1}{[(n+1)/n]^n} = \frac{1}{\displaystyle\lim_{n\to\infty}(1+1/n)^n} = \frac{1}{e} < 1$ (see Equation 7.4.9

[ET 3.8.6]), so the series converges by the Root Test.

37. $\displaystyle\lim_{n\to\infty} \sqrt[n]{|a_n|} = \lim_{n\to\infty} \left(2^{1/n} - 1\right) = 1 - 1 = 0$, so the series converges by the Root Test.

12.8 Power Series   ET 11.8

1. A power series is a series of the form $\sum_{n=0}^{\infty} c_n x^n = c_0 + c_1 x + c_2 x^2 + c_3 x^3 + \cdots$, where x is a variable and the

c_n's are constants called the coefficients of the series.

More generally, a series of the form $\sum_{n=0}^{\infty} c_n (x-a)^n = c_0 + c_1 (x-a) + c_2 (x-a)^2 + \cdots$ is called a power

series in $(x-a)$ or a power series centered at a or a power series about a.

3. $\displaystyle\lim_{n\to\infty} \left|\frac{a_{n+1}}{a_n}\right| = \lim_{n\to\infty} \left|\frac{x^{n+1}}{\sqrt{n+1}} \cdot \frac{\sqrt{n}}{x}\right| = \lim_{n\to\infty} \left|\frac{x}{\sqrt{n+1}/\sqrt{n}}\right| = \lim_{n\to\infty} \frac{|x|}{\sqrt{1+1/n}} = |x|$. By the Ratio Test, the

series converges when $|x| < 1$, so the radius of convergence $R = 1$. When $x = 1$, the series $\sum_{n=1}^{\infty} 1/\sqrt{n}$ diverges

because it is a p-series with $p = \frac{1}{2} \le 1$, but when $x = -1$, it converges by the Alternating Series Test. So the

interval of convergence is $I = [-1, 1)$.

5. If $a_n = nx^n$, then $\displaystyle\lim_{n\to\infty} \left|\frac{a_{n+1}}{a_n}\right| = \lim_{n\to\infty} \left|\frac{(n+1)x^{n+1}}{nx^n}\right| = |x| \lim_{n\to\infty} \frac{n+1}{n} = |x| < 1$ for convergence (by the

Ratio Test). So $R = 1$. When $x = 1$ or -1, $\displaystyle\lim_{n\to\infty} nx^n$ does not exist, so $\sum_{n=0}^{\infty} nx^n$ diverges for $x = \pm 1$. So

$I = (-1, 1)$.

7. If $a_n = \dfrac{x^n}{n!}$, then $\displaystyle\lim_{n\to\infty} \left|\frac{a_{n+1}}{a_n}\right| = \lim_{n\to\infty} \left|\frac{x^{n+1}/(n+1)!}{x^n/n!}\right| = |x| \lim_{n\to\infty} \frac{1}{n+1} = 0 < 1$ for all x. So, by the Ratio

Test, $R = \infty$, and $I = (-\infty, \infty)$.

9. $\lim\limits_{n\to\infty}\left|\dfrac{a_{n+1}}{a_n}\right| = \lim\limits_{n\to\infty}\dfrac{(n+1)\,4^{n+1}\,|x|^{n+1}}{n4^n\,|x|^n} = \lim\limits_{n\to\infty}\left(1+\dfrac{1}{n}\right)4\,|x| = 4\,|x|$. Now $4\,|x| < 1 \iff |x| < \frac{1}{4}$, so by

the Ratio Test, $R = \frac{1}{4}$. When $x = \frac{1}{4}$, we get the divergent series $\sum_{n=1}^{\infty}(-1)^n\,n$, and when $x = -\frac{1}{4}$, we get the

divergent series $\sum_{n=1}^{\infty}n$. Thus, $I = \left(-\frac{1}{4},\frac{1}{4}\right)$.

11. If $a_n = \dfrac{3^n x^n}{(n+1)^2}$, then $\lim\limits_{n\to\infty}\left|\dfrac{a_{n+1}}{a_n}\right| = \lim\limits_{n\to\infty}\left|\dfrac{3^{n+1}x^{n+1}}{(n+2)^2}\cdot\dfrac{(n+1)^2}{3^n x^n}\right| = 3\,|x|\lim\limits_{n\to\infty}\left(\dfrac{n+1}{n+2}\right)^2 = 3\,|x| < 1$ for

convergence, so $|x| < \frac{1}{3}$ and $R = \frac{1}{3}$. When $x = \frac{1}{3}$, $\sum\limits_{n=0}^{\infty}\dfrac{3^n x^n}{(n+1)^2} = \sum\limits_{n=0}^{\infty}\dfrac{1}{(n+1)^2} = \sum\limits_{n=1}^{\infty}\dfrac{1}{n^2}$ which is a

convergent p-series ($p = 2 > 1$). When $x = -\frac{1}{3}$, $\sum\limits_{n=0}^{\infty}\dfrac{3^n x^n}{(n+1)^2} = \sum\limits_{n=0}^{\infty}\dfrac{(-1)^n}{(n+1)^2}$ which converges by the

Alternating Series Test, so $I = \left[-\frac{1}{3},\frac{1}{3}\right]$.

13. If $a_n = \dfrac{x^n}{\ln n}$, then $\lim\limits_{n\to\infty}\left|\dfrac{a_{n+1}}{a_n}\right| = \lim\limits_{n\to\infty}\left|\dfrac{x^{n+1}}{\ln(n+1)}\cdot\dfrac{\ln n}{x^n}\right| = |x|\lim\limits_{n\to\infty}\dfrac{\ln n}{\ln(n+1)} \overset{\text{H}}{=} |x|$, so $R = 1$. When

$x = 1$, $\sum\limits_{n=2}^{\infty}\dfrac{x^n}{\ln n} = \sum\limits_{n=2}^{\infty}\dfrac{1}{\ln n}$ which diverges because $\dfrac{1}{\ln n} > \dfrac{1}{n}$ and $\sum\limits_{n=2}^{\infty}\dfrac{1}{n}$ is the divergent harmonic series. When

$x = -1$, $\sum\limits_{n=2}^{\infty}\dfrac{x^n}{\ln n} = \sum\limits_{n=2}^{\infty}\dfrac{(-1)^n}{\ln n}$ which converges by the Alternating Series Test. So $I = [-1,1)$.

15. $\lim\limits_{n\to\infty}\left|\dfrac{a_{n+1}}{a_n}\right| = \lim\limits_{n\to\infty}\dfrac{\sqrt{n+1}\,|x-1|^{n+1}}{\sqrt{n}\,|x-1|^n} = \lim\limits_{n\to\infty}\sqrt{1+\dfrac{1}{n}}\,|x-1| = |x-1|$, so by the Ratio Test, the series

converges when $|x-1| < 1 \iff -1 < x - 1 < 1 \iff 0 < x < 2$. $R = 1$. When $x = 2$, the series becomes

$\sum_{n=0}^{\infty}\sqrt{n}$, which diverges by the Test for Divergence. When $x = 0$, the series becomes $\sum_{n=0}^{\infty}(-1)^n\sqrt{n}$, which

also diverges by the Test for Divergence. Thus, $I = (0,2)$.

17. $\lim\limits_{n\to\infty}\left|\dfrac{a_{n+1}}{a_n}\right| = \lim\limits_{n\to\infty}\left[\dfrac{|x+2|^{n+1}}{(n+1)\,2^{n+1}}\cdot\dfrac{n2^n}{|x+2|^n}\right] = \lim\limits_{n\to\infty}\dfrac{n}{n+1}\cdot\dfrac{|x+2|}{2} = \dfrac{|x+2|}{2}$, so by the Ratio Test, the

series converges when $\dfrac{|x+2|}{2} < 1 \iff |x+2| < 2 \iff -2 < x + 2 < 2 \iff -4 < x < 0$. $R = 2$. When

$x = -4$, the series becomes $\sum\limits_{n=1}^{\infty}(-1)^n\dfrac{(-2)^n}{n2^n} = \sum\limits_{n=1}^{\infty}\dfrac{2^n}{n2^n} = \sum\limits_{n=1}^{\infty}\dfrac{1}{n}$, which is the divergent harmonic series.

When $x = 0$, the series is $\sum\limits_{n=1}^{\infty}\dfrac{(-1)^n}{n}$, the alternating harmonic series, which converges by the Alternating Series

Test. Thus, $I = (-4,0]$.

19. If $a_n = \dfrac{(x-2)^n}{n^n}$, then $\lim\limits_{n\to\infty}\sqrt[n]{|a_n|} = \lim\limits_{n\to\infty}\dfrac{|x-2|}{n} = 0$, so the series converges for all x (by the Root Test).

$R = \infty$ and $I = (-\infty,\infty)$.

21. If $a_n = \dfrac{2^n (x-3)^n}{n+3}$, then

$$\lim_{n\to\infty} \left| \frac{a_{n+1}}{a_n} \right| = \lim_{n\to\infty} \left| \frac{2^{n+1}(x-3)^{n+1}}{n+4} \cdot \frac{n+3}{2^n(x-3)^n} \right| = 2|x-3| \lim_{n\to\infty} \frac{n+3}{n+4} = 2|x-3| < 1 \text{ for}$$

convergence, or $|x-3| < \frac{1}{2} \Leftrightarrow \frac{5}{2} < x < \frac{7}{2}$, and $R = \frac{1}{2}$. When $x = \frac{5}{2}$, $\displaystyle\sum_{n=0}^{\infty} \frac{2^n(x-3)^n}{n+3} = \sum_{n=0}^{\infty} \frac{(-1)^n}{n+3}$ which

converges by the Alternating Series Test. When $x = \frac{7}{2}$, $\displaystyle\sum_{n=0}^{\infty} \frac{2^n(x-3)^n}{n+3} = \sum_{n=0}^{\infty} \frac{1}{n+3}$, similar to the harmonic

series, which diverges. So $I = \left[\frac{5}{2}, \frac{7}{2}\right)$.

23. If $a_n = n!\,(2x-1)^n$, then $\displaystyle\lim_{n\to\infty} \left| \frac{a_{n+1}}{a_n} \right| = \lim_{n\to\infty} \left| \frac{(n+1)!\,(2x-1)^{n+1}}{n!\,(2x-1)^n} \right| = \lim_{n\to\infty} (n+1)\,|2x-1| \to \infty$ as

$n \to \infty$ for all $x \neq \frac{1}{2}$. Since the series diverges for all $x \neq \frac{1}{2}$, $R = 0$ and $I = \left\{\frac{1}{2}\right\}$.

25. $\displaystyle\lim_{n\to\infty} \left| \frac{a_{n+1}}{a_n} \right| = \lim_{n\to\infty} \left[\frac{|4x+1|^{n+1}}{(n+1)^2} \cdot \frac{n^2}{|4x+1|^n} \right] = \lim_{n\to\infty} \frac{|4x+1|}{(1+1/n)^2} = |4x+1|$, so by the Ratio Test, the

series converges when $|4x+1| < 1 \Leftrightarrow -1 < 4x+1 < 1 \Leftrightarrow -2 < 4x < 0 \Leftrightarrow -\frac{1}{2} < x < 0$, so

$R = \frac{1}{4}$. When $x = -\frac{1}{2}$, the series becomes $\displaystyle\sum_{n=1}^{\infty} \frac{(-1)^n}{n^2}$, which converges by the Alternating Series Test. When

$x = 0$, the series becomes $\displaystyle\sum_{n=1}^{\infty} \frac{1}{n^2}$, a convergent p-series $(p = 2 > 1)$. $I = \left[-\frac{1}{2}, 0\right]$.

27. If $a_n = \dfrac{x^n}{(\ln n)^n}$ then $\displaystyle\lim_{n\to\infty} \sqrt[n]{|a_n|} = \lim_{n\to\infty} \frac{|x|}{\ln n} = 0 < 1$ for all x, so $R = \infty$ and $I = (-\infty, \infty)$ by the Root

Test.

29. (a) We are given that the power series $\sum_{n=0}^{\infty} c_n x^n$ is convergent for $x = 4$. So by Theorem 3, it must converge for

at least $-4 < x \leq 4$. In particular, it converges when $x = -2$, that is, $\sum_{n=0}^{\infty} c_n (-2)^n$ is convergent.

(b) It does not follow that $\sum_{n=0}^{\infty} c_n (-4)^n$ is necessarily convergent. [See the comments after Theorem 3 about

convergence at the endpoint of an interval. An example is $c_n = (-1)^n / (n4^n)$.]

31. If $a_n = \dfrac{(n!)^k}{(kn)!} x^n$, then

$$\lim_{n\to\infty} \left| \frac{a_{n+1}}{a_n} \right| = \lim_{n\to\infty} \frac{[(n+1)!]^k (kn)!}{(n!)^k [k(n+1)]!} |x| = \lim_{n\to\infty} \frac{(n+1)^k}{(kn+k)(kn+k-1)\cdots(kn+2)(kn+1)} |x|$$

$$= \lim_{n\to\infty} \left[\frac{(n+1)}{(kn+1)} \frac{(n+1)}{(kn+2)} \cdots \frac{(n+1)}{(kn+k)} \right] |x|$$

$$= \lim_{n\to\infty} \left[\frac{n+1}{kn+1} \right] \lim_{n\to\infty} \left[\frac{n+1}{kn+2} \right] \cdots \lim_{n\to\infty} \left[\frac{n+1}{kn+k} \right] |x| = \left(\frac{1}{k} \right)^k |x| < 1 \Leftrightarrow$$

$|x| < k^k$ for convergence, and the radius of convergence is $R = k^k$.

33. (a) If $a_n = \dfrac{(-1)^n \, x^{2n+1}}{n! \, (n+1)! \, 2^{2n+1}}$, then $\displaystyle\lim_{n \to \infty} \left| \dfrac{a_{n+1}}{a_n} \right| = \left(\dfrac{x}{2} \right)^2 \lim_{n \to \infty} \dfrac{1}{(n+1)(n+2)} = 0$ for all x. So $J_1(x)$

converges for all x; the domain is $(-\infty, \infty)$.

(b), (c) The initial terms of $J_1(x)$ up to $n = 5$ are

$$a_0 = \frac{x}{2}, \; a_1 = -\frac{x^3}{16}, \; a_2 = \frac{x^5}{384},$$

$$a_3 = -\frac{x^7}{18{,}432}, \; a_4 = \frac{x^9}{1{,}474{,}560}, \text{ and}$$

$$a_5 = -\frac{x^{11}}{176{,}947{,}200}. \text{ The partial sums seem to}$$

approximate $J_1(x)$ well near the origin, but as $|x|$ increases, we need to take a large number of terms to get a good approximation.

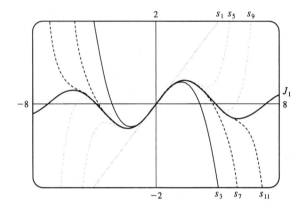

35. $s_{2n-1} = 1 + 2x + x^2 + 2x^3 + \cdots + x^{2n-2} + 2x^{2n-1} = (1 + 2x)\left(1 + x^2 + x^4 + \cdots + x^{2n-2}\right)$

$= (1 + 2x)\dfrac{1 - x^{2n}}{1 - x^2}$ [by (12.2.3 [ET 11.2.3]) with $r = x^2$]$\to \dfrac{1 + 2x}{1 - x^2}$ as $n \to \infty$ [by (12.2.4 [ET 11.2.4])],

when $|x| < 1$. Also $s_{2n} = s_{2n-1} + x^{2n} \to \dfrac{1 + 2x}{1 - x^2}$ since $x^{2n} \to 0$ for $|x| < 1$. Therefore, $s_n \to \dfrac{1 + 2x}{1 - x^2}$

since s_{2n} and s_{2n-1} both approach $\dfrac{1 + 2x}{1 - x^2}$ as $n \to \infty$. Thus, the interval of convergence is $(-1, 1)$ and

$f(x) = \dfrac{1 + 2x}{1 - x^2}$.

37. We use the Root Test on the series $\sum c_n x^n$. $\displaystyle\lim_{n \to \infty} \sqrt[n]{|c_n x^n|} = |x| \lim_{n \to \infty} \sqrt[n]{|c_n|} = c \, |x| < 1$ for convergence, or

$|x| < 1/c$, so $R = 1/c$.

39. For $2 < x < 3$, $\sum c_n x^n$ diverges and $\sum d_n x^n$ converges. By Exercise 12.2.61 [ET 11.2.61], $\sum (c_n + d_n) \, x^n$

diverges. Since both series converge for $|x| < 2$, the radius of convergence of $\sum (c_n + d_n) \, x^n$ is 2.

12.9 Representations of Functions as Power Series　　　　ET 11.9

1. If $f(x) = \displaystyle\sum_{n=0}^{\infty} c_n x^n$ has radius of convergence 10, then $f'(x) = \displaystyle\sum_{n=1}^{\infty} n c_n x^{n-1}$ also has radius of convergence 10

by Theorem 2.

3. $f(x) = \dfrac{1}{1 + x} = \dfrac{1}{1 - (-x)} = \displaystyle\sum_{n=0}^{\infty} (-x)^n = \sum_{n=0}^{\infty} (-1)^n \, x^n$ with $|-x| < 1 \;\Leftrightarrow\; |x| < 1$, so $R = 1$ and

$I = (-1, 1)$.

5. Replacing x with x^3 in (1) gives $f(x) = \dfrac{1}{1 - x^3} = \displaystyle\sum_{n=0}^{\infty} (x^3)^n = \sum_{n=0}^{\infty} x^{3n}$. The series converges when $|x^3| < 1$;

that is, when $|x| < 1$, so $I = (-1, 1)$.

7. $f(x) = \dfrac{1}{4 + x^2} = \dfrac{1}{4}\left(\dfrac{1}{1 + x^2/4}\right) = \dfrac{1}{4}\sum\limits_{n=0}^{\infty}(-1)^n\left(\dfrac{x^2}{4}\right)^n$ (using Exercise 3) $= \sum\limits_{n=0}^{\infty}\dfrac{(-1)^n\, x^{2n}}{4^{n+1}}$, with

$\left|\dfrac{x^2}{4}\right| < 1 \iff x^2 < 4 \iff |x| < 2$, so $R = 2$ and $I = (-2, 2)$.

9. $f(x) = \dfrac{1}{x - 5} = -\dfrac{1}{5}\left(\dfrac{1}{1 - x/5}\right) = -\dfrac{1}{5}\sum\limits_{n=0}^{\infty}\left(\dfrac{x}{5}\right)^n$. The series converges when $\left|\dfrac{x}{5}\right| < 1$; that is, when $|x| < 5$,

so $I = (-5, 5)$.

11. $f(x) = \dfrac{3}{x^2 + x - 2} = \dfrac{3}{(x + 2)(x - 1)} = \dfrac{A}{x + 2} + \dfrac{B}{x - 1} \Rightarrow 3 = A(x - 1) + B(x + 2)$. Taking $x = -2$,

we get $A = -1$. Taking $x = 1$, we get $B = 1$. Thus,

$$\dfrac{3}{x^2 + x - 2} = \dfrac{1}{x - 1} - \dfrac{1}{x + 2} = -\dfrac{1}{1 - x} - \dfrac{1}{2}\dfrac{1}{1 + x/2} = -\sum\limits_{n=0}^{\infty} x^n - \dfrac{1}{2}\sum\limits_{n=0}^{\infty}\left(-\dfrac{x}{2}\right)^n$$

$$= \sum\limits_{n=0}^{\infty}\left[-1 - \tfrac{1}{2}\left(-\tfrac{1}{2}\right)^n\right] x^n = \sum\limits_{n=0}^{\infty}\left[-1 + \left(-\tfrac{1}{2}\right)^{n+1}\right] x^n = \sum\limits_{n=0}^{\infty}\left[\dfrac{(-1)^{n+1}}{2^{n+1}} - 1\right] x^n$$

We represented the given function as the sum of two geometric series; the first converges for $x \in (-1, 1)$ and the second converges for $x \in (-2, 2)$. Thus, the sum converges for $x \in (-1, 1) = I$.

13. $f(x) = \dfrac{1}{(1 + x)^2} = -\dfrac{d}{dx}\left(\dfrac{1}{1 + x}\right) = -\dfrac{d}{dx}\left[\sum_{n=0}^{\infty}(-1)^n x^n\right]$ (from Exercise 3)

$= \sum_{n=1}^{\infty}(-1)^{n+1} n x^{n-1}$ [from Theorem 2(a)] $= \sum_{n=0}^{\infty}(-1)^n(n + 1)x^n$ with $R = 1$.

15. $f(x) = \dfrac{1}{(1 + x)^3} = -\dfrac{1}{2}\dfrac{d}{dx}\left[\dfrac{1}{(1 + x)^2}\right] = -\dfrac{1}{2}\dfrac{d}{dx}\left[\sum_{n=0}^{\infty}(-1)^n(n + 1)x^n\right]$ (from Exercise 13)

$= -\tfrac{1}{2}\sum_{n=1}^{\infty}(-1)^n(n + 1)n x^{n-1} = \tfrac{1}{2}\sum_{n=0}^{\infty}(-1)^n(n + 2)(n + 1)x^n$ with $R = 1$.

17. $f(x) = \ln(5 - x) = -\displaystyle\int\dfrac{dx}{5 - x} = -\dfrac{1}{5}\displaystyle\int\dfrac{dx}{1 - x/5}$

$= -\dfrac{1}{5}\displaystyle\int\left[\sum\limits_{n=0}^{\infty}\left(\dfrac{x}{5}\right)^n\right]dx = C - \dfrac{1}{5}\sum\limits_{n=0}^{\infty}\dfrac{x^{n+1}}{5^n(n + 1)} = C - \sum\limits_{n=1}^{\infty}\dfrac{x^n}{n 5^n}$

Putting $x = 0$, we get $C = \ln 5$. The series converges for $|x/5| < 1 \iff |x| < 5$, so $R = 5$.

19. $\dfrac{1}{2 - x} = \dfrac{1}{2(1 - x/2)} = \dfrac{1}{2}\sum\limits_{n=0}^{\infty}\left(\dfrac{x}{2}\right)^n = \sum\limits_{n=0}^{\infty}\dfrac{1}{2^{n+1}}x^n$ for $\left|\dfrac{x}{2}\right| < 1 \iff |x| < 2$. Now

$\dfrac{1}{(x - 2)^2} = \left(\dfrac{1}{2 - x}\right)' = \left(\sum\limits_{n=0}^{\infty}\dfrac{1}{2^{n+1}}x^n\right)' = \sum\limits_{n=0}^{\infty}\left(\dfrac{1}{2^{n+1}}x^n\right)' = \sum\limits_{n=1}^{\infty}\dfrac{n}{2^{n+1}}x^{n-1} = \sum\limits_{n=0}^{\infty}\dfrac{n + 1}{2^{n+2}}x^n$. So

$f(x) = \dfrac{x^3}{(x - 2)^2} = x^3\sum\limits_{n=0}^{\infty}\dfrac{n + 1}{2^{n+2}}x^n = \sum\limits_{n=0}^{\infty}\dfrac{n + 1}{2^{n+2}}x^{n+3}$ or $\sum\limits_{n=3}^{\infty}\dfrac{n - 2}{2^{n-1}}x^n$ for $|x| < 2$. Thus, $R = 2$ and

$I = (-2, 2)$.

21. $f(x) = \ln(3 + x) = \displaystyle\int \frac{dx}{3+x} = \frac{1}{3}\int \frac{dx}{1+x/3} = \frac{1}{3}\int \sum_{n=0}^{\infty} (-1)^n \left(\frac{x}{3}\right)^n dx$ (from Exercise 3)

$= C + \dfrac{1}{3} \displaystyle\sum_{n=0}^{\infty} \frac{(-1/3)^n}{n+1} x^{n+1} = \ln 3 + \frac{1}{3} \sum_{n=1}^{\infty} \frac{(-1/3)^{n-1}}{n} x^n$ $[C = f(0) = \ln 3]$

$= \ln 3 + \displaystyle\sum_{n=1}^{\infty} \frac{(-1)^{n-1}}{n 3^n} x^n$ with $R = 3$.

The terms of the series are $a_0 = \ln 3$, $a_1 = \dfrac{x}{3}$, $a_2 = -\dfrac{x^2}{18}$, $a_3 = \dfrac{x^3}{81}$, $a_4 = -\dfrac{x^4}{324}$, $a_5 = \dfrac{x^5}{1215}, \ldots$.

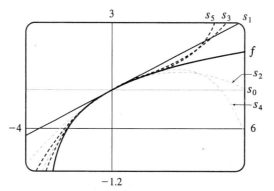

As n increases, $s_n(x)$ approximates f better on the interval of convergence, which is $(-3, 3)$.

23. $f(x) = \ln\left(\dfrac{1+x}{1-x}\right) = \ln(1+x) - \ln(1-x) = \displaystyle\int \frac{dx}{1+x} + \int \frac{dx}{1-x}$

$= \displaystyle\int \left[\sum_{n=0}^{\infty} (-1)^n x^n + \sum_{n=0}^{\infty} x^n\right] dx = \int \sum_{n=0}^{\infty} 2x^{2n} dx = \sum_{n=0}^{\infty} \frac{2x^{2n+1}}{2n+1} + C$

But $f(0) = \ln \dfrac{1}{1} = 0$, so $C = 0$ and we have $f(x) = \displaystyle\sum_{n=0}^{\infty} \frac{2x^{2n+1}}{2n+1}$ with $R = 1$. If $x = \pm 1$, then

$f(x) = \pm 2 \displaystyle\sum_{n=0}^{\infty} \frac{1}{2n+1}$, which both diverge by the Limit Comparison Test with $b_n = \dfrac{1}{n}$.

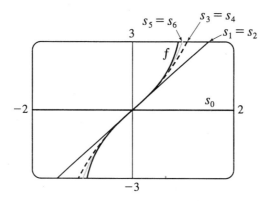

As n increases, $s_n(x)$ approximates f better on the interval of convergence, which is $(-1, 1)$.

25. $\displaystyle\int \frac{dx}{1+x^4} = \int \sum_{n=0}^{\infty} (-1)^n x^{4n} dx = C + \sum_{n=0}^{\infty} \frac{(-1)^n x^{4n+1}}{4n+1}$ with $R = 1$.

27. By Example 7, $\arctan x = \sum_{n=0}^{\infty} (-1)^n \dfrac{x^{2n+1}}{2n+1}$, so

$$\int \frac{\arctan x}{x}\, dx = \int \sum_{n=0}^{\infty} (-1)^n \frac{x^{2n}}{2n+1}\, dx = C + \sum_{n=0}^{\infty} (-1)^n \frac{x^{2n+1}}{(2n+1)^2} \quad \text{with } R = 1.$$

29. We use the representation $\displaystyle\int \frac{dx}{1+x^4} = C + \sum_{n=0}^{\infty} \frac{(-1)^n\, x^{4n+1}}{4n+1}$ from Exercise 25 with $C = 0$. So

$$\int_0^{0.2} \frac{dx}{1+x^4} = \left[x - \frac{x^5}{5} + \frac{x^9}{9} - \frac{x^{13}}{13} + \cdots \right]_0^{0.2} = 0.2 - \frac{0.2^5}{5} + \frac{0.2^9}{9} - \frac{0.2^{13}}{13} + \cdots$$

Since the series is alternating, the error in the nth-order approximation is less than the first neglected term, by The Alternating Series Estimation Theorem. If we use only the first two terms of the series, then the error is at most $0.2^9/9 \approx 5.7 \times 10^{-8}$. So, to six decimal places, $\displaystyle\int_0^{0.2} \frac{dx}{1+x^4} \approx 0.2 - \frac{0.2^5}{5} = 0.199936.$

31. We substitute x^4 for x in Example 7, and find that

$$\int x^2 \tan^{-1}(x^4)\, dx = \int x^2 \sum_{n=0}^{\infty} (-1)^n \frac{(x^4)^{2n+1}}{2n+1}\, dx$$

$$= \int \sum_{n=0}^{\infty} (-1)^n \frac{x^{8n+6}}{2n+1}\, dx = C + \sum_{n=0}^{\infty} (-1)^n \frac{x^{8n+7}}{(2n+1)(8n+7)}$$

So $\displaystyle\int_0^{1/3} x^2 \tan^{-1}(x^4)\, dx = \left[\frac{x^7}{7} - \frac{x^{15}}{45} + \cdots \right]_0^{1/3} = \frac{1}{7 \cdot 3^7} - \frac{1}{45 \cdot 3^{15}} + \cdots$. The series is alternating, so if we use only one term, the error is at most $1/(45 \cdot 3^{15}) \approx 1.5 \times 10^{-9}$. So $\displaystyle\int_0^{1/3} x^2 \tan^{-1}(x^4)\, dx \approx 1/(7 \cdot 3^7) \approx 0.000065$ to six decimal places.

33. Using the result of Example 6, $\ln(1-x) = -\sum_{n=1}^{\infty} x^n/n$, with $x = -0.1$, we have
$\ln 1.1 = \ln[1 - (-0.1)] = 0.1 - \frac{0.01}{2} + \frac{0.001}{3} - \frac{0.0001}{4} + \frac{0.00001}{5} - \cdots$. The series is alternating, so if we use only the first four terms, the error is at most $\frac{0.00001}{5} = 0.000002$. So
$\ln 1.1 \approx 0.1 - \frac{0.01}{2} + \frac{0.001}{3} - \frac{0.0001}{4} \approx 0.09531.$

35. (a) $J_0(x) = \sum_{n=0}^{\infty} \dfrac{(-1)^n\, x^{2n}}{2^{2n}\, (n!)^2}$, $J_0'(x) = \sum_{n=1}^{\infty} \dfrac{(-1)^n\, 2n\, x^{2n-1}}{2^{2n}\, (n!)^2}$, and $J_0''(x) = \sum_{n=1}^{\infty} \dfrac{(-1)^n\, 2n(2n-1)\, x^{2n-2}}{2^{2n}\, (n!)^2}$, so

$$x^2 J_0''(x) + x J_0'(x) + x^2 J_0(x) = \sum_{n=1}^{\infty} \frac{(-1)^n\, 2n(2n-1)\, x^{2n}}{2^{2n}\, (n!)^2} + \sum_{n=1}^{\infty} \frac{(-1)^n\, 2n x^{2n}}{2^{2n}\, (n!)^2} + \sum_{n=0}^{\infty} \frac{(-1)^n\, x^{2n+2}}{2^{2n}\, (n!)^2}$$

$$= \sum_{n=1}^{\infty} \frac{(-1)^n\, 2n(2n-1)\, x^{2n}}{2^{2n}\, (n!)^2} + \sum_{n=1}^{\infty} \frac{(-1)^n\, 2n x^{2n}}{2^{2n}\, (n!)^2} + \sum_{n=1}^{\infty} \frac{(-1)^{n-1}\, x^{2n}}{2^{2n-2}\, [(n-1)!]^2}$$

$$= \sum_{n=1}^{\infty} (-1)^n \left[\frac{2n(2n-1) + 2n - 2^2 n^2}{2^{2n}\, (n!)^2} \right] x^{2n} = \sum_{n=1}^{\infty} (-1)^n \left[\frac{4n^2 - 2n + 2n - 4n^2}{2^{2n}\, (n!)^2} \right] x^{2n} = 0$$

(b) $\displaystyle\int_0^1 J_0(x)\, dx = \int_0^1 \left[\sum_{n=0}^{\infty} \frac{(-1)^n\, x^{2n}}{2^{2n}\, (n!)^2} \right] dx = \int_0^1 \left(1 - \frac{x^2}{4} + \frac{x^4}{64} - \frac{x^6}{2304} + \cdots \right) dx$

$$= \left[x - \frac{x^3}{3 \cdot 4} + \frac{x^5}{5 \cdot 64} - \frac{x^7}{7 \cdot 2304} + \cdots \right]_0^1 = 1 - \frac{1}{12} + \frac{1}{320} - \frac{1}{16{,}128} + \cdots$$

Since $\frac{1}{16{,}128} \approx 0.000062$, it follows from The Alternating Series Estimation Theorem that, correct to three decimal places, $\int_0^1 J_0(x)\, dx \approx 1 - \frac{1}{12} + \frac{1}{320} \approx 0.920.$

37. (a) $f(x) = \sum\limits_{n=0}^{\infty} \dfrac{x^n}{n!} \Rightarrow f'(x) = \sum\limits_{n=1}^{\infty} \dfrac{nx^{n-1}}{n!} = \sum\limits_{n=1}^{\infty} \dfrac{x^{n-1}}{(n-1)!} = \sum\limits_{n=0}^{\infty} \dfrac{x^n}{n!} = f(x)$

(b) By Theorem 10.4.2 [ET 9.4.2], the only solution to the differential equation $df(x)/dx = f(x)$ is
$f(x) = Ke^x$, but $f(0) = 1$, so $K = 1$ and $f(x) = e^x$.
Or: We could solve the equation $df(x)/dx = f(x)$ as a separable differential equation.

39. If $a_n = \dfrac{x^n}{n^2}$, then by the Ratio Test, $\lim\limits_{n\to\infty} \left| \dfrac{a_{n+1}}{a_n} \right| = |x| \lim\limits_{n\to\infty} \left(\dfrac{n}{n+1} \right)^2 = |x| < 1$ for convergence, so $R = 1$.

When $x = \pm 1$, $\sum\limits_{n=1}^{\infty} \left| \dfrac{x^n}{n^2} \right| = \sum\limits_{n=1}^{\infty} \dfrac{1}{n^2}$ which is a convergent p-series ($p = 2 > 1$), so the interval of convergence for

f is $[-1, 1]$. By Theorem 2, the radii of convergence of f' and f'' are both 1, so we need only check the endpoints.

$f(x) = \sum\limits_{n=1}^{\infty} \dfrac{x^n}{n^2} \Rightarrow f'(x) = \sum\limits_{n=1}^{\infty} \dfrac{nx^{n-1}}{n^2} = \sum\limits_{n=0}^{\infty} \dfrac{x^n}{n+1}$, and this series diverges for $x = 1$ (harmonic series)

and converges for $x = -1$ (Alternating Series Test), so the interval of convergence is $[-1, 1)$.

$f''(x) = \sum\limits_{n=1}^{\infty} \dfrac{nx^{n-1}}{n+1}$ diverges at both 1 and -1 (Test for Divergence) since $\lim\limits_{n\to\infty} \dfrac{n}{n+1} = 1 \neq 0$, so its interval of

convergence is $(-1, 1)$.

☰12.10 Taylor and Maclaurin Series ET 11.10

1. Using Theorem 5 with $\sum\limits_{n=0}^{\infty} b_n (x-5)^n$, $b_n = \dfrac{f^{(n)}(a)}{n!}$, so $b_8 = \dfrac{f^{(8)}(5)}{8!}$.

3.

n	$f^{(n)}(x)$	$f^{(n)}(0)$
0	$\cos x$	1
1	$-\sin x$	0
2	$-\cos x$	-1
3	$\sin x$	0
4	$\cos x$	1
...	...	...

$\cos x = f(0) + f'(0)x + \dfrac{f''(0)}{2!}x^2 + \dfrac{f^{(3)}(0)}{3!}x^3 + \dfrac{f^{(4)}(0)}{4!}x^4 + \cdots$

$= 1 - \dfrac{x^2}{2!} + \dfrac{x^4}{4!} - \cdots = \sum\limits_{n=0}^{\infty} \dfrac{(-1)^n x^{2n}}{(2n)!}$

If $a_n = \dfrac{(-1)^n x^{2n}}{(2n)!}$, then

$\lim\limits_{n\to\infty} \left| \dfrac{a_{n+1}}{a_n} \right| = x^2 \lim\limits_{n\to\infty} \dfrac{1}{(2n+2)(2n+1)} = 0 < 1$ for all x. So

$R = \infty$ (Ratio Test).

5.

n	$f^{(n)}(x)$	$f^{(n)}(0)$
0	$(1+x)^{-3}$	1
1	$-3(1+x)^{-4}$	-3
2	$12(1+x)^{-5}$	12
3	$-60(1+x)^{-6}$	-60
4	$360(1+x)^{-7}$	360
...	...	...

$(1+x)^{-3} = f(0) + f'(0)x + \dfrac{f''(0)}{2!}x^2 + \dfrac{f'''(0)}{3!}x^3$

$+ \dfrac{f^{(4)}(0)}{4!}x^4 + \cdots$

$= 1 - 3x + \dfrac{12}{2}x^2 - \dfrac{60}{6}x^3 + \dfrac{360}{24}x^4 + \cdots$

$= \sum\limits_{n=0}^{\infty} \dfrac{(-1)^n (n+2)! x^n}{2(n!)} = \sum\limits_{n=0}^{\infty} \dfrac{(-1)^n (n+2)(n+1) x^n}{2}$

$\lim\limits_{n\to\infty} \left| \dfrac{a_{n+1}}{a_n} \right| = \lim\limits_{n\to\infty} \dfrac{(n+3)(n+2)|x|^{n+1}}{(n+2)(n+1)|x|} = |x| < 1$ for

convergence, so $R = 1$.

7.

n	$f^{(n)}(x)$	$f^{(n)}(0)$
0	$\sinh x$	0
1	$\cosh x$	1
2	$\sinh x$	0
3	$\cosh x$	1
4	$\sinh x$	0
$\cdots$	$\cdots$	$\cdots$

So $f^{(n)}(0) = \begin{cases} 0 & \text{if } n \text{ is even} \\ 1 & \text{if } n \text{ is odd} \end{cases}$ and $\sinh x = \sum\limits_{n=0}^{\infty} \dfrac{x^{2n+1}}{(2n+1)!}$. If

$a_n = \dfrac{x^{2n+1}}{(2n+1)!}$ then

$\lim\limits_{n\to\infty} \left| \dfrac{a_{n+1}}{a_n} \right| = x^2 \lim\limits_{n\to\infty} \dfrac{1}{(2n+3)(2n+2)} = 0 < 1$ for all x, so

$R = \infty$.

9.

n	$f^{(n)}(x)$	$f^{(n)}(2)$
0	$1 + x + x^2$	7
1	$1 + 2x$	5
2	2	2
3	0	0
4	0	0
$\cdots$	$\cdots$	$\cdots$

$f(x) = 7 + 5(x - 2) + \dfrac{2}{2!}(x-2)^2 + \sum\limits_{n=3}^{\infty} \dfrac{0}{n!}(x-2)^n$

$= 7 + 5(x-2) + (x-2)^2$

Since $a_n = 0$ for large n, $R = \infty$.

11. Clearly, $f^{(n)}(x) = e^x$, so $f^{(n)}(3) = e^3$ and $e^x = \sum\limits_{n=0}^{\infty} \dfrac{e^3}{n!}(x-3)^n$. If $a_n = \dfrac{e^3}{n!}(x-3)^n$, then

$\lim\limits_{n\to\infty} \left| \dfrac{a_{n+1}}{a_n} \right| = \lim\limits_{n\to\infty} \dfrac{|x-3|}{n+1} = 0$ for all x, so $R = \infty$.

13.

n	$f^{(n)}(x)$	$f^{(n)}(1)$
0	x^{-1}	1
1	$-x^{-2}$	-1
2	$2x^{-3}$	2
3	$-3 \cdot 2x^{-4}$	$-3 \cdot 2$
4	$4 \cdot 3 \cdot 2x^{-5}$	$4 \cdot 3 \cdot 2$
$\cdots$	$\cdots$	$\cdots$

So $f^{(n)}(1) = (-1)^n n!$, and $\dfrac{1}{x} = \sum\limits_{n=0}^{\infty} \dfrac{(-1)^n n!}{n!}(x-1)^n = \sum\limits_{n=0}^{\infty} (-1)^n (x-1)^n$. If $a_n = (-1)^n (x-1)^n$ then

$\lim\limits_{n\to\infty} \left| \dfrac{a_{n+1}}{a_n} \right| = |x-1| < 1$ for convergence, so $0 < x < 2$ and $R = 1$.

15.

n	$f^{(n)}(x)$	$f^{(n)}\left(\frac{\pi}{4}\right)$
0	$\sin x$	$\sqrt{2}/2$
1	$\cos x$	$\sqrt{2}/2$
2	$-\sin x$	$-\sqrt{2}/2$
3	$-\cos x$	$-\sqrt{2}/2$
4	$\sin x$	$\sqrt{2}/2$
...	...	...

$$\sin x = f\left(\tfrac{\pi}{4}\right) + f'\left(\tfrac{\pi}{4}\right)\left(x - \tfrac{\pi}{4}\right) + \frac{f''\left(\tfrac{\pi}{4}\right)}{2!}\left(x - \tfrac{\pi}{4}\right)^2$$

$$+ \frac{f^{(3)}\left(\tfrac{\pi}{4}\right)}{3!}\left(x - \tfrac{\pi}{4}\right)^3 + \frac{f^{(4)}\left(\tfrac{\pi}{4}\right)}{4!}\left(x - \tfrac{\pi}{4}\right)^4 + \cdots$$

$$= \frac{\sqrt{2}}{2}\left[1 + \left(x - \tfrac{\pi}{4}\right) - \tfrac{1}{2!}\left(x - \tfrac{\pi}{4}\right)^2\right.$$

$$\left. - \tfrac{1}{3!}\left(x - \tfrac{\pi}{4}\right)^3 + \tfrac{1}{4!}\left(x - \tfrac{\pi}{4}\right)^4 + \cdots\right]$$

$$= \frac{\sqrt{2}}{2}\left[1 - \tfrac{1}{2!}\left(x - \tfrac{\pi}{4}\right)^2 + \tfrac{1}{4!}\left(x - \tfrac{\pi}{4}\right)^4 - \cdots\right]$$

$$+ \frac{\sqrt{2}}{2}\left[\left(x - \tfrac{\pi}{4}\right) - \tfrac{1}{3!}\left(x - \tfrac{\pi}{4}\right)^3 + \cdots\right]$$

$$= \frac{\sqrt{2}}{2}\sum_{n=0}^{\infty}(-1)^n\left[\tfrac{1}{(2n)!}\left(x - \tfrac{\pi}{4}\right)^{2n} + \tfrac{1}{(2n+1)!}\left(x - \tfrac{\pi}{4}\right)^{2n+1}\right]$$

The series can also be written in the more elegant form $\sin x = \dfrac{\sqrt{2}}{2}\displaystyle\sum_{n=0}^{\infty}\dfrac{(-1)^{n(n-1)/2}\left(x - \tfrac{\pi}{4}\right)^n}{n!}$. If

$a_n = \dfrac{(-1)^{n(n-1)/2}\left(x - \tfrac{\pi}{4}\right)^n}{n!}$, then $\displaystyle\lim_{n\to\infty}\left|\dfrac{a_{n+1}}{a_n}\right| = \lim_{n\to\infty}\dfrac{\left|x - \tfrac{\pi}{4}\right|}{n+1} = 0 < 1$ for all x, so $R = \infty$.

17. If $f(x) = \cos x$, then by Formula 9 with $a = 0$, $|R_n(x)| \leq \dfrac{\left|f^{(n+1)}(x)\right|}{(n+1)!}|x|^{n+1}$. But $f^{(n+1)}(x) = \pm\sin x$ or

$\pm\cos x$. In each case, $\left|f^{(n+1)}(x)\right| \leq 1$, so $|R_n(x)| \leq \dfrac{1}{(n+1)!}|x|^{n+1} \to 0$ as $n \to \infty$ by Equation 10. So

$\displaystyle\lim_{n\to\infty} R_n(x) = 0$ and, by Theorem 8, the series in Exercise 3 represents $\cos x$ for all x.

19. If $f(x) = \sinh x$, then $R_n(x) = \dfrac{f^{(n+1)}(z)}{(n+1)!}x^{n+1}$, where $0 < |z| < |x|$. But for all n,

$\left|f^{(n+1)}(z)\right| \leq \cosh z \leq \cosh x$ (since all derivatives are either sinh or cosh, $|\sinh z| < |\cosh z|$ for all z, and

$|z| < |x| \Rightarrow \cosh z < \cosh x$), so $|R_n(z)| \leq \dfrac{\cosh x}{(n+1)!}x^{n+1} \to 0$ as $n \to \infty$ (by Equation 10). So by

Theorem 8, the series represents $\sinh x$ for all x.

21. $\cos x = \displaystyle\sum_{n=0}^{\infty}(-1)^n\frac{x^{2n}}{(2n)!} \Rightarrow f(x) = \cos(\pi x) = \displaystyle\sum_{n=0}^{\infty}\frac{(-1)^n(\pi x)^{2n}}{(2n)!} = \displaystyle\sum_{n=0}^{\infty}\frac{(-1)^n\pi^{2n}x^{2n}}{(2n)!}$, $R = \infty$

23. $\tan^{-1}x = \displaystyle\sum_{n=0}^{\infty}(-1)^n\frac{x^{2n+1}}{2n+1} \Rightarrow f(x) = x\tan^{-1}x = x\displaystyle\sum_{n=0}^{\infty}(-1)^n\frac{x^{2n+1}}{2n+1} = \displaystyle\sum_{n=0}^{\infty}(-1)^n\frac{x^{2n+2}}{2n+1}$, $R = 1$

25. $e^x = \displaystyle\sum_{n=0}^{\infty}\frac{x^n}{n!} \Rightarrow f(x) = x^2 e^{-x} = x^2\displaystyle\sum_{n=0}^{\infty}\frac{(-x)^n}{n!} = \displaystyle\sum_{n=0}^{\infty}\frac{(-1)^n x^{n+2}}{n!}$, $R = \infty$

27. $\sin^2 x = \tfrac{1}{2}[1 - \cos 2x] = \dfrac{1}{2}\left[1 - \displaystyle\sum_{n=0}^{\infty}\frac{(-1)^n(2x)^{2n}}{(2n)!}\right] = 2^{-1}\left[1 - 1 - \displaystyle\sum_{n=1}^{\infty}\frac{(-1)^n(2x)^{2n}}{(2n)!}\right]$

$$= \displaystyle\sum_{n=1}^{\infty}\frac{(-1)^{n+1}2^{2n-1}x^{2n}}{(2n)!}, \quad R = \infty$$

29. $\dfrac{\sin x}{x} = \dfrac{1}{x}\displaystyle\sum_{n=0}^{\infty}\frac{(-1)^n x^{2n+1}}{(2n+1)!} = \displaystyle\sum_{n=0}^{\infty}\frac{(-1)^n x^{2n}}{(2n+1)!}$ and this series also gives the required value at $x = 0$ (namely 1),

so $R = \infty$.

31.

n	$f^{(n)}(x)$	$f^{(n)}(0)$
0	$(1+x)^{1/2}$	1
1	$\frac{1}{2}(1+x)^{-1/2}$	$\frac{1}{2}$
2	$-\frac{1}{4}(1+x)^{-3/2}$	$-\frac{1}{4}$
3	$\frac{3}{8}(1+x)^{-5/2}$	$\frac{3}{8}$
4	$-\frac{15}{16}(1+x)^{-7/2}$	$-\frac{15}{16}$
...	...	...

So $f^{(n)}(0) = \dfrac{(-1)^{n-1}\, 1 \cdot 3 \cdot 5 \cdot \cdots \cdot (2n-3)}{2^n}$ for $n \geq 2$,

and

$$\sqrt{1+x} = 1 + \frac{x}{2} + \sum_{n=2}^{\infty} \frac{(-1)^{n-1}\, 1 \cdot 3 \cdot 5 \cdot \cdots \cdot (2n-3)}{2^n n!} x^n.$$

If $a_n = \dfrac{(-1)^{n-1}\, 1 \cdot 3 \cdot 5 \cdot \cdots \cdot (2n-3)}{2^n n!} x^n$, then

$$\lim_{n \to \infty} \left| \frac{a_{n+1}}{a_n} \right| = \frac{|x|}{2} \lim_{n \to \infty} \frac{2n-1}{n+1} = |x| < 1 \text{ for}$$

convergence, so $R = 1$.

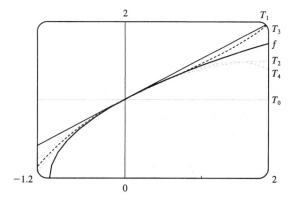

33. $\cos x = \sum_{n=0}^{\infty} (-1)^n \dfrac{x^{2n}}{(2n)!} \quad \Rightarrow \quad f(x) = \cos(x^2) = \sum_{n=0}^{\infty} \dfrac{(-1)^n (x^2)^{2n}}{(2n)!} = \sum_{n=0}^{\infty} \dfrac{(-1)^n x^{4n}}{(2n)!}, \; R = \infty$

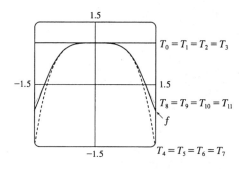

35. $\ln(1+x) = \displaystyle\int \frac{dx}{1+x} = \int \sum_{n=0}^{\infty} (-1)^n x^n \, dx = C + \sum_{n=0}^{\infty} (-1)^n \frac{x^{n+1}}{n+1} = \sum_{n=1}^{\infty} \frac{(-1)^{n-1} x^n}{n}$ with $C = 0$ and

$R = 1$, so $\ln(1.1) = \displaystyle\sum_{n=1}^{\infty} \frac{(-1)^{n-1}(0.1)^n}{n}$. This is an alternating series with $b_5 = \dfrac{(0.1)^5}{5} = 0.000002$, so to five

decimal places, $\ln(1.1) \approx \displaystyle\sum_{n=1}^{4} \frac{(-1)^{n-1}(0.1)^n}{n} \approx 0.09531$.

37. $\displaystyle\int \sin(x^2)\, dx = \int \sum_{n=0}^{\infty} (-1)^n \frac{(x^2)^{2n+1}}{(2n+1)!}\, dx = \int \sum_{n=0}^{\infty} \frac{(-1)^n x^{4n+2}}{(2n+1)!}\, dx = C + \sum_{n=0}^{\infty} \frac{(-1)^n x^{4n+3}}{(4n+3)(2n+1)!}$

39. Using the series from Exercise 31 and substituting x^3 for x, we get

$$\int \sqrt{x^3 + 1}\, dx = \int \left[1 + \frac{x^3}{2} + \sum_{n=2}^{\infty} \frac{(-1)^{n-1} \, 1 \cdot 3 \cdot 5 \cdots (2n-3)}{2^n n!} x^{3n} \right] dx$$

$$= C + x + \frac{x^4}{8} + \sum_{n=2}^{\infty} \frac{(-1)^{n-1} \, 1 \cdot 3 \cdot 5 \cdots (2n-3)}{2^n n! \, (3n+1)} x^{3n+1}$$

41. Using our series from Exercise 37, we get

$$\int_0^1 \sin\left(x^2\right) dx = \sum_{n=0}^{\infty} \left[\frac{(-1)^n \, x^{4n+3}}{(4n+3)(2n+1)!} \right]_0^1 = \sum_{n=0}^{\infty} \frac{(-1)^n}{(4n+3)(2n+1)!} \text{ and } |c_3| = \frac{1}{75{,}600} < 0.000014, \text{ so by}$$

the Alternating Series Estimation Theorem, we have $\displaystyle\sum_{n=0}^{2} \frac{(-1)^n}{(4n+3)(2n+1)!} = \frac{1}{3} - \frac{1}{42} + \frac{1}{1320} \approx 0.310$ (correct

to three decimal places).

43. We first find a series representation for $f(x) = (1+x)^{-1/2}$, and then substitute.

n	$f^{(n)}(x)$	$f^{(n)}(0)$
0	$(1+x)^{-1/2}$	1
1	$-\frac{1}{2}(1+x)^{-3/2}$	$-\frac{1}{2}$
2	$\frac{3}{4}(1+x)^{-5/2}$	$\frac{3}{4}$
3	$-\frac{15}{8}(1+x)^{-7/2}$	$-\frac{15}{8}$
…	…	…

$$\frac{1}{\sqrt{1+x}} = 1 - \frac{x}{2} + \frac{3}{4}\left(\frac{x^2}{2!}\right) - \frac{15}{8}\left(\frac{x^3}{3!}\right) + \cdots \quad \Rightarrow \quad \frac{1}{\sqrt{1+x^3}} = 1 - \frac{1}{2}x^3 + \frac{3}{8}x^6 - \frac{5}{16}x^9 + \cdots \quad \Rightarrow$$

$$\int_0^{0.1} \frac{dx}{\sqrt{1+x^3}} = \left[x - \frac{1}{8}x^4 + \frac{3}{56}x^7 - \frac{1}{32}x^{10} + \cdots \right]_0^{0.1} \approx (0.1) - \frac{1}{8}(0.1)^4 \text{, by the Alternating Series}$$

Estimation Theorem, since $\frac{3}{56}(0.1)^7 \approx 0.0000000054 < 10^{-8}$, which is the maximum desired error. Therefore,

$$\int_0^{0.1} \frac{dx}{\sqrt{1+x^3}} \approx 0.09998750.$$

45. $\displaystyle\lim_{x \to 0} \frac{x - \tan^{-1}x}{x^3} = \lim_{x \to 0} \frac{x - \left(x - \frac{1}{3}x^3 + \frac{1}{5}x^5 - \frac{1}{7}x^7 + \cdots \right)}{x^3} = \lim_{x \to 0} \frac{\frac{1}{3}x^3 - \frac{1}{5}x^5 + \frac{1}{7}x^7 - \cdots}{x^3}$

$$= \lim_{x \to 0} \left(\frac{1}{3} - \frac{1}{5}x^2 + \frac{1}{7}x^4 - \cdots \right) = \frac{1}{3}$$

since power series are continuous functions.

47. $\displaystyle\lim_{x \to 0} \frac{\sin x - x + \frac{1}{6}x^3}{x^5} = \lim_{x \to 0} \frac{\left(x - \frac{1}{3!}x^3 + \frac{1}{5!}x^5 - \frac{1}{7!}x^7 + \cdots \right) - x + \frac{1}{6}x^3}{x^5}$

$$= \lim_{x \to 0} \frac{\frac{1}{5!}x^5 - \frac{1}{7!}x^7 + \cdots}{x^5} = \lim_{x \to 0} \left(\frac{1}{5!} - \frac{x^2}{7!} + \frac{x^4}{9!} - \cdots \right) = \frac{1}{5!} = \frac{1}{120}$$

since power series are continuous functions.

49. As in Example 8(a), we have $e^{-x^2} = 1 - \dfrac{x^2}{1!} + \dfrac{x^4}{2!} - \dfrac{x^6}{3!} + \cdots$ and we know that $\cos x = 1 - \dfrac{x^2}{2!} + \dfrac{x^4}{4!} - \cdots$

from Equation 16. Therefore, $e^{-x^2} \cos x = \left(1 - x^2 + \tfrac{1}{2}x^4 - \cdots\right)\left(1 - \tfrac{1}{2}x^2 + \tfrac{1}{24}x^4 - \cdots\right)$. Writing only the

terms with degree ≤ 4, we get $e^{-x^2} \cos x = 1 - \tfrac{1}{2}x^2 + \tfrac{1}{24}x^4 - x^2 + \tfrac{1}{2}x^4 + \tfrac{1}{2}x^4 + \cdots = 1 - \tfrac{3}{2}x^2 + \tfrac{25}{24}x^4 + \cdots$.

51.

$$
\begin{array}{r}
-x + \tfrac{1}{2}x^2 - \tfrac{1}{3}x^3 + \cdots \\[4pt]
1 + x + \tfrac{1}{2}x^2 + \tfrac{1}{6}x^3 + \cdots \overline{\big)\; -x - \tfrac{1}{2}x^2 - \tfrac{1}{3}x^3 - \cdots} \\[4pt]
\underline{-x - x^2 - \tfrac{1}{2}x^3 - \cdots} \\[4pt]
\tfrac{1}{2}x^2 + \tfrac{1}{6}x^3 - \cdots \\[4pt]
\underline{\tfrac{1}{2}x^2 + \tfrac{1}{2}x^3 + \cdots} \\[4pt]
-\tfrac{1}{3}x^3 + \cdots \\[4pt]
\underline{-\tfrac{1}{3}x^3 + \cdots} \\[4pt]
\cdots
\end{array}
$$

From Example 6 in Section 12.9 [ET 11.9], we have $\ln(1-x) = -x - \tfrac{1}{2}x^2 - \tfrac{1}{3}x^3 - \cdots$, $|x| < 1$. Therefore,

$y = \dfrac{\ln(1-x)}{e^x} = \dfrac{-x - \tfrac{1}{2}x^2 - \tfrac{1}{3}x^3 - \cdots}{1 + x + \tfrac{1}{2}x^2 + \tfrac{1}{6}x^3 + \cdots}$. So by the long division above,

$\dfrac{\ln(1-x)}{e^x} = -x + \dfrac{x^2}{2} - \dfrac{x^3}{3} + \cdots$, $|x| < 1$.

53. $\displaystyle\sum_{n=0}^{\infty} (-1)^n \frac{x^{4n}}{n!} = \sum_{n=0}^{\infty} \frac{\left(-x^4\right)^n}{n!} = e^{-x^4}$, by (11).

55. $\displaystyle\sum_{n=0}^{\infty} \frac{(-1)^n \, \pi^{2n+1}}{4^{2n+1}(2n+1)!} = \sum_{n=0}^{\infty} \frac{(-1)^n \left(\frac{\pi}{4}\right)^{2n+1}}{(2n+1)!} = \sin\frac{\pi}{4} = \frac{1}{\sqrt{2}}$, by (15).

57. $3 + \dfrac{9}{2!} + \dfrac{27}{3!} + \dfrac{81}{4!} + \cdots = \dfrac{3^1}{1!} + \dfrac{3^2}{2!} + \dfrac{3^3}{3!} + \dfrac{3^4}{4!} + \cdots = \displaystyle\sum_{n=1}^{\infty} \frac{3^n}{n!} = \sum_{n=0}^{\infty} \frac{3^n}{n!} - 1 = e^3 - 1$, by (11).

59. Assume that $|f'''(x)| \le M$, so $f'''(x) \le M$ for $a \le x \le d$. Now $\int_a^x f'''(t)\,dt \le \int_a^x M\,dt$

$\Rightarrow\quad f''(x) - f''(a) \le M(x - a) \quad\Rightarrow\quad f''(x) \le f''(a) + M(x - a)$. Thus,

$\int_a^x f''(t)\,dt \le \int_a^x [f''(a) + M(t - a)]\,dt \quad\Rightarrow\quad f'(x) - f'(a) \le f''(a)(x - a) + \tfrac{1}{2}M(x - a)^2$

$\Rightarrow\quad f'(x) \le f'(a) + f''(a)(x - a) + \tfrac{1}{2}M(x - a)^2 \quad\Rightarrow$

$\int_a^x f'(t)\,dt \le \int_a^x \left[f'(a) + f''(a)(t - a) + \tfrac{1}{2}M(t - a)^2\right]dt$

$\Rightarrow\quad f(x) - f(a) \le f'(a)(x - a) + \tfrac{1}{2}f''(a)(x - a)^2 + \tfrac{1}{6}M(x - a)^3$. So

$f(x) - f(a) - f'(a)(x - a) - \tfrac{1}{2}f''(a)(x - a)^2 \le \tfrac{1}{6}M(x - a)^3$. But

$R_2(x) = f(x) - T_2(x) = f(x) - f(a) - f'(a)(x - a) - \tfrac{1}{2}f''(a)(x - a)^2$, so $R_2(x) \le \tfrac{1}{6}M(x - a)^3$. A

similar argument using $f'''(x) \ge -M$ shows that $R_2(x) \ge -\tfrac{1}{6}M(x - a)^3$. So $|R_2(x_2)| \le \tfrac{1}{6}M\,|x - a|^3$.

Although we have assumed that $x > a$, a similar calculation shows that this inequality is also true if $x < a$.

12.11 The Binomial Series ET 11.11

1. The general binomial series in (2) is

$$(1+x)^k = \sum_{n=0}^{\infty} \binom{k}{n} x^n = 1 + kx + \frac{k(k-1)}{2!}x^2 + \frac{k(k-1)(k-2)}{3!}x^3 + \cdots.$$

$$(1+x)^{1/2} = \sum_{n=0}^{\infty} \binom{\frac{1}{2}}{n} x^n = 1 + \left(\frac{1}{2}\right)x + \frac{\left(\frac{1}{2}\right)\left(-\frac{1}{2}\right)}{2!}x^2 + \frac{\left(\frac{1}{2}\right)\left(-\frac{1}{2}\right)\left(-\frac{3}{2}\right)}{3!}x^3 + \cdots$$

$$= 1 + \frac{x}{2} - \frac{x^2}{2^2 \cdot 2!} + \frac{1 \cdot 3 \cdot x^3}{2^3 \cdot 3!} - \frac{1 \cdot 3 \cdot 5 \cdot x^4}{2^4 \cdot 4!} + \cdots$$

$$= 1 + \frac{x}{2} + \sum_{n=2}^{\infty} \frac{(-1)^{n-1} 1 \cdot 3 \cdot 5 \cdots (2n-3) x^n}{2^n \cdot n!}, \quad R = 1$$

3. $\dfrac{1}{(2+x)^3} = \dfrac{1}{[2(1+x/2)]^3} = \dfrac{1}{8}\left(1+\dfrac{x}{2}\right)^{-3} = \dfrac{1}{8}\sum_{n=0}^{\infty} \binom{-3}{n}\left(\dfrac{x}{2}\right)^n$. The binomial coefficient is

$$\binom{-3}{n} = \frac{(-3)(-4)(-5)\cdots(-3-n+1)}{n!} = \frac{(-1)^n \cdot 2 \cdot 3 \cdot 4 \cdot 5 \cdots (n+1)(n+2)}{2 \cdot n!}$$

$$= \frac{(-1)^n (n+1)(n+2)}{2}$$

so $\dfrac{1}{(2+x)^3} = \dfrac{1}{8}\sum_{n=0}^{\infty} \dfrac{(-1)^n (n+1)(n+2)}{2} \dfrac{x^n}{2^n} = \sum_{n=0}^{\infty} \dfrac{(-1)^n (n+1)(n+2) x^n}{2^{n+4}}$ for $\left|\dfrac{x}{2}\right| < 1 \Leftrightarrow |x| < 2$,

so $R = 2$.

5. $\sqrt[4]{1-8x} = (1-8x)^{1/4} = \sum_{n=0}^{\infty} \binom{\frac{1}{4}}{n}(-8x)^n$

$$= 1 + \frac{1}{4}(-8x) + \frac{\frac{1}{4}\left(-\frac{3}{4}\right)}{2!}(-8x)^2 + \frac{\left(\frac{1}{4}\right)\left(-\frac{3}{4}\right)\left(-\frac{7}{4}\right)}{3!}(-8x)^3 + \cdots$$

$$= 1 - 2x + \sum_{n=2}^{\infty} \frac{(-1)^n (-1)^{n-1} \cdot 3 \cdot 7 \cdots (4n-5) 8^n}{4^n \cdot n!}x^n$$

$$= 1 - 2x - \sum_{n=2}^{\infty} \frac{3 \cdot 7 \cdots (4n-5) 2^n}{n!}x^n$$

and $|-8x| < 1 \Leftrightarrow |x| < \frac{1}{8}$, so $R = \frac{1}{8}$.

7. $\dfrac{x}{\sqrt{4+x^2}} = \dfrac{x}{2\sqrt{1+x^2/4}} = \dfrac{x}{2}\left(1+\dfrac{x^2}{4}\right)^{-1/2} = \dfrac{x}{2}\sum_{n=0}^{\infty} \binom{-\frac{1}{2}}{n}\left(\dfrac{x^2}{4}\right)^n$

$$= \frac{x}{2}\left[1 + \left(-\frac{1}{2}\right)\frac{x^2}{4} + \frac{\left(-\frac{1}{2}\right)\left(-\frac{3}{2}\right)}{2!}\left(\frac{x^2}{4}\right)^2 + \frac{\left(-\frac{1}{2}\right)\left(-\frac{3}{2}\right)\left(-\frac{5}{2}\right)}{3!}\left(\frac{x^2}{4}\right)^3 + \cdots\right]$$

$$= \frac{x}{2} + \sum_{n=1}^{\infty} (-1)^n \frac{1 \cdot 3 \cdot 5 \cdots (2n-1)}{n! \, 2^{3n+1}}x^{2n+1} \text{ and } \frac{x^2}{4} < 1 \Leftrightarrow \frac{|x|}{2} < 1 \Leftrightarrow |x| < 2, \text{ so } R = 2.$$

9. $\dfrac{1}{\sqrt[3]{8+x}} = (8+x)^{-1/3} = 8^{-1/3}\left(1+\dfrac{x}{8}\right)^{-1/3} = \dfrac{1}{2}\left(1+\dfrac{x}{8}\right)^{-1/3}$

$$= \dfrac{1}{2}\left[1+\left(-\dfrac{1}{3}\right)\left(\dfrac{x}{8}\right)+\dfrac{\left(-\frac{1}{3}\right)\left(-\frac{4}{3}\right)}{2!}\left(\dfrac{x}{8}\right)^2+\cdots\right]$$

$$= \dfrac{1}{2}\left[1+\sum_{n=1}^{\infty}\dfrac{(-1)^n\,1\cdot 4\cdot 7\cdots\cdots(3n-2)}{3^n\cdot n!\,8^n}x^n\right] \text{ and } \left|\dfrac{x}{8}\right|<1 \;\Leftrightarrow\; |x|<8, \text{ so } R=8.$$

The three Taylor polynomials are $T_1(x) = \dfrac{1}{2}-\dfrac{1}{48}x$, $T_2(x)=\dfrac{1}{2}-\dfrac{1}{48}x+\dfrac{1}{576}x^2$, and

$T_3(x) = \dfrac{1}{2}-\dfrac{1}{48}x+\dfrac{1}{576}x^2-\dfrac{4\cdot 7}{2\cdot 27\cdot 6\cdot 512}x^3 = \dfrac{1}{2}-\dfrac{1}{48}x+\dfrac{1}{576}x^2-\dfrac{7}{41{,}472}x^3.$

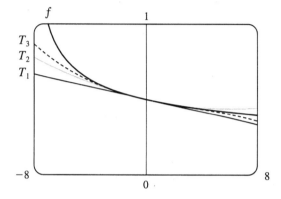

11. (a) $\left[1+\left(-x^2\right)\right]^{-1/2} = 1+\left(-\dfrac{1}{2}\right)\left(-x^2\right)+\dfrac{\left(-\frac{1}{2}\right)\left(-\frac{3}{2}\right)}{2!}\left(-x^2\right)^2+\dfrac{\left(-\frac{1}{2}\right)\left(-\frac{3}{2}\right)\left(-\frac{5}{2}\right)}{3!}\left(-x^2\right)^3+\cdots$

$$= 1+\sum_{n=1}^{\infty}\dfrac{1\cdot 3\cdot 5\cdots\cdots(2n-1)}{2^n\cdot n!}x^{2n}$$

(b) $\sin^{-1}x = \displaystyle\int\dfrac{1}{\sqrt{1-x^2}}\,dx = C+x+\sum_{n=1}^{\infty}\dfrac{1\cdot 3\cdot 5\cdots\cdots(2n-1)}{(2n+1)\,2^n\cdot n!}x^{2n+1}$

$$= x+\sum_{n=1}^{\infty}\dfrac{1\cdot 3\cdot 5\cdots\cdots(2n-1)}{(2n+1)\,2^n\cdot n!}x^{2n+1} \text{ since } 0=\sin^{-1}0=C.$$

13. (a) $(1+x)^{-1/2} = 1+\left(-\dfrac{1}{2}\right)x+\dfrac{\left(-\frac{1}{2}\right)\left(-\frac{3}{2}\right)}{2!}x^2+\dfrac{\left(-\frac{1}{2}\right)\left(-\frac{3}{2}\right)\left(-\frac{5}{2}\right)}{3!}x^3+\cdots$

$$= 1+\sum_{n=1}^{\infty}\dfrac{(-1)^n\,1\cdot 3\cdot 5\cdots\cdots(2n-1)}{2^n\cdot n!}x^n$$

(b) Take $x=0.1$ in the above series. $\dfrac{1\cdot 3\cdot 5\cdot 7}{2^4 4!}(0.1)^4 < 0.00003$, so

$$\dfrac{1}{\sqrt{1.1}} \approx 1-\dfrac{0.1}{2}+\dfrac{1\cdot 3}{2^2\cdot 2!}(0.1)^2-\dfrac{1\cdot 3\cdot 5}{2^3\cdot 3!}(0.1)^3 \approx 0.953.$$

15. (a) $[1 + (-x)]^{-2} = 1 + (-2)(-x) + \dfrac{(-2)(-3)}{2!}(-x)^2 + \dfrac{(-2)(-3)(-4)}{3!}(-x)^3 + \cdots$

$$= 1 + 2x + 3x^2 + 4x^3 + \cdots = \sum_{n=0}^{\infty} (n+1) x^n,$$

so $\dfrac{x}{(1-x)^2} = \sum_{n=0}^{\infty} (n+1) x^{n+1} = \sum_{n=1}^{\infty} n x^n.$

(b) With $x = \frac{1}{2}$ in part (a), we have $\displaystyle\sum_{n=1}^{\infty} \frac{n}{2^n} = \dfrac{\frac{1}{2}}{\left(1 - \frac{1}{2}\right)^2} = 2.$

17. (a) $(1 + x^2)^{1/2} = 1 + \left(\frac{1}{2}\right)x^2 + \dfrac{\left(\frac{1}{2}\right)\left(-\frac{1}{2}\right)}{2!}(x^2)^2 + \dfrac{\left(\frac{1}{2}\right)\left(-\frac{1}{2}\right)\left(-\frac{3}{2}\right)}{3!}(x^2)^3 + \cdots$

$$= 1 + \frac{x^2}{2} + \sum_{n=2}^{\infty} \frac{(-1)^{n-1} \, 1 \cdot 3 \cdot 5 \cdot \cdots \cdot (2n-3)}{2^n \cdot n!} x^{2n}$$

(b) The coefficient of x^{10} (corresponding to $n = 5$) in the above Maclaurin series is $\dfrac{f^{(10)}(0)}{10!}$, so

$$\frac{f^{(10)}(0)}{10!} = \frac{(-1)^4 \cdot 1 \cdot 3 \cdot 5 \cdot 7}{2^5 \cdot 5!} \quad\Rightarrow\quad f^{(10)}(0) = 10!\left(\frac{1 \cdot 3 \cdot 5 \cdot 7}{2^5 \cdot 5!}\right) = 99{,}225.$$

19. (a) $g(x) = \displaystyle\sum_{n=0}^{\infty} \binom{k}{n} x^n \quad\Rightarrow\quad g'(x) = \sum_{n=1}^{\infty} \binom{k}{n} n x^{n-1}$, so

$$(1+x)g'(x) = (1+x)\sum_{n=1}^{\infty} \binom{k}{n} n x^{n-1} = \sum_{n=1}^{\infty} \binom{k}{n} n x^{n-1} + \sum_{n=1}^{\infty} \binom{k}{n} n x^n$$

$$= \sum_{n=0}^{\infty} \binom{k}{n+1}(n+1)x^n + \sum_{n=0}^{\infty} \binom{k}{n} n x^n \qquad \begin{bmatrix}\text{Replace } n \text{ with } n+1 \\ \text{in the first series}\end{bmatrix}$$

$$= \sum_{n=0}^{\infty} (n+1) \frac{k(k-1)(k-2)\cdots(k-n+1)(k-n)}{(n+1)!} x^n$$

$$\qquad + \sum_{n=0}^{\infty} \left[(n)\frac{k(k-1)(k-2)\cdots(k-n+1)}{n!}\right] x^n$$

$$= \sum_{n=0}^{\infty} \frac{(n+1)k(k-1)(k-2)\cdots(k-n+1)}{(n+1)!}[(k-n)+n]x^n$$

$$= k\sum_{n=0}^{\infty} \frac{k(k-1)(k-2)\cdots(k-n+1)}{n!} x^n = k\sum_{n=0}^{\infty} \binom{k}{n} x^n = kg(x)$$

Thus, $g'(x) = \dfrac{kg(x)}{1+x}.$

(b) $h(x) = (1+x)^{-k} g(x) \quad\Rightarrow$

$$h'(x) = -k(1+x)^{-k-1} g(x) + (1+x)^{-k} g'(x) \quad \text{[Product Rule]}$$

$$= -k(1+x)^{-k-1} g(x) + (1+x)^{-k} \frac{kg(x)}{1+x} \quad \text{[from part (a)]}$$

$$= -k(1+x)^{-k-1} g(x) + k(1+x)^{-k-1} g(x) = 0$$

(c) From part (b) we see that $h(x)$ must be constant for $x \in (-1, 1)$, so $h(x) = h(0) = 1$ for $x \in (-1, 1)$. Thus, $h(x) = 1 = (1+x)^{-k} g(x) \quad\Leftrightarrow\quad g(x) = (1+x)^k$ for $x \in (-1, 1)$.

▀12.12 Applications of Taylor Polynomials   **ET 11.12**

1. (a)

n	$f^{(n)}(x)$	$f^{(n)}(0)$	$T_n(x)$
0	$\cos x$	1	1
1	$-\sin x$	0	1
2	$-\cos x$	-1	$1 - \frac{1}{2}x^2$
3	$\sin x$	0	$1 - \frac{1}{2}x^2$
4	$\cos x$	1	$1 - \frac{1}{2}x^2 + \frac{1}{24}x^4$
5	$-\sin x$	0	$1 - \frac{1}{2}x^2 + \frac{1}{24}x^4$
6	$-\cos x$	-1	$1 - \frac{1}{2}x^2 + \frac{1}{24}x^4 - \frac{1}{720}x^6$

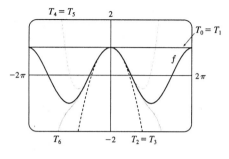

(b)

x	f	$T_0 = T_1$	$T_2 = T_3$	$T_4 = T_5$	T_6
$\frac{\pi}{4}$	0.7071	1	0.6916	0.7074	0.7071
$\frac{\pi}{2}$	0	1	-0.2337	0.0200	-0.0009
π	-1	1	-3.9348	0.1239	-1.2114

(c) As n increases, $T_n(x)$ is a good approximation to $f(x)$ on a larger and larger interval.

3.

n	$f^{(n)}(x)$	$f^{(n)}(1)$
0	$\ln x$	0
1	$1/x$	1
2	$-1/x^2$	-1
3	$2/x^3$	2
4	$-6/x^4$	-6

$$T_4(x) = \sum_{n=0}^{4} \frac{f^{(n)}(1)}{n!}(x-1)^n = 0 + (x-1) - \frac{1}{2}(x-1)^2 + \frac{1}{3}(x-1)^3 - \frac{1}{4}(x-1)^4$$

5.

n	$f^{(n)}(x)$	$f^{(n)}\left(\frac{\pi}{6}\right)$
0	$\sin x$	$\frac{1}{2}$
1	$\cos x$	$\frac{\sqrt{3}}{2}$
2	$-\sin x$	$-\frac{1}{2}$
3	$-\cos x$	$-\frac{\sqrt{3}}{2}$

$$T_3(x) = \sum_{n=0}^{3} \frac{f^{(n)}\left(\frac{\pi}{6}\right)}{n!}\left(x-\frac{\pi}{6}\right)^n = \frac{1}{2} + \frac{\sqrt{3}}{2}\left(x-\frac{\pi}{6}\right) - \frac{1}{4}\left(x-\frac{\pi}{6}\right)^2 - \frac{\sqrt{3}}{12}\left(x-\frac{\pi}{6}\right)^3$$

7.

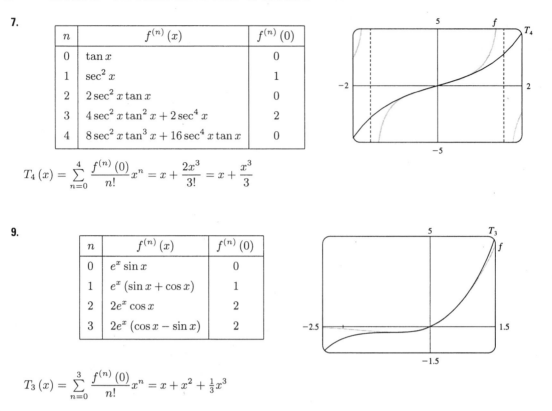

n	$f^{(n)}(x)$	$f^{(n)}(0)$
0	$\tan x$	0
1	$\sec^2 x$	1
2	$2\sec^2 x \tan x$	0
3	$4\sec^2 x \tan^2 x + 2\sec^4 x$	2
4	$8\sec^2 x \tan^3 x + 16\sec^4 x \tan x$	0

$$T_4(x) = \sum_{n=0}^{4} \frac{f^{(n)}(0)}{n!} x^n = x + \frac{2x^3}{3!} = x + \frac{x^3}{3}$$

9.

n	$f^{(n)}(x)$	$f^{(n)}(0)$
0	$e^x \sin x$	0
1	$e^x (\sin x + \cos x)$	1
2	$2e^x \cos x$	2
3	$2e^x (\cos x - \sin x)$	2

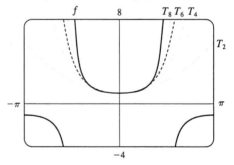

$$T_3(x) = \sum_{n=0}^{3} \frac{f^{(n)}(0)}{n!} x^n = x + x^2 + \tfrac{1}{3}x^3$$

11. In Maple, we can find the Taylor polynomials by the following method: first define `f:=sec(x);` and then set `T2:=convert(taylor(f,x=0,3),polynom);`, `T4:=convert(taylor(f,x=0,5),polynom);`, etc. (The third argument in the `taylor` function is one more than the degree of the desired polynomial). We must convert to the type `polynom` because the output of the `taylor` function contains an error term which we do not want. In Mathematica, we use `Tn:=Normal[Series[f,{x,0,n}]]`, with n=2, 4, etc. Note that in Mathematica, the "degree" argument is the same as the degree of the desired polynomial. In Derive, author sec x, then enter `Calculus,Taylor,8,0`; and then simplify the expression. The eighth Taylor polynomial is

$$T_8(x) = 1 + \tfrac{1}{2}x^2 + \tfrac{5}{24}x^4 + \tfrac{61}{720}x^6 + \tfrac{277}{8064}x^8.$$

13.

$$f(x) = \sqrt{x} \qquad\qquad f(4) = 2$$
$$f'(x) = \tfrac{1}{2}x^{-1/2} \qquad\qquad f'(4) = \tfrac{1}{4}$$
$$f''(x) = -\tfrac{1}{4}x^{-3/2} \qquad\qquad f''(4) = -\tfrac{1}{32}$$
$$f'''(x) = \tfrac{3}{8}x^{-5/2}$$

(a) $\sqrt{x} \approx T_2(x) = 2 + \frac{1}{4}(x - 4) - \frac{1/32}{2!}(x - 4)^2$

$\qquad = 2 + \frac{1}{4}(x - 4) - \frac{1}{64}(x - 4)^2$

(b) $|R_2(x)| \le \frac{M}{3!}|x - 4|^3$, where $|f'''(x)| \le M$. Now

$\quad 4 \le x \le 4.2 \quad \Rightarrow \quad |x - 4| \le 0.2 \quad \Rightarrow \quad |x - 4|^3 \le 0.008.$

Since $f'''(x)$ is decreasing on $[4, 4.2]$, we can take

$M = |f'''(4)| = \frac{3}{8}4^{-5/2} = \frac{3}{256}$, so

$|R_2(x)| \le \frac{3/256}{6}(0.008) = \frac{0.008}{512} = 0.000015625.$

(c)   0.00002

From the graph of

$|R_2(x)| = |\sqrt{x} - T_2(x)|$, it seems that

the error is less than 1.52×10^{-5} on

$[4, 4.2]$.

15.

$f(x) = \sin x \qquad\qquad f\left(\frac{\pi}{4}\right) = \frac{\sqrt{2}}{2} \qquad\qquad f^{(4)}(x) = \sin x \qquad\qquad f^{(4)}\left(\frac{\pi}{4}\right) = \frac{\sqrt{2}}{2}$

$f'(x) = \cos x \qquad\qquad f'\left(\frac{\pi}{4}\right) = \frac{\sqrt{2}}{2} \qquad\qquad f^{(5)}(x) = \cos x \qquad\qquad f^{(5)}\left(\frac{\pi}{4}\right) = \frac{\sqrt{2}}{2}$

$f''(x) = -\sin x \qquad\qquad f''\left(\frac{\pi}{4}\right) = -\frac{\sqrt{2}}{2} \qquad\qquad f^{(6)}(x) = -\sin x$

$f'''(x) = -\cos x \qquad\qquad f'''\left(\frac{\pi}{4}\right) = -\frac{\sqrt{2}}{2}$

(a) $\sin x \approx T_5(x)$

$\quad = \frac{\sqrt{2}}{2} + \frac{\sqrt{2}}{2}\left(x - \frac{\pi}{4}\right) - \frac{\sqrt{2}}{4}\left(x - \frac{\pi}{4}\right)^2 - \frac{\sqrt{2}}{12}\left(x - \frac{\pi}{4}\right)^3$

$\qquad + \frac{\sqrt{2}}{48}\left(x - \frac{\pi}{4}\right)^4 + \frac{\sqrt{2}}{240}\left(x - \frac{\pi}{4}\right)^5$

(b) $|R_5(x)| \le \frac{M}{6!}\left|x - \frac{\pi}{4}\right|^6$, where $\left|f^{(6)}(x)\right| \le M$. Now

$\quad 0 \le x \le \frac{\pi}{2} \quad \Rightarrow \quad \left(x - \frac{\pi}{4}\right)^6 \le \left(\frac{\pi}{4}\right)^6$, and letting $x = \frac{\pi}{2}$

gives $M = 1$, so

$|R_5(x)| \le \frac{1}{6!}\left(\frac{\pi}{4}\right)^6 = \frac{1}{720}\left(\frac{\pi}{4}\right)^6 \approx 0.00033.$

(c)   0.0003

From the graph of

$|R_5(x)| = |\sin x - T_5(x)|$, it seems that

the error is less than 0.00026 on $\left[0, \frac{\pi}{2}\right]$.

17. $f(x) = \tan x \qquad\qquad f(0) = 0 \qquad\qquad f'''(x) = 4\sec^2 x \tan^2 x + 2\sec^4 x \qquad\qquad f'''(0) = 2$

$f'(x) = \sec^2 x \qquad\qquad f'(0) = 1 \qquad\qquad f^{(4)}(x) = 8\sec^2 x \tan^3 x + 16\sec^4 x \tan x$

$f''(x) = 2\sec^2 x \tan x \qquad\qquad f''(0) = 0$

(a) $\tan x \approx T_3(x) = x + \frac{1}{3}x^3$

(b) $|R_3(x)| \le \frac{M}{4!}|x|^4$, where $\left|f^{(4)}(x)\right| \le M$. Now

$\quad 0 \le x \le \frac{\pi}{6} \quad \Rightarrow \quad x^4 \le \left(\frac{\pi}{6}\right)^4$, and letting $x = \frac{\pi}{6}$ gives

$|R_3(x)| \le \dfrac{8\left(\frac{2}{\sqrt{3}}\right)^2\left(\frac{1}{\sqrt{3}}\right)^3 + 16\left(\frac{2}{\sqrt{3}}\right)^4\left(\frac{1}{\sqrt{3}}\right)}{4!}\left(\frac{\pi}{6}\right)^4$

$\qquad = \frac{4\sqrt{3}}{9}\left(\frac{\pi}{6}\right)^4 \approx 0.057859$

(c)   0.01

From the graph, it seems that the error is

less than 0.006 on $[0, \pi]$.

19. $f(x) = e^{x^2}$ $\qquad$ $f(0) = 1$ $\qquad$ $f'''(x) = e^{x^2}(12x + 8x^3)$ $\qquad$ $f'''(0) = 0$

$\qquad$ $f'(x) = e^{x^2}(2x)$ $\qquad$ $f'(0) = 0$ $\qquad$ $f^{(4)}(x) = e^{x^2}(12 + 48x^2 + 16x^4)$

$\qquad$ $f''(x) = e^{x^2}(2 + 4x^2)$ $\qquad$ $f''(0) = 2$

(a) $e^{x^2} \approx T_3(x) = 1 + \frac{2}{2!}x^2 = 1 + x^2$ $\qquad$ (c)

(b) $|R_3(x)| \leq \dfrac{M}{4!}|x|^4$, where $\left|f^{(4)}(x)\right| \leq M$. Now

$\qquad$ $0 \leq x \leq 0.1 \;\Rightarrow\; x^4 \leq (0.1)^4$, and letting $x = 0.1$ gives

$\qquad$ $|R_3(x)| \leq \dfrac{e^{0.01}(12 + 0.48 + 0.0016)}{24}(0.1)^4 \approx 0.00006.$

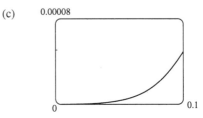

0.00008

0 $\qquad$ 0.1

From the graph of

$$|R_3(x)| = \left|e^{x^2} - \left(1 + x^2\right)\right|,$$ it appears

that the error is less than 0.000051 on

$[0, 0.1]$.

21. $\quad f(x) = x^{3/4}$ $\qquad$ $f(16) = 8$ $\qquad$ $f'''(x) = \frac{15}{64}x^{-9/4}$ $\qquad$ $f'''(16) = \frac{15}{32,768}$

$\qquad$ $f'(x) = \frac{3}{4}x^{-1/4}$ $\qquad$ $f'(16) = \frac{3}{8}$ $\qquad$ $f^{(4)}(x) = -\frac{135}{256}x^{-13/4}$

$\qquad$ $f''(x) = -\frac{3}{16}x^{-5/4}$ $\qquad$ $f''(16) = -\frac{3}{512}$

(a) $x^{3/4} \approx T_3(x) = 8 + \frac{3}{8}(x - 16) - \frac{3}{1024}(x - 16)^2$ $\qquad$ (c) $\qquad$ 5×10^{-6}

$\qquad\qquad\qquad\qquad\qquad + \frac{5}{65,536}(x - 16)^3$

(b) $|R_3(x)| \leq \dfrac{M}{4!}|x - 16|^4$, where $\left|f^{(4)}(x)\right| \leq M$. Now

$\qquad$ $15 \leq x \leq 17 \;\Rightarrow\; |x - 16|^4 \leq 1^4 = 1$, and letting $x = 15$

$\qquad$ to minimize the denominator of $f^{(4)}(x)$ gives

$\qquad$ 15 $\qquad\qquad\qquad$ 17

$\qquad$ 0

$\qquad$ $|R_3(x)| \leq \dfrac{135/\left[256\,(15)^{13/4}\right]}{4!}(1) \approx 0.0000033.$ $\qquad$ It appears that the error is less than

$\qquad\qquad\qquad\qquad\qquad\qquad\qquad\qquad\qquad$ 3×10^{-6} on $(15, 17)$.

23. From Exercise 5, $\sin x = \frac{1}{2} + \frac{\sqrt{3}}{2}\left(x - \frac{\pi}{6}\right) - \frac{1}{4}\left(x - \frac{\pi}{6}\right)^2 - \frac{\sqrt{3}}{12}\left(x - \frac{\pi}{6}\right)^3 + R_3(x)$, where

$R_3(x) \leq \dfrac{M}{4!}\left|x - \frac{\pi}{6}\right|^4$ with $\left|f^{(4)}(x)\right| = |\sin x| \leq M = 1$. Now $35° = \left(\frac{\pi}{6} + \frac{\pi}{36}\right)$ radians, so

the error is $\left|R_3\left(\frac{\pi}{36}\right)\right| \leq \dfrac{\left(\frac{\pi}{36}\right)^4}{4!} < 0.000003.$ Therefore, to five decimal places,

$\sin 35° \approx \frac{1}{2} + \frac{\sqrt{3}}{2}\left(\frac{\pi}{36}\right) - \frac{1}{4}\left(\frac{\pi}{36}\right)^2 - \frac{\sqrt{3}}{12}\left(\frac{\pi}{36}\right)^3 \approx 0.57358.$

25. All derivatives of e^x are e^x, so $|R_n(x)| \leq \dfrac{e^x}{(n+1)!}|x|^{n+1}$, where $0 < x < 0.1$. Letting $x = 0.1$,

$R_n(0.1) \leq \dfrac{e^{0.1}}{(n+1)!}(0.1)^{n+1} < 0.00001$, and by trial and error we find that $n = 3$ satisfies this inequality since

$R_3(0.1) < 0.0000046$. Thus, by adding the three terms of the Maclaurin series for e^x corresponding to $n = 0, 1$,

and 2, we can estimate $e^{0.1}$ to within 0.00001.

27. $\sin x = x - \frac{1}{3!}x^3 + \frac{1}{5!}x^5 - \cdots$. By the Alternating

Series Estimation Theorem, the error in the

approximation $\sin x = x - \frac{1}{3!}x^3$ is less than

$\left|\frac{1}{5!}x^5\right| < 0.01 \;\Leftrightarrow\; \left|x^5\right| < 120\,(0.01) \;\Leftrightarrow$

$|x| < (1.2)^{1/5} \approx 1.037$. The curves intersect at

$x \approx 1.043$, so the graph confirms our estimate. Since

both the sine function and the given approximation are

odd functions, we need to check the estimate only for

$x > 0$.

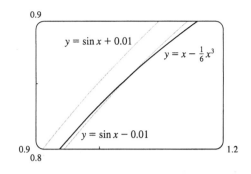

29. Let $s\,(t)$ be the position function of the car, and for convenience set $s\,(0) = 0$. The velocity of the car is

$v\,(t) = s'\,(t)$ and the acceleration is $a\,(t) = s''\,(t)$, so the second degree Taylor polynomial is

$T_2\,(t) = s\,(0) + v\,(0)\,t + \dfrac{a\,(0)}{2}t^2 = 20t + t^2$. We estimate the distance travelled during the next second to be

$s\,(1) \approx T_2\,(1) = 20 + 1 = 21$ m. The function $T_2\,(t)$ would not be accurate over a full minute, since the car could

not possibly maintain an acceleration of 2 m/s^2 for that long (if it did, its final speed would be

140 m/s ≈ 315 mi/h!)

31. $E = \dfrac{q}{D^2} - \dfrac{q}{(D+d)^2} = \dfrac{q}{D^2} - \dfrac{q}{D^2\,(1+d/D)^2} = \dfrac{q}{D^2}\left[1 - \left(1 + \dfrac{d}{D}\right)^{-2}\right].$

We use the Binomial Series to expand $(1+d/D)^{-2}$:

$$E = \frac{q}{D^2}\left[1 - \left(1 - 2\left(\frac{d}{D}\right) + \frac{2\cdot 3}{2!}\left(\frac{d}{D}\right)^2 - \frac{2\cdot 3\cdot 4}{3!}\left(\frac{d}{D}\right)^3 + \cdots\right)\right]$$

$$= \frac{q}{D^2}\left[2\left(\frac{d}{D}\right) - 3\left(\frac{d}{D}\right)^2 + 4\left(\frac{d}{D}\right)^3 - \cdots\right] \approx 2qd\cdot\frac{1}{D^3}$$

when D is much larger than d, that is, when P is far away from the dipole.

33. (a) If the water is deep, then $2\pi d/L$ is large, and we know that $\tanh x \to 1$ as $x \to \infty$. So we can approximate

$\tanh(2\pi d/L) \approx 1$, and so $v^2 \approx gL/\,(2\pi) \;\Leftrightarrow\; v \approx \sqrt{gL/\,(2\pi)}$.

(b) From the calculations at right, the first term in the
Maclaurin series of $\tanh x$ is x, so if the water is
shallow, we can approximate $\tanh\dfrac{2\pi d}{L} \approx \dfrac{2\pi d}{L}$,
and so $v^2 \approx \dfrac{gL}{2\pi}\cdot\dfrac{2\pi d}{L} \;\Leftrightarrow\; v \approx \sqrt{gd}$.

$f\,(x) = \tanh x$ $f\,(0) = 0$
$f'\,(x) = \text{sech}^2\,x$ $f'\,(0) = 1$
$f''\,(x) = -2\,\text{sech}^2\,x\tanh x$ $f''\,(0) = 0$
$f'''\,(x) = 2\,\text{sech}^2\,x\,(3\tanh^2 x - 1)$ $f'''\,(0) = -2$

(c) Since $\tanh x$ is an odd function, its Maclaurin series is alternating, so the error in the approximation

$\tanh\dfrac{2\pi d}{L} \approx \dfrac{2\pi d}{L}$ is less than the first neglected term, which is $\dfrac{|f'''\,(0)|}{3!}\left(\dfrac{2\pi d}{L}\right)^3 = \dfrac{1}{3}\left(\dfrac{2\pi d}{L}\right)^3$. If

$L > 10d$, then $\dfrac{1}{3}\left(\dfrac{2\pi d}{L}\right)^3 < \dfrac{1}{3}\left(2\pi\cdot\dfrac{1}{10}\right)^3 = \dfrac{\pi^3}{375}$, so the error in the approximation $v^2 = gd$ is less than

$\dfrac{gL}{2\pi}\cdot\dfrac{\pi^3}{375} \approx 0.0132gL$.

35. Using $f(x) = T_n(x) + R_n(x)$ with $n = 1$ and $x = r$, we have $f(r) = T_1(r) + R_1(r)$, where T_1 is the first-degree Taylor polynomial of f at a. Because $a = x_n$, $f(r) = f(x_n) + f'(x_n)(r - x_n) + R_1(r)$. But r is a root of f, so $f(r) = 0$ and we have $0 = f(x_n) + f'(x_n)(r - x_n) + R_1(r)$. Taking the first two terms to the left side and dividing by $f'(x_n)$, we have $f'(x_n)(x_n - r) - f(x_n) = R_1(r)$ $\Rightarrow$ $x_n - r - \dfrac{f(x_n)}{f'(x_n)} = \dfrac{R_1(r)}{f'(x_n)}$.

By the formula for Newton's method, the left side of the preceding equation is $x_{n+1} - r$, so

$$|x_{n+1} - r| = \left| \frac{R_1(r)}{f'(x_n)} \right|. \text{ Taylor's Inequality gives us } |R_1(r)| \leq \frac{|f''(r)|}{2!} |r - x_n|^2. \text{ Combining this inequality}$$

with the facts $|f''(x)| \leq M$ and $|f'(x)| \geq K$ gives us $|x_{n+1} - r| \leq \dfrac{M}{2K} |x_n - r|^2$.

▤12 Review

CONCEPT CHECK

1. (a) See Definition 12.1.1 [ET 11.1.1].

(b) See Definition 12.2.2 [ET 11.2.2].

(c) The terms of the sequence $\{a_n\}$ approach 3 as n becomes large.

(d) By adding sufficiently many terms of the series, we can make the partial sums as close to 3 as we like.

2. (a) See Definition 12.1.9 [ET 11.1.9].

(b) A sequence is monotonic if it is either increasing or decreasing.

(c) By Theorem 12.1.10 [ET 11.1.10], every bounded, monotonic sequence is convergent.

3. (a) See (4) in Section 12.2 [ET 11.2].

(b) See (1) in Section 12.3 [ET 11.3].

4. If $\sum a_n = 3$, then $\lim\limits_{n \to \infty} a_n = 0$ and $\lim\limits_{n \to \infty} s_n = 3$.

5. (a) See the Test for Divergence (12.2.7 [ET 11.2.7]).

(b) See the Integral Test on page 749 [ET 715].

(c) See the Comparison Test on page 756 [ET 722].

(d) See the Limit Comparison Test on page 757 [ET 723].

(e) See the Alternating Series Test on page 761 [ET 727].

(f) See the Ratio Test on page 767 [ET 733].

(g) See the Root Test on page 769 [ET 735].

6. (a) See Definition 12.6.1 [ET 11.6.1].

(b) By (12.6.3 [ET 11.6.3]), it is convergent.

(c) See Definition 12.6.2 [ET 11.6.2].

7. (a) Use either (2) or (3) in Section 12.3 [ET 11.3].

(b) See Example 5 in Section 12.4 [ET 11.4].

(c) By adding terms until you reach the desired accuracy given by the Alternating Series Estimation Theorem on page 763 [ET 729].

8. (a) $\sum_{n=0}^{\infty} c_n (x - a)^n$

(b) Given the power series $\sum_{n=0}^{\infty} c_n (x - a)^n$, the radius of convergence is:

(i) 0 if the series converges only when $x = a$

(ii) ∞ if the series converges for all x, or

(iii) a positive number R such that the series converges if $|x - a| < R$ and diverges if $|x - a| > R$.

(c) The interval of convergence of a power series is the interval that consists of all values of x for which the series converges. Corresponding to the cases in part (b), the interval of convergence is: (i) the single point $\{a\}$, (ii) all real numbers, that is, the real number line $(-\infty, \infty)$, or (iii) an interval with endpoints $a - R$ and $a + R$ which can contain neither, either, or both of the endpoints. In this case, we must test the series for convergence at each endpoint to determine the interval of convergence.

9. (a), (b) See Theorem 12.9.2 [ET 11.9.2].

10. (a) $T_n (x) = \sum_{i=0}^{n} \dfrac{f^{(i)} (a)}{i!} (x - a)^i$

(b) $\sum_{n=0}^{\infty} \dfrac{f^{(n)} (a)}{n!} (x - a)^n$

(c) $\sum_{n=0}^{\infty} \dfrac{f^{(n)} (0)}{n!} x^n$ $[a = 0$ in part (b)$]$

(d) See Theorem 12.10.8 [ET 11.10.8].

(e) See Taylor's Inequality (12.10.9 [ET 11.10.9]).

11. (a) – (e) See the table on page 792 [ET 758].

12. See the Binomial Series (12.11.2 [ET 11.11.2]) for the expansion. The radius of convergence for the binomial series is 1.

TRUE-FALSE QUIZ

1. False. See Note 2 after Theorem 12.2.6 [ET 11.2.6].

3. False. For example, take $c_n = (-1)^n / (n6^n)$.

5. False, since $\lim\limits_{n \to \infty} \left| \dfrac{a_{n+1}}{a_n} \right| = \lim\limits_{n \to \infty} \left| \dfrac{n^3}{(n + 1)^3} \right| = \lim\limits_{n \to \infty} \dfrac{1}{(1 + 1/n)^3} = 1.$

7. False. See the note after Example 2 in Section 12.4 [ET 11.4].

9. True. See (7) in Section 12.1 [ET 11.1].

11. True. By Theorem 12.10.5 [ET 11.10.5] the coefficient of x^3 is $\dfrac{f''' (0)}{3!} = \dfrac{1}{3}$ $\Rightarrow$ $f''' (0) = 2.$

Or: Use Theorem 12.9.2 [ET 11.9.2] to differentiate f three times.

13. False. For example, let $a_n = b_n = (-1)^n$. Then $\{a_n\}$ and $\{b_n\}$ are divergent, but $a_n b_n = 1$, so $\{a_n b_n\}$ is convergent.

15. True by Theorem 12.6.3 [ET 11.6.3]. $[\sum (-1)^n a_n$ is absolutely convergent and hence convergent.$]$

EXERCISES

1. $\left\{\dfrac{2+n^3}{1+2n^3}\right\}$ converges since $\lim\limits_{n\to\infty}\dfrac{2+n^3}{1+2n^3}=\lim\limits_{n\to\infty}\dfrac{2/n^3+1}{1/n^3+2}=\dfrac{1}{2}$.

3. $\lim\limits_{n\to\infty}a_n=\lim\limits_{n\to\infty}\dfrac{n^3}{1+n^2}=\lim\limits_{n\to\infty}\dfrac{n}{1/n^2+1}=\infty$, so the sequence diverges.

5. $\{\sin n\}$ is divergent since $\lim\limits_{n\to\infty}\sin n$ does not exist.

7. $\left\{\left(1+\dfrac{3}{n}\right)^{4n}\right\}$ is convergent. Let $y=\left(1+\dfrac{3}{x}\right)^{4x}$. Then

$$\lim_{x\to\infty}\ln y=\lim_{x\to\infty}4x\ln(1+3/x)=\lim_{x\to\infty}\dfrac{\ln(1+3/x)}{1/(4x)}\overset{\text{H}}{=}\lim_{x\to\infty}\dfrac{\dfrac{1}{1+3/x}\left(-\dfrac{3}{x^2}\right)}{-1/(4x^2)}=\lim_{x\to\infty}\dfrac{12}{1+3/x}=12$$

so $\lim\limits_{x\to\infty}y=\lim\limits_{n\to\infty}\left(1+\dfrac{3}{n}\right)^{4n}=e^{12}$.

9. We use induction, hypothesizing that $a_{n-1}<a_n<2$. Note first that $1<a_2=\frac{1}{3}(1+5)=\frac{5}{3}<2$, so the hypothesis holds for $n=2$. Now assume that $a_{k-1}<a_k<2$. Then $a_k=\frac{1}{3}(a_{k-1}+4)<\frac{1}{3}(a_k+4)<\frac{1}{3}(2+4)=2$. So $a_k<a_{k+1}<2$, and the induction is complete. To find the limit of the sequence, we note that $L=\lim\limits_{n\to\infty}a_n=\lim\limits_{n\to\infty}a_{n+1}\ \Rightarrow\ L=\frac{1}{3}(L+4)\ \Rightarrow\ L=2$.

11. $\dfrac{n}{n^3+1}<\dfrac{n}{n^3}=\dfrac{1}{n^2}$, so $\sum\limits_{n=1}^{\infty}\dfrac{n}{n^3+1}$ converges by the Comparison Test with the convergent p-series $\sum\limits_{n=1}^{\infty}\dfrac{1}{n^2}$ $(p=2>1)$.

13. $\lim\limits_{n\to\infty}\left|\dfrac{a_{n+1}}{a_n}\right|=\lim\limits_{n\to\infty}\left[\dfrac{(n+1)^3}{5^{n+1}}\cdot\dfrac{5^n}{n^3}\right]=\lim\limits_{n\to\infty}\left(1+\dfrac{1}{n}\right)^3\cdot\dfrac{1}{5}=\dfrac{1}{5}<1$, so $\sum\limits_{n=1}^{\infty}\dfrac{n^3}{5^n}$ converges by the Ratio Test.

15. $\lim\limits_{n\to\infty}\sqrt[n]{|a_n|}=\lim\limits_{n\to\infty}\dfrac{n}{3n+1}=\dfrac{1}{3}<1$, so the series converges by the Root Test.

17. $\left|\dfrac{\sin n}{1+n^2}\right|\le\dfrac{1}{1+n^2}<\dfrac{1}{n^2}$ and since $\sum\limits_{n=1}^{\infty}\dfrac{1}{n^2}$ converges (p-series with $p=2>1$), so does $\sum\limits_{n=1}^{\infty}\left|\dfrac{\sin n}{1+n^2}\right|$ by the Comparison Test, and so does $\sum\limits_{n=1}^{\infty}\dfrac{\sin n}{1+n^2}$ by Theorem 12.6.3 [ET 11.6.3].

19. $\lim\limits_{n\to\infty}\left|\dfrac{a_{n+1}}{a_n}\right|=\lim\limits_{n\to\infty}\dfrac{1\cdot3\cdot5\cdots\cdots(2n-1)(2n+1)}{5^{n+1}(n+1)!}\cdot\dfrac{5^n n!}{1\cdot3\cdot5\cdots\cdots(2n-1)}=\lim\limits_{n\to\infty}\dfrac{2n+1}{5(n+1)}=\dfrac{2}{5}<1$, so the series converges by the Ratio Test.

21. Let $b_n=\dfrac{\sqrt{n}}{n+1}>0$. Then $0\le\lim\limits_{n\to\infty}b_n=\lim\limits_{n\to\infty}\dfrac{\sqrt{n}}{n+1}\le\lim\limits_{n\to\infty}\dfrac{\sqrt{n}}{n}=\lim\limits_{n\to\infty}\dfrac{1}{\sqrt{n}}=0$, so $\lim\limits_{n\to\infty}b_n=0$. If $f(x)=\dfrac{\sqrt{x}}{x+1}$ for $x>0$, then $f'(x)=\dfrac{(x+1)\cdot\frac{1}{2\sqrt{x}}-\sqrt{x}\cdot1}{(x+1)^2}=\dfrac{(x+1)-2x}{2\sqrt{x}(x+1)^2}=\dfrac{1-x}{2\sqrt{x}(x+1)^2}$, so $f'(x)<0$ for $x>1$. It follows that $f(1)>f(2)>f(3)>\cdots$; that is, $b_n>b_{n+1}$ for all n. Thus, $\sum\limits_{n=1}^{\infty}(-1)^{n-1}\dfrac{\sqrt{n}}{n+1}$ converges by the Alternating Series Test.

23. Consider the series of absolute values: $\sum_{n=1}^{\infty} n^{-1/3}$ is a p-series with $p = \frac{1}{3} < 1$ and is therefore divergent. But if we apply the Alternating Series Test we see that $a_{n+1} < a_n$ and $\lim_{n \to \infty} n^{-1/3} = 0$. Therefore $\sum_{n=1}^{\infty} (-1)^{n-1} n^{-1/3}$ is conditionally convergent.

25. $\left| \dfrac{a_{n+1}}{a_n} \right| = \left| \dfrac{(-1)^{n+1} (n+2) 3^{n+1}}{2^{2n+3}} \cdot \dfrac{2^{2n+1}}{(-1)^n (n+1) 3^n} \right| = \dfrac{n+2}{n+1} \cdot \dfrac{3}{4} = \dfrac{1 + (2/n)}{1 + (1/n)} \cdot \dfrac{3}{4} \to \dfrac{3}{4} < 1$ as $n \to \infty$, so

by the Ratio Test, $\displaystyle\sum_{n=1}^{\infty} \dfrac{(-1)^n (n+1) 3^n}{2^{2n+1}}$ is absolutely convergent.

27. Convergent geometric series. $\displaystyle\sum_{n=1}^{\infty} \dfrac{2^{2n+1}}{5^n} = \sum_{n=1}^{\infty} \dfrac{(2^2)^n \cdot 2^1}{5^n} = 2 \sum_{n=1}^{\infty} \dfrac{4^n}{5^n} = 2 \left(\dfrac{\frac{4}{5}}{1 - \frac{4}{5}} \right) = 8.$

29. $\displaystyle\sum_{n=1}^{\infty} \left[\tan^{-1} (n+1) - \tan^{-1} n \right] = \lim_{n \to \infty} \left[(\tan^{-1} 2 - \tan^{-1} 1) + (\tan^{-1} 3 - \tan^{-1} 2) + \cdots \right.$

$$\left. + (\tan^{-1} (n+1) - \tan^{-1} n) \right]$$

$$= \lim_{n \to \infty} \left[\tan^{-1} (n+1) - \tan^{-1} 1 \right] = \frac{\pi}{2} - \frac{\pi}{4} = \frac{\pi}{4}$$

31. $1.2 + 0.0\overline{345} = \dfrac{12}{10} + \dfrac{345/10{,}000}{1 - 1/1000} = \dfrac{12}{10} + \dfrac{345}{9990} = \dfrac{4111}{3330}$

33. $\displaystyle\sum_{n=1}^{\infty} \dfrac{(-1)^{n+1}}{n^5} = 1 - \dfrac{1}{32} + \dfrac{1}{243} - \dfrac{1}{1024} + \dfrac{1}{3125} - \dfrac{1}{7776} + \dfrac{1}{16{,}807} - \dfrac{1}{32{,}768} + \cdots$. Since $\dfrac{1}{32{,}768} < 0.000031$,

$\displaystyle\sum_{n=1}^{\infty} \dfrac{(-1)^{n+1}}{n^5} \approx \sum_{n=1}^{7} \dfrac{(-1)^{n+1}}{n^5} \approx 0.9721.$

35. $\displaystyle\sum_{n=1}^{\infty} \dfrac{1}{2 + 5^n} \approx \sum_{n=1}^{8} \dfrac{1}{2 + 5^n} \approx 0.18976224$. To estimate the error, note that $\dfrac{1}{2 + 5^n} < \dfrac{1}{5^n}$, so the remainder term is

$R_8 = \displaystyle\sum_{n=9}^{\infty} \dfrac{1}{2 + 5^n} < \sum_{n=9}^{\infty} \dfrac{1}{5^n} = \dfrac{1/5^9}{1 - 1/5} = 6.4 \times 10^{-7}$ (geometric series with $a = \frac{1}{5^9}$ and $r = \frac{1}{5}$).

37. Use the Limit Comparison Test. $\displaystyle\lim_{n \to \infty} \left| \dfrac{\left(\frac{n+1}{n} \right) a_n}{a_n} \right| = \lim_{n \to \infty} \dfrac{n+1}{n} = \lim_{n \to \infty} \left(1 + \dfrac{1}{n} \right) = 1 > 0$. Since $\sum |a_n|$ is

convergent, so is $\displaystyle\sum \left| \left(\dfrac{n+1}{n} \right) a_n \right|$, by the Limit Comparison Test.

39. $\displaystyle\lim_{n \to \infty} \left| \dfrac{a_{n+1}}{a_n} \right| = \lim_{n \to \infty} \left[\dfrac{|x+2|^{n+1}}{(n+1) 4^{n+1}} \cdot \dfrac{n 4^n}{|x+2|^n} \right] = \lim_{n \to \infty} \left[\dfrac{n}{n+1} \cdot \dfrac{|x+2|}{4} \right] = \dfrac{|x+2|}{4} < 1 \iff |x+2| < 4,$

so $R = 4$. $|x + 2| < 4 \iff -4 < x + 2 < 4 \iff -6 < x < 2$. If $x = -6$, then the series becomes

$\displaystyle\sum_{n=1}^{\infty} \dfrac{(-4)^n}{n 4^n} = \sum_{n=1}^{\infty} \dfrac{(-1)^n}{n}$, the alternating harmonic series, which converges by the Alternating Series Test. When

$x = 2$, the series becomes the harmonic series $\displaystyle\sum_{n=1}^{\infty} \dfrac{1}{n}$, which diverges. Thus, $I = [-6, 2)$.

41. $\lim_{n\to\infty} \left| \dfrac{a_{n+1}}{a_n} \right| = \lim_{n\to\infty} \left| \dfrac{2^{n+1} (x-3)^{n+1}}{\sqrt{n+4}} \cdot \dfrac{\sqrt{n+3}}{2^n (x-3)^n} \right| = 2\,|x-3| \lim_{n\to\infty} \sqrt{\dfrac{n+3}{n+4}} = 2\,|x-3| < 1 \quad \Leftrightarrow$

$|x-3| < \frac{1}{2}$, so $R = \frac{1}{2}$. For $x = \frac{7}{2}$, the series becomes $\displaystyle\sum_{n=0}^{\infty} \dfrac{1}{\sqrt{n+3}} = \sum_{n=3}^{\infty} \dfrac{1}{n^{1/2}}$, which diverges $(p = \frac{1}{2} \le 1)$,

but for $x = \frac{5}{2}$, we get $\displaystyle\sum_{n=0}^{\infty} \dfrac{(-1)^n}{\sqrt{n+3}}$, which is a convergent alternating series, so $I = \left[\frac{5}{2}, \frac{7}{2}\right)$.

43.

$$f(x) = \sin x \qquad f\left(\tfrac{\pi}{6}\right) = \tfrac{1}{2} \qquad f'''(x) = -\cos x \qquad f'''\left(\tfrac{\pi}{6}\right) = -\tfrac{\sqrt{3}}{2}$$

$$f'(x) = \cos x \qquad f'\left(\tfrac{\pi}{6}\right) = \tfrac{\sqrt{3}}{2} \qquad f^{(4)}(x) = \sin x \qquad f^{(4)}\left(\tfrac{\pi}{6}\right) = \tfrac{1}{2}$$

$$f''(x) = -\sin x \qquad f''\left(\tfrac{\pi}{6}\right) = -\tfrac{1}{2} \qquad \cdots \qquad \cdots$$

$$f^{(2n)}\left(\tfrac{\pi}{6}\right) = (-1)^n \cdot \tfrac{1}{2} \text{ and } f^{(2n+1)}\left(\tfrac{\pi}{6}\right) = (-1)^n \cdot \tfrac{\sqrt{3}}{2}.$$

$$\sin x = \sum_{n=0}^{\infty} \dfrac{f^{(n)}\left(\tfrac{\pi}{6}\right)}{n!} \left(x - \tfrac{\pi}{6}\right)^n = \sum_{n=0}^{\infty} \dfrac{(-1)^n}{2\,(2n)!} \left(x - \tfrac{\pi}{6}\right)^{2n} + \sum_{n=0}^{\infty} \dfrac{(-1)^n \sqrt{3}}{2\,(2n+1)!} \left(x - \tfrac{\pi}{6}\right)^{2n+1}$$

45. $\dfrac{1}{1+x} = \dfrac{1}{1-(-x)} = \displaystyle\sum_{n=0}^{\infty} (-1)^n x^n$ for $|x| < 1 \quad \Rightarrow \quad \dfrac{x^2}{1+x} = \displaystyle\sum_{n=0}^{\infty} (-1)^n x^{n+2}$ with $R = 1$.

47. $\dfrac{1}{1-x} = \displaystyle\sum_{n=0}^{\infty} x^n$ for $|x| < 1 \quad \Rightarrow \quad \ln(1-x) = -\displaystyle\int \dfrac{dx}{1-x} = -\int \sum_{n=0}^{\infty} x^n \, dx = C - \sum_{n=0}^{\infty} \dfrac{x^{n+1}}{n+1}$.

$\ln(1-0) = C - 0 \quad \Rightarrow \quad C = 0 \quad \Rightarrow \quad \ln(1-x) = -\displaystyle\sum_{n=0}^{\infty} \dfrac{x^{n+1}}{n+1} = \sum_{n=1}^{\infty} \dfrac{-x^n}{n}$ with $R = 1$.

49. $\sin x = \displaystyle\sum_{n=0}^{\infty} \dfrac{(-1)^n x^{2n+1}}{(2n+1)!} \quad \Rightarrow \quad \sin(x^4) = \displaystyle\sum_{n=0}^{\infty} \dfrac{(-1)^n (x^4)^{2n+1}}{(2n+1)!} = \sum_{n=0}^{\infty} \dfrac{(-1)^n x^{8n+4}}{(2n+1)!}$ for all x, so the radius of

convergence is ∞.

51. $f(x) = 1 / \sqrt[4]{16-x} = (16-x)^{-1/4} = \tfrac{1}{2} \left(1 - \tfrac{1}{16}x\right)^{-1/4}$

$$= \tfrac{1}{2} \left[1 + \left(-\tfrac{1}{4}\right)\left(-\tfrac{x}{16}\right) + \dfrac{\left(-\tfrac{1}{4}\right)\left(-\tfrac{5}{4}\right)}{2!} \left(-\tfrac{x}{16}\right)^2 + \cdots \right]$$

$$= \dfrac{1}{2} + \sum_{n=1}^{\infty} \dfrac{1 \cdot 5 \cdot 9 \cdots \cdots (4n-3)}{2 \cdot 4^n \cdot n! \cdot 16^n} x^n = \dfrac{1}{2} + \sum_{n=1}^{\infty} \dfrac{1 \cdot 5 \cdot 9 \cdots \cdots (4n-3)}{2^{6n+1} n!} x^n$$

for $\left| -\dfrac{x}{16} \right| < 1 \quad \Rightarrow \quad R = 16$.

53. $e^x = \displaystyle\sum_{n=0}^{\infty} \dfrac{x^n}{n!}$ so $\dfrac{e^x}{x} = \dfrac{1}{x} + \sum_{n=1}^{\infty} \dfrac{x^{n-1}}{n!}$ and $\displaystyle\int \dfrac{e^x}{x} \, dx = C + \ln|x| + \sum_{n=1}^{\infty} \dfrac{x^n}{n \cdot n!}$.

55. (a)

$$f(x) = x^{1/2} \qquad f(1) = 1 \qquad f'''(x) = \tfrac{3}{8}x^{-5/2} \qquad f'''(1) = \tfrac{3}{8}$$

$$f'(x) = \tfrac{1}{2}x^{-1/2} \qquad f'(1) = \tfrac{1}{2} \qquad f^{(4)}(x) = -\tfrac{15}{16}x^{-7/2}$$

$$f''(x) = -\tfrac{1}{4}x^{-3/2} \qquad f''(1) = -\tfrac{1}{4}$$

$$\sqrt{x} \approx T_3(x) = 1 + \frac{1/2}{1!}(x-1) - \frac{1/4}{2!}(x-1)^2 + \frac{3/8}{3!}(x-1)^3$$

$$= 1 + \tfrac{1}{2}(x-1) - \tfrac{1}{8}(x-1)^2 + \tfrac{1}{16}(x-1)^3$$

(b)

(c) $|R_3(x)| \le \dfrac{M}{4!}|x-1|^4$, where $\left|f^{(4)}(x)\right| \le M$ with

$f^{(4)}(x) = -\tfrac{15}{16}x^{-7/2}$. Now $0.9 \le x \le 1.1 \Rightarrow$

$(x-1)^4 \le (0.1)^4$, and letting $x = 0.9$ gives

$$M = \frac{15}{16(0.9)^{7/2}}, \text{ so}$$

$$|R_3(x)| \le \frac{15}{16(0.9)^{7/2}\, 4!}(0.1)^4 \approx 0.000005648.$$

(d)

From the graph of $|R_3(x)| = |\sqrt{x} - T_3(x)|$, it appears that the error is less than 5×10^{-6} on $[0.9, 1.1]$.

57. $\sin x = \displaystyle\sum_{n=0}^{\infty}(-1)^n \frac{x^{2n+1}}{(2n+1)!} = x - \frac{x^3}{3!} + \frac{x^5}{5!} - \frac{x^7}{7!} + \cdots$, so $\sin x - x = -\dfrac{x^3}{3!} + \dfrac{x^5}{5!} - \dfrac{x^7}{7!} + \cdots$ and

$\dfrac{\sin x - x}{x^3} = -\dfrac{1}{3!} + \dfrac{x^2}{5!} - \dfrac{x^4}{7!} + \cdots$ and $\displaystyle\lim_{x\to 0}\frac{\sin x - x}{x^3} = \lim_{x\to 0}\left(-\frac{1}{6} + \frac{x^2}{120} - \frac{x^4}{5040} + \cdots\right) = -\frac{1}{6}$.

59. $f(x) = \sum_{n=0}^{\infty} c_n x^n \Rightarrow f(-x) = \sum_{n=0}^{\infty} c_n(-x)^n = \sum_{n=0}^{\infty}(-1)^n c_n x^n$

(a) If f is an odd function, then $f(-x) = -f(x) \Rightarrow \sum_{n=0}^{\infty}(-1)^n c_n x^n = \sum_{n=0}^{\infty} -c_n x^n$. The coefficients of any power series are uniquely determined (by Theorem 12.10.5 [ET 11.10.5]), so $(-1)^n c_n = -c_n$. If n is even, then $(-1)^n = 1$, so $c_n = -c_n \Rightarrow 2c_n = 0 \Rightarrow c_n = 0$. Thus, all even coefficients are 0.

(b) If f is even, then $f(-x) = f(x) \Rightarrow \sum_{n=0}^{\infty}(-1)^n c_n x^n = \sum_{n=0}^{\infty} c_n x^n \Rightarrow (-1)^n c_n = c_n$. If n is odd, then $(-1)^n = -1$, so $-c_n = c_n \Rightarrow 2c_n = 0 \Rightarrow c_n = 0$. Thus, all odd coefficients are 0.

Problems Plus

1. It would be far too much work to compute 15 derivatives of f. The key idea is to remember that $f^{(n)}(0)$ occurs in the coefficient of x^n in the Maclaurin series of f. We start with the Maclaurin series for sin:

$\sin x = x - \dfrac{x^3}{3!} + \dfrac{x^5}{5!} - \cdots$. Then $\sin(x^3) = x^3 - \dfrac{x^9}{3!} + \dfrac{x^{15}}{5!} - \cdots$ and so the coefficient of x^{15} is

$\dfrac{f^{(15)}(0)}{15!} = \dfrac{1}{5!}$. Therefore, $f^{(15)}(0) = \dfrac{15!}{5!} = 6 \cdot 7 \cdot 8 \cdot 9 \cdot 10 \cdot 11 \cdot 12 \cdot 13 \cdot 14 \cdot 15 = 10{,}897{,}286{,}400$.

3. (a) From Formula 14a in Appendix D, with $x = y = \theta$, we get $\tan 2\theta = \dfrac{2 \tan \theta}{1 - \tan^2 \theta}$, so $\cot 2\theta = \dfrac{1 - \tan^2 \theta}{2 \tan \theta}$ $\Rightarrow$

$2 \cot 2\theta = \dfrac{1 - \tan^2 \theta}{\tan \theta} = \cot \theta - \tan \theta$. Replacing θ by $\frac{1}{2}x$, we get $2 \cot x = \cot \frac{1}{2}x - \tan \frac{1}{2}x$,

or $\tan \frac{1}{2}x = \cot \frac{1}{2}x - 2 \cot x$.

(b) From part (a), $\tan \dfrac{x}{2^n} = \cot \dfrac{x}{2^n} - 2 \cot \dfrac{x}{2^{n-1}}$, so the nth partial sum of $\displaystyle\sum_{n=1}^{\infty} \dfrac{1}{2^n} \tan \dfrac{x}{2^n}$ is

$$s_n = \frac{\tan(x/2)}{2} + \frac{\tan(x/4)}{4} + \frac{\tan(x/8)}{8} + \cdots + \frac{\tan(x/2^n)}{2^n}$$

$$= \left[\frac{\cot(x/2)}{2} - \cot x\right] + \left[\frac{\cot(x/4)}{4} - \frac{\cot(x/2)}{2}\right] + \left[\frac{\cot(x/8)}{8} - \frac{\cot(x/4)}{4}\right] + \cdots$$

$$+ \left[\frac{\cot(x/2^n)}{2^n} - \frac{\cot(x/2^{n-1})}{2^{n-1}}\right] = -\cot x + \frac{\cot(x/2^n)}{2^n} \quad \text{(telescoping sum)}$$

Now $\dfrac{\cot(x/2^n)}{2^n} = \dfrac{\cos(x/2^n)}{2^n \sin(x/2^n)} = \dfrac{\cos(x/2^n)}{x} \cdot \dfrac{x/2^n}{\sin(x/2^n)} \to \dfrac{1}{x} \cdot 1 = \dfrac{1}{x}$ as

$n \to \infty$ since $x/2^n \to 0$ for $x \neq 0$. Therefore, if $x \neq 0$ and $x \neq n\pi$, then

$\displaystyle\sum_{n=1}^{\infty} \frac{1}{2^n} \tan \frac{x}{2^n} = \lim_{n \to \infty} \left(-\cot x + \frac{1}{2^n} \cot \frac{x}{2^n}\right) = -\cot x + \frac{1}{x}$. If $x = 0$, then all terms in the series are 0,

so the sum is 0.

5. (a) At each stage, each side is replaced by four shorter sides, each of
length $\frac{1}{3}$ of the side length at the preceding stage. Writing s_0 and ℓ_0
for the number of sides and the length of the side of the initial
triangle, we generate the table at right. In general, we have
$s_n = 3 \cdot 4^n$ and $\ell_n = \left(\frac{1}{3}\right)^n$, so the length of the perimeter at the nth
stage of construction is $p_n = s_n \ell_n = 3 \cdot 4^n \cdot \left(\frac{1}{3}\right)^n = 3 \cdot \left(\frac{4}{3}\right)^n$.

$s_0 = 3$	$\ell_0 = 1$
$s_1 = 3 \cdot 4$	$\ell_1 = 1/3$
$s_2 = 3 \cdot 4^2$	$\ell_2 = 1/3^2$
$s_3 = 3 \cdot 4^3$	$\ell_3 = 1/3^3$
$\cdots$	$\cdots$

(b) $p_n = \dfrac{4^n}{3^{n-1}} = 4 \left(\dfrac{4}{3}\right)^{n-1}$. Since $\frac{4}{3} > 1$, $p_n \to \infty$ as $n \to \infty$.

(c) The area of each of the small triangles added at a given stage is one-ninth of the area of the triangle added at the preceding stage. Let a be the area of the original triangle. Then the area a_n of each of the small triangles added at stage n is $a_n = a \cdot \dfrac{1}{9^n} = \dfrac{a}{9^n}$. Since a small triangle is added to each side at every stage, it follows that the total area A_n added to the figure at the nth stage is

$$A_n = s_{n-1} \cdot a_n = 3 \cdot 4^{n-1} \cdot \frac{a}{9^n} = a \cdot \frac{4^{n-1}}{3^{2n-1}}.$$ Then the total area enclosed by the snowflake curve is

$$A = a + A_1 + A_2 + A_3 + \cdots = a + a \cdot \frac{1}{3} + a \cdot \frac{4}{3^3} + a \cdot \frac{4^2}{3^5} + a \cdot \frac{4^3}{3^7} + \cdots.$$ After the first term, this is a

geometric series with common ratio $\frac{4}{9}$, so $A = a + \dfrac{a/3}{1 - \frac{4}{9}} = a + \dfrac{a}{3} \cdot \dfrac{9}{5} = \dfrac{8a}{5}$. But the area of the original

equilateral triangle with side 1 is $a = \frac{1}{2} \cdot 1 \cdot \sin \frac{\pi}{3} = \frac{\sqrt{3}}{4}$. So the area enclosed by the snowflake curve is

$\frac{8}{5} \cdot \frac{\sqrt{3}}{4} = \frac{2\sqrt{3}}{5}$.

7. (a) Let $a = \arctan x$ and $b = \arctan y$. Then, from Formula 14b in Appendix D,

$$\tan(a - b) = \frac{\tan a - \tan b}{1 + \tan a \tan b} = \frac{\tan(\arctan x) - \tan(\arctan y)}{1 + \tan(\arctan x)\tan(\arctan y)} = \frac{x - y}{1 + xy} \quad \Rightarrow$$

$$\arctan x - \arctan y = a - b = \arctan \frac{x - y}{1 + xy} \quad \text{since} \quad -\frac{\pi}{2} < \arctan x - \arctan y < \frac{\pi}{2}$$

(b) From part (a) we have

$$\arctan \tfrac{120}{119} - \arctan \tfrac{1}{239} = \arctan \frac{\frac{120}{119} - \frac{1}{239}}{1 + \frac{120}{119} \cdot \frac{1}{239}} = \arctan \frac{\frac{28,561}{28,441}}{\frac{28,561}{28,441}} = \arctan 1 = \frac{\pi}{4}$$

(c) Replacing y by $-y$ in the formula of part (a), we get $\arctan x + \arctan y = \arctan \dfrac{x + y}{1 - xy}$. So

$$4\arctan \tfrac{1}{5} = 2\left(\arctan \tfrac{1}{5} + \arctan \tfrac{1}{5}\right) = 2\arctan \frac{\frac{1}{5} + \frac{1}{5}}{1 - \frac{1}{5} \cdot \frac{1}{5}} = 2\arctan \tfrac{5}{12} = \arctan \tfrac{5}{12} + \arctan \tfrac{5}{12}$$

$$= \arctan \frac{\frac{5}{12} + \frac{5}{12}}{1 - \frac{5}{12} \cdot \frac{5}{12}} = \arctan \tfrac{120}{119}$$

Thus, from part (b), we have $4\arctan \tfrac{1}{5} - \arctan \tfrac{1}{239} = \arctan \tfrac{120}{119} - \arctan \tfrac{1}{239} = \tfrac{\pi}{4}$.

(d) From Example 7 in Section 12.9 [ET 11.9] we have $\arctan x = x - \dfrac{x^3}{3} + \dfrac{x^5}{5} - \dfrac{x^7}{7} + \dfrac{x^9}{9} - \dfrac{x^{11}}{11} + \cdots$, so

$$\arctan \frac{1}{5} = \frac{1}{5} - \frac{1}{3 \cdot 5^3} + \frac{1}{5 \cdot 5^5} - \frac{1}{7 \cdot 5^7} + \frac{1}{9 \cdot 5^9} - \frac{1}{11 \cdot 5^{11}} + \cdots$$

This is an alternating series and the size of the terms decreases to 0, so by the Alternating Series Estimation Theorem, the sum lies between s_5 and s_6, that is, $0.197395560 < \arctan \tfrac{1}{5} < 0.197395562$.

(e) From the series in part (d) we get $\arctan \dfrac{1}{239} = \dfrac{1}{239} - \dfrac{1}{3 \cdot 239^3} + \dfrac{1}{5 \cdot 239^5} - \cdots$. The third term is less than 2.6×10^{-13}, so by the Alternating Series Estimation Theorem we have, to nine decimal places, $\arctan \tfrac{1}{239} \approx s_2 \approx 0.004184076$. Thus, $0.004184075 < \arctan \tfrac{1}{239} < 0.004184077$.

(f) From part (c) we have $\pi = 16 \arctan \frac{1}{5} - 4 \arctan \frac{1}{239}$, so from parts (d) and (e) we have

$16\,(0.197395560) - 4\,(0.004184077) < \pi < 16\,(0.197395562) - 4\,(0.004184075) \quad \Rightarrow$

$3.141592652 < \pi < 3.141592692$. So, to 7 decimal places, $\pi \approx 3.1415927$.

9. We start with the geometric series $\sum\limits_{n=0}^{\infty} x^n = \dfrac{1}{1-x}$, $|x| < 1$, and differentiate:

$$\sum_{n=1}^{\infty} nx^{n-1} = \frac{d}{dx}\left(\sum_{n=0}^{\infty} x^n\right) = \frac{d}{dx}\left(\frac{1}{1-x}\right) = \frac{1}{(1-x)^2} \text{ for } |x| < 1 \quad \Rightarrow$$

$$\sum_{n=1}^{\infty} nx^n = x \sum_{n=1}^{\infty} nx^{n-1} = \frac{x}{(1-x)^2} \text{ for } |x| < 1. \text{ Differentiate again:}$$

$$\sum_{n=1}^{\infty} n^2 x^{n-1} = \frac{d}{dx}\frac{x}{(1-x)^2} = \frac{(1-x)^2 - x \cdot 2(1-x)(-1)}{(1-x)^4} = \frac{x+1}{(1-x)^3} \quad \Rightarrow \quad \sum_{n=1}^{\infty} n^2 x^n = \frac{x^2+x}{(1-x)^3} \quad \Rightarrow$$

$$\sum_{n=1}^{\infty} n^3 x^{n-1} = \frac{d}{dx}\frac{x^2+x}{(1-x)^3} = \frac{(1-x)^3(2x+1) - (x^2+x)\,3(1-x)^2(-1)}{(1-x)^6} = \frac{x^2+4x+1}{(1-x)^4} \quad \Rightarrow$$

$$\sum_{n=1}^{\infty} n^3 x^n = \frac{x^3+4x^2+x}{(1-x)^4}, \ |x| < 1.$$ The radius of convergence is 1 because that is the radius of convergence for

the geometric series we started with. If $x = \pm 1$, the series is $\sum n^3 (\pm 1)^n$, which diverges by the Test For

Divergence, so the interval of convergence is $(-1, 1)$.

11. $a_{n+1} = \dfrac{a_n + b_n}{2}$, $b_{n+1} = \sqrt{b_n a_{n+1}}$. So $a_1 = \cos\theta$, $b_1 = 1 \quad \Rightarrow \quad a_2 = \dfrac{1 + \cos\theta}{2} = \cos^2\frac{\theta}{2}$,

$b_2 = \sqrt{b_1 a_2} = \sqrt{\cos^2\frac{\theta}{2}} = \cos\frac{\theta}{2}$ since $-\frac{\pi}{2} \le \theta \le \frac{\pi}{2}$. Then

$a_3 = \frac{1}{2}\left(\cos\frac{\theta}{2} + \cos^2\frac{\theta}{2}\right) = \cos\frac{\theta}{2} \cdot \frac{1}{2}\left(1 + \cos\frac{\theta}{2}\right) = \cos\frac{\theta}{2}\cos^2\frac{\theta}{4} \quad \Rightarrow$

$b_3 = \sqrt{b_2 a_3} = \sqrt{\cos\frac{\theta}{2}\cos\frac{\theta}{2}\cos^2\frac{\theta}{4}} = \cos\frac{\theta}{2}\cos\frac{\theta}{4} \quad \Rightarrow$

$a_4 = \frac{1}{2}\left(\cos\frac{\theta}{2}\cos^2\frac{\theta}{4} + \cos\frac{\theta}{2}\cos\frac{\theta}{4}\right) = \cos\frac{\theta}{2}\cos\frac{\theta}{4} \cdot \frac{1}{2}\left(1 + \cos\frac{\theta}{4}\right) = \cos\frac{\theta}{2}\cos\frac{\theta}{4}\cos^2\frac{\theta}{8} \quad \Rightarrow$

$b_4 = \sqrt{\cos\frac{\theta}{2}\cos\frac{\theta}{4}\cos\frac{\theta}{2}\cos\frac{\theta}{4}\cos^2\frac{\theta}{8}} = \cos\frac{\theta}{2}\cos\frac{\theta}{4}\cos\frac{\theta}{8}$. By now we see the pattern:

$b_n = \cos\frac{\theta}{2}\cos\frac{\theta}{2^2}\cos\frac{\theta}{2^3} \cdots \cdots \cos\frac{\theta}{2^{n-1}}$ and $a_n = b_n \cos\frac{\theta}{2^{n-1}}$. (This could be proved by mathematical

induction.) By Exercise 10(a), $\sin\theta = 2^{n-1}\sin\frac{\theta}{2^{n-1}}\cos\frac{\theta}{2}\cos\frac{\theta}{4}\cdots\cos\frac{\theta}{2^{n-1}}$. So

$b_n = \cos\frac{\theta}{2}\cos\frac{\theta}{2^2}\cos\frac{\theta}{2^3}\cdots\cos\frac{\theta}{2^{n-1}} \to \frac{\sin\theta}{\theta}$ as $n \to \infty$ by Exercise 10(b), and

$a_n = b_n \cos\frac{\theta}{2^{n-1}} \to \frac{\sin\theta}{\theta} \cdot 1 = \frac{\sin\theta}{\theta}$ as $n \to \infty$. So $\lim\limits_{n\to\infty} a_n = \lim\limits_{n\to\infty} b_n = \frac{\sin\theta}{\theta}$.

13. Let $f(x) = \sum_{m=0}^{\infty} c_m x^m$ and $g(x) = e^{f(x)} = \sum_{n=0}^{\infty} d_n x^n$. Then $g'(x) = \sum_{n=0}^{\infty} nd_n x^{n-1}$, so nd_n occurs as

the coefficient of x^{n-1}. But also

$$g'(x) = e^{f(x)} f'(x) = \left(\sum_{n=0}^{\infty} d_n x^n\right)\left(\sum_{m=1}^{\infty} mc_m x^{m-1}\right)$$

$$= \left(d_0 + d_1 x + d_2 x^2 + \cdots + d_{n-1}x^{n-1} + \cdots\right)\left(c_1 + 2c_2 x + 3c_3 x^2 + \cdots + nc_n x^{n-1} + \cdots\right)$$

so the coefficient of x^{n-1} is $c_1 d_{n-1} + 2c_2 d_{n-2} + 3c_3 d_{n-3} + \cdots + nc_n d_0 = \sum_{i=1}^{n} ic_i d_{n-i}$. Therefore,

$nd_n = \sum_{i=1}^{n} ic_i d_{n-i}$.

15. $u = 1 + \dfrac{x^3}{3!} + \dfrac{x^6}{6!} + \dfrac{x^9}{9!} + \cdots,\ v = x + \dfrac{x^4}{4!} + \dfrac{x^7}{7!} + \dfrac{x^{10}}{10!} + \cdots,\ w = \dfrac{x^2}{2!} + \dfrac{x^5}{5!} + \dfrac{x^8}{8!} + \cdots.$ The key idea is to

differentiate: $\dfrac{du}{dx} = \dfrac{3x^2}{3!} + \dfrac{6x^5}{6!} + \dfrac{9x^8}{9!} + \cdots = \dfrac{x^2}{2!} + \dfrac{x^5}{5!} + \dfrac{x^8}{8!} + \cdots = w.$ Similarly,

$\dfrac{dv}{dx} = 1 + \dfrac{x^3}{3!} + \dfrac{x^6}{6!} + \dfrac{x^9}{9!} + \cdots = u,$ and $\dfrac{dw}{dx} = x + \dfrac{x^4}{4!} + \dfrac{x^7}{7!} + \dfrac{x^{10}}{10!} + \cdots = v.$ So $u' = w,\ v' = u,$ and

$w' = v.$ Now differentiate the left hand side of the desired equation:

$$\frac{d}{dx}\left(u^3 + v^3 + w^3 - 3uvw\right) = 3u^2 u' + 3v^2 v' + 3w^2 w' - 3\left(u'vw + uv'w + uvw'\right)$$

$$= 3u^2 w + 3v^2 u + 3w^2 v - 3\left(vw^2 + u^2 w + uv^2\right) = 0 \quad\Rightarrow$$

$u^3 + v^3 + w^3 - 3uvw = C.$ To find the value of the constant C, we put $x = 0$ in the last equation and get

$1^3 + 0^3 + 0^3 - 3\left(1 \cdot 0 \cdot 0\right) = C \quad\Rightarrow\quad C = 1,$ so $u^3 + v^3 + w^3 - 3uvw = 1.$

17. If L is the length of a side of the equilateral triangle, then the area is $A = \frac{1}{2}L \cdot \frac{\sqrt{3}}{2}L = \frac{\sqrt{3}}{4}L^2$ and so $L^2 = \frac{4}{\sqrt{3}}A.$

Let r be the radius of one of the circles. When there are n rows of circles, the figure shows that

$$L = \sqrt{3}r + r + (n-2)(2r) + r + \sqrt{3}r = r\left(2n - 2 + 2\sqrt{3}\right),\ \text{so } r = \frac{L}{2\left(n + \sqrt{3} - 1\right)}.\ \text{The number of circles}$$

is $1 + 2 + \cdots + n = \dfrac{n(n+1)}{2}$ and so the total area of the circles is

$$A_n = \frac{n(n+1)}{2}\pi r^2 = \frac{n(n+1)}{2}\pi \frac{L^2}{4\left(n + \sqrt{3} - 1\right)^2} = \frac{n(n+1)}{2}\pi \frac{4A/\sqrt{3}}{4\left(n + \sqrt{3} - 1\right)^2}$$

$$= \frac{n(n+1)}{\left(n + \sqrt{3} - 1\right)^2}\frac{\pi A}{2\sqrt{3}} \quad\Rightarrow$$

$$\frac{A_n}{A} = \frac{n(n+1)}{\left(n + \sqrt{3} - 1\right)^2}\frac{\pi}{2\sqrt{3}}$$

$$= \frac{1 + 1/n}{\left[1 + \left(\sqrt{3} - 1\right)/n\right]^2}\frac{\pi}{2\sqrt{3}} \to \frac{\pi}{2\sqrt{3}}\ \text{as } n \to \infty$$

$\sqrt{3}r \quad r \quad 2r \quad 2r \quad r \quad \sqrt{3}r$

$\longmapsto\! L \!\longmapsto$

19. Call the series S. We group the terms according to the number of digits in their denominators:

$$S = \underbrace{\left(1 + \tfrac{1}{2} + \cdots + \tfrac{1}{8} + \tfrac{1}{9}\right)}_{g_1} + \underbrace{\left(\tfrac{1}{11} + \cdots + \tfrac{1}{99}\right)}_{g_2} + \underbrace{\left(\tfrac{1}{111} + \cdots + \tfrac{1}{999}\right)}_{g_3} + \cdots$$

Now in the group g_n, there are 9^n terms, since we have 9 choices for each of the n digits in the denominator.

Furthermore, each term in g_n is less than $\dfrac{1}{10^{n-1}}$. So $g_n < 9^n \cdot \dfrac{1}{10^{n-1}} = 9\left(\tfrac{9}{10}\right)^{n-1}$. Now $\sum_{n=1}^{\infty} 9\left(\tfrac{9}{10}\right)^{n-1}$ is a

geometric series with $a = 9$ and $r = \tfrac{9}{10} < 1$. Therefore, by the Comparison Test,

$$S = \sum_{n=1}^{\infty} g_n < \sum_{n=1}^{\infty} 9\left(\tfrac{9}{10}\right)^{n-1} = \frac{9}{1 - 9/10} = 90.$$

Vectors and the Geometry of Space

13.1 Three-Dimensional Coordinate Systems

1. We start at the origin, which has coordinates $(0, 0, 0)$. First we move 4 units along the positive x-axis, affecting only the x-coordinate, bringing us to the point $(4, 0, 0)$. We then move 3 units straight downward, in the negative z-direction. Thus only the z-coordinate is affected, and we arrive at $(4, 0, -3)$.

3. The distance from a point to the xz-plane is the absolute value of the y-coordinate of the point. $Q(-5, -1, 4)$ has the y-coordinate with the smallest absolute value, so Q is the point closest to the xz-plane. $R(0, 3, 8)$ must lie in the yz-plane since the distance from R to the yz-plane, given by the x-coordinate of R, is 0.

5. The equation $x + y = 2$ represents the set of all points in $\mathbb{R}^3$ whose x- and y-coordinates have a sum of 2, or equivalently where $y = 2 - x$. This is the set $\{(x, 2 - x, z) \mid x \in \mathbb{R}, z \in \mathbb{R}\}$ which is a vertical plane that intersects the xy-plane in the line $y = 2 - x$, $z = 0$.

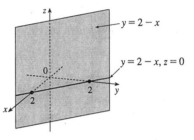

7. We first find the lengths of the sides of the triangle by using the distance formula between pairs of vertices:

$$|PQ| = \sqrt{[1 - (-2)]^2 + (2 - 4)^2 + (-1 - 0)^2} = \sqrt{9 + 4 + 1} = \sqrt{14}$$

$$|QR| = \sqrt{(-1 - 1)^2 + (1 - 2)^2 + [2 - (-1)]^2} = \sqrt{4 + 1 + 9} = \sqrt{14}$$

$$|PR| = \sqrt{[-1 - (-2)]^2 + (1 - 4)^2 + (2 - 0)^2} = \sqrt{1 + 9 + 4} = \sqrt{14}$$

Since all three sides have the same length, PQR is an equilateral triangle.

9. (a) First we find the distances between points:

$$|AB| = \sqrt{(7 - 5)^2 + (9 - 1)^2 + (-1 - 3)^2} = \sqrt{84} = 2\sqrt{21}$$

$$|BC| = \sqrt{(1 - 7)^2 + (-15 - 9)^2 + [11 - (-1)]^2} = \sqrt{756} = 6\sqrt{21}$$

$$|AC| = \sqrt{(1 - 5)^2 + (-15 - 1)^2 + (11 - 3)^2} = \sqrt{336} = 4\sqrt{21}$$

In order for the points to lie on a straight line, the sum of the two shortest distances must be equal to the longest distance. Since $|AB| + |AC| = |BC|$, the three points lie on a straight line.

(b) The distances between points are

$$|KL| = \sqrt{(1-0)^2 + (2-3)^2 + [-2-(-4)]^2} = \sqrt{6}$$

$$|LM| = \sqrt{(3-1)^2 + (0-2)^2 + [1-(-2)]^2} = \sqrt{17}$$

$$|KM| = \sqrt{(3-0)^2 + (0-3)^2 + [1-(-4)]^2} = \sqrt{43}$$

Since $\sqrt{6} + \sqrt{17} \neq \sqrt{43}$, the three points do not lie on a straight line.

11. An equation of the sphere with center $(0, 1, -1)$ and radius 4 is $(x-0)^2 + (y-1)^2 + [z-(-1)]^2 = 4^2$ or $x^2 + (y-1)^2 + (z+1)^2 = 16$. The intersection of this sphere with the yz-plane is the set of points on the sphere whose x-coordinate is 0. Putting $x = 0$ in the equation, we have $(y-1)^2 + (z+1)^2 = 16$, $x = 0$, which represents a circle in the yz-plane with center $(0, 1, -1)$ and radius 4.

13. The radius of the sphere is the distance between $(4, 3, -1)$ and $(3, 8, 1)$:

$r = \sqrt{(3-4)^2 + (8-3)^2 + [1-(-1)]^2} = \sqrt{30}$. Thus, an equation of the sphere is

$(x-3)^2 + (y-8)^2 + (z-1)^2 = 30$.

15. Completing squares in the equation gives

$(x^2 + 2x + 1) + (y^2 + 8y + 16) + (z^2 - 4z + 4) = 28 + 1 + 16 + 4 \Rightarrow$

$(x+1)^2 + (y+4)^2 + (z-2)^2 = 49$ which we recognize as an equation of a sphere with center $(-1, -4, 2)$ and radius 7.

17. Completing squares in the equation gives $\left(x^2 - x + \frac{1}{4}\right) + \left(y^2 - y + \frac{1}{4}\right) + \left(z^2 - z + \frac{1}{4}\right) = \frac{1}{4} + \frac{1}{4} + \frac{1}{4} \Rightarrow$

$\left(x - \frac{1}{2}\right)^2 + \left(y - \frac{1}{2}\right)^2 + \left(z - \frac{1}{2}\right)^2 = \frac{3}{4}$ which we recognize as an equation of a sphere with center $\left(\frac{1}{2}, \frac{1}{2}, \frac{1}{2}\right)$ and

radius $\sqrt{\frac{3}{4}} = \frac{\sqrt{3}}{2}$.

19. (a) If the midpoint of the line segment from $P_1(x_1, y_1, z_1)$ to $P_2(x_2, y_2, z_2)$ is

$Q = \left(\dfrac{x_1 + x_2}{2}, \dfrac{y_1 + y_2}{2}, \dfrac{z_1 + z_2}{2}\right)$, then the distances $|P_1 Q|$ and $|Q P_2|$ are equal, and each is half of $|P_1 P_2|$.

We verify that this is the case:

$$|P_1 P_2| = \sqrt{(x_2 - x_1)^2 + (y_2 - y_1)^2 + (z_2 - z_1)^2}$$

$$|P_1 Q| = \sqrt{\left[\frac{1}{2}(x_1 + x_2) - x_1\right]^2 + \left[\frac{1}{2}(y_1 + y_2) - y_1\right]^2 + \left[\frac{1}{2}(z_1 + z_2) - z_1\right]^2}$$

$$= \sqrt{\left(\frac{1}{2}x_2 - \frac{1}{2}x_1\right)^2 + \left(\frac{1}{2}y_2 - \frac{1}{2}y_1\right)^2 + \left(\frac{1}{2}z_2 - \frac{1}{2}z_1\right)^2}$$

$$= \sqrt{\left(\frac{1}{2}\right)^2 \left[(x_2 - x_1)^2 + (y_2 - y_1)^2 + (z_2 - z_1)^2\right]}$$

$$= \frac{1}{2}\sqrt{(x_2 - x_1)^2 + (y_2 - y_1)^2 + (z_2 - z_1)^2}$$

$$= \frac{1}{2}|P_1 P_2|$$

$$|QP_2| = \sqrt{\left[x_2 - \tfrac{1}{2}(x_1 + x_2)\right]^2 + \left[y_2 - \tfrac{1}{2}(y_1 + y_2)\right]^2 + \left[z_2 - \tfrac{1}{2}(z_1 + z_2)\right]^2}$$

$$= \sqrt{\left(\tfrac{1}{2}x_2 - \tfrac{1}{2}x_1\right)^2 + \left(\tfrac{1}{2}y_2 - \tfrac{1}{2}y_1\right)^2 + \left(\tfrac{1}{2}z_2 - \tfrac{1}{2}z_1\right)^2}$$

$$= \sqrt{\left(\tfrac{1}{2}\right)^2 \left[(x_2 - x_1)^2 + (y_2 - y_1)^2 + (z_2 - z_1)^2\right]}$$

$$= \tfrac{1}{2}\sqrt{(x_2 - x_1)^2 + (y_2 - y_1)^2 + (z_2 - z_1)^2}$$

$$= \tfrac{1}{2}|P_1 P_2|$$

So Q is indeed the midpoint of $P_1 P_2$.

(b) By part (a), the midpoints of sides AB, BC and CA are $P_1\left(-\tfrac{1}{2}, 1, 4\right)$, $P_2\left(1, \tfrac{1}{2}, 5\right)$ and $P_3\left(\tfrac{5}{2}, \tfrac{3}{2}, 4\right)$. (Recall that a median of a triangle is a line segment from a vertex to the mipoint of the opposite side.) Then the lengths of the medians are:

$$|AP_2| = \sqrt{0^2 + \left(\tfrac{1}{2} - 2\right)^2 + (5 - 3)^2} = \sqrt{\tfrac{9}{4} + 4} = \sqrt{\tfrac{25}{4}} = \tfrac{5}{2}$$

$$|BP_3| = \sqrt{\left(\tfrac{5}{2} + 2\right)^2 + \left(\tfrac{3}{2}\right)^2 + (4 - 5)^2} = \sqrt{\tfrac{81}{4} + \tfrac{9}{4} + 1} = \sqrt{\tfrac{94}{4}} = \tfrac{1}{2}\sqrt{94}$$

$$|CP_1| = \sqrt{\left(-\tfrac{1}{2} - 4\right)^2 + (1 - 1)^2 + (4 - 5)^2} = \sqrt{\tfrac{81}{4} + 1} = \tfrac{1}{2}\sqrt{85}$$

21. (a) Since the sphere touches the xy-plane, its radius is the distance from its center, $(2, -3, 6)$, to the xy-plane, namely 6. Therefore $r = 6$ and an equation of the sphere is $(x - 2)^2 + (y + 3)^2 + (z - 6)^2 = 6^2 = 36$.

(b) The radius of this sphere is the distance from its center $(2, -3, 6)$ to the yz-plane, which is 2. Therefore, an equation is $(x - 2)^2 + (y + 3)^2 + (z - 6)^2 = 4$.

(c) Here the radius is the distance from the center $(2, -3, 6)$ to the xz-plane, which is 3. Therefore, an equation is $(x - 2)^2 + (y + 3)^2 + (z - 6)^2 = 9$.

23. The equation $x = 9$ represents a plane parallel to the yz-plane and 9 units in front of it.

25. The inequality $y > 2$ represents a half-space containing all points to the right of the plane $y = 2$.

27. The inequality $|z| \le 2$ is equivalent to $-2 \le z \le 2$, so it represents all points on and between the two horizontal planes $z = 2$ and $z = -2$.

29. The inequality $x^2 + y^2 + z^2 > 1$ is equivalent to $\sqrt{x^2 + y^2 + z^2} > 1$, so the region consists of those points whose distance from the origin is greater than 1. This is the set of all points outside the sphere with radius 1 and center $(0, 0, 0)$.

31. Completing the square in z gives $x^2 + y^2 + \left(z^2 - 2z + 1\right) < 3 + 1$ or $x^2 + y^2 + (z - 1)^2 < 4$, which is equivalent to $\sqrt{x^2 + y^2 + (z - 1)^2} < 2$. Thus the region consists of those points whose distance from the point $(0, 0, 1)$ is less than 2. This is the set of all points inside the sphere with radius 2 and center $(0, 0, 1)$.

33. Here $x^2 + z^2 \le 9$ or equivalently $\sqrt{x^2 + z^2} \le 3$ which describes the set of all points in $\mathbb{R}^3$ whose distance from the y-axis is at most 3. Thus, the inequality represents the region consisting of all points on and inside a circular cylinder of radius 3 with axis the y-axis.

35. This describes all points with negative y-coordinates, that is, $y < 0$.

37. This describes a region all of whose points have a distance to the origin which is greater than r, but smaller than R. So inequalities describing the region are $r < \sqrt{x^2 + y^2 + z^2} < R$, or $r^2 < x^2 + y^2 + z^2 < R^2$.

39. (a) To find the x- and y-coordinates of the point P, we project it onto L_2 and project the resulting point Q onto the x- and y-axes. To find the z-coordinate, we project P onto either the xz-plane or the yz-plane (using our knowledge of its x- or y-coordinate) and then project the resulting point onto the z-axis. (Or, we could draw a line parallel to QO from P to the z-axis.) The coordinates of P are $(2, 1, 4)$.

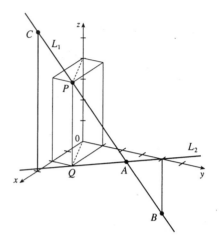

(b) A is the intersection of L_1 and L_2, B is directly below the y-intercept of L_2, and C is directly above the x-intercept of L_2.

41. We need to find a set of points $\{P(x, y, z) \,|\, |AP| = |BP|\}$.

$\sqrt{(x + 1)^2 + (y - 5)^2 + (z - 3)^2} = \sqrt{(x - 6)^2 + (y - 2)^2 + (z + 2)^2} \Rightarrow$

$(x + 1)^2 + (y - 5) + (z - 3)^2 = (x - 6)^2 + (y - 2)^2 + (z + 2)^2 \Rightarrow$

$x^2 + 2x + 1 + y^2 - 10y + 25 + z^2 - 6z + 9 = x^2 - 12x + 36 + y^2 - 4y + 4 + z^2 + 4z + 4 \Rightarrow$

$14x - 6y - 10z = 9$. Thus the set of points is a plane perpendicular to the line segment joining A and B (since this plane must contain the perpendicular bisector of the line segment AB).

13.2 Vectors   ET 12.2

1. (a) The cost of a theater ticket is a scalar, because it has only magnitude.

(b) The current in a river is a vector, because it has both magnitude (the speed of the current) and direction at any given location.

(c) If we assume that the initial path is linear, the initial flight path from Houston to Dallas is a vector, because it has both magnitude (distance) and direction.

(d) The population of the world is a scalar, because it has only magnitude.

3. Vectors are equal when they share the same length and direction (but not necessarily location). Using the symmetry of the parallelogram as a guide, we see that $\overrightarrow{AB} = \overrightarrow{DC}, \overrightarrow{DA} = \overrightarrow{CB}, \overrightarrow{DE} = \overrightarrow{EB}$, and $\overrightarrow{EA} = \overrightarrow{CE}$.

5. (a) By the Triangle Law, $\overrightarrow{AB} + \overrightarrow{BC}$ is the vector with initial point A and terminal point C, namely $\overrightarrow{AC}$.

(b) By the Triangle Law, $\overrightarrow{CD} + \overrightarrow{DA}$ is the vector with initial point C and terminal point A, namely $\overrightarrow{CA}$.

(c) First we consider $\overrightarrow{BC} - \overrightarrow{DC}$ as $\overrightarrow{BC} + \left(-\overrightarrow{DC}\right)$. Then since $-\overrightarrow{DC}$ has the same length as $\overrightarrow{CD}$ but points in the opposite direction, we have $-\overrightarrow{DC} = \overrightarrow{CD}$ and so $\overrightarrow{BC} - \overrightarrow{DC} = \overrightarrow{BC} + \overrightarrow{CD} = \overrightarrow{BD}$.

(d) We use the Triangle Law twice: $\overrightarrow{BC} + \overrightarrow{CD} + \overrightarrow{DA} = \left(\overrightarrow{BC} + \overrightarrow{CD}\right) + \overrightarrow{DA} = \overrightarrow{BD} + \overrightarrow{DA} = \overrightarrow{BA}$.

7. $\mathbf{a} = \langle 4 - 1, 4 - 3 \rangle = \langle 3, 1 \rangle$

9. $\mathbf{a} = \langle -3 - (-1), 4 - (-1) \rangle = \langle -2, 5 \rangle$

11. $\mathbf{a} = \langle 2 - 0, 3 - 3, -1 - 1 \rangle = \langle 2, 0, -2 \rangle$

13. $\langle 3, -1 \rangle + \langle -2, 4 \rangle = \langle 3 + (-2), -1 + 4 \rangle = \langle 1, 3 \rangle$

15. $\langle 1, 0, 1 \rangle + \langle 0, 0, 1 \rangle = \langle 1 + 0, 0 + 0, 1 + 1 \rangle$
$$= \langle 1, 0, 2 \rangle$$

17. $|\mathbf{a}| = \sqrt{(-4)^2 + 3^2} = \sqrt{25} = 5$

$\mathbf{a} + \mathbf{b} = \langle -4 + 6, 3 + 2 \rangle = \langle 2, 5 \rangle$

$\mathbf{a} - \mathbf{b} = \langle -4 - 6, 3 - 2 \rangle = \langle -10, 1 \rangle$

$2\mathbf{a} = \langle 2(-4), 2(3) \rangle = \langle -8, 6 \rangle$

$3\mathbf{a} + 4\mathbf{b} = \langle -12, 9 \rangle + \langle 24, 8 \rangle = \langle 12, 17 \rangle$

19. $|\mathbf{a}| = \sqrt{6^2 + 2^2 + 3^2} = \sqrt{49} = 7$

$\mathbf{a} + \mathbf{b} = \langle 6 + (-1), 2 + 5, 3 + (-2) \rangle = \langle 5, 7, 1 \rangle$

$\mathbf{a} - \mathbf{b} = \langle 6 - (-1), 2 - 5, 3 - (-2) \rangle = \langle 7, -3, 5 \rangle$

$2\mathbf{a} = \langle 2(6), 2(2), 2(3) \rangle = \langle 12, 4, 6 \rangle$

$3\mathbf{a} + 4\mathbf{b} = \langle 18, 6, 9 \rangle + \langle -4, 20, -8 \rangle = \langle 14, 26, 1 \rangle$

21. $|\mathbf{a}| = \sqrt{1^2 + (-2)^2 + 1^2} = \sqrt{6}$

$\mathbf{a} + \mathbf{b} = (\mathbf{i} - 2\mathbf{j} + \mathbf{k}) + (\mathbf{j} + 2\mathbf{k}) = \mathbf{i} - \mathbf{j} + 3\mathbf{k}$

$\mathbf{a} - \mathbf{b} = (\mathbf{i} - 2\mathbf{j} + \mathbf{k}) - (\mathbf{j} + 2\mathbf{k}) = \mathbf{i} - 3\mathbf{j} - \mathbf{k}$

$2\mathbf{a} = 2(\mathbf{i} - 2\mathbf{j} + \mathbf{k}) = 2\mathbf{i} - 4\mathbf{j} + 2\mathbf{k}$

$3\mathbf{a} + 4\mathbf{b} = 3(\mathbf{i} - 2\mathbf{j} + \mathbf{k}) + 4(\mathbf{j} + 2\mathbf{k})$
$$= 3\mathbf{i} - 6\mathbf{j} + 3\mathbf{k} + 4\mathbf{j} + 8\mathbf{k} = 3\mathbf{i} - 2\mathbf{j} + 11\mathbf{k}$$

23. $|\langle 9, -5 \rangle| = \sqrt{9^2 + (-5)^2} = \sqrt{106}$, so $\mathbf{u} = \frac{1}{\sqrt{106}} \langle 9, -5 \rangle = \left\langle \frac{9}{\sqrt{106}}, \frac{-5}{\sqrt{106}} \right\rangle$.

25. $|8\mathbf{i} - \mathbf{j} + 4\mathbf{k}| = \sqrt{8^2 + (-1)^2 + 4^2} = \sqrt{81} = 9$, so $\mathbf{u} = \frac{1}{9}(8\mathbf{i} - \mathbf{j} + 4\mathbf{k}) = \frac{8}{9}\mathbf{i} - \frac{1}{9}\mathbf{j} + \frac{4}{9}\mathbf{k}$.

27. By the Triangle Law, $\overrightarrow{AB} + \overrightarrow{BC} = \overrightarrow{AC}$. Then $\overrightarrow{AB} + \overrightarrow{BC} + \overrightarrow{CA} = \overrightarrow{AC} + \overrightarrow{CA}$, but

$\overrightarrow{AC} + \overrightarrow{CA} = \overrightarrow{AC} + \left(-\overrightarrow{AC}\right) = \mathbf{0}$. So $\overrightarrow{AB} + \overrightarrow{BC} + \overrightarrow{CA} = \mathbf{0}$.

29. (a), (b)

(c) From the sketch, we estimate that $s \approx 1.3$ and $t \approx 1.6$.

(d) $\mathbf{c} = s\,\mathbf{a} + t\,\mathbf{b} \quad \Leftrightarrow \quad 7 = 3s + 2t$ and $1 = 2s - t$.

Solving these equations gives $s = \frac{9}{7}$ and $t = \frac{11}{7}$.

31. $|\mathbf{F}_1| = 10$ lb and $|\mathbf{F}_2| = 12$ lb.

$\mathbf{F}_1 = -|\mathbf{F}_1| \cos 45^\circ\,\mathbf{i} + |\mathbf{F}_1| \sin 45^\circ\,\mathbf{j} = -10 \cos 45^\circ\,\mathbf{i} + 10 \sin 45^\circ\,\mathbf{j}$

$= -5\sqrt{2}\,\mathbf{i} + 5\sqrt{2}\,\mathbf{j}$

$\mathbf{F}_2 = |\mathbf{F}_2| \cos 30^\circ\,\mathbf{i} + |\mathbf{F}_2| \sin 30^\circ\,\mathbf{j} = 12 \cos 30^\circ\,\mathbf{i} + 12 \sin 30^\circ\,\mathbf{j} = 6\sqrt{3}\,\mathbf{i} + 6\,\mathbf{j}$

$\mathbf{F} = \mathbf{F}_1 + \mathbf{F}_2 = \left(6\sqrt{3} - 5\sqrt{2}\right)\mathbf{i} + \left(6 + 5\sqrt{2}\right)\mathbf{j} \approx 3.32\mathbf{i} + 13.07\mathbf{j}$

$|\mathbf{F}| \approx \sqrt{(3.32)^2 + (13.07)^2} \approx 13.5$ lb. $\tan\theta = \dfrac{6 + 5\sqrt{2}}{6\sqrt{3} - 5\sqrt{2}} \quad \Rightarrow \quad \theta = \tan^{-1}\dfrac{6 + 5\sqrt{2}}{6\sqrt{3} - 5\sqrt{2}} \approx 76^\circ.$

33. With respect to the water's surface, the woman's velocity is the vector sum of the velocity of the ship with respect to the water, and the woman's velocity with respect to the ship. If we let north be the positive y-direction, then $\mathbf{v} = \langle 0, 22 \rangle + \langle -3, 0 \rangle = \langle -3, 22 \rangle$. The woman's speed is $|\mathbf{v}| = \sqrt{9 + 484} \approx 22.2$ mi/h. The vector $\mathbf{v}$ makes an angle θ with the east, where $\theta = \tan^{-1}\frac{22}{-3} \approx 98^\circ$. Therefore, the woman's direction is about $N(98 - 90)^\circ\,W = N\,8^\circ\,W$.

35. Let $\mathbf{T}_1$ and $\mathbf{T}_2$ represent the tension vectors in each side of the clothesline as shown in the figure. $\mathbf{T}_1$ and $\mathbf{T}_2$ have equal vertical components and opposite horizontal components, so we can write $\mathbf{T}_1 = -a\,\mathbf{i} + b\,\mathbf{j}$ and $\mathbf{T}_2 = a\,\mathbf{i} + b\,\mathbf{j}\,(a, b > 0)$.

By similar triangles, $\dfrac{b}{a} = \dfrac{0.08}{4} \quad \Rightarrow \quad a = 50b$. The force due to gravity acting on the shirt has magnitude $0.8g \approx (0.8)(9.8) = 7.84$ N, hence we have $\mathbf{w} = -7.84\,\mathbf{j}$. The resultant $\mathbf{T}_1 + \mathbf{T}_2$ of the tensile forces counterbalances $\mathbf{w}$, so $\mathbf{T}_1 + \mathbf{T}_2 = -\mathbf{w} \quad \Rightarrow \quad (-a\,\mathbf{i} + b\,\mathbf{j}) + (a\,\mathbf{i} + b\,\mathbf{j}) = 7.84\,\mathbf{j} \quad \Rightarrow$ $(-50b\,\mathbf{i} + b\,\mathbf{j}) + (50b\,\mathbf{i} + b\,\mathbf{j}) = 2b\,\mathbf{j} = 7.84\,\mathbf{j} \quad \Rightarrow \quad b = \frac{7.84}{2} = 3.92$ and $a = 50b = 196$. Thus the tensions are $\mathbf{T}_1 = -a\,\mathbf{i} + b\,\mathbf{j} = -196\,\mathbf{i} + 3.92\,\mathbf{j}$ and $\mathbf{T}_2 = a\,\mathbf{i} + b\,\mathbf{j} = 196\,\mathbf{i} + 3.92\,\mathbf{j}$.

Alternatively, we can find the value of θ and proceed as in Example 5.

37. $|\mathbf{r} - \mathbf{r}_0|$ is the distance between the points (x, y, z) and (x_0, y_0, z_0), so the set of points is a sphere with radius 1 and center (x_0, y_0, z_0).

Alternate Method: $|\mathbf{r} - \mathbf{r}_0| = 1 \quad \Leftrightarrow \quad \sqrt{(x - x_0)^2 + (y - y_0)^2 + (z - z_0)^2} = 1 \quad \Leftrightarrow$

$(x - x_0)^2 + (y - y_0)^2 + (z - z_0)^2 = 1$, which is the equation of a sphere with radius 1 and center (x_0, y_0, z_0).

39. $\mathbf{a} + (\mathbf{b} + \mathbf{c}) = \langle a_1, a_2 \rangle + (\langle b_1, b_2 \rangle + \langle c_1, c_2 \rangle) = \langle a_1, a_2 \rangle + \langle b_1 + c_1, b_2 + c_2 \rangle$

$\qquad = \langle a_1 + b_1 + c_1, a_2 + b_2 + c_2 \rangle = \langle (a_1 + b_1) + c_1, (a_2 + b_2) + c_2 \rangle$

$\qquad = \langle a_1 + b_1, a_2 + b_2 \rangle + \langle c_1, c_2 \rangle = (\langle a_1, a_2 \rangle + \langle b_1, b_2 \rangle) + \langle c_1, c_2 \rangle$

$\qquad = (\mathbf{a} + \mathbf{b}) + \mathbf{c}$

41. Consider triangle ABC, where D and E are the midpoints of AB and BC. We know that $\overrightarrow{AB} + \overrightarrow{BC} = \overrightarrow{AC}$ (1)

and $\overrightarrow{DB} + \overrightarrow{BE} = \overrightarrow{DE}$ (2). However, $\overrightarrow{DB} = \frac{1}{2}\overrightarrow{AB}$, and $\overrightarrow{BE} = \frac{1}{2}\overrightarrow{BC}$. Substituting these expressions for $\overrightarrow{DB}$ and

$\overrightarrow{BE}$ into (2) gives $\frac{1}{2}\overrightarrow{AB} + \frac{1}{2}\overrightarrow{BC} = \overrightarrow{DE}$. Comparing this with (1) gives $\overrightarrow{DE} = \frac{1}{2}\overrightarrow{AC}$. Therefore $\overrightarrow{AC}$ and $\overrightarrow{DE}$ are

parallel and $\left|\overrightarrow{DE}\right| = \frac{1}{2}\left|\overrightarrow{AC}\right|$.

13.3 The Dot Product   ET 12.3

1. (a) $\mathbf{a} \cdot \mathbf{b}$ is a scalar, and the dot product is defined only for vectors, so $(\mathbf{a} \cdot \mathbf{b}) \cdot \mathbf{c}$ has no meaning.

(b) $(\mathbf{a} \cdot \mathbf{b})\,\mathbf{c}$ is a scalar multiple of a vector, so it does have meaning.

(c) Both $|\mathbf{a}|$ and $\mathbf{b} \cdot \mathbf{c}$ are scalars, so $|\mathbf{a}|\,(\mathbf{b} \cdot \mathbf{c})$ is an ordinary product of real numbers, and has meaning.

(d) Both $\mathbf{a}$ and $\mathbf{b} + \mathbf{c}$ are vectors, so the dot product $\mathbf{a} \cdot (\mathbf{b} + \mathbf{c})$ has meaning.

(e) $\mathbf{a} \cdot \mathbf{b}$ is a scalar, but $\mathbf{c}$ is a vector, and so the two quantities cannot be added and this expression has no meaning.

(f) $|\mathbf{a}|$ is a scalar, and the dot product is defined only for vectors, so $|\mathbf{a}| \cdot (\mathbf{b} + \mathbf{c})$ has no meaning.

3. $\mathbf{a} \cdot \mathbf{b} = \langle 4, -1 \rangle \cdot \langle 3, 6 \rangle = (4)(3) + (-1)(6) = 6$

5. $\mathbf{a} \cdot \mathbf{b} = \langle 5, 0, -2 \rangle \cdot \langle 3, -1, 10 \rangle = (5)(3) + (0)(-1) + (-2)(10) = -5$

7. $\mathbf{a} \cdot \mathbf{b} = (\mathbf{i} - 2\mathbf{j} + 3\mathbf{k}) \cdot (5\mathbf{i} + 9\mathbf{k}) = (1)(5) + (-2)(0) + (3)(9) = 32$

9. Use Theorem 3: $\mathbf{a} \cdot \mathbf{b} = |\mathbf{a}|\,|\mathbf{b}|\cos\theta = (12)(15)\cos\frac{\pi}{6} = 180 \cdot \frac{\sqrt{3}}{2} = 90\sqrt{3} \approx 155.9$

11. $\mathbf{u}$, $\mathbf{v}$, and $\mathbf{w}$ are all unit vectors, so the triangle is an equilateral triangle. Thus the angle between $\mathbf{u}$ and $\mathbf{v}$ is $60°$ and

$\mathbf{u} \cdot \mathbf{v} = |\mathbf{u}|\,|\mathbf{v}|\cos 60° = (1)(1)\left(\frac{1}{2}\right) = \frac{1}{2}$. If $\mathbf{w}$ is moved so it has the same initial point as $\mathbf{u}$, we can see that the

angle between them is $120°$ and we have $\mathbf{u} \cdot \mathbf{w} = |\mathbf{u}|\,|\mathbf{w}|\cos 120° = (1)(1)\left(-\frac{1}{2}\right) = -\frac{1}{2}$.

13. (a) $\mathbf{i} \cdot \mathbf{j} = \langle 1, 0, 0 \rangle \cdot \langle 0, 1, 0 \rangle = (1)(0) + (0)(1) + (0)(0) = 0$. Similarly $\mathbf{j} \cdot \mathbf{k} = (0)(0) + (1)(0) + (0)(1) = 0$

and $\mathbf{k} \cdot \mathbf{i} = (0)(1) + (0)(0) + (1)(0) = 0$.

Another Method: Because $\mathbf{i}$, $\mathbf{j}$, and $\mathbf{k}$ are mutually perpendicular, the cosine factor in each dot product is

$\cos\frac{\pi}{2} = 0$.

(b) By Property #1 of the Dot Product, $\mathbf{i} \cdot \mathbf{i} = |\mathbf{i}|^2 = 1^2 = 1$ since $\mathbf{i}$ is a unit vector. Similarly, $\mathbf{j} \cdot \mathbf{j} = |\mathbf{j}|^2 = 1$ and

$\mathbf{k} \cdot \mathbf{k} = |\mathbf{k}|^2 = 1$.

15. $|\mathbf{a}| = \sqrt{3^2 + 4^2} = 5$, $|\mathbf{b}| = \sqrt{5^2 + 12^2} = 13$, and $\mathbf{a} \cdot \mathbf{b} = (3)(5) + (4)(12) = 63$. Using Corollary 6, we have

$\cos\theta = \dfrac{\mathbf{a} \cdot \mathbf{b}}{|\mathbf{a}|\,|\mathbf{b}|} = \dfrac{63}{5 \cdot 13} = \dfrac{63}{65}$. So the angle between $\mathbf{a}$ and $\mathbf{b}$ is $\theta = \cos^{-1}\frac{63}{65} \approx 14°$.

17. $|\mathbf{a}| = \sqrt{1^2 + 2^2 + 3^2} = \sqrt{14}$, $|\mathbf{b}| = \sqrt{4^2 + 0^2 + (-1)^2} = \sqrt{17}$, and $\mathbf{a} \cdot \mathbf{b} = (1)(4) + (2)(0) + (3)(-1) = 1$.

Then $\cos\theta = \dfrac{\mathbf{a} \cdot \mathbf{b}}{|\mathbf{a}|\,|\mathbf{b}|} = \dfrac{1}{\sqrt{14} \cdot \sqrt{17}} = \dfrac{1}{\sqrt{238}}$ and the angle between $\mathbf{a}$ and $\mathbf{b}$ is $\theta = \cos^{-1}\frac{1}{\sqrt{238}} \approx 86°$.

19. $|\mathbf{a}| = \sqrt{0^2 + 1^2 + 1^2} = \sqrt{2}$, $|\mathbf{b}| = \sqrt{1^2 + 2^2 + (-3)^2} = \sqrt{14}$, and $\mathbf{a} \cdot \mathbf{b} = (0)(1) + (1)(2) + (1)(-3) = -1$.

Then $\cos\theta = \dfrac{\mathbf{a} \cdot \mathbf{b}}{|\mathbf{a}|\,|\mathbf{b}|} = \dfrac{-1}{\sqrt{2} \cdot \sqrt{14}} = \dfrac{-1}{2\sqrt{7}}$ and $\theta = \cos^{-1}\left(-\frac{1}{2\sqrt{7}}\right) \approx 101°$.

21. Let a, b and c be the angles at vertices A, B and C respectively. Then a is the angle between vectors $\overrightarrow{AB}$ and $\overrightarrow{AC}$, b is the angle between vectors $\overrightarrow{BA}$ and $\overrightarrow{BC}$, and c is the angle between vectors $\overrightarrow{CA}$ and $\overrightarrow{CB}$.

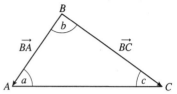

Thus

$$\cos a = \frac{\overrightarrow{AB} \cdot \overrightarrow{AC}}{\left|\overrightarrow{AB}\right|\left|\overrightarrow{AC}\right|} = \frac{1}{\sqrt{30 \cdot 29}}\langle 5, -1, 2\rangle \cdot \langle -2, -4, -3\rangle = \frac{1}{\sqrt{870}}(-10 + 4 - 6) = -\frac{12}{\sqrt{870}} \text{ and}$$

$a = \cos^{-1}\left(-\dfrac{12}{\sqrt{870}}\right) \approx 114°$. Similarly

$$\cos b = \frac{\overrightarrow{BA} \cdot \overrightarrow{BC}}{\left|\overrightarrow{BA}\right|\left|\overrightarrow{BC}\right|} = \frac{1}{\sqrt{30 \cdot 83}}\langle 5, -1, 2\rangle \cdot \langle -7, -3, -5\rangle = \frac{1}{\sqrt{2490}}(35 - 3 + 10) = \frac{42}{\sqrt{2490}}, \text{ so}$$

$b = \cos^{-1}\dfrac{42}{\sqrt{2490}} \approx 33°$, and

$$\cos c = \frac{\overrightarrow{CA} \cdot \overrightarrow{CB}}{\left|\overrightarrow{CA}\right|\left|\overrightarrow{CB}\right|} = \frac{1}{\sqrt{29 \cdot 83}}\langle 2, 4, 3\rangle \cdot \langle 7, 3, 5\rangle = \frac{1}{\sqrt{2407}}(14 + 12 + 15) = \frac{41}{\sqrt{2407}}, \text{ so}$$

$c = \cos^{-1}\dfrac{41}{\sqrt{2407}} \approx 33°$.

Alternate Solution: Apply the Law of Cosines three times as follows: $\cos a = \dfrac{\left|\overrightarrow{BC}\right|^2 - \left|\overrightarrow{AB}\right|^2 - \left|\overrightarrow{AC}\right|^2}{2\left|\overrightarrow{AB}\right|\left|\overrightarrow{AC}\right|}$,

$\cos b = \dfrac{\left|\overrightarrow{AC}\right|^2 - \left|\overrightarrow{AB}\right|^2 - \left|\overrightarrow{BC}\right|^2}{2\left|\overrightarrow{AB}\right|\left|\overrightarrow{BC}\right|}$, and $\cos c = \dfrac{\left|\overrightarrow{AB}\right|^2 - \left|\overrightarrow{AC}\right|^2 - \left|\overrightarrow{BC}\right|^2}{2\left|\overrightarrow{AC}\right|\left|\overrightarrow{BC}\right|}$.

23. Since $\mathbf{a} = -2\mathbf{b}$, $\mathbf{a}$ and $\mathbf{b}$ are parallel vectors (and thus not orthogonal).

25. $\mathbf{a} \cdot \mathbf{b} = (-5)(6) + (3)(-8) + (7)(2) = -40 \neq 0$, so $\mathbf{a}$ and $\mathbf{b}$ are not orthogonal. Also, since $\mathbf{a}$ is not a scalar multiple of $\mathbf{b}$, $\mathbf{a}$ and $\mathbf{b}$ are not parallel.

27. $\mathbf{a} \cdot \mathbf{b} = (-1)(3) + (2)(4) + (5)(-1) = 0$, so $\mathbf{a}$ and $\mathbf{b}$ are orthogonal (and not parallel).

29. For the two vectors to be orthogonal, we need $\langle -3x, 2x\rangle \cdot \langle 4, x\rangle = 0$ $\Leftrightarrow$ $(-3x)(4) + (2x)(x) = 0$ $\Leftrightarrow$ $-12x + 2x^2 = 0$ $\Leftrightarrow$ $2x(x - 6) = 0$ $\Leftrightarrow$ $x = 0$ or $x = 6$.

31. Let $\mathbf{a} = a_1\mathbf{i} + a_2\mathbf{j} + a_3\mathbf{k}$ be a vector orthogonal to both $\mathbf{i} + \mathbf{j}$ and $\mathbf{i} + \mathbf{k}$. Then $\mathbf{a} \cdot (\mathbf{i} + \mathbf{j}) = 0$ $\Leftrightarrow$ $a_1 + a_2 = 0$ and $\mathbf{a} \cdot (\mathbf{i} + \mathbf{k}) = 0$ $\Leftrightarrow$ $a_1 + a_3 = 0$, so $a_1 = -a_2 = -a_3$. Furthermore $\mathbf{a}$ is to be a unit vector, so $1 = a_1^2 + a_2^2 + a_3^2 = 3a_1^2$ implies $a_1 = \pm\frac{1}{\sqrt{3}}$. Thus $\mathbf{a} = \frac{1}{\sqrt{3}}\mathbf{i} - \frac{1}{\sqrt{3}}\mathbf{j} - \frac{1}{\sqrt{3}}\mathbf{k}$ and $\mathbf{a} = -\frac{1}{\sqrt{3}}\mathbf{i} + \frac{1}{\sqrt{3}}\mathbf{j} + \frac{1}{\sqrt{3}}\mathbf{k}$ are two such unit vectors.

33. Since $|\langle 1, 2, 2\rangle| = \sqrt{1 + 4 + 4} = 3$, using (8) and (9) we have $\cos\alpha = \frac{1}{3}$, $\cos\beta = \frac{2}{3}$ and $\cos\gamma = \frac{2}{3}$, while $\alpha = \cos^{-1}\frac{1}{3} \approx 71°$ and $\beta = \gamma = \cos^{-1}\frac{2}{3} \approx 48°$.

35. $|-8\mathbf{i} + 3\mathbf{j} + 2\mathbf{k}| = \sqrt{64 + 9 + 4} = \sqrt{77}$, so $\cos\alpha = -\frac{8}{\sqrt{77}}$, $\cos\beta = \frac{3}{\sqrt{77}}$ and $\cos\gamma = \frac{2}{\sqrt{77}}$, while

$\alpha = \cos^{-1}\frac{-8}{\sqrt{77}} \approx 156°$, $\beta = \cos^{-1}\frac{3}{\sqrt{77}} \approx 70°$ and $\gamma = \cos^{-1}\frac{2}{\sqrt{77}} \approx 77°$.

37. $|\langle 2, 1.2, 0.8\rangle| = \sqrt{4 + 1.44 + 0.64} = \frac{5}{5}\sqrt{6.08} = \frac{\sqrt{152}}{5}$, so $\cos\alpha = \frac{10}{\sqrt{152}} = \frac{5}{\sqrt{38}}$, $\cos\beta = \frac{6}{\sqrt{152}} = \frac{3}{\sqrt{38}}$ and

$\cos\gamma = \frac{4}{\sqrt{152}} = \frac{2}{\sqrt{38}}$, while $\alpha = \cos^{-1}\frac{5}{\sqrt{38}} \approx 36°$, $\beta = \cos^{-1}\frac{3}{\sqrt{38}} \approx 61°$ and $\gamma = \cos^{-1}\frac{2}{\sqrt{38}} \approx 71°$.

39. $|\mathbf{a}| = \sqrt{4 + 9} = \sqrt{13}$. The scalar projection of $\mathbf{b}$ onto $\mathbf{a}$ is $\text{comp}_{\mathbf{a}}\,\mathbf{b} = \dfrac{\mathbf{a}\cdot\mathbf{b}}{|\mathbf{a}|} = \dfrac{2\cdot4 + 3\cdot1}{\sqrt{13}} = \dfrac{11}{\sqrt{13}}$.

The vector projection of $\mathbf{b}$ onto $\mathbf{a}$ is $\text{proj}_{\mathbf{a}}\,\mathbf{b} = \dfrac{11}{\sqrt{13}}\dfrac{\mathbf{a}}{|\mathbf{a}|} = \dfrac{11}{\sqrt{13}}\cdot\dfrac{1}{\sqrt{13}}\langle 2, 3\rangle = \dfrac{11}{13}\langle 2, 3\rangle = \left\langle\dfrac{22}{13}, \dfrac{33}{13}\right\rangle$.

41. $|\mathbf{a}| = \sqrt{16 + 4 + 0} = 2\sqrt{5}$ so the scalar projection of $\mathbf{b}$ onto $\mathbf{a}$ is $\text{comp}_{\mathbf{a}}\mathbf{b} = \dfrac{\mathbf{a}\cdot\mathbf{b}}{|\mathbf{a}|} = \dfrac{1}{2\sqrt{5}}(4 + 2 + 0) = \dfrac{3}{\sqrt{5}}$.

The vector projection of $\mathbf{b}$ onto $\mathbf{a}$ is $\text{proj}_{\mathbf{a}}\mathbf{b} = \dfrac{3}{\sqrt{5}}\dfrac{\mathbf{a}}{|\mathbf{a}|} = \dfrac{3}{\sqrt{5}}\cdot\dfrac{1}{2\sqrt{5}}\langle 4, 2, 0\rangle = \dfrac{1}{5}\langle 6, 3, 0\rangle = \left\langle\dfrac{6}{5}, \dfrac{3}{5}, 0\right\rangle$.

43. $|\mathbf{a}| = \sqrt{1 + 0 + 1} = \sqrt{2}$ so the scalar projection of $\mathbf{b}$ onto $\mathbf{a}$ is $\text{comp}_{\mathbf{a}}\,\mathbf{b} = \dfrac{\mathbf{a}\cdot\mathbf{b}}{|\mathbf{a}|} = \dfrac{1}{\sqrt{2}}(1 + 0 + 0) = \dfrac{1}{\sqrt{2}}$ while

the vector projection of $\mathbf{b}$ onto $\mathbf{a}$ is $\text{proj}_{\mathbf{a}}\,\mathbf{b} = \dfrac{1}{\sqrt{2}}\dfrac{\mathbf{a}}{|\mathbf{a}|} = \dfrac{1}{\sqrt{2}}\cdot\dfrac{1}{\sqrt{2}}(\mathbf{i} + \mathbf{k}) = \dfrac{1}{2}(\mathbf{i} + \mathbf{k})$.

45. $(\text{orth}_{\mathbf{a}}\,\mathbf{b})\cdot\mathbf{a} = (\mathbf{b} - \text{proj}_{\mathbf{a}}\,\mathbf{b})\cdot\mathbf{a} = \mathbf{b}\cdot\mathbf{a} - (\text{proj}_{\mathbf{a}}\,\mathbf{b})\cdot\mathbf{a} = \mathbf{b}\cdot\mathbf{a} - \dfrac{\mathbf{a}\cdot\mathbf{b}}{|\mathbf{a}|^2}\mathbf{a}\cdot\mathbf{a}$

$= \mathbf{b}\cdot\mathbf{a} - \dfrac{\mathbf{a}\cdot\mathbf{b}}{|\mathbf{a}|^2}|\mathbf{a}|^2 = \mathbf{b}\cdot\mathbf{a} - \mathbf{a}\cdot\mathbf{b} = 0$

So they are orthogonal by (7).

47. $\text{comp}_{\mathbf{a}}\,\mathbf{b} = \dfrac{\mathbf{a}\cdot\mathbf{b}}{|\mathbf{a}|} = 2 \iff \mathbf{a}\cdot\mathbf{b} = 2\,|\mathbf{a}| = 2\sqrt{10}$. If $\mathbf{b} = \langle b_1, b_2, b_3\rangle$, then we need $3b_1 + 0b_2 - 1b_3 = 2\sqrt{10}$.

One possible solution is obtained by taking $b_1 = 0$, $b_2 = 0$, $b_3 = -2\sqrt{10}$.

In general, $\mathbf{b} = \langle s, t, 3s - 2\sqrt{10}\rangle$, $s, t \in \mathbb{R}$.

49. Here $\mathbf{D} = (4 - 2)\,\mathbf{i} + (9 - 3)\,\mathbf{j} + (15 - 0)\,\mathbf{k} = 2\mathbf{i} + 6\mathbf{j} + 15\mathbf{k}$ so using (12) we have

$W = \mathbf{F}\cdot\mathbf{D} = 20 + 108 - 90 = 38$ joules.

51. $W = |\mathbf{F}|\,|\mathbf{D}|\cos\theta = (25)\,(10)\cos 20°$

≈ 235 ft-lb

53. First note that $\mathbf{n} = \langle a, b\rangle$ is perpendicular to the line, because if $Q_1 = (a_1, b_1)$ and $Q_2 = (a_2, b_2)$ lie on the line,

then $\mathbf{n}\cdot\overrightarrow{Q_1Q_2} = aa_2 - aa_1 + bb_2 - bb_1 = 0$, since $aa_2 + bb_2 = -c = aa_1 + bb_1$ from the equation of the line.

Let $P_2 = (x_2, y_2)$ lie on the line. Then the distance from P_1 to the line is the absolute value of the scalar projection

of $\overrightarrow{P_1P_2}$ onto $\mathbf{n}$. $\text{comp}_{\mathbf{n}}\left(\overrightarrow{P_1P_2}\right) = \dfrac{|\mathbf{n}\cdot\langle x_2 - x_1, y_2 - y_1\rangle|}{|\mathbf{n}|} = \dfrac{|ax_2 - ax_1 + by_2 - by_1|}{\sqrt{a^2 + b^2}} = \dfrac{|ax_1 + by_1 + c|}{\sqrt{a^2 + b^2}}$

since $ax_2 + by_2 = -c$. The required distance is $\dfrac{|3\cdot-2 + -4\cdot3 + 5|}{\sqrt{3^2 + 4^2}} = \dfrac{13}{5}$.

55. For convenience, consider the unit cube positioned so that its back left corner is at the origin, and its edges lie along the coordinate axes. The diagonal of the cube that begins at the origin and ends at $(1, 1, 1)$ has vector representation $\langle 1, 1, 1 \rangle$. The angle θ between this vector and the vector of the edge which also begins at the origin and runs along the x-axis [that is, $\langle 1, 0, 0 \rangle$] is given by $\cos \theta = \dfrac{\langle 1,1,1 \rangle \cdot \langle 1,0,0 \rangle}{|\langle 1,1,1 \rangle| \, |\langle 1,0,0 \rangle|} = \dfrac{1}{\sqrt{3}}$ $\Rightarrow$ $\theta = \cos^{-1} \frac{1}{\sqrt{3}} \approx 55\,°$.

57. Consider the H-C-H combination consisting of the sole carbon atom and the two hydrogen atoms that are at $(1, 0, 0)$ and $(0, 1, 0)$ (or any H-C-H combination, for that matter). Vector representations of the line segments emanating from the carbon atom and extending to these two hydrogen atoms are $\langle 1 - \frac{1}{2}, 0 - \frac{1}{2}, 0 - \frac{1}{2} \rangle = \langle \frac{1}{2}, -\frac{1}{2}, -\frac{1}{2} \rangle$ and $\langle 0 - \frac{1}{2}, 1 - \frac{1}{2}, 0 - \frac{1}{2} \rangle = \langle -\frac{1}{2}, \frac{1}{2}, -\frac{1}{2} \rangle$. The bond angle, θ, is therefore given by

$$\cos \theta = \frac{\langle \frac{1}{2}, -\frac{1}{2}, -\frac{1}{2} \rangle \cdot \langle -\frac{1}{2}, \frac{1}{2}, -\frac{1}{2} \rangle}{|\langle \frac{1}{2}, -\frac{1}{2}, -\frac{1}{2} \rangle| \, |\langle -\frac{1}{2}, \frac{1}{2}, -\frac{1}{2} \rangle|} = \frac{-\frac{1}{4} - \frac{1}{4} + \frac{1}{4}}{\sqrt{\frac{3}{4}}\sqrt{\frac{3}{4}}} = -\frac{1}{3} \quad \Rightarrow \quad \theta = \cos^{-1}\left(-\frac{1}{3}\right) \approx 109.5\,°.$$

59. Let $\mathbf{a} = \langle a_1, a_2, a_3 \rangle$ and $= \langle b_1, b_2, b_3 \rangle$.

Property 2: $\mathbf{a} \cdot \mathbf{b} = \langle a_1, a_2, a_3 \rangle \cdot \langle b_1, b_2, b_3 \rangle = a_1 b_1 + a_2 b_2 + a_3 b_3$

$\qquad\qquad = b_1 a_1 + b_2 a_2 + b_3 a_3 = \langle b_1, b_2, b_3 \rangle \cdot \langle a_1, a_2, a_3 \rangle = \mathbf{b} \cdot \mathbf{a}$

Property 4: $(c\mathbf{a}) \cdot \mathbf{b} = \langle ca_1, ca_2, ca_3 \rangle \cdot \langle b_1, b_2, b_3 \rangle = (ca_1) \, b_1 + (ca_2) \, b_2 + (ca_3) \, b_3$

$\qquad\qquad = c\,(a_1 b_1 + a_2 b_2 + a_3 b_3) = c\,(\mathbf{a} \cdot \mathbf{b}) = a_1\,(cb_1) + a_2\,(cb_2) + a_3\,(cb_3)$

$\qquad\qquad = \langle a_1, a_2, a_3 \rangle \cdot \langle cb_1, cb_2, cb_3 \rangle = \mathbf{a} \cdot (c\mathbf{b})$

Property 5: $\mathbf{0} \cdot \mathbf{a} = \langle 0, 0, 0 \rangle \cdot \langle a_1, a_2, a_3 \rangle = (0)\,(a_1) + (0)\,(a_2) + (0)\,(a_3) = 0$

61. $|\mathbf{a} \cdot \mathbf{b}| = \big|\, |\mathbf{a}|\,|\mathbf{b}| \cos \theta \,\big| = |\mathbf{a}|\,|\mathbf{b}|\,|\cos \theta|$. Since $|\cos \theta| \leq 1$, $|\mathbf{a} \cdot \mathbf{b}| = |\mathbf{a}|\,|\mathbf{b}|\,|\cos \theta| \leq |\mathbf{a}|\,|\mathbf{b}|$.

Note: We have equality in the case of $\cos \theta = \pm 1$, so $\theta = 0$ or $\theta = \pi$, thus equality when $\mathbf{a}$ and $\mathbf{b}$ are parallel.

63. (a)

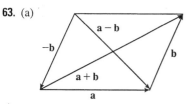

The Parallelogram Law states that the sum of the squares of the lengths of the diagonals of a parallelogram equals the sum of the squares of its (four) sides.

(b) $|\mathbf{a} + \mathbf{b}|^2 = (\mathbf{a} + \mathbf{b}) \cdot (\mathbf{a} + \mathbf{b}) = |\mathbf{a}|^2 + 2\,(\mathbf{a} \cdot \mathbf{b}) + |\mathbf{b}|^2$ and

$|\mathbf{a} - \mathbf{b}|^2 = (\mathbf{a} - \mathbf{b}) \cdot (\mathbf{a} - \mathbf{b}) = |\mathbf{a}|^2 - 2\,(\mathbf{a} \cdot \mathbf{b}) + |\mathbf{b}|^2$. Adding these two equations gives

$|\mathbf{a} + \mathbf{b}|^2 + |\mathbf{a} - \mathbf{b}|^2 = 2\,|\mathbf{a}|^2 + 2\,|\mathbf{b}|^2$.

13.4 The Cross Product　　　　　　　　　　　　　　　　　　　　　ET 12.4

1. $\mathbf{a} \times \mathbf{b} = \begin{vmatrix} \mathbf{i} & \mathbf{j} & \mathbf{k} \\ 1 & 2 & 0 \\ 0 & 3 & 1 \end{vmatrix} = \begin{vmatrix} 2 & 0 \\ 3 & 1 \end{vmatrix} \mathbf{i} - \begin{vmatrix} 1 & 0 \\ 0 & 1 \end{vmatrix} \mathbf{j} + \begin{vmatrix} 1 & 2 \\ 0 & 3 \end{vmatrix} \mathbf{k} = (2 - 0)\,\mathbf{i} - (1 - 0)\,\mathbf{j} + (3 - 0)\,\mathbf{k} = 2\mathbf{i} - \mathbf{j} + 3\mathbf{k}$

3. $\mathbf{a} \times \mathbf{b} = \begin{vmatrix} \mathbf{i} & \mathbf{j} & \mathbf{k} \\ 1 & -1 & 0 \\ 3 & 2 & 1 \end{vmatrix} = \begin{vmatrix} -1 & 0 \\ 2 & 1 \end{vmatrix} \mathbf{i} - \begin{vmatrix} 1 & 0 \\ 3 & 1 \end{vmatrix} \mathbf{j} + \begin{vmatrix} 1 & -1 \\ 3 & 2 \end{vmatrix} \mathbf{k} = (-1 - 0)\,\mathbf{i} - (1 - 0)\,\mathbf{j} + [2 - (-3)]\,\mathbf{k} = -\mathbf{i} - \mathbf{j} + 5\mathbf{k}$

5. $\mathbf{a} \times \mathbf{b} = \begin{vmatrix} \mathbf{i} & \mathbf{j} & \mathbf{k} \\ 2 & 1 & -1 \\ 0 & 1 & 2 \end{vmatrix} = \begin{vmatrix} 1 & -1 \\ 1 & 2 \end{vmatrix} \mathbf{i} - \begin{vmatrix} 2 & -1 \\ 0 & 2 \end{vmatrix} \mathbf{j} + \begin{vmatrix} 2 & 1 \\ 0 & 1 \end{vmatrix} \mathbf{k} = [2 - (-1)]\,\mathbf{i} - (4 - 0)\,\mathbf{j} + (2 - 0)\,\mathbf{k} = 3\mathbf{i} - 4\mathbf{j} + 2\mathbf{k}$

7. $\mathbf{a} \times \mathbf{b} = \begin{vmatrix} \mathbf{i} & \mathbf{j} & \mathbf{k} \\ 3 & 2 & 4 \\ 1 & -2 & -3 \end{vmatrix} = \begin{vmatrix} 2 & 4 \\ -2 & -3 \end{vmatrix} \mathbf{i} - \begin{vmatrix} 3 & 4 \\ 1 & -3 \end{vmatrix} \mathbf{j} + \begin{vmatrix} 3 & 2 \\ 1 & -2 \end{vmatrix} \mathbf{k}$

$= [-6 - (-8)]\,\mathbf{i} - (-9 - 4)\,\mathbf{j} + (-6 - 2)\,\mathbf{k} = 2\mathbf{i} + 13\mathbf{j} - 8\mathbf{k}$

9. (a) Since $\mathbf{b} \times \mathbf{c}$ is a vector, the dot product $\mathbf{a} \cdot (\mathbf{b} \times \mathbf{c})$ is meaningful and is a scalar.

(b) $\mathbf{b} \cdot \mathbf{c}$ is a scalar, so $\mathbf{a} \times (\mathbf{b} \cdot \mathbf{c})$ is meaningless, as the cross product is defined only for two *vectors*.

(c) Since $\mathbf{b} \times \mathbf{c}$ is a vector, the cross product $\mathbf{a} \times (\mathbf{b} \times \mathbf{c})$ is meaningful and results in another vector.

(d) $\mathbf{a} \cdot \mathbf{b}$ is a scalar, so the cross product $(\mathbf{a} \cdot \mathbf{b}) \times \mathbf{c}$ is meaningless.

(e) Since $(\mathbf{a} \cdot \mathbf{b})$ and $(\mathbf{c} \cdot \mathbf{d})$ are both scalars, the cross product $(\mathbf{a} \cdot \mathbf{b}) \times (\mathbf{c} \cdot \mathbf{d})$ is meaningless.

(f) $\mathbf{a} \times \mathbf{b}$ and $\mathbf{c} \times \mathbf{d}$ are both vectors, so the dot product $(\mathbf{a} \times \mathbf{b}) \cdot (\mathbf{c} \times \mathbf{d})$ is meaningful and is a scalar.

11. If we sketch $\mathbf{u}$ and $\mathbf{v}$ starting from the same initial point, we see that the angle between them is $30\,°$. Using Theorem 6, we have $|\mathbf{u} \times \mathbf{v}| = |\mathbf{u}|\,|\mathbf{v}| \sin 30\,° = (6)\,(8)\left(\frac{1}{2}\right) = 24$. By the right-hand rule, $\mathbf{u} \times \mathbf{v}$ is directed into the page.

13. $\mathbf{a} \times \mathbf{b} = \begin{vmatrix} \mathbf{i} & \mathbf{j} & \mathbf{k} \\ 0 & 1 & 2 \\ 3 & 1 & 0 \end{vmatrix} = \begin{vmatrix} 1 & 2 \\ 1 & 0 \end{vmatrix} \mathbf{i} - \begin{vmatrix} 0 & 2 \\ 3 & 0 \end{vmatrix} \mathbf{j} + \begin{vmatrix} 0 & 1 \\ 3 & 1 \end{vmatrix} \mathbf{k} = -2\mathbf{i} + 6\mathbf{j} - 3\mathbf{k}$

$\mathbf{b} \times \mathbf{a} = \begin{vmatrix} \mathbf{i} & \mathbf{j} & \mathbf{k} \\ 3 & 1 & 0 \\ 0 & 1 & 2 \end{vmatrix} = \begin{vmatrix} 1 & 0 \\ 1 & 2 \end{vmatrix} \mathbf{i} - \begin{vmatrix} 3 & 0 \\ 0 & 2 \end{vmatrix} \mathbf{j} + \begin{vmatrix} 3 & 1 \\ 0 & 1 \end{vmatrix} \mathbf{k} = 2\mathbf{i} - 6\mathbf{j} + 3\mathbf{k}$

(and notice $\mathbf{a} \times \mathbf{b} = -\mathbf{b} \times \mathbf{a}$ here, as we know is always true by Theorem 8.)

15. We know that the cross product of two vectors is orthogonal to both. So we calculate

$\langle 1, -1, 1 \rangle \times \langle 0, 4, 4 \rangle = \begin{vmatrix} \mathbf{i} & \mathbf{j} & \mathbf{k} \\ 1 & -1 & 1 \\ 0 & 4 & 4 \end{vmatrix} = \begin{vmatrix} -1 & 1 \\ 4 & 4 \end{vmatrix} \mathbf{i} - \begin{vmatrix} 1 & 1 \\ 0 & 4 \end{vmatrix} \mathbf{j} + \begin{vmatrix} 1 & -1 \\ 0 & 4 \end{vmatrix} \mathbf{k} = -8\mathbf{i} - 4\mathbf{j} + 4\mathbf{k}$. So two unit vectors

orthogonal to both are $\pm \dfrac{\langle -8, -4, 4 \rangle}{\sqrt{64 + 16 + 16}} = \pm \dfrac{\langle -8, -4, 4 \rangle}{4\sqrt{6}}$, that is, $\left\langle -\frac{2}{\sqrt{6}}, -\frac{1}{\sqrt{6}}, \frac{1}{\sqrt{6}} \right\rangle$ and $\left\langle \frac{2}{\sqrt{6}}, \frac{1}{\sqrt{6}}, -\frac{1}{\sqrt{6}} \right\rangle$.

17. Let $\mathbf{a} = \langle a_1, a_2, a_3 \rangle$. Then $\mathbf{0} \times \mathbf{a} = \begin{vmatrix} \mathbf{i} & \mathbf{j} & \mathbf{k} \\ 0 & 0 & 0 \\ a_1 & a_2 & a_3 \end{vmatrix} = \begin{vmatrix} 0 & 0 \\ a_2 & a_3 \end{vmatrix} \mathbf{i} - \begin{vmatrix} 0 & 0 \\ a_1 & a_3 \end{vmatrix} \mathbf{j} + \begin{vmatrix} 0 & 0 \\ a_1 & a_2 \end{vmatrix} \mathbf{k} = \mathbf{0}$,

$\mathbf{a} \times \mathbf{0} = \begin{vmatrix} \mathbf{i} & \mathbf{j} & \mathbf{k} \\ a_1 & a_2 & a_3 \\ 0 & 0 & 0 \end{vmatrix} = \begin{vmatrix} a_2 & a_3 \\ 0 & 0 \end{vmatrix} \mathbf{i} - \begin{vmatrix} a_1 & a_3 \\ 0 & 0 \end{vmatrix} \mathbf{j} + \begin{vmatrix} a_1 & a_2 \\ 0 & 0 \end{vmatrix} \mathbf{k} = \mathbf{0}$.

19. $\mathbf{a} \times \mathbf{b} = \langle a_2b_3 - a_3b_2, a_3b_1 - a_1b_3, a_1b_2 - a_2b_1 \rangle$

$$= \langle (-1)(b_2a_3 - b_3a_2), (-1)(b_3a_1 - b_1a_3), (-1)(b_1a_2 - b_2a_1) \rangle$$

$$= - \langle b_2a_3 - b_3a_2, b_3a_1 - b_1a_3, b_1a_2 - b_2a_1 \rangle = -\mathbf{b} \times \mathbf{a}$$

21. $\mathbf{a} \times (\mathbf{b} + \mathbf{c}) = \mathbf{a} \times \langle b_1 + c_1, b_2 + c_2, b_3 + c_3 \rangle$

$$= \langle a_2(b_3 + c_3) - a_3(b_2 + c_2), a_3(b_1 + c_1) - a_1(b_3 + c_3), a_1(b_2 + c_2) - a_2(b_1 + c_1) \rangle$$

$$= \langle a_2b_3 + a_2c_3 - a_3b_2 - a_3c_2, a_3b_1 + a_3c_1 - a_1b_3 - a_1c_3, a_1b_2 + a_1c_2 - a_2b_1 - a_2c_1 \rangle$$

$$= \langle (a_2b_3 - a_3b_2) + (a_2c_3 - a_3c_2), (a_3b_1 - a_1b_3) + (a_3c_1 - a_1c_3),$$

$$(a_1b_2 - a_2b_1) + (a_1c_2 - a_2c_1) \rangle$$

$$= \langle a_2b_3 - a_3b_2, a_3b_1 - a_1b_3, a_1b_2 - a_2b_1 \rangle + \langle a_2c_3 - a_3c_2, a_3c_1 - a_1c_3, a_1c_2 - a_2c_1 \rangle$$

$$= (\mathbf{a} \times \mathbf{b}) + (\mathbf{a} \times \mathbf{c})$$

23. We know that the area of the parallelogram determined by two vectors is equal to the length of the cross product of these vectors. The vectors corresponding to $\overrightarrow{AB}$ and $\overrightarrow{AD}$ are $\mathbf{a} = \langle 3, -1, 0 \rangle$ and $\mathbf{b} = \langle 2, -2, 0 \rangle$, so the area of

parallelogram $ABCD$ is $|\mathbf{a} \times \mathbf{b}| = \begin{vmatrix} \mathbf{i} & \mathbf{j} & \mathbf{k} \\ 3 & -1 & 0 \\ 2 & -2 & 0 \end{vmatrix} = |(0)\,\mathbf{i} - (0)\,\mathbf{j} + (-6 + 2)\,\mathbf{k}| = |-4\mathbf{k}| = 4.$

25. (a) Because the plane through P, Q, and R contains the vectors $\overrightarrow{PQ}$ and $\overrightarrow{PR}$, a vector orthogonal to both of these vectors (such as their cross product) is also orthogonal to the plane. Here $\overrightarrow{PQ} = \langle -1, 2, 0 \rangle$ and $\overrightarrow{PR} = \langle -1, 0, 3 \rangle$, so

$$\overrightarrow{PQ} \times \overrightarrow{PR} = \langle (2)(3) - (0)(0), (0)(-1) - (-1)(3), (-1)(0) - (2)(-1) \rangle = \langle 6, 3, 2 \rangle$$

Therefore, $\langle 6, 3, 2 \rangle$ (or any scalar multiple thereof) is orthogonal to the plane through P, Q, and R.

(b) Note that the area of the triangle determined by P, Q, and R is equal to half of the area of the parallelogram determined by the three points. From part (a), the area of the parallelogram is

$$\left| \overrightarrow{PQ} \times \overrightarrow{PR} \right| = |\langle 6, 3, 2 \rangle| = \sqrt{36 + 9 + 4} = 7, \text{ so the area of the triangle is } \tfrac{1}{2}(7) = \tfrac{7}{2}.$$

27. (a) $\overrightarrow{PQ} = \langle 1, -1, 1 \rangle$ and $\overrightarrow{PR} = \langle 4, 3, 7 \rangle$, so a vector orthogonal to the plane through P, Q, and R is

$\overrightarrow{PQ} \times \overrightarrow{PR} = \langle (-1)(7) - (1)(3), (1)(4) - (1)(7), (1)(3) - (-1)(4) \rangle = \langle -10, -3, 7 \rangle$ (or any scalar multiple thereof).

(b) The area of the parallelogram determined by $\overrightarrow{PQ}$ and $\overrightarrow{PR}$ is

$$\left| \overrightarrow{PQ} \times \overrightarrow{PR} \right| = |\langle -10, -3, 7 \rangle| = \sqrt{100 + 9 + 49} = \sqrt{158}, \text{ so the area of triangle } PQR \text{ is } \tfrac{1}{2}\sqrt{158}.$$

29. We know that the volume of the parallelepiped determined by $\mathbf{a}$, $\mathbf{b}$ and $\mathbf{c}$ is the magnitude of their scalar triple product, which is

$$\mathbf{a} \cdot (\mathbf{b} \times \mathbf{c}) = \begin{vmatrix} 1 & 0 & 6 \\ 2 & 3 & -8 \\ 8 & -5 & 6 \end{vmatrix} = 1 \begin{vmatrix} 3 & -8 \\ -5 & 6 \end{vmatrix} - 0 + 6 \begin{vmatrix} 2 & 3 \\ 8 & -5 \end{vmatrix}$$

$$= (18 - 40) + 6(-10 - 24) = -226$$

Thus the volume of the parallelepiped is $|-226| = 226$ cubic units.

31. $\mathbf{a} = \overrightarrow{PQ} = \langle 1, -1, 2 \rangle$, $\mathbf{b} = \overrightarrow{PR} = \langle 3, 0, 6 \rangle$ and $\mathbf{c} = \overrightarrow{PS} = \langle 2, -2, -3 \rangle$.

$$\mathbf{a} \cdot (\mathbf{b} \times \mathbf{c}) = \begin{vmatrix} 1 & -1 & 2 \\ 3 & 0 & 6 \\ 2 & -2 & -3 \end{vmatrix} = 1 \begin{vmatrix} 0 & 6 \\ -2 & -3 \end{vmatrix} - (-1) \begin{vmatrix} 3 & 6 \\ 2 & -3 \end{vmatrix} + 2 \begin{vmatrix} 3 & 0 \\ 2 & -2 \end{vmatrix} = 12 - 21 - 12 = -21, \text{ so the volume}$$

of the parallelepiped is 21 cubic units.

33. $\mathbf{a} \cdot (\mathbf{b} \times \mathbf{c}) = \begin{vmatrix} 2 & 3 & 1 \\ 1 & -1 & 0 \\ 7 & 3 & 2 \end{vmatrix} = 2 \begin{vmatrix} -1 & 0 \\ 3 & 2 \end{vmatrix} - 3 \begin{vmatrix} 1 & 0 \\ 7 & 2 \end{vmatrix} + 1 \begin{vmatrix} 1 & -1 \\ 7 & 3 \end{vmatrix} = -4 - 6 + 10 = 0$, which says that the volume

of the parallelepiped determined by $\mathbf{a}$, $\mathbf{b}$ and $\mathbf{c}$ is 0, and thus these three vectors are coplanar.

35. The magnitude of the torque is

$|\boldsymbol{\tau}| = |\mathbf{r} \times \mathbf{F}| = |\mathbf{r}| \, |\mathbf{F}| \sin \theta = (0.18 \text{ m}) \, (60 \text{ N}) \sin (70 + 10)\,° = 10.8 \sin 80\,° \approx 10.6 \text{ J.}$

37. Using the notation of the text, $\mathbf{r} = \langle 0, 0.3, 0 \rangle$ and $\mathbf{F}$ has direction $\langle 0, 3, -4 \rangle$. The angle θ between them can be

determined by $\cos \theta = \dfrac{\langle 0, 0.3, 0 \rangle \cdot \langle 0, 3, -4 \rangle}{|\langle 0, 0.3, 0 \rangle| \, |\langle 0, 3, -4 \rangle|} \ \Rightarrow \ \cos \theta = \dfrac{0.9}{(0.3)\,(5)} \ \Rightarrow \ \cos \theta = 0.6 \ \Rightarrow \ \theta \approx 53.1\,°.$

Then $|\boldsymbol{\tau}| = |\mathbf{r}| \, |\mathbf{F}| \sin \theta \ \Rightarrow \ 100 = 0.3 \, |\mathbf{F}| \sin 53.1\,° \ \Rightarrow \ |\mathbf{F}| \approx 417 \text{ N.}$

39. (a)

The distance between a point and a line is the length of the perpendicular

from the point to the line, here $\left| \overrightarrow{PS} \right| = d$. But referring to triangle

PQS, $d = \left| \overrightarrow{PS} \right| = \left| \overrightarrow{QP} \right| \sin \theta = |\mathbf{b}| \sin \theta.$ But θ is the angle between

$\overrightarrow{QP} = \mathbf{b}$ and $\overrightarrow{QR} = \mathbf{a}.$ Thus by Theorem 6, $\sin \theta = \dfrac{|\mathbf{a} \times \mathbf{b}|}{|\mathbf{a}| \, |\mathbf{b}|}$ and so

$$d = |\mathbf{b}| \sin \theta = \frac{|\mathbf{b}| \, |\mathbf{a} \times \mathbf{b}|}{|\mathbf{a}| \, |\mathbf{b}|} = \frac{|\mathbf{a} \times \mathbf{b}|}{|\mathbf{a}|}.$$

(b) $\mathbf{a} = \overrightarrow{QR} = \langle -1, -2, -1 \rangle$ and $\mathbf{b} = \overrightarrow{QP} = \langle 1, -5, -7 \rangle$. Then

$\mathbf{a} \times \mathbf{b} = \langle (-2)\,(-7) - (-1)\,(-5), (-1)\,(1) - (-1)\,(-7), (-1)\,(-5) - (-2)\,(1) \rangle = \langle 9, -8, 7 \rangle.$ Thus the

distance is $d = \dfrac{|\mathbf{a} \times \mathbf{b}|}{|\mathbf{a}|} = \frac{1}{\sqrt{6}} \sqrt{81 + 64 + 49} = \sqrt{\frac{194}{6}} = \sqrt{\frac{97}{3}}.$

41. $(\mathbf{a} - \mathbf{b}) \times (\mathbf{a} + \mathbf{b}) = (\mathbf{a} - \mathbf{b}) \times \mathbf{a} + (\mathbf{a} - \mathbf{b}) \times \mathbf{b}$ by Theorem 8 #3

$\qquad\qquad\qquad\qquad = \mathbf{a} \times \mathbf{a} + (-\mathbf{b}) \times \mathbf{a} + \mathbf{a} \times \mathbf{b} + (-\mathbf{b}) \times \mathbf{b}$ by Theorem 8 #4

$\qquad\qquad\qquad\qquad = (\mathbf{a} \times \mathbf{a}) - (\mathbf{b} \times \mathbf{a}) + (\mathbf{a} \times \mathbf{b}) - (\mathbf{b} \times \mathbf{b})$ by Theorem 8 #2

$\qquad\qquad\qquad\qquad = \mathbf{0} - (\mathbf{b} \times \mathbf{a}) + (\mathbf{a} \times \mathbf{b}) - \mathbf{0}$ by Example 2

$\qquad\qquad\qquad\qquad = (\mathbf{a} \times \mathbf{b}) + (\mathbf{a} \times \mathbf{b})$ by Theorem 8 #1

$\qquad\qquad\qquad\qquad = 2\,(\mathbf{a} \times \mathbf{b})$

43. $\mathbf{a} \times (\mathbf{b} \times \mathbf{c}) + \mathbf{b} \times (\mathbf{c} \times \mathbf{a}) + \mathbf{c} \times (\mathbf{a} \times \mathbf{b})$

$\qquad = [(\mathbf{a} \cdot \mathbf{c})\, \mathbf{b} - (\mathbf{a} \cdot \mathbf{b})\, \mathbf{c}] + [(\mathbf{b} \cdot \mathbf{a})\, \mathbf{c} - (\mathbf{b} \cdot \mathbf{c})\, \mathbf{a}] + [(\mathbf{c} \cdot \mathbf{b})\, \mathbf{a} - (\mathbf{c} \cdot \mathbf{a})\, \mathbf{b}]$ by Exercise 42

$\qquad = (\mathbf{a} \cdot \mathbf{c})\, \mathbf{b} - (\mathbf{a} \cdot \mathbf{b})\, \mathbf{c} + (\mathbf{a} \cdot \mathbf{b})\, \mathbf{c} - (\mathbf{b} \cdot \mathbf{c})\, \mathbf{a} + (\mathbf{b} \cdot \mathbf{c})\, \mathbf{a} - (\mathbf{a} \cdot \mathbf{c})\, \mathbf{b} = \mathbf{0}$

45. (a) No. If $\mathbf{a} \cdot \mathbf{b} = \mathbf{a} \cdot \mathbf{c}$, then $\mathbf{a} \cdot (\mathbf{b} - \mathbf{c}) = 0$, so $\mathbf{a}$ is perpendicular to $\mathbf{b} - \mathbf{c}$, which can happen if $\mathbf{b} \neq \mathbf{c}$. For

example, let $\mathbf{a} = \langle 1, 1, 1 \rangle$, $\mathbf{b} = \langle 1, 0, 0 \rangle$ and $\mathbf{c} = \langle 0, 1, 0 \rangle$.

(b) No. If $\mathbf{a} \times \mathbf{b} = \mathbf{a} \times \mathbf{c}$ then $\mathbf{a} \times (\mathbf{b} - \mathbf{c}) = \mathbf{0}$, which implies that $\mathbf{a}$ is parallel to $\mathbf{b} - \mathbf{c}$, which of course can happen if $\mathbf{b} \neq \mathbf{c}$.

(c) Yes. Since $\mathbf{a} \cdot \mathbf{c} = \mathbf{a} \cdot \mathbf{b}$, $\mathbf{a}$ is perpendicular to $\mathbf{b} - \mathbf{c}$, by part (a). From part (b), $\mathbf{a}$ is also parallel to $\mathbf{b} - \mathbf{c}$. Thus since $\mathbf{a} \neq \mathbf{0}$ but is both parallel and perpendicular to $\mathbf{b} - \mathbf{c}$, we have $\mathbf{b} - \mathbf{c} = \mathbf{0}$, so $\mathbf{b} = \mathbf{c}$.

13.5 Equations of Lines and Planes $\qquad$ ET 12.5

1. (a) True; each of the first two lines has a direction vector parallel to the direction vector of the third line, so these vectors are each scalar multiples of the third direction vector. Then the first two direction vectors are also scalar multiples of each other, so these vectors, and hence the two lines, are parallel.

(b) False; for example, the x- and y-axes are both perpendicular to the z-axis, yet the x- and y-axes are not parallel.

(c) True; each of the first two planes has a normal vector parallel to the normal vector of the third plane, so these two normal vectors are parallel to each other and the planes are parallel.

(d) False; for example, the xy- and yz-planes are not parallel, yet they are both perpendicular to the xz-plane.

(e) False; the x- and y-axes are not parallel, yet they are both parallel to the plane $z = 1$.

(f) True; if each line is perpendicular to a plane, then the lines' direction vectors are both parallel to a normal vector for the plane. Thus, the direction vectors are parallel to each other and the lines are parallel.

(g) False; the planes $y = 1$ and $z = 1$ are not parallel, yet they are both parallel to the x-axis.

(h) True; if each plane is perpendicular to a line, then any normal vector for each plane is parallel to a direction vector for the line. Thus, the normal vectors are parallel to each other and the planes are parallel.

(i) True; see Figure 9 and the accompanying discussion.

(j) False; they can be skew, as in Example 3.

(k) True. Consider any normal vector for the plane and any direction vector for the line. If the normal vector is perpendicular to the direction vector, the line and plane are parallel. Otherwise, the vectors meet at an angle θ, $0° \leq \theta < 90°$, and the line will intersect the plane at an angle $90° - \theta$.

3. For this line, we have $\mathbf{r}_0 = -2\mathbf{i} + 4\mathbf{j} + 10\mathbf{k}$ and $\mathbf{v} = 3\mathbf{i} + \mathbf{j} - 8\mathbf{k}$, so a vector equation is
$\mathbf{r} = \mathbf{r}_0 + t\mathbf{v} = (-2\mathbf{i} + 4\mathbf{j} + 10\mathbf{k}) + t(3\mathbf{i} + \mathbf{j} - 8\mathbf{k}) = (-2 + 3t)\mathbf{i} + (4 + t)\mathbf{j} + (10 - 8t)\mathbf{k}$ and parametric equations are $x = -2 + 3t$, $y = 4 + t$, $z = 10 - 8t$.

5. A line perpendicular to the given plane has the same direction as a normal vector to the plane, such as $\mathbf{n} = \langle 1, 3, 1 \rangle$. So $\mathbf{r}_0 = \mathbf{i} + 6\mathbf{k}$, and we can take $\mathbf{v} = \mathbf{i} + 3\mathbf{j} + \mathbf{k}$. Then a vector equation is $\mathbf{r} = (\mathbf{i} + 6\mathbf{k}) + t(\mathbf{i} + 3\mathbf{j} + \mathbf{k}) = (1 + t)\mathbf{i} + 3t\mathbf{j} + (6 + t)\mathbf{k}$, and parametric equations are $x = 1 + t$, $y = 3t$, $z = 6 + t$.

7. $\mathbf{v} = \langle 3 - 3, 2 - 1, -6 - (-1) \rangle = \langle 0, 1, -5 \rangle$, and letting $P_0 = (3, 1, -1)$, parametric equations are $x = 3$, $y = 1 + t$, $z = -1 - 5t$, while symmetric equations are $x = 3$, $y - 1 = (z + 1)/(-5)$. Notice here that the direction number $a = 0$, so rather than writing $(x - 3)/0$ in the symmetric equation we must write the equation $x = 3$ separately.

9. $\mathbf{v} = \langle 2 - 0, 1 - \frac{1}{2}, -3 - 1 \rangle = \langle 2, \frac{1}{2}, -4 \rangle$, and letting $P_0 = (2, 1, -3)$, parametric equations are $x = 2 + 2t$, $y = 1 + \frac{1}{2}t$, $z = -3 - 4t$, while symmetric equations are $\dfrac{x - 2}{2} = \dfrac{y - 1}{1/2} = \dfrac{z + 3}{-4}$ or $\dfrac{x - 2}{2} = 2y - 2 = \dfrac{z + 3}{-4}$.

11. Direction vectors of the lines are $\mathbf{v}_1 = \langle 6, 9, 12 \rangle$ and $\mathbf{v}_2 = \langle 4, 6, 8 \rangle$, and since $\mathbf{v}_1 = \frac{3}{2}\mathbf{v}_2$, the direction vectors and thus the lines are parallel.

13. (a) A direction vector of the line with parametric equations $x = 1 + 2t$, $y = 3t$, $z = 5 - 7t$ is $\mathbf{v} = \langle 2, 3, -7 \rangle$ and the desired parallel line must also have $\mathbf{v}$ as a direction vector. Here $P_0 = (0, 2, -1)$, so symmetric equations for the line are $\dfrac{x}{2} = \dfrac{y-2}{3} = \dfrac{z+1}{-7}$.

(b) The line intersects the xy-plane when $z = 0$, so we need $\dfrac{x}{2} = \dfrac{y-2}{3} = \dfrac{1}{-7}$ or $x = -\frac{2}{7}$, $y = \frac{11}{7}$. Thus the point of intersection with the xy-plane is $\left(-\frac{2}{7}, \frac{11}{7}, 0\right)$. Similarly for the yz-plane, we need $x = 0$ $\Leftrightarrow$ $0 = \dfrac{y-2}{3} = \dfrac{z+1}{-7}$ $\Leftrightarrow$ $y = 2$, $z = -1$. Thus the line intersects the yz-plane at $(0, 2, -1)$. For the xz-plane, we need $y = 0$ $\Leftrightarrow$ $\dfrac{x}{2} = -\dfrac{2}{3} = \dfrac{z+1}{-7}$ $\Leftrightarrow$ $x = -\frac{4}{3}$, $z = \frac{11}{3}$. So the line intersects the xz-plane at $\left(-\frac{4}{3}, 0, \frac{11}{3}\right)$.

15. The lines aren't parallel since the direction vectors $\langle 2, 4, -3 \rangle$ and $\langle 1, 3, 2 \rangle$ aren't parallel, so we check to see if the lines intersect. The parametric equations of the lines are L_1: $x = 4 + 2t$, $y = -5 + 4t$, $z = 1 - 3t$ and L_2: $x = 2 + s$, $y = -1 + 3s$, $z = 2s$. For the lines to intersect we must be able to find one value of t and one value of s satisfying the following three equations: $4 + 2t = 2 + s$, $-5 + 4t = -1 + 3s$, $1 - 3t = 2s$. Solving the first two equations we get $t = -5$, $s = -8$ and checking, we see that these values don't satisfy the third equation. Thus L_1 and L_2 aren't parallel and don't intersect, so they must be skew lines.

17. Since the direction vectors are $\mathbf{v}_1 = \langle -6, 9, -3 \rangle$ and $\mathbf{v}_2 = \langle 2, -3, 1 \rangle$, we have $\mathbf{v}_1 = -3\mathbf{v}_2$ so the lines are parallel.

19. Since the plane is perpendicular to the vector $\langle -2, 1, 5 \rangle$, we can take $\langle -2, 1, 5 \rangle$ as a normal vector to the plane. $(6, 3, 2)$ is a point on the plane, so setting $a = -2$, $b = 1$, $c = 5$ and $x_0 = 6$, $y_0 = 3$, $z_0 = 2$ in Equation 6 gives $-2(x - 6) + 1(y - 3) + 5(z - 2) = 0$ or $-2x + y + 5z = 1$ to be an equation of the plane.

21. $\mathbf{i} + \mathbf{j} - \mathbf{k} = \langle 1, 1, -1 \rangle$ is a normal vector to the plane and $(1, -1, 1)$ is a point on the plane, so setting $a = 1$, $b = 1$, $c = -1$, $x_0 = 1$, $y_0 = -1$, $z_0 = 1$ in Equation 6 gives $1(x - 1) + 1[y - (-1)] - 1(z - 1) = 0$ or $x + y - z = -1$ to be an equation of the plane.

23. Since the two planes are parallel, they will have the same normal vectors. So we can take $\mathbf{n} = \langle 2, -1, 3 \rangle$, and an equation of the plane is $2(x - 0) - 1(y - 0) + 3(z - 0) = 0$ or $2x - y + 3z = 0$.

25. Since the two planes are parallel, they will have the same normal vectors. So we can take $\mathbf{n} = \langle 3, 0, -7 \rangle$, and an equation of the plane is $3(x - 4) + 0[y - (-2)] - 7(z - 3) = 0$ or $3x - 7z = -9$.

27. Here the vectors $\mathbf{a} = \langle 1 - 0, 0 - 1, 1 - 1 \rangle = \langle 1, -1, 0 \rangle$ and $\mathbf{b} = \langle 1 - 0, 1 - 1, 0 - 1 \rangle = \langle 1, 0, -1 \rangle$ lie in the plane, so $\mathbf{a} \times \mathbf{b}$ is a normal vector to the plane. Thus, we can take $\mathbf{n} = \mathbf{a} \times \mathbf{b} = \langle 1 - 0, 0 + 1, 0 + 1 \rangle = \langle 1, 1, 1 \rangle$. If P_0 is the point $(0, 1, 1)$, an equation of the plane is $1(x - 0) + 1(y - 1) + 1(z - 1) = 0$ or $x + y + z = 2$.

29. Here the vectors $\mathbf{a} = \langle 8 - 3, 2 - (-1), 4 - 2 \rangle = \langle 5, 3, 2 \rangle$ and $\mathbf{b} = \langle -1 - 3, -2 - (-1), -3 - 2 \rangle = \langle -4, -1, -5 \rangle$ lie in the plane, so a normal vector to the plane is $\mathbf{n} = \mathbf{a} \times \mathbf{b} = \langle -15 + 2, -8 + 25, -5 + 12 \rangle = \langle -13, 17, 7 \rangle$ and an equation of the plane is $-13(x - 3) + 17[y - (-1)] + 7(z - 2) = 0$ or $-13x + 17y + 7z = -42$.

31. If we first find two nonparallel vectors in the plane, their cross product will be a normal vector to the plane. Since the given line lies in the plane, its direction vector $\mathbf{a} = \langle -2, 5, 4 \rangle$ is one vector in the plane. We can verify that the given point $(6, 0, -2)$ does not lie on this line, so to find another nonparallel vector $\mathbf{b}$ which lies in the plane, we can pick any point on the line and find a vector connecting the points. If we put $t = 0$, we see that $(4, 3, 7)$ is on the line, so $\mathbf{b} = \langle 6 - 4, 0 - 3, -2 - 7 \rangle = \langle 2, -3, -9 \rangle$ and $\mathbf{n} = \mathbf{a} \times \mathbf{b} = \langle -45 + 12, 8 - 18, 6 - 10 \rangle = \langle -33, -10, -4 \rangle$. Thus, an equation of the plane is $-33(x - 6) - 10(y - 0) - 4[z - (-2)] = 0$ or $33x + 10y + 4z = 190$.

33. A direction vector for the line of intersection is $\mathbf{a} = \mathbf{n}_1 \times \mathbf{n}_2 = \langle 1, 1, -1 \rangle \times \langle 2, -1, 3 \rangle = \langle 2, -5, -3 \rangle$, and $\mathbf{a}$ is parallel to the desired plane. Another vector parallel to the plane is the vector connecting any point on the line of intersection to the given point $(-1, 2, 1)$ in the plane. Setting $x = 0$, the equations of the planes reduce to $y - z = 2$ and $-y + 3z = 1$ with simultaneous solution $y = \frac{7}{2}$ and $z = \frac{3}{2}$. So a point on the line is $\left(0, \frac{7}{2}, \frac{3}{2} \right)$ and another vector parallel to the plane is $\left\langle -1, -\frac{3}{2}, -\frac{1}{2} \right\rangle$. Then a normal vector to the plane is $\mathbf{n} = \langle 2, -5, -3 \rangle \times \left\langle -1, -\frac{3}{2}, -\frac{1}{2} \right\rangle = \langle -2, 4, -8 \rangle$ and an equation of the plane is $-2(x + 1) + 4(y - 2) - 8(z - 1) = 0$ or $x - 2y + 4z = -1$.

35. Substituting the parametric equations of the line into the equation of the plane gives $x + y + z = 1 + t + 2t + 3t = 1 \implies t = 0$. This value of t corresponds to the point of intersection $(1, 0, 0)$, obtained by substitution of $t = 0$ into the equations of the line.

37. Substituting the parametric equations of the line into the equation of the plane gives $2x + y - z + 5 = 2(1 + 2t) + (-1) - t + 5 = 0 \implies 3t + 6 = 0 \implies t = -2$. Therefore, the point of intersection is $x = 1 + 2(-2) = -3$, $y = -1$ and $z = -2$ and the point of intersection is $(-3, -1, -2)$.

39. Setting $x = 0$, we see that $(0, 1, 0)$ satisfies the equations of both planes, so that they do in fact have a line of intersection. $\mathbf{v} = \mathbf{n}_1 \times \mathbf{n}_2 = \langle 1, 1, 1 \rangle \times \langle 1, 0, 1 \rangle = \langle 1, 0, -1 \rangle$ is the direction of this line. Therefore, direction numbers of the intersecting line are $1, 0, -1$.

41. The normal vectors to the planes are $\mathbf{n}_1 = \langle 1, 0, 1 \rangle$ and $\mathbf{n}_2 = \langle 0, 1, 1 \rangle$. Thus the normal vectors (and consequently the planes) aren't parallel. Furthermore, $\mathbf{n}_1 \cdot \mathbf{n}_2 = 1 \neq 0$ so the planes aren't perpendicular. Letting θ be the angle between the two planes, we have $\cos \theta = \dfrac{\mathbf{n}_1 \cdot \mathbf{n}_2}{|\mathbf{n}_1| \, |\mathbf{n}_2|} = \dfrac{1}{\sqrt{2}\sqrt{2}} = \dfrac{1}{2}$ and $\theta = \cos^{-1} \frac{1}{2} = 60°$.

43. The normals are $\mathbf{n}_1 = \langle 1, 4, -3 \rangle$ and $\mathbf{n}_2 = \langle -3, 6, 7 \rangle$, so the normals (and thus the planes) aren't parallel. But $\mathbf{n}_1 \cdot \mathbf{n}_2 = -3 + 24 - 21 = 0$, so the normals (and thus the planes) are perpendicular.

45. The normals are $\mathbf{n}_1 = \langle 2, 4, -2 \rangle$ and $\mathbf{n}_2 = \langle -3, -6, 3 \rangle$. Since $\mathbf{n}_1 = -\frac{3}{2}\mathbf{n}_2$, the normals (and thus the planes) are parallel.

47. (a) To find a point on the line of intersection, set one of the variables equal to a constant, say $z = 0$. (This will only work if the line of intersection crosses the xy-plane; otherwise, try setting x or y equal to 0.) Then the equations of the planes reduce to $x + y = 2$ and $3x - 4y = 6$. Solving these two equations gives $x = 2$, $y = 0$. So a point on the line of intersection is $(2, 0, 0)$. The direction of the line is $\mathbf{v} = \mathbf{n}_1 \times \mathbf{n}_2 = \langle 5 - 4, -3 - 5, -4 - 3 \rangle = \langle 1, -8, -7 \rangle$, and symmetric equations for the line are $x - 2 = \dfrac{y}{-8} = \dfrac{z}{-7}$.

(b) The angle between the planes satisfies $\cos \theta = \dfrac{\mathbf{n}_1 \cdot \mathbf{n}_2}{|\mathbf{n}_1| \, |\mathbf{n}_2|} = \dfrac{3 - 4 - 5}{\sqrt{3}\sqrt{50}} = -\dfrac{\sqrt{6}}{5}$. Therefore $\theta = \cos^{-1} \left(-\dfrac{\sqrt{6}}{5} \right) \approx 119°$ (or $61°$).

49. Setting $x = 0$, the equations of the two planes become $z = y$ and $5y + z = -1$, which intersect at $y = -\frac{1}{6}$ and $z = -\frac{1}{6}$. Thus we can choose $(x_0, y_0, z_0) = \left(0, -\frac{1}{6}, -\frac{1}{6}\right)$. The vector giving the direction of this intersecting line, $\mathbf{v}$, is perpendicular to the normal vectors of both planes. So $\mathbf{v} = \mathbf{n}_1 \times \mathbf{n}_2 = \langle 2, -5, -1 \rangle \times \langle 1, 1, -1 \rangle = \langle 6, 1, 7 \rangle$. Therefore, by Equations 2, parametric equations for this line are $x = 6t$, $y = -\frac{1}{6} + t$, $z = -\frac{1}{6} + 7t$.

51. The plane contains all perpendicular bisectors of the line segment joining $(1, 1, 0)$ and $(0, 1, 1)$. All of these bisectors pass through the midpoint of this segment $\left(\frac{1}{2}, \frac{1+1}{2}, \frac{1}{2}\right) = \left(\frac{1}{2}, 1, \frac{1}{2}\right)$. The direction of this line segment $\langle 1 - 0, 1 - 1, 0 - 1 \rangle = \langle 1, 0, -1 \rangle$ is perpendicular to the plane so that we can choose this to be $\mathbf{n}$. Therefore the equation of the plane is $1\left(x - \frac{1}{2}\right) + 0\left(y - 1\right) - 1\left(z - \frac{1}{2}\right) = 0 \iff x = z$.

53. The plane contains the points $(a, 0, 0)$, $(0, b, 0)$ and $(0, 0, c)$. Thus the vectors $\mathbf{a} = \langle -a, b, 0 \rangle$ and $\mathbf{b} = \langle -a, 0, c \rangle$ lie in the plane, and $\mathbf{n} = \mathbf{a} \times \mathbf{b} = \langle bc - 0, 0 + ac, 0 + ab \rangle = \langle bc, ac, ab \rangle$ is a normal vector to the plane. The equation of the plane is therefore $bcx + acy + abz = abc + 0 + 0$ or $bcx + acy + abz = abc$. Notice that if $a \neq 0$, $b \neq 0$ and $c \neq 0$ then we can rewrite the equation as $\dfrac{x}{a} + \dfrac{y}{b} + \dfrac{z}{c} = 1$. This is a good equation to remember!

55. Two vectors which are perpendicular to the required line are the normal of the given plane, $\langle 1, 1, 1 \rangle$, and a direction vector for the given line, $\langle 1, -1, 2 \rangle$. So a direction vector for the required line is $\langle 1, 1, 1 \rangle \times \langle 1, -1, 2 \rangle = \langle 3, -1, -2 \rangle$. Thus L is given by $\langle x, y, z \rangle = \langle 0, 1, 2 \rangle + t \langle 3, -1, -2 \rangle$, or in parametric form, $x = 3t$, $y = 1 - t$, $z = 2 - 2t$.

57. Let P_i have normal vector $\mathbf{n}_i$. Then $\mathbf{n}_1 = \langle 4, -2, 6 \rangle$, $\mathbf{n}_2 = \langle 4, -2, -2 \rangle$, $\mathbf{n}_3 = \langle -6, 3, -9 \rangle$, $\mathbf{n}_4 = \langle 2, -1, -1 \rangle$. Now $\mathbf{n}_1 = -\frac{2}{3}\mathbf{n}_3$, so $\mathbf{n}_1$ and $\mathbf{n}_3$ are parallel, and hence P_1 and P_3 are parallel; similarly P_2 and P_4 are parallel because $\mathbf{n}_2 = 2\mathbf{n}_4$. However, $\mathbf{n}_1$ and $\mathbf{n}_2$ are not parallel. $\left(0, 0, \frac{1}{2}\right)$ lies on P_1, but not on P_3, so they are not the same plane, but both P_2 and P_4 contain the point $(0, 0, -3)$, so these two planes are identical.

59. Let $Q = (2, 2, 0)$ and $R = (3, -1, 5)$, points on the line corresponding to $t = 0$ and $t = 1$. Let $P = (1, 2, 3)$. Then $\mathbf{a} = \overrightarrow{QR} = \langle 1, -3, 5 \rangle$, $\mathbf{b} = \overrightarrow{QP} = \langle -1, 0, 3 \rangle$. The distance is

$$d = \frac{|\mathbf{a} \times \mathbf{b}|}{|\mathbf{a}|} = \frac{|\langle 1, -3, 5 \rangle \times \langle -1, 0, 3 \rangle|}{|\langle 1, -3, 5 \rangle|} = \frac{|\langle -9, -8, -3 \rangle|}{|\langle 1, -3, 5 \rangle|} = \frac{\sqrt{9^2 + 8^2 + 3^2}}{\sqrt{1^2 + 3^2 + 5^2}} = \frac{\sqrt{154}}{\sqrt{35}} = \sqrt{\frac{22}{5}}.$$

61. By Equation 8, the distance is $D = \dfrac{1}{\sqrt{1 + 4 + 4}} \left[(1)(2) + (-2)(8) + (-2)(5) - 1\right] = \dfrac{25}{3}$.

63. Put $y = z = 0$ in the equation of the first plane, to get the point $(-1, 0, 0)$ on the plane. Because the planes are parallel, the distance D between them is the distance from $(-1, 0, 0)$ to the second plane. By Equation 8,

$$D = \frac{|3(-1) + 6(0) - 3(0) - 4|}{\sqrt{3^2 + 6^2 + (-3)^2}} = \frac{7}{3\sqrt{6}}.$$

65. The distance between two parallel planes is the same as the distance between a point on one of the planes and the other plane. Let $P_0 = (x_0, y_0, z_0)$ be a point on the plane given by $ax + by + cz + d_1 = 0$. Then $ax_0 + by_0 + cz_0 + d_1 = 0$ and the distance between P_0 and the plane given by $ax + by + cz + d_2 = 0$ is, from Equation 8, $D = \dfrac{|ax_0 + by_0 + cz_0 + d_2|}{\sqrt{a^2 + b^2 + c^2}} = \dfrac{|-d_1 + d_2|}{\sqrt{a^2 + b^2 + c^2}} = \dfrac{|d_1 - d_2|}{\sqrt{a^2 + b^2 + c^2}}$.

67. $L_1\colon x = y = z \quad \Rightarrow \quad x = y$ (1). $L_2\colon x + 1 = y/2 = z/3 \quad \Rightarrow \quad x + 1 = y/2$ (2). The solution of (1) and (2) is

$x = y = -2$. However, when $x = -2$, $x = z \quad \Rightarrow \quad z = -2$, but $x + 1 = z/3 \quad \Rightarrow \quad z = -3$, a contradiction.

Hence the lines do not intersect. For L_1, $\mathbf{v}_1 = \langle 1, 1, 1 \rangle$, and for L_2, $\mathbf{v}_2 = \langle 1, 2, 3 \rangle$, so the lines are not parallel.

Thus the lines are skew lines. If two lines are skew, they can be viewed as lying in two parallel planes and so the

distance between the skew lines would be the same as the distance between these parallel planes. The common

normal vector to the planes must be perpendicular to both $\langle 1, 1, 1 \rangle$ and $\langle 1, 2, 3 \rangle$, the direction vectors of the two

lines. So set $\mathbf{n} = \langle 1, 1, 1 \rangle \times \langle 1, 2, 3 \rangle = \langle 3 - 2, -3 + 1, 2 - 1 \rangle = \langle 1, -2, 1 \rangle$. From above, we know that

$(-2, -2, -2)$ and $(-2, -2, -3)$ are points of L_1 and L_2 respectively. So in the notation of Equation 7,

$1(-2) - 2(-2) + 1(-2) + d_1 = 0 \quad \Rightarrow \quad d_1 = 0$ and $1(-2) - 2(-2) + 1(-3) + d_2 = 0 \quad \Rightarrow \quad d_2 = 1$. By

Exercise 65, the distance between these two skew lines is $D = \dfrac{|0 - 1|}{\sqrt{1 + 4 + 1}} = \dfrac{1}{\sqrt{6}}$.

Alternate solution (without reference to planes): A vector which is perpendicular to both of the lines is

$\mathbf{n} = \langle 1, 1, 1 \rangle \times \langle 1, 2, 3 \rangle = \langle 1, -2, 1 \rangle$. Pick any point on each of the lines, say $(-2, -2, -2)$ and $(-2, -2, -3)$,

and form the vector $\mathbf{b} = \langle 0, 0, 1 \rangle$ connecting the two points. The distance between the two skew lines is the

absolute value of the scalar projection of $\mathbf{b}$ along $\mathbf{n}$, that is, $D = \dfrac{|\mathbf{n} \cdot \mathbf{b}|}{|\mathbf{n}|} = \dfrac{|1 \cdot 0 - 2 \cdot 0 + 1 \cdot 1|}{\sqrt{1 + 4 + 1}} = \dfrac{1}{\sqrt{6}}$.

69. If $a \neq 0$, then $ax + by + cz + d = 0 \quad \Rightarrow \quad a(x + d/a) + b(y - 0) + c(z - 0) = 0$ which by (6) is the scalar

equation of the plane through the point $(-d/a, 0, 0)$ with normal vector $\langle a, b, c \rangle$. Similarly, if $b \neq 0$ (or if $c \neq 0$)

the equation of the plane can be rewritten as $a(x - 0) + b(y + d/b) + c(z - 0) = 0$ [or as

$a(x - 0) + b(y - 0) + c(z + d/c) = 0$] which by (6) is the scalar equation of a plane through the point

$(0, -d/b, 0)$ [or the point $(0, 0, -d/c)$] with normal vector $\langle a, b, c \rangle$.

13.6 Cylinders and Quadric Surfaces

1. (a) In $\mathbb{R}^2$, the equation $y = x^2$ represents a parabola.

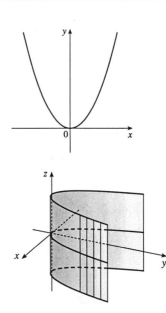

(b) In $\mathbb{R}^3$, the equation $y = x^2$ doesn't involve z, so any

horizontal plane with equation $z = k$ intersects the graph

in a curve with equation $y = x^2$. Thus, the surface is a

parabolic cylinder, made up of infinitely many shifted

copies of the same parabola. The rulings are parallel to

the z-axis.

(c) In $\mathbb{R}^3$, the equation $z = y^2$ also represents a parabolic cylinder. Since x doesn't appear, the graph is formed by moving the parabola $z = y^2$ in the direction of the x-axis. Thus, the rulings of the cylinder are parallel to the x-axis.

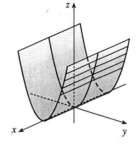

3. Since x is missing from the equation, the vertical traces $y^2 + 4z^2 = 4$, $x = k$, are copies of the same ellipse in the plane $x = k$. Thus, the surface $y^2 + 4z^2 = 4$ is an elliptic cylinder with rulings parallel to the x-axis.

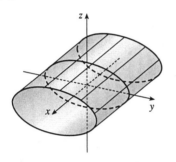

5. Since z is missing, each horizontal trace $x = y^2$, $z = k$, is a copy of the same parabola in the plane $z = k$. Thus, the surface $x - y^2 = 0$ is a parabolic cylinder with rulings parallel to the z-axis.

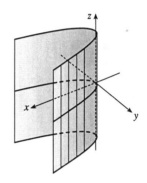

7. Since y is missing, each vertical trace $z = \cos x$, $y = k$ is a copy of a cosine curve in the plane $y = k$. Thus, the surface $z = \cos x$ is a cylindrical surface with rulings parallel to the y-axis.

9. The trace in any plane $x = k$ is given by $z^2 - y^2 = 1 - k^2$, $x = k$ whose graph is a hyperbola. The trace in any plane $y = k$ is the circle given by $x^2 + z^2 = 1 + k^2$, $y = k$, and the trace in any plane $z = k$ is the hyperbola given by $x^2 - y^2 = 1 - k^2$, $z = k$. Thus the surface is a hyperboloid of one sheet with axis the y-axis.

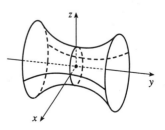

11. Traces: $x = k$, $9y^2 + 36z^2 = 36 - 4k^2$, an ellipse for $|k| < 3$; $y = k$, $4x^2 + 36z^2 = 36 - 9k^2$, an ellipse for $|k| < 2$; $z = k$, $4x^2 + 9y^2 = 36\left(1 - k^2\right)$, an ellipse for $|k| < 1$. Thus the surface is an ellipsoid with center at the origin and axes along the x-, y- and z-axes.

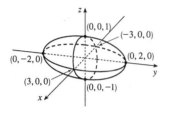

13. Traces: $x = k$, $4z^2 - y^2 = 1 + k^2$, a hyperbola; $y = k$, $4z^2 - x^2 = 1 + k^2$, a hyperbola; $z = k$, $-x^2 - y^2 = 1 - 4k^2$ or $x^2 + y^2 = 4k^2 - 1$, a circle for $k > \frac{1}{2}$ or $k < -\frac{1}{2}$. Thus the surface is a hyperboloid of two sheets with axis the z-axis.

15. Traces: $x = k$, $y^2 = k^2 + z^2$ or $y^2 - z^2 = k^2$, a hyperbola for $k \neq 0$ and two intersecting lines for $k = 0$; $y = k$, $x^2 + z^2 = k^2$, a circle for $k \neq 0$; $z = k$, $y^2 = x^2 + k^2$ or $y^2 - x^2 = k^2$, a hyperbola for $k \neq 0$ and two intersecting lines for $k = 0$. Thus the surface is a cone (right circular) with axis the y-axis and vertex the origin.

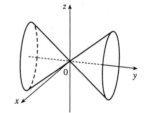

17. Traces: $x = k$, $k^2 + 4z^2 - y = 0$ or $y - k^2 = 4z^2$, a parabola; $y = k$, $x^2 + 4z^2 = k$, an ellipse for $k > 0$; $z = k$, $x^2 + 4k^2 - y = 0$ or $y - 4k^2 = x^2$, a parabola. Thus the surface is an elliptic paraboloid with axis the y-axis and vertex the origin.

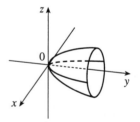

19. Traces: $x = k \Rightarrow y = z^2 - k^2$, parabolas; $y = k \Rightarrow k = z^2 - x^2$, hyperbolas on the z-axis for $k > 0$, and hyperbolas on the x-axis for $k < 0$; $z = k \Rightarrow y = k^2 - x^2$, parabolas. Thus, $\dfrac{y}{1} = \dfrac{z^2}{1^2} - \dfrac{x^2}{1^2}$ is a hyperbolic paraboloid.

21. This is the equation of an ellipsoid: $x^2 + 4y^2 + 9z^2 = x^2 + \dfrac{y^2}{(1/2)^2} + \dfrac{z^2}{(1/3)^2} = 1$, with x-intercepts ± 1, y-intercepts $\pm\frac{1}{2}$ and z-intercepts $\pm\frac{1}{3}$. So the major axis is the x-axis and the only possible graph is VII.

23. This is the equation of a hyperboloid of one sheet, with $a = b = c = 1$. Since the coefficient of y^2 is negative, the axis of the hyperboloid is the y-axis, hence the correct graph is II.

25. There are no real values of x and z that satisfy this equation for $y < 0$, so this surface does not extend to the left of the xz-plane. The surface intersects the plane $y = k > 0$ in an ellipse. Notice that y occurs to the first power whereas x and z occur to the second power. So the surface is an elliptic paraboloid with axis the y-axis. Its graph is VI.

27. This surface is a cylinder because the variable y is missing from the equation. The intersection of the surface and the xz-plane is an ellipse. So the graph is VIII.

29. $z^2 = 3x^2 + 4y^2 - 12$ or $3x^2 + 4y^2 - z^2 = 12$ or

$$\frac{x^2}{4} + \frac{y^2}{3} - \frac{z^2}{12} = 1 \text{ or } \frac{x^2}{2^2} + \frac{y^2}{\left(\sqrt{3}\right)^2} - \frac{z^2}{\left(\sqrt{12}\right)^2} = 1$$

represents a hyperboloid of one sheet with axis the z-axis.

31. $z = x^2 + y^2 + 1$ or $z - 1 = x^2 + y^2$, a circular paraboloid with axis the z-axis and vertex $(0, 0, 1)$.

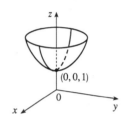

33. Completing the square in all three variables gives

$$(x + 2)^2 + (y - 3)^2 - 4(z + 1)^2 = 13 + 9 \text{ or}$$

$$\frac{(x + 2)^2}{\left(\sqrt{22}\right)^2} + \frac{(y - 3)^2}{\left(\sqrt{22}\right)^2} - \frac{(z + 1)^2}{\left(\frac{1}{2}\sqrt{22}\right)^2} = 1,$$

a hyperboloid of one sheet with center $(-2, 3, -1)$ and axis the vertical line $y = 3$, $x = -2$.

35. Completing the square in y gives

$$x^2 - (y - 2)^2 + z = 4 - 4 = 0 \text{ or}$$

$$z = (y - 2)^2 - x^2, \text{ a hyperbolic paraboloid}$$

with center at $(0, 2, 0)$.

37.

39.

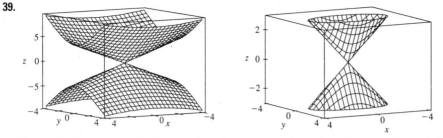

To restrict the z-range as in the second graph, we can use the option `view` = `-2..2` in Maple's `plot3d` command, or `PlotRange` -> `{-2,2}` in Mathematica's `Plot3D` command.

41.

43. The surface is a paraboloid of revolution (circular paraboloid) with vertex at the origin, axis the y-axis and opens to the right. Thus the trace in the yz-plane is also a parabola: $y = z^2$, $x = 0$. The equation is $y = x^2 + z^2$.

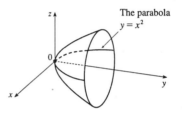

The parabola $y = x^2$

45. Let $P = (x, y, z)$ be an arbitrary point equidistant from $(-1, 0, 0)$ and the plane $x = 1$. Then the distance from P to $(-1, 0, 0)$ is $\sqrt{(x+1)^2 + y^2 + z^2}$ and the distance from P to the plane $x = 1$ is $|x - 1| / \sqrt{1^2} = |x - 1|$ (by Equation 13.5.8 [ET 12.5.8]). So $|x - 1| = \sqrt{(x+1)^2 + y^2 + z^2}$ $\Leftrightarrow$ $(x - 1)^2 = (x + 1)^2 + y^2 + z^2$ $\Leftrightarrow$ $x^2 - 2x + 1 = x^2 + 2x + 1 + y^2 + z^2$ $\Leftrightarrow$ $-4x = y^2 + z^2$. Thus the collection of all such points P is a circular paraboloid with vertex at the origin, axis the x-axis, which opens in the negative direction.

47. If (a, b, c) satisfies $z = y^2 - x^2$, then $c = b^2 - a^2$. L_1: $x = a + t$, $y = b + t$, $z = c + 2(b - a)t$, L_2: $x = a + t$, $y = b - t$, $z = c - 2(b + a)t$. Substitute the parametric equations of L_1 into the equation of the hyperbolic paraboloid in order to find the points of intersection: $z = y^2 - x^2$ $\Rightarrow$ $c + 2(b - a)t = (b + t)^2 - (a + t)^2 = b^2 - a^2 + 2(b - a)t$ $\Rightarrow$ $c = b^2 - a^2$. As this is true for all values of t, L_1 lies on $z = y^2 - x^2$. Performing similar operations with L_2 gives: $z = y^2 - x^2$ $\Rightarrow$ $c - 2(b + a)t = (b - t)^2 - (a + t)^2 = b^2 - a^2 - 2(b + a)t$ $\Rightarrow$ $c = b^2 - a^2$. This tells us that all of L_2 also lies on $z = y^2 - x^2$.

49.

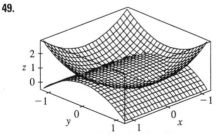

The curve of intersection looks like a bent ellipse. The projection of this curve onto the xy-plane is the set of points $(x, y, 0)$ which satisfy $x^2 + y^2 = 1 - y^2$ $\Leftrightarrow$ $x^2 + 2y^2 = 1$ $\Leftrightarrow$ $x^2 + \dfrac{y^2}{\left(1/\sqrt{2}\right)^2} = 1$. This is an equation of an ellipse.

13.7 Cylindrical and Spherical Coordinates

ET 12.7

1. See Figure 1 and the accompanying discussion; see the paragraph accompanying Figure 3.

3.

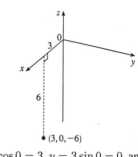

$x = 3\cos\frac{\pi}{2} = 0$, $y = 3\sin\frac{\pi}{2} = 3$, and $z = 1$, so the point is $(0, 3, 1)$ in rectangular coordinates.

5.

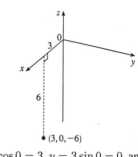

$x = 3\cos 0 = 3$, $y = 3\sin 0 = 0$, and $z = -6$, so the point is $(3, 0, -6)$ in rectangular coordinates.

7.

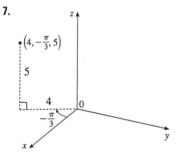

$x = 4\cos\left(-\frac{\pi}{3}\right) = 2$, $y = 4\sin\left(-\frac{\pi}{3}\right) = -2\sqrt{3}$, and $z = 5$, so the point is $\left(2, -2\sqrt{3}, 5\right)$ in rectangular coordinates.

9. $r^2 = x^2 + y^2 = 1^2 + (-1)^2 = 2$ so $r = \sqrt{2}$; $\tan\theta = \frac{y}{x} = \frac{-1}{1} = -1$ and the point $(1, -1)$ is in the fourth quadrant of the xy-plane, so $\theta = \frac{7\pi}{4} + 2n\pi$; $z = 4$. Thus, one set of cylindrical coordinates is $\left(\sqrt{2}, \frac{7\pi}{4}, 4\right)$.

11. $r^2 = (-1)^2 + \left(-\sqrt{3}\right)^2 = 4$ so $r = 2$; $\tan\theta = \frac{-\sqrt{3}}{-1} = \sqrt{3}$ and the point $\left(-1, -\sqrt{3}\right)$ is in the third quadrant of the xy-plane, so $\theta = \frac{4\pi}{3} + 2n\pi$; $z = 2$. Thus, one set of cylindrical coordinates is $\left(2, \frac{4\pi}{3}, 2\right)$.

13.

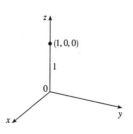

$x = \rho\sin\phi\cos\theta = (1)\sin 0\cos 0 = 0$,
$y = \rho\sin\phi\sin\theta = (1)\sin 0\sin 0 = 0$, and
$z = \rho\cos\phi = (1)\cos 0 = 1$ so the point is
$(0, 0, 1)$ in rectangular coordinates.

15.

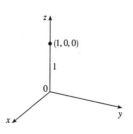

$x = \sin\frac{\pi}{6}\cos\frac{\pi}{6} = \frac{\sqrt{3}}{4}$, $y = \sin\frac{\pi}{6}\sin\frac{\pi}{6} = \frac{1}{4}$, and
$z = \cos\frac{\pi}{6} = \frac{\sqrt{3}}{2}$, so the point is $\left(\frac{\sqrt{3}}{4}, \frac{1}{4}, \frac{\sqrt{3}}{2}\right)$ in rectangular coordinates.

17.

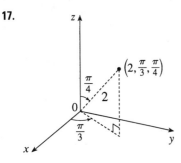

$x = 2\sin\frac{\pi}{4}\cos\frac{\pi}{3} = \frac{\sqrt{2}}{2}$, $y = 2\sin\frac{\pi}{4}\sin\frac{\pi}{3} = \frac{\sqrt{6}}{2}$,

$z = 2\cos\frac{\pi}{4} = \sqrt{2}$ so the point is $\left(\frac{\sqrt{2}}{2}, \frac{\sqrt{6}}{2}, \sqrt{2}\right)$ in rectangular

coordinates.

19. $\rho = \sqrt{9+0+0} = 3$, $\cos\phi = \frac{0}{3} = 0$ so $\phi = \frac{\pi}{2}$, and $\cos\theta = \dfrac{-3}{3\sin\frac{\pi}{2}} = -1$ so $\theta = \pi$, thus spherical coordinates

are $\left(3, \pi, \frac{\pi}{2}\right)$.

21. $\rho = \sqrt{3+1} = 2$, $\cos\phi = \frac{1}{2}$ so $\phi = \frac{\pi}{3}$, and $\cos\theta = \dfrac{\sqrt{3}}{2\sin\frac{\pi}{3}} = \dfrac{\sqrt{3}\cdot 2}{2\cdot\sqrt{3}} = 1$ so $\theta = 0$, thus the point is $\left(2, 0, \frac{\pi}{3}\right)$ in

spherical coordinates.

Note: It is also apparent that $\theta = 0$ since the point is in the xz-plane and $x > 0$.

23. $\rho = \sqrt{x^2+y^2+z^2} = \sqrt{r^2+z^2} = \sqrt{2+0} = \sqrt{2}$; $\theta = \frac{\pi}{4}$; $z = \rho\cos\phi = \sqrt{2}\cos\phi = 0$ so $\phi = \frac{\pi}{2}$ and the point

is $\left(\sqrt{2}, \frac{\pi}{4}, \frac{\pi}{2}\right)$.

25. $\rho = \sqrt{r^2+z^2} = \sqrt{4^2+4^2} = 4\sqrt{2}$; $\theta = \frac{\pi}{3}$; $z = 4 = 4\sqrt{2}\cos\phi$ so $\cos\phi = \frac{1}{\sqrt{2}}$ $\Rightarrow$ $\phi = \frac{\pi}{4}$ and the point is

$\left(4\sqrt{2}, \frac{\pi}{3}, \frac{\pi}{4}\right)$.

27. $z = \rho\cos\phi = 2\cos 0 = 2$, $\rho^2 = x^2+y^2+z^2 = r^2+z^2$ $\Rightarrow$ $r = \sqrt{\rho^2-z^2} = \sqrt{2^2-2^2} = 0$,

(or $r = 2\sin 0 = 0$), $\theta = 0$ and the point is $(0,0,2)$.

29. $z = 8\cos\frac{\pi}{2} = 0$, $r = 8\sin\frac{\pi}{2} = 8$, $\theta = \frac{\pi}{6}$ and the point is $\left(8, \frac{\pi}{6}, 0\right)$.

31. Since $r = 3$, $x^2+y^2 = 9$ and the surface is a cylinder with radius 3 and axis the z-axis.

33. Since $\phi = 0$, $x = 0$ and $y = 0$ while $z = \rho \geq 0$. Thus the "surface" is the positive z-axis including the origin.

35. Since $\phi = \frac{\pi}{3}$, the surface is one frustum of the right circular cone with vertex at the origin and axis the positive

z-axis.

37. $z = r^2 = x^2+y^2$, so the surface is a circular paraboloid with vertex at the origin and axis the positive z-axis.

39. $2 = \rho\cos\phi = z$ is a plane through the point $(0,0,2)$ and parallel to the xy-plane.

41. $r = 2\cos\theta$ $\Rightarrow$ $r^2 = x^2+y^2 = 2r\cos\theta = 2x$ $\Leftrightarrow$ $(x-1)^2+y^2 = 1$, which is the equation of a circular

cylinder with radius 1, whose axis is the vertical line $x = 1$, $y = 0$, $z = z$.

43. Since $r^2+z^2 = 25$ and $r^2 = x^2+y^2$, we have $x^2+y^2+z^2 = 25$, a sphere with radius 5 and center at the origin.

45. Since $x^2 = \rho^2\sin^2\phi\cos^2\theta$ and $z^2 = \rho^2\cos^2\phi$, the equation of the surface in rectangular coordinates is

$x^2+z^2 = 4$. Thus the surface is a circular cylinder of radius 2 about the y-axis.

47. Since $r^2-r = 0$, $r = 0$ or $r = 1$. But $x^2+y^2 = r^2$. Thus the surface consists of the right circular cylinder of

radius 1 and axis the z-axis along with the surface given by $x^2+y^2 = 0$, that is, the z-axis.

49. (a) $r^2 = x^2+y^2$, so $r^2+z^2 = 16$.

(b) $\rho^2 = x^2+y^2+z^2$, so $\rho^2 = 16$ or $\rho = 4$.

51. (a) $r \cos\theta + 2r \sin\theta + 3z = 6$

(b) $\rho \sin\phi \cos\theta + 2\rho \sin\phi \sin\theta + 3\rho \cos\phi = 6$ or $\rho (\sin\phi \cos\theta + 2\sin\phi \sin\theta + 3\cos\phi) = 6$.

53. (a) $r^2 (\cos^2\theta - \sin^2\theta) - 2z^2 = 4$ or $2z^2 = r^2 \cos 2\theta - 4$.

(b) $\rho^2 (\sin^2\phi \cos^2\theta - \sin^2\phi \sin^2\theta - 2\cos^2\phi) = 4$ or $\rho^2 (\sin^2\phi \cos 2\theta - 2\cos^2\phi) = 4$.

55. (a) $r^2 = 2r \sin\theta$ or $r = 2\sin\theta$.

(b) $\rho^2 \sin^2\phi (\cos^2\theta + \sin^2\theta) = 2\rho \sin\phi \sin\theta$ or $\rho \sin^2\phi = 2\sin\phi \sin\theta$ or $\rho \sin\phi = 2\sin\theta$.

57.

$z = r^2 = x^2 + y^2$ is a circular paraboloid with vertex $(0,0,0)$, opening upward. $z = 2 - r^2 \Rightarrow z - 2 = -(x^2 + y^2)$ is a circular paraboloid with vertex $(0,0,2)$ opening downward. Thus $r^2 \le z \le 2 - r^2$ is the solid region enclosed by these two surfaces.

59.

$-\frac{\pi}{2} \le \theta \le \frac{\pi}{2}$ restricts the solid to the 4 octants in which x is positive. $\rho = \sec\phi \Rightarrow \rho \cos\phi = z = 1$, which is the equation of a horizontal plane. $0 \le \phi \le \frac{\pi}{6}$ describes a cone, opening upward. So the solid lies above the cone $\phi = \frac{\pi}{6}$ and below the plane $z = 1$.

61. $z \ge \sqrt{x^2 + y^2}$ because the solid lies above the cone. Squaring both sides of this inequality gives $z^2 \ge x^2 + y^2$ $\Rightarrow 2z^2 \ge x^2 + y^2 + z^2 = \rho^2 \Rightarrow z^2 = \rho^2 \cos^2\phi \ge \frac{1}{2}\rho^2 \Rightarrow \cos^2\phi \ge \frac{1}{2}$. The cone opens upward so that the inequality is $\cos\phi \ge \frac{1}{\sqrt{2}}$, or equivalently $0 \le \phi \le \frac{\pi}{4}$. In spherical coordinates the sphere $z = x^2 + y^2 + z^2$ is $\rho \cos\phi = \rho^2 \Rightarrow \rho = \cos\phi$. $0 \le \rho \le \cos\phi$ because the solid lies below the sphere. The solid can therefore be described as the region in spherical coordinates satisfying $0 \le \rho \le \cos\phi$, $0 \le \phi \le \frac{\pi}{4}$.

63. In cylindrical coordinates, the equation of the cylinder is $r = 3$, $0 \le z \le 10$. The hemisphere is the upper part of the sphere radius 3, center $(0,0,10)$, equation $r^2 + (z - 10)^2 = 3^2$, $z \ge 10$. In Maple, we can use either the `coords=cylindrical` option in a regular `plot` command, or the `plots[cylinderplot]` command. In Mathematica, we can use `ParametricPlot3d`.

▅13 Review

CONCEPT CHECK

1. A scalar is a real number, while a vector is a quantity that has both a real-valued magnitude and a direction.

2. To add two vectors geometrically, we can use either the Triangle Law or the Parallelogram Law, as illustrated in Figures 4 and 5 in Section 13.2 [ET 12.2]. Algebraically, we add the corresponding components of the vectors.

3. For $c > 0$, $c\mathbf{a}$ is a vector with the same direction as $\mathbf{a}$ and length c times the length of $\mathbf{a}$. If $c < 0$, $c\mathbf{a}$ points in the opposite direction as $\mathbf{a}$ and has length $|c|$ times the length of $\mathbf{a}$. (See Figures 6 and 7 in Section 13.2 [ET 12.2].) Algebraically, to find $c\mathbf{a}$ we multiply each component of $\mathbf{a}$ by c.

4. See (1) in Section 13.2 [ET 12.2].

5. See Theorem 13.3.3 [ET 12.3.3] and Definition 13.3.1 [ET 12.3.1].

6. The dot product can be used to find the angle between two vectors and the scalar projection of one vector onto another. In particular, the dot product can determine if two vectors are orthogonal. Also, the dot product can be used to determine the work done moving an object given the force and displacement vectors.

7. See the boxed equations on page 835 [ET 801] as well as Figures 3 and 4 and the accompanying discussion on page 834 [ET 800].

8. See Theorem 13.4.6 [ET 12.4.6] and the preceding discussion; use either (1) or (4) in Section 13.4 [ET 12.4].

9. The cross product can be used to create a vector orthogonal to two given vectors as well as to determine if two vectors are parallel. The cross product can also be used to find the area of a parallelogram determined by two vectors. In addition, the cross product can be used to determine torque if the force and position vectors are known.

10. (a) The area of the parallelogram determined by $\mathbf{a}$ and $\mathbf{b}$ is the length of the cross product: $|\mathbf{a} \times \mathbf{b}|$.

    (b) The volume of the parallelepiped determined by $\mathbf{a}$, $\mathbf{b}$, and $\mathbf{c}$ is the magnitude of their scalar triple product: $|\mathbf{a} \cdot (\mathbf{b} \times \mathbf{c})|$.

11. If an equation of the plane is known, it can be written as $ax + by + cz + d = 0$. A normal vector, which is perpendicular to the plane, is $\langle a, b, c \rangle$ (or any scalar multiple of $\langle a, b, c \rangle$). If an equation is not known, we can use points on the plane to find two non-parallel vectors which lie in the plane. The cross product of these vectors is a vector perpendicular to the plane.

12. The angle between two intersecting planes is defined as the acute angle between their normal vectors. We can find this angle using Corollary 13.3.6 [ET 12.3.6].

13. See (1), (2), and (3) in Section 13.5 [ET 12.5].

14. See (4), (5), and (6) in Section 13.5 [ET 12.5].

15. (a) Determine the vectors $\overrightarrow{PQ} = \langle a_1, a_2, a_3 \rangle$ and $\overrightarrow{PR} = \langle b_1, b_2, b_3 \rangle$. If there is a scalar t such that $\langle a_1, a_2, a_3 \rangle = t \langle b_1, b_2, b_3 \rangle$, then the vectors are parallel and the points must all lie on the same line.
    Alternatively, if $\overrightarrow{PQ} \times \overrightarrow{PR} = \mathbf{0}$, then $\overrightarrow{PQ}$ and $\overrightarrow{PR}$ are parallel, so P, Q, and R are collinear.
    Thirdly, an algebraic method is to determine an equation of the line joining two of the points, and then check whether or not the third point satisfies this equation.

    (b) Find the vectors $\overrightarrow{PQ} = \mathbf{a}$, $\overrightarrow{PR} = \mathbf{b}$, $\overrightarrow{PS} = \mathbf{c}$. $\mathbf{a} \times \mathbf{b}$ is normal to the plane formed by P, Q and R, and so S lies on this plane if $\mathbf{a} \times \mathbf{b}$ and $\mathbf{c}$ are orthogonal, that is, if $(\mathbf{a} \times \mathbf{b}) \cdot \mathbf{c} = 0$. (Or use the reasoning in Example 4 in Section 13.4 [ET 12.4].)
    Alternatively, find an equation for the plane determined by three of the points and check whether or not the fourth point satisfies this equation.

16. (a) See Exercise 13.4.39 [ET 12.4.39].

(b) See Example 8 in Section 13.5 [ET 12.5].

(c) See Example 10 in Section 13.5 [ET 12.5].

17. The traces of a surface are the curves of intersection of the surface with planes parallel to the coordinate planes. We can find the trace in the plane $x = k$ (parallel to the yz-plane) by setting $x = k$ and determining the curve represented by the resulting equation. Traces in the planes $y = k$ (parallel to the xz-plane) and $z = k$ (parallel to the xy-plane) are found similarly.

18. See Table 1 in Section 13.6 [ET 12.6].

19. (a) See (1) and the discussion accompanying Figure 3 in Section 13.7 [ET 12.7].

(b) See (3) and Figures 6-8, and the accompanying discussion, in Section 13.7 [ET 12.7].

TRUE-FALSE QUIZ

1. True, by Theorem 13.3.2 [ET 12.3.2] #2.

3. True. If θ is the angle between **u** and **v**, then by Theorem 13.4.6 [ET 12.4.6],
$|\mathbf{u} \times \mathbf{v}| = |\mathbf{u}|\,|\mathbf{v}|\sin\theta = |\mathbf{v}|\,|\mathbf{u}|\sin\theta = |\mathbf{v} \times \mathbf{u}|$.
(Or, by Theorem 13.4.8 [ET 12.4.8], $|\mathbf{u} \times \mathbf{v}| = |-\mathbf{v} \times \mathbf{u}| = |-1|\,|\mathbf{v} \times \mathbf{u}| = |\mathbf{v} \times \mathbf{u}|$.)

5. Theorem 13.4.8 [ET 12.4.8] #2 tells us that this is true.

7. This is true by Theorem 13.4.8 [ET 12.4.8] #5.

9. This is true because $\mathbf{u} \times \mathbf{v}$ is orthogonal to **u** (see Theorem 13.4.5 [ET 12.4.5]), and the dot product of two orthogonal vectors is 0.

11. If $|\mathbf{u}| = 1$, $|\mathbf{v}| = 1$ and θ is the angle between these two vectors (so $0 \le \theta \le \pi$), then by Theorem 13.4.6 [ET 12.4.6], $|\mathbf{u} \times \mathbf{v}| = |\mathbf{u}|\,|\mathbf{v}|\sin\theta = \sin\theta$, which is equal to 1 if and only if $\theta = \frac{\pi}{2}$ (that is, if and only if the two vectors are orthogonal). Therefore, the assertion that the cross product of two unit vectors is a unit vector is false.

13. This is false. In $\mathbb{R}^2$, $x^2 + y^2 = 1$ represents a circle, but $\{(x, y, z) \mid x^2 + y^2 = 1\}$ represents a *three-dimensional surface*, namely, a circular cylinder with axis the z-axis.

EXERCISES

1. (a) By the formula for an equation of a sphere (see Section 13.1 [ET 12.1]), an equation of the sphere with center $(1, -1, 2)$ and radius 3 is $(x - 1)^2 + (y + 1)^2 + (z - 2)^2 = 9$.

(b) Completing squares gives $(x + 2)^2 + (y + 3)^2 + (z - 5)^2 = -2 + 4 + 9 + 25 = 36$. Thus, the sphere is centered at $(-2, -3, 5)$ and has radius 6.

3. $\mathbf{u} \cdot \mathbf{v} = |\mathbf{u}|\,|\mathbf{v}|\cos 45° = (2)(3)\frac{\sqrt{2}}{2} = 3\sqrt{2}$. $|\mathbf{u} \times \mathbf{v}| = |\mathbf{u}|\,|\mathbf{v}|\sin 45° = (2)(3)\frac{\sqrt{2}}{2} = 3\sqrt{2}$. By the right-hand rule, $\mathbf{u} \times \mathbf{v}$ is directed out of the page.

5. For the two vectors to be orthogonal, we need $\langle 3, 2, x \rangle \cdot \langle 2x, 4, x \rangle = 0$ ⇔
$(3)(2x) + (2)(4) + (x)(x) = 0$ ⇔ $x^2 + 6x + 8 = 0$ ⇔ $(x + 2)(x + 4) = 0$ ⇔ $x = -2$ or $x = -4$.

7. (a) $(\mathbf{u} \times \mathbf{v}) \cdot \mathbf{w} = \mathbf{u} \cdot (\mathbf{v} \times \mathbf{w}) = 2$

(b) $\mathbf{u} \cdot (\mathbf{w} \times \mathbf{v}) = \mathbf{u} \cdot [-(\mathbf{v} \times \mathbf{w})] = -\mathbf{u} \cdot (\mathbf{v} \times \mathbf{w}) = -2$

(c) $\mathbf{v} \cdot (\mathbf{u} \times \mathbf{w}) = (\mathbf{v} \times \mathbf{u}) \cdot \mathbf{w} = -(\mathbf{u} \times \mathbf{v}) \cdot \mathbf{w} = -2$

(d) $(\mathbf{u} \times \mathbf{v}) \cdot \mathbf{v} = \mathbf{u} \cdot (\mathbf{v} \times \mathbf{v}) = \mathbf{u} \cdot \mathbf{0} = 0$

9. For simplicity, consider a unit cube positioned with its back left corner at the origin. Vector representations of the diagonals joining the points $(0, 0, 0)$ to $(1, 1, 1)$ and $(1, 0, 0)$ to $(0, 1, 1)$ are $\langle 1, 1, 1 \rangle$ and $\langle -1, 1, 1 \rangle$. Let θ be the angle between these two vectors. $\langle 1, 1, 1 \rangle \cdot \langle -1, 1, 1 \rangle = -1 + 1 + 1 = 1 = |\langle 1, 1, 1 \rangle| |\langle -1, 1, 1 \rangle| \cos \theta = 3 \cos \theta$
$\Rightarrow \quad \cos \theta = \frac{1}{3} \quad \Rightarrow \quad \theta = \cos^{-1} \frac{1}{3} \approx 71°$.

11. $\overrightarrow{AB} = \langle 1, 0, -1 \rangle$, $\overrightarrow{AC} = \langle 0, 4, 3 \rangle$, so

(a) a vector perpendicular to the plane is $\overrightarrow{AB} \times \overrightarrow{AC} = \langle 0 + 4, - (3 + 0), 4 - 0 \rangle = \langle 4, -3, 4 \rangle$.

(b) $\frac{1}{2} \left| \overrightarrow{AB} \times \overrightarrow{AC} \right| = \frac{1}{2} \sqrt{16 + 9 + 16} = \frac{\sqrt{41}}{2}$.

13. Let F_1 be the magnitude of the force directed $20°$ away from the direction of shore, and let F_2 be the magnitude of the other force. Separating these forces into components parallel to the direction of the resultant force and perpendicular to it gives $F_1 \cos 20° + F_2 \cos 30° = 255$ (1), and $F_1 \sin 20° - F_2 \sin 30° = 0 \Rightarrow$
$F_1 = F_2 \dfrac{\sin 30°}{\sin 20°}$ (2). Substituting (2) into (1) gives $F_2 \left(\sin 30° \cot 20° + \cos 30° \right) = 255 \Rightarrow F_2 \approx 114$ N.
Substituting this into (2) gives $F_1 \approx 166$ N.

15. $x = 1 + 2t, y = 2 - t, z = 4 + 3t$

17. $\mathbf{v} = \langle 4, -3, 5 \rangle$, so $x = 1 + 4t, y = -3t, z = 1 + 5t$.

19. Since the two planes are parallel, they will have the same normal vectors. So se can take $\mathbf{n} = \langle 1, 2, 5 \rangle$ and an equation of the plane is $1 \left[x - (-4) \right] + 2 (y - 1) + 5 (z - 2) = 0$ or $x + 2y + 5z = 8$.

21. Substitution of the parametric equations into the equation of the plane gives
$2x - y + z = 2 (2 - t) - (1 + 3t) + 4t = 2 \Rightarrow -t + 3 = 2 \Rightarrow t = 1$. When $t = 1$, the parametric equations give $x = 2 - 1 = 1, y = 1 + 3 = 4$ and $z = 4$. Therefore, the point of intersection is $(1, 4, 4)$.

23. Since the direction vectors $\langle 2, 3, 4 \rangle$ and $\langle 6, -1, 2 \rangle$ aren't parallel, neither are the lines. For the lines to intersect, the three equations $1 + 2t = -1 + 6s, 2 + 3t = 3 - s, 3 + 4t = -5 + 2s$ must be satisfied simultaneously. Solving the first two equations gives $t = \frac{1}{5}, s = \frac{2}{5}$ and checking we see these values don't satisfy the third equation. Thus the lines aren't parallel and they don't intersect, so they must be skew.

25. By Exercise 13.6.65 [ET 12.6.65], $D = \dfrac{|2 - 24|}{\sqrt{26}} = \dfrac{22}{\sqrt{26}}$.

27. The equation $x = z$ represents a plane perpendicular to the xz-plane and intersecting the xz-plane in the line $x = z, y = 0$.

29. A circular paraboloid with vertex the origin and axis the y-axis.

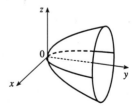

31. A (right elliptical) cone with vertex at the origin and axis the x-axis.

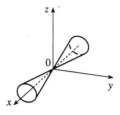

33. An equivalent equation is $-x^2 + \dfrac{y^2}{4} - z^2 = 1$, a hyperboloid of two sheets with axis the y-axis. For $|y| > 2$, traces parallel to the xz-plane are circles.

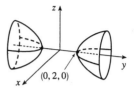

35. $4x^2 + y^2 = 16 \iff \dfrac{x^2}{4} + \dfrac{y^2}{16} = 1$. The equation of the ellipsoid is $\dfrac{x^2}{4} + \dfrac{y^2}{16} + \dfrac{z^2}{c^2} = 1$, since the horizontal trace in the plane $z = 0$ must be the original ellipse. The traces of the ellipsoid in the yz-plane must be circles since the surface is obtained by rotation about the x-axis. Therefore, $c^2 = 16$ and the equation of the ellipsoid is

$$\dfrac{x^2}{4} + \dfrac{y^2}{16} + \dfrac{z^2}{16} = 1 \iff 4x^2 + y^2 + z^2 = 16.$$

37. $x = 2\cos\frac{\pi}{6} = \sqrt{3}$, $y = 2\sin\frac{\pi}{6} = 1$, $z = 2$, so in rectangular coordinates the point is $\left(\sqrt{3}, 1, 2\right)$.
$\rho = \sqrt{3+1+4} = 2\sqrt{2}$, $\theta = \frac{\pi}{6}$, and $\cos\phi = z/\rho = \frac{1}{\sqrt{2}}$, so $\phi = \frac{\pi}{4}$ and the spherical coordinates are $\left(2\sqrt{2}, \frac{\pi}{6}, \frac{\pi}{4}\right)$.

39. $x = 4\sin\frac{\pi}{6}\cos\frac{\pi}{3} = 1$, $y = 4\sin\frac{\pi}{6}\sin\frac{\pi}{3} = \sqrt{3}$, $z = 4\cos\frac{\pi}{6} = 2\sqrt{3}$ so in rectangular coordinates the point is $\left(1, \sqrt{3}, 2\sqrt{3}\right)$. $r^2 = x^2 + y^2 = 4$, $r = 2$, so the cylindrical coordinates are $\left(2, \frac{\pi}{3}, 2\sqrt{3}\right)$.

41. $\theta = \frac{\pi}{4}$. In spherical coordinates, this is a half-plane including the z-axis and intersecting the xy-plane in the half-line $x = y$, $x > 0$.

43. Since $\rho = 3\sec\phi$, $\rho\cos\phi = 3$ or $z = 3$. Thus the surface is a plane parallel to the xy-plane and through the point $(0, 0, 3)$.

45. $x^2 + y^2 + z^2 = 4$. In cylindrical coordinates, this becomes $r^2 + z^2 = 4$. In spherical coordinates, it becomes $\rho^2 = 4$ or $\rho = 2$.

47. The resulting surface is a circular paraboloid with equation $z = 4x^2 + 4y^2$. Changing to cylindrical coordinates we have $z = 4\left(x^2 + y^2\right) = 4r^2$.

Problems Plus

1. Since three-dimensional situations are often difficult to visualize and work with, let us first try to find an analogous problem in two dimensions. The analogue of a cube is a square and the analogue of a sphere is a circle. Thus a similar problem in two dimensions is the following: if five circles with the same radius r are contained in a square of side 1 m so that the circles touch each other and four of the circles touch two sides of the square, find r.

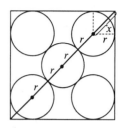

The diagonal of the square is $\sqrt{2}$. The diagonal is also $4r + 2x$. But x is the diagonal of a smaller square of side r.

Therefore $x = \sqrt{2}\,r \quad \Rightarrow \quad \sqrt{2} = 4r + 2x = 4r + 2\sqrt{2}\,r = \left(4 + 2\sqrt{2}\right)r \quad \Rightarrow \quad r = \frac{\sqrt{2}}{4 + 2\sqrt{2}}$.

Let us use these ideas to solve the original three-dimensional problem. The diagonal of the cube is $\sqrt{1^2 + 1^2 + 1^2} = \sqrt{3}$. The diagonal of the cube is also $4r + 2x$ where x is the diagonal of a smaller cube with edge r. Therefore $x = \sqrt{r^2 + r^2 + r^2} = \sqrt{3}\,r \quad \Rightarrow \quad \sqrt{3} = 4r + 2x = 4r + 2\sqrt{3}\,r = \left(4 + 2\sqrt{3}\right)r$. Thus

$r = \dfrac{\sqrt{3}}{4 + 2\sqrt{3}} = \dfrac{2\sqrt{3} - 3}{2}$. The radius of each ball is $\left(\sqrt{3} - \frac{3}{2}\right)$ m.

3. (a) We find the line of intersection L as in Example 13.5.7(b) [ET 12.5.7(b)]. Observe that the point $(-1, c, c)$ lies on both planes. Now since L lies in both planes, it is perpendicular to both of the normal vectors $\mathbf{n}_1$ and $\mathbf{n}_2$, and thus parallel to their cross product $\mathbf{n}_1 \times \mathbf{n}_2 = \begin{vmatrix} \mathbf{i} & \mathbf{j} & \mathbf{k} \\ c & 1 & 1 \\ 1 & -c & c \end{vmatrix} = \langle 2c,\, -c^2 + 1,\, -c^2 - 1 \rangle$. So symmetric equations of L can be written as $\dfrac{x + 1}{-2c} = \dfrac{y - c}{c^2 - 1} = \dfrac{z - c}{c^2 + 1}$, provided that $c \neq 0, \pm 1$.

If $c = 0$, then the two planes are given by $y + z = 0$ and $x = -1$, so symmetric equations of L are $x = -1$, $y = -z$. If $c = -1$, then the two planes are given by $-x + y + z = -1$ and $x + y + z = -1$, and they intersect in the line $x = 0$, $y = -z - 1$. If $c = 1$, then the two planes are given by $x + y + z = 1$ and $x - y + z = 1$, and they intersect in the line $y = 0$, $x = 1 - z$.

(b) If we set $z = t$ in the symmetric equations and solve for x and y separately, we get $x + 1 = \dfrac{(t - c)(-2c)}{c^2 + 1}$,

$y - c = \dfrac{(t - c)(c^2 - 1)}{c^2 + 1} \quad \Rightarrow \quad x = \dfrac{-2ct + (c^2 - 1)}{c^2 + 1}, \quad y = \dfrac{(c^2 - 1)t + 2c}{c^2 + 1}$. Eliminating c from these equations, we have $x^2 + y^2 = t^2 + 1$. So the curve traced out by L in the plane $z = t$ is a circle with center at $(0, 0, t)$ and radius $\sqrt{t^2 + 1}$.

(c) The area of a horizontal cross-section of the solid is $A(z) = \pi(z^2 + 1)$, so

$V = \int_0^1 A(z)\,dz = \pi\left[\frac{1}{3}z^3 + z\right]_0^1 = \frac{4\pi}{3}$.

5. (a) When $\theta = \theta_s$, the block is not moving, so the sum of the forces on the

block must be $\mathbf{0}$, thus $\mathbf{N} + \mathbf{F} + \mathbf{W} = \mathbf{0}$. This relationship is illustrated

geometrically in the figure. Since the vectors form a right triangle, we have

$$\tan(\theta_s) = \frac{|\mathbf{F}|}{|\mathbf{N}|} = \frac{\mu_s n}{n} = \mu_s.$$

(b) We place the block at the origin and sketch the force vectors acting on the block, including the additional
horizontal force $\mathbf{H}$, with initial points at the origin. We then rotate this system so that $\mathbf{F}$ lies along the positive
x-axis and the inclined plane is parallel to the x-axis.

$|\mathbf{F}|$ is maximal, so $|\mathbf{F}| = \mu_s n$ for $\theta > \theta_s$. Then the vectors, in terms of components parallel and perpendicular
to the inclined plane, are

$$\mathbf{N} = n\mathbf{j} \qquad \mathbf{F} = (\mu_s n)\,\mathbf{i}$$

$$\mathbf{W} = (-mg\sin\theta)\,\mathbf{i} + (-mg\cos\theta)\,\mathbf{j}$$
$$\mathbf{H} = (h_{\min}\cos\theta)\,\mathbf{i} + (-h_{\min}\sin\theta)\,\mathbf{j}$$

Equating components, we have

$$\mu_s n - mg\sin\theta + h_{\min}\cos\theta = 0 \quad\Rightarrow\quad h_{\min}\cos\theta + \mu_s n = mg\sin\theta \tag{1}$$
$$n - mg\cos\theta - h_{\min}\sin\theta = 0 \quad\Rightarrow\quad h_{\min}\sin\theta + mg\cos\theta = n \tag{2}$$

(c) Since (2) is solved for n, we substitute into (1):

$$h_{\min}\cos\theta + \mu_s\,(h_{\min}\sin\theta + mg\cos\theta) = mg\sin\theta \quad\Rightarrow$$
$$h_{\min}\cos\theta + h_{\min}\mu_s\sin\theta = mg\sin\theta - mg\mu_s\cos\theta \quad\Rightarrow$$

$$h_{\min} = mg\left(\frac{\sin\theta - \mu_s\cos\theta}{\cos\theta + \mu_s\sin\theta}\right) = mg\left(\frac{\tan\theta - \mu_s}{1 + \mu_s\tan\theta}\right)$$

From part (a) we know $\mu_s = \tan\theta_s$, so this becomes $h_{\min} = mg\left(\dfrac{\tan\theta - \tan\theta_s}{1 + \tan\theta_s\tan\theta}\right)$ and using a trigonometric

identity, this is $mg\tan(\theta - \theta_s)$ as desired.

Note for $\theta = \theta_s$, $h_{min} = mg \tan 0 = 0$, which makes sense since the block is at rest for θ_s, thus no additional force $\mathbf{H}$ is necessary to prevent it from moving. As θ increases, the factor $\tan(\theta - \theta_s)$, and hence the value of h_{min}, increases slowly for small values of $\theta - \theta_s$ but much more rapidly as $\theta - \theta_s$ becomes significant. This seems reasonable, as the steeper the inclined plane, the less the horizontal components of the various forces affect the movement of the block, so we would need a much larger magnitude of horizontal force to keep the block motionless. If we allow $\theta \to 90°$, corresponding to the inclined plane being placed vertically, the value of h_{min} is quite large; this is to be expected, as it takes a great amount of horizontal force to keep an object from moving vertically. In fact, without friction (so $\theta_s = 0$), we would have $\theta \to 90°$ $\Rightarrow$ $h_{min} \to \infty$, and it would be impossible to keep the block from slipping.

(d) Since h_{max} is the largest value of h that keeps the block from slipping, the force of friction is keeping the block from moving *up* the inclined plane; thus, $\mathbf{F}$ is directed *down* the plane. Our system of forces is similar to that in part (b), then, except that we have $\mathbf{F} = -(\mu_s n)\,\mathbf{i}$. (Note that $|\mathbf{F}|$ is again maximal.) Following our procedure in parts (b) and (c), we equate components:

$$-\mu_s n - mg \sin \theta + h_{max} \cos \theta = 0 \quad \Rightarrow \quad h_{max} \cos \theta - \mu_s n = mg \sin \theta$$

$$n - mg \cos \theta - h_{max} \sin \theta = 0 \quad \Rightarrow \quad h_{max} \sin \theta + mg \cos \theta = n$$

Then substituting,

$$h_{max} \cos \theta - \mu_s (h_{max} \sin \theta + mg \cos \theta) = mg \sin \theta \quad \Rightarrow$$

$$h_{max} \cos \theta - h_{max}\mu_s \sin \theta = mg \sin \theta + mg\mu_s \cos \theta \quad \Rightarrow$$

$$h_{max} = mg \left(\frac{\sin \theta + \mu_s \cos \theta}{\cos \theta - \mu_s \sin \theta} \right) = mg \left(\frac{\tan \theta + \mu_s}{1 - \mu_s \tan \theta} \right)$$

$$= mg \left(\frac{\tan \theta + \tan \theta_s}{1 - \tan \theta_s \tan \theta} \right) = mg \tan(\theta + \theta_s)$$

We would expect h_{max} to increase as θ increases, with similar behavior as we established for h_{min}, but with h_{max} values always larger than h_{min}. We can see that this is the case if we graph h_{max} as a function of θ, as the curve is the graph of h_{min} translated $2\theta_s$ to the left, so the equation does seem reasonable. Notice that the equation predicts $h_{max} \to \infty$ as $\theta \to (90° - \theta_s)$. In fact, as h_{max} increases, the normal force increases as well. When $(90° - \theta_s) \le \theta \le 90°$, the horizontal force is completely counteracted by the sum of the normal and frictional forces, so no part of the horizontal force contributes to moving the block up the plane no matter how large its magnitude.

Vector Functions

14.1 Vector Functions and Space Curves

1. The component functions t^2, $\sqrt{t-1}$, and $\sqrt{5-t}$ are all defined when $t-1 \geq 0 \quad \Rightarrow \quad t \geq 1$ and $5-t \geq 0 \quad \Rightarrow$ $t \leq 5$, so the domain of $\mathbf{r}\,(t)$ is $[1,5]$.

3. $\displaystyle\lim_{t\to0+}\cos t = \cos 0 = 1$, $\displaystyle\lim_{t\to0+}\sin t = \sin 0 = 0$, $\displaystyle\lim_{t\to0+} t \ln t = \lim_{t\to0+}\frac{\ln t}{1/t} = \lim_{t\to0+}\frac{1/t}{-1/t^2} = \lim_{t\to0+} -t = 0$ (by

l'Hospital's Rule). Thus $\displaystyle\lim_{t\to0+}\langle\cos t, \sin t, t \ln t\rangle = \Big\langle \lim_{t\to0+}\cos t, \lim_{t\to0+}\sin t, \lim_{t\to0+} t \ln t \Big\rangle = \langle 1, 0, 0\rangle$.

5. $\displaystyle\lim_{t\to1}\sqrt{t+3} = 2$, $\displaystyle\lim_{t\to1}\frac{t-1}{t^2-1} = \lim_{t\to1}\frac{1}{t+1} = \frac{1}{2}$, $\displaystyle\lim_{t\to1}\left(\frac{\tan t}{t}\right) = \tan 1$

Thus the given limit equals $\left\langle 2, \frac{1}{2}, \tan 1\right\rangle$.

7. $x = \cos 4t$, $y = t$, $z = \sin 4t$. At any point (x, y, z) on the curve, $x^2 + z^2 = \cos^2 4t + \sin^2 4t = 1$. So the curve lies on a circular cylinder with axis the y-axis. Since $y = t$, this is a helix. So the graph is V.

9. $x = t$, $y = 1/\left(1+t^2\right)$, $z = t^2$. Note that y and z are positive for all t. The curve passes through $(0, 1, 0)$ when $t = 0$. As $t \to \infty$, $(x, y, z) \to (\infty, 0, \infty)$, and as $t \to -\infty$, $(x, y, z) \to (-\infty, 0, \infty)$. So the graph is I.

11. $x = \cos t$, $y = \sin t$, $z = \sin 5t$. $\quad x^2 + y^2 = \cos^2 t + \sin^2 t = 1$, so the curve lies on a circular cylinder with axis the z-axis. Each of x, y and z is periodic, and at $t = 0$ and $t = 2\pi$ the curve passes through the same point, so the curve repeats itself and the graph is IV.

13. The parametric equations for this curve are $x = t^4 + 1$, $y = t$. We can make a table of values, or we can eliminate the parameter: $t = y \quad \Rightarrow \quad x = y^4 + 1$, with $y \in \mathbb{R}$. By comparing different values of t, we find the direction in which t increases as indicated in the graph.

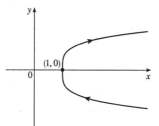

15. The corresponding parametric equations are $x = t$, $y = -t$, $z = 2t$, which are parametric equations of a line through the origin and with direction vector $\langle 1, -1, 2\rangle$.

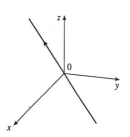

17. The parametric equations give

$x^2 + z^2 = \sin^2 t + \cos^2 t = 1$, $y = 3$, which is a circle of

radius 1, center $(0, 3, 0)$ in the plane $y = 3$.

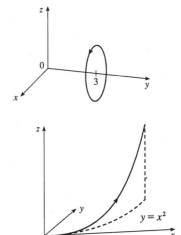

19. The parametric equations are $x = t^2$, $y = t^4$, $z = t^6$. These

are positive for $t \neq 0$ and 0 when $t = 0$. So the curve lies

entirely in the first quadrant. The projection of the graph onto

the xy-plane is $y = x^2$, $y > 0$, a half parabola. On the

xz-plane $z = x^3$, $z > 0$, a half cubic, and the yz-plane,

$y^3 = z^2$.

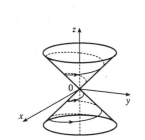

21. If $x = t \cos t$, $y = t \sin t$, and $z = t$, then

$x^2 + y^2 = t^2 \cos^2 t + t^2 \sin^2 t = t^2 = z^2$, so the curve lies

on the cone $z^2 = x^2 + y^2$. Since $z = t$, the curve is a spiral

on this cone.

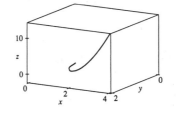

23. $\mathbf{r}(t) = \langle \sin t, \cos t, t^2 \rangle$

25. $\mathbf{r}(t) = \langle \sqrt{t}, t, t^2 - 2 \rangle$

27.

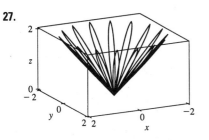

$x = (1 + \cos 16t) \cos t$, $y = (1 + \cos 16t) \sin t$,

$z = 1 + \cos 16t$. At any point on the graph,

$$x^2 + y^2 = (1 + \cos 16t)^2 \cos^2 t + (1 + \cos 16t)^2 \sin^2 t$$
$$= (1 + \cos 16t)^2 = z^2, \text{ so the graph lies on the cone}$$

$x^2 + y^2 = z^2$. From the graph at left, we see that this curve looks

like the projection of a leaved two-dimensional curve onto a cone.

29. If $t = -1$, then $x = 1, y = 4, z = 0$, so the curve passes through the point $(1, 4, 0)$. If $t = 3$, then

$x = 9, y = -8, z = 28$, so the curve passes through the point $(9, -8, 28)$. For the point $(4, 7, -6)$ to be on the

curve, we require $y = 1 - 3t = 7 \implies t = -2$. But then $z = 1 + (-2)^3 = -7 \neq -6$, so $(4, 7, -6)$ is not on the

curve.

31. Both equations are solved for z, so we can substitute to eliminate z: $\sqrt{x^2 + y^2} = 1 + y \Rightarrow$

$x^2 + y^2 = 1 + 2y + y^2 \Rightarrow x^2 = 1 + 2y \Rightarrow y = \frac{1}{2}(x^2 - 1)$. We can form parametric equations for the

curve C of intersection by choosing a parameter $x = t$, then $y = \frac{1}{2}(t^2 - 1)$ and

$z = 1 + y = 1 + \frac{1}{2}(t^2 - 1) = \frac{1}{2}(t^2 + 1)$. Thus a vector function representing C is

$\mathbf{r}(t) = t\,\mathbf{i} + \frac{1}{2}(t^2 - 1)\,\mathbf{j} + \frac{1}{2}(t^2 + 1)\,\mathbf{k}$.

33.

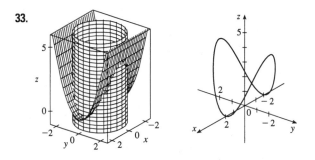

The projection of the curve C of intersection onto

the xy-plane is the circle $x^2 + y^2 = 4$, $z = 0$.

Then we can write $x = 2\cos t$, $y = 2\sin t$,

$0 \le t \le 2\pi$. Since C also lies on the surface

$z = x^2$, we have $z = x^2 = (2\cos t)^2 = 4\cos^2 t$.

Then parametric equations for C are $x = 2\cos t$,

$y = 2\sin t$, $z = 4\cos^2 t$, $0 \le t \le 2\pi$.

35. Let $\mathbf{u}(t) = \langle u_1(t), u_2(t), u_3(t) \rangle$ and $\mathbf{v}(t) = \langle v_1(t), v_2(t), v_3(t) \rangle$. In each part of this problem the basic
procedure is to use Equation 1 and then analyze the individual component functions using the limit properties we
have already developed for real-valued functions.

(a) $\lim\limits_{t \to a} \mathbf{u}(t) + \lim\limits_{t \to a} \mathbf{v}(t) = \left\langle \lim\limits_{t \to a} u_1(t), \lim\limits_{t \to a} u_2(t), \lim\limits_{t \to a} u_3(t) \right\rangle + \left\langle \lim\limits_{t \to a} v_1(t), \lim\limits_{t \to a} v_2(t), \lim\limits_{t \to a} v_3(t) \right\rangle$ and the

limits of these component functions must each exist since the vector functions both possess limits as $t \to a$.

Then adding the two vectors and using the addition property of limits for real-valued functions, we have that

$$\lim\limits_{t \to a} \mathbf{u}(t) + \lim\limits_{t \to a} \mathbf{v}(t) = \left\langle \lim\limits_{t \to a} u_1(t) + \lim\limits_{t \to a} v_1(t), \lim\limits_{t \to a} u_2(t) + \lim\limits_{t \to a} v_2(t), \lim\limits_{t \to a} u_3(t) + \lim\limits_{t \to a} v_3(t) \right\rangle$$

$$= \left\langle \lim\limits_{t \to a} [u_1(t) + v_1(t)], \lim\limits_{t \to a} [u_2(t) + v_2(t)], \lim\limits_{t \to a} [u_3(t) + v_3(t)] \right\rangle$$

$$= \lim\limits_{t \to a} \langle u_1(t) + v_1(t), u_2(t) + v_2(t), u_3(t) + v_3(t) \rangle \quad \text{[using (1) backward]}$$

$$= \lim\limits_{t \to a} [\mathbf{u}(t) + \mathbf{v}(t)]$$

(b) $\lim\limits_{t \to a} c\,\mathbf{u}(t) = \lim\limits_{t \to a} \langle cu_1(t), cu_2(t), cu_3(t) \rangle = \left\langle \lim\limits_{t \to a} cu_1(t), \lim\limits_{t \to a} cu_2(t), \lim\limits_{t \to a} cu_3(t) \right\rangle$

$$= \left\langle c \lim\limits_{t \to a} u_1(t), c \lim\limits_{t \to a} u_2(t), c \lim\limits_{t \to a} u_3(t) \right\rangle = c \left\langle \lim\limits_{t \to a} u_1(t), \lim\limits_{t \to a} u_2(t), \lim\limits_{t \to a} u_3(t) \right\rangle$$

$$= c \lim\limits_{t \to a} \langle u_1(t), u_2(t), u_3(t) \rangle = c \lim\limits_{t \to a} \mathbf{u}(t)$$

(c) $\lim\limits_{t \to a} \mathbf{u}(t) \cdot \lim\limits_{t \to a} \mathbf{v}(t) = \left\langle \lim\limits_{t \to a} u_1(t), \lim\limits_{t \to a} u_2(t), \lim\limits_{t \to a} u_3(t) \right\rangle \cdot \left\langle \lim\limits_{t \to a} v_1(t), \lim\limits_{t \to a} v_2(t), \lim\limits_{t \to a} v_3(t) \right\rangle$

$$= \left[\lim\limits_{t \to a} u_1(t) \right] \left[\lim\limits_{t \to a} v_1(t) \right] + \left[\lim\limits_{t \to a} u_2(t) \right] \left[\lim\limits_{t \to a} v_2(t) \right] + \left[\lim\limits_{t \to a} u_3(t) \right] \left[\lim\limits_{t \to a} v_3(t) \right]$$

$$= \lim\limits_{t \to a} u_1(t)\, v_1(t) + \lim\limits_{t \to a} u_2(t)\, v_2(t) + \lim\limits_{t \to a} u_3(t)\, v_3(t)$$

$$= \lim\limits_{t \to a} [u_1(t)\, v_1(t) + u_2(t)\, v_2(t) + u_3(t)\, v_3(t)] = \lim\limits_{t \to a} [\mathbf{u}(t) \cdot \mathbf{v}(t)]$$

(d) $\lim\limits_{t \to a} \mathbf{u}(t) \times \lim\limits_{t \to a} \mathbf{v}(t) = \left\langle \lim\limits_{t \to a} u_1(t), \lim\limits_{t \to a} u_2(t), \lim\limits_{t \to a} u_3(t) \right\rangle \times \left\langle \lim\limits_{t \to a} v_1(t), \lim\limits_{t \to a} v_2(t), \lim\limits_{t \to a} v_3(t) \right\rangle$

$$= \left\langle \left[\lim\limits_{t \to a} u_2(t)\right] \left[\lim\limits_{t \to a} v_3(t)\right] - \left[\lim\limits_{t \to a} u_3(t)\right] \left[\lim\limits_{t \to a} v_2(t)\right], \right.$$

$$\left[\lim\limits_{t \to a} u_3(t)\right] \left[\lim\limits_{t \to a} v_1(t)\right] - \left[\lim\limits_{t \to a} u_1(t)\right] \left[\lim\limits_{t \to a} v_3(t)\right],$$

$$\left.\left[\lim\limits_{t \to a} u_1(t)\right] \left[\lim\limits_{t \to a} v_2(t)\right] - \left[\lim\limits_{t \to a} u_2(t)\right] \left[\lim\limits_{t \to a} v_1(t)\right] \right\rangle$$

$$= \left\langle \lim\limits_{t \to a} [u_2(t) v_3(t) - u_3(t) v_2(t)], \lim\limits_{t \to a} [u_3(t) v_1(t) - u_1(t) v_3(t)], \right.$$

$$\left. \lim\limits_{t \to a} [u_1(t) v_2(t) - u_2(t) v_1(t)] \right\rangle$$

$$= \lim\limits_{t \to a} \langle u_2(t) v_3(t) - u_3(t) v_2(t), u_3(t) v_1(t) - u_1(t) v_3(t),$$

$$u_1(t) v_2(t) - u_2(t) v_1(t) \rangle$$

$$= \lim\limits_{t \to a} [\mathbf{u}(t) \times \mathbf{v}(t)]$$

37. Let $\mathbf{r}(t) = \langle f(t), g(t), h(t) \rangle$ and $\mathbf{b} = \langle b_1, b_2, b_3 \rangle$. If $\lim\limits_{t \to a} \mathbf{r}(t) = \mathbf{b}$, then $\lim\limits_{t \to a} \mathbf{r}(t)$ exists, so by (1),

$\mathbf{b} = \lim\limits_{t \to a} \mathbf{r}(t) = \left\langle \lim\limits_{t \to a} f(t), \lim\limits_{t \to a} g(t), \lim\limits_{t \to a} h(t) \right\rangle$. By the definition of equal vectors we have $\lim\limits_{t \to a} f(t) = b_1$,

$\lim\limits_{t \to a} g(t) = b_2$ and $\lim\limits_{t \to a} h(t) = b_3$. But these are limits of real-valued functions, so by the definition of limits, for

every $\epsilon > 0$ there exists $\delta_1 > 0$, $\delta_2 > 0$, $\delta_3 > 0$ so $|f(t) - b_1| < \epsilon/3$ whenever $0 < |t - a| < \delta_1$,

$|g(t) - b_2| < \epsilon/3$ whenever $0 < |t - a| < \delta_2$, and $|h(t) - b_3| < \epsilon/3$ whenever $0 < |t - a| < \delta_3$. Letting

$\delta = $ minimum of $\{\delta_1, \delta_2, \delta_3\}$, we have $|f(t) - b_1| + |g(t) - b_2| + |h(t) - b_3| < \epsilon/3 + \epsilon/3 + \epsilon/3 = \epsilon$ whenever

$0 < |t - a| < \delta$. But $|\mathbf{r}(t) - \mathbf{b}| = |\langle f(t) - b_1, g(t) - b_2, h(t) - b_3 \rangle|$

$= \sqrt{(f(t) - b_1)^2 + (g(t) - b_2)^2 + (h(t) - b_3)^2} \le \sqrt{[f(t) - b_1]^2} + \sqrt{[g(t) - b_2]^2} + \sqrt{[h(t) - b_3]^2}$

$= |f(t) - b_1| + |g(t) - b_2| + |h(t) - b_3|$. Thus for every $\epsilon > 0$ there exists $\delta > 0$ such that

$|\mathbf{r}(t) - \mathbf{b}| \le |f(t) - b_1| + |g(t) - b_2| + |h(t) - b_3| < \epsilon$ whenever $0 < |t - a| < \delta$. Conversely, if for every

$\epsilon > 0$, there exists $\delta > 0$ such that $|\mathbf{r}(t) - \mathbf{b}| < \epsilon$ whenever $0 < |t - a| < \delta$, then

$|\langle f(t) - b_1, g(t) - b_2, h(t) - b_3 \rangle| < \epsilon \quad \Leftrightarrow \quad \sqrt{[f(t) - b_1]^2 + [g(t) - b_2]^2 + [h(t) - b_3]^2} < \epsilon \quad \Leftrightarrow$

$[f(t) - b_1]^2 + [g(t) - b_2]^2 + [h(t) - b_3]^2 < \epsilon^2$ whenever $0 < |t - a| < \delta$. But each term on the left side of this

inequality is positive so $[f(t) - b_1]^2 < \epsilon^2$, $[g(t) - b_2]^2 < \epsilon^2$ and $[h(t) - b_3]^2 < \epsilon^2$ whenever $0 < |t - a| < \delta$, or

taking the square root of both sides in each of the above we have $|f(t) - b_1| < \epsilon$, $|g(t) - b_2| < \epsilon$ and

$|h(t) - b_3| < \epsilon$ whenever $0 < |t - a| < \delta$. And by definition of limits of real-valued functions we have

$\lim\limits_{t \to a} f(t) = b_1$, $\lim\limits_{t \to a} g(t) = b_2$ and $\lim\limits_{t \to a} h(t) = b_3$. But by (1), $\lim\limits_{t \to a} \mathbf{r}(t) = \left\langle \lim\limits_{t \to a} f(t), \lim\limits_{t \to a} g(t), \lim\limits_{t \to a} h(t) \right\rangle$, so

$\lim\limits_{t \to a} \mathbf{r}(t) = \langle b_1, b_2, b_3 \rangle = \mathbf{b}$.

14.2 Derivatives and Integrals of Vector Functions

ET 13.2

1. (a)

(b) $\dfrac{\mathbf{r}(4.5) - \mathbf{r}(4)}{0.5} = 2\left[\mathbf{r}(4.5) - \mathbf{r}(4)\right]$, so we draw a vector in the

same direction but with twice the length of the vector

$\mathbf{r}(4.5) - \mathbf{r}(4)$. $\dfrac{\mathbf{r}(4.2) - \mathbf{r}(4)}{0.2} = 5\left[\mathbf{r}(4.2) - \mathbf{r}(4)\right]$, so we draw

a vector in the same direction but with 5 times the length of the

vector $\mathbf{r}(4.2) - \mathbf{r}(4)$.

(c) By Definition 1, $\mathbf{r}'(4) = \lim\limits_{h \to 0} \dfrac{\mathbf{r}(4 + h) - \mathbf{r}(4)}{h}$.

$\mathbf{T}(4) = \dfrac{\mathbf{r}'(4)}{|\mathbf{r}'(4)|}$.

(d) $\mathbf{T}(4)$ is a unit vector in the same direction as $\mathbf{r}'(4)$, that is,

parallel to the tangent line to the curve at $\mathbf{r}(4)$ with length 1.

3. (a), (c)

(b) $\mathbf{r}'(t) = \langle -\sin t, \cos t \rangle$

5. Since $(x - 1)^2 = t^2 = y$, the curve is a

parabola.

(a), (c)

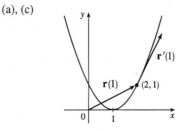

(b) $\mathbf{r}'(t) = \mathbf{i} + 2t\mathbf{j}$

7. $x^{-2} = e^{-2t} = y$, so $y = 1/x^2$, $x > 0$.

(a), (c)

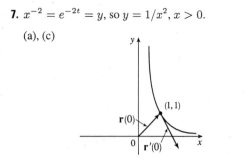

(b) $\mathbf{r}'(t) = e^t\mathbf{i} - 2e^{-2t}\mathbf{j}$

9. $\mathbf{r}'(t) = \left\langle \dfrac{d}{dt}\left[t^2\right], \dfrac{d}{dt}\left[1-t\right], \dfrac{d}{dt}\left[\sqrt{t}\right] \right\rangle = \left\langle 2t, -1, \dfrac{1}{2\sqrt{t}} \right\rangle$

11. $\mathbf{r}(t) = \mathbf{i} - \mathbf{j} + e^{4t}\mathbf{k} \quad \Rightarrow \quad \mathbf{r}'(t) = 0\mathbf{i} + 0\mathbf{j} + 4e^{4t}\mathbf{k} = 4e^{4t}\mathbf{k}$

13. $\mathbf{r}'(t) = -\dfrac{2t}{4-t^2}\mathbf{i} + \dfrac{1}{2\sqrt{1+t}}\mathbf{j} - 12e^{3t}\mathbf{k}$

15. $\mathbf{r}'(t) = \mathbf{0} + \mathbf{b} + 2\mathbf{c}\,t = \mathbf{b} + 2t\,\mathbf{c}$ by Formulas 1 and 3 of Theorem 3.

17. $\mathbf{r}'(t) = \langle 30t^4, 12t^2, 2 \rangle \quad \Rightarrow \quad \mathbf{r}'(1) = \langle 30, 12, 2 \rangle$. So $|\mathbf{r}'(1)| = \sqrt{30^2 + 12^2 + 2^2} = \sqrt{1048} = 2\sqrt{262}$ and

$\mathbf{T}(1) = \dfrac{\mathbf{r}'(1)}{|\mathbf{r}'(1)|} = \frac{1}{2\sqrt{262}}\langle 30, 12, 2 \rangle = \left\langle \frac{15}{\sqrt{262}}, \frac{6}{\sqrt{262}}, \frac{1}{\sqrt{262}} \right\rangle$.

19. $\mathbf{r}'(t) = \mathbf{i} + 2\cos t\,\mathbf{j} - 3\sin t\,\mathbf{k}$, $\mathbf{r}'\left(\frac{\pi}{6}\right) = \mathbf{i} + \sqrt{3}\mathbf{j} - \frac{3}{2}\mathbf{k}$. Thus

$\mathbf{T}\left(\frac{\pi}{6}\right) = \dfrac{1}{\sqrt{1^2 + \left(\sqrt{3}\right)^2 + (-3/2)^2}}\left(\mathbf{i} + \sqrt{3}\mathbf{j} - \frac{3}{2}\mathbf{k}\right) = \frac{1}{5/2}\left(\mathbf{i} + \sqrt{3}\mathbf{j} - \frac{3}{2}\mathbf{k}\right) = \frac{2}{5}\mathbf{i} + \frac{2\sqrt{3}}{5}\mathbf{j} - \frac{3}{5}\mathbf{k}$.

21. $\mathbf{r}(t) = \langle t, t^2, t^3 \rangle \quad \Rightarrow \quad \mathbf{r}'(t) = \langle 1, 2t, 3t^2 \rangle$. Then $\mathbf{r}'(1) = \langle 1, 2, 3 \rangle$ and $|\mathbf{r}'(1)| = \sqrt{1^2 + 2^2 + 3^2} = \sqrt{14}$, so

$\mathbf{T}(1) = \dfrac{\mathbf{r}'(1)}{|\mathbf{r}'(1)|} = \frac{1}{\sqrt{14}}\langle 1, 2, 3 \rangle = \left\langle \frac{1}{\sqrt{14}}, \frac{2}{\sqrt{14}}, \frac{3}{\sqrt{14}} \right\rangle$. $\mathbf{r}''(t) = \langle 0, 2, 6t \rangle$, so

$\mathbf{r}'(t) \times \mathbf{r}''(t) = \begin{vmatrix} \mathbf{i} & \mathbf{j} & \mathbf{k} \\ 1 & 2t & 3t^2 \\ 0 & 2 & 6t \end{vmatrix} = \begin{vmatrix} 2t & 3t^2 \\ 2 & 6t \end{vmatrix}\mathbf{i} - \begin{vmatrix} 1 & 3t^2 \\ 0 & 6t \end{vmatrix}\mathbf{j} + \begin{vmatrix} 1 & 2t \\ 0 & 2 \end{vmatrix}\mathbf{k}$

$\qquad = \left(12t^2 - 6t^2\right)\mathbf{i} - (6t - 0)\mathbf{j} + (2 - 0)\mathbf{k} = \langle 6t^2, -6t, 2 \rangle$.

23. The vector equation for the curve is $\mathbf{r}(t) = \langle t^5, t^4, t^3 \rangle$, so $\mathbf{r}'(t) = \langle 5t^4, 4t^3, 3t^2 \rangle$. The point $(1, 1, 1)$ corresponds to $t = 1$, so the tangent vector there is $\mathbf{r}'(1) = \langle 5, 4, 3 \rangle$. Thus, the tangent line goes through the point $(1, 1, 1)$ and is parallel to the vector $\langle 5, 4, 3 \rangle$. Parametric equations are $x = 1 + 5t$, $y = 1 + 4t$, $z = 1 + 3t$.

25. $\mathbf{r}(t) = \langle t\cos 2\pi t, t\sin 2\pi t, 4t \rangle$, $\mathbf{r}'(t) = \langle \cos 2\pi t - 2\pi t\sin 2\pi t, \sin 2\pi t + 2\pi t\cos 2\pi t, 4 \rangle$. At $\left(0, \frac{1}{4}, 1\right)$, $t = \frac{1}{4}$ and $\mathbf{r}'\left(\frac{1}{4}\right) = \left\langle 0 - \frac{\pi}{2}, 1 + 0, 4 \right\rangle = \left\langle -\frac{\pi}{2}, 1, 4 \right\rangle$. Thus, parametric equations of the tangent line are $x = -\frac{\pi}{2}t$, $y = \frac{1}{4} + t$, $z = 1 + 4t$.

27. $r(t) = \langle t, \sqrt{2}\cos t, \sqrt{2}\sin t \rangle \Rightarrow$

$r'(t) = \langle 1, -\sqrt{2}\sin t, \sqrt{2}\cos t \rangle$. At $(\frac{\pi}{4}, 1, 1)$, $t = \frac{\pi}{4}$ and

$r'(\frac{\pi}{4}) = \langle 1, -1, 1 \rangle$. Thus, parametric equations of the tangent line

are $x = \frac{\pi}{4} + t$, $y = 1 - t$, $z = 1 + t$.

29. (a) $r(t) = \langle t^3, t^4, t^5 \rangle \Rightarrow r'(t) = \langle 3t^2, 4t^3, 5t^4 \rangle$, and since $r'(0) = \langle 0, 0, 0 \rangle = 0$, the curve is not smooth.

(b) $r(t) = \langle t^3 + t, t^4, t^5 \rangle \Rightarrow r'(t) = \langle 3t^2 + 1, 4t^3, 5t^4 \rangle$. $r'(t)$ is continuous since its component functions

are continuous. Also, $r'(t) \neq 0$, as the y- and z-components are 0 only for $t = 0$, but $r'(0) = \langle 1, 0, 0 \rangle \neq 0$.

Thus, the curve is smooth.

(c) $r(t) = \langle \cos^3 t, \sin^3 t \rangle \Rightarrow r'(t) = \langle -3\cos^2 t \sin t, 3\sin^2 t \cos t \rangle$. Since

$r'(0) = \langle -3\cos^2 0 \sin 0, 3\sin^2 0 \cos 0 \rangle = \langle 0, 0 \rangle = 0$, the curve is not smooth.

31. The angle of intersection of the two curves is the angle between the two tangent vectors to the curves at the point of

intersection. Since $r'_1(t) = \langle 1, 2t, 3t^2 \rangle$ and $t = 0$ at $(0, 0, 0)$, $r'_1(0) = \langle 1, 0, 0 \rangle$ is a tangent vector to r_1 at

$(0, 0, 0)$. Similarly, $r'_2(t) = \langle \cos t, 2\cos 2t, 1 \rangle$ and since $r_2(0) = \langle 0, 0, 0 \rangle$, $r'_2(0) = \langle 1, 2, 1 \rangle$ is a tangent vector to

r_2 at $(0, 0, 0)$. If θ is the angle between these two tangent vectors, then $\cos\theta = \frac{1}{\sqrt{1}\sqrt{6}}\langle 1, 0, 0 \rangle \cdot \langle 1, 2, 1 \rangle = \frac{1}{\sqrt{6}}$ and

$\theta = \cos^{-1}\frac{1}{\sqrt{6}} \approx 66°$.

33. $\int_0^1 (t\,i + t^2\,j + t^3\,k)\,dt = \left(\int_0^1 t\,dt\right)i + \left(\int_0^1 t^2\,dt\right)j + \left(\int_0^1 t^3\,dt\right)k$

$= \left[\frac{t^2}{2}\right]_0^1 i + \left[\frac{t^3}{3}\right]_0^1 j + \left[\frac{t^4}{4}\right]_0^1 k$

$= \frac{1}{2}i + \frac{1}{3}j + \frac{1}{4}k$

35. $\int_0^{\pi/4} (\cos 2t\,i + \sin 2t\,j + t\sin t\,k)\,dt$

$= \left[\frac{1}{2}\sin 2t\,i - \frac{1}{2}\cos 2t\,j\right]_0^{\pi/4} + \left[[-t\cos t]_0^{\pi/4} + \int_0^{\pi/4}\cos t\,dt\right]k$

$= \frac{1}{2}i + \frac{1}{2}j + \left[-\frac{\pi}{4}\cos\frac{\pi}{4} + \sin\frac{\pi}{4}\right]k$

$= \frac{1}{2}i + \frac{1}{2}j + \frac{1}{\sqrt{2}}\left(1 - \frac{\pi}{4}\right)k$

$= \frac{1}{2}i + \frac{1}{2}j + \frac{4-\pi}{4\sqrt{2}}k$

37. $\int (e^t\,i + 2t\,j + \ln t\,k)\,dt = \left(\int e^t\,dt\right)i + \left(\int 2t\,dt\right)j + \left(\int \ln t\,dt\right)k = e^t\,i + t^2\,j + (t\ln t - t)k + C$, where C is

a vector constant of integration.

39. $r'(t) = t^2\,i + 4t^3\,j - t^2\,k \Rightarrow r(t) = \frac{1}{3}t^3\,i + t^4\,j - \frac{1}{3}t^3\,k + C$, where C is a constant vector. But

$j = r(0) = (0)i + (0)j - (0)k + C$. Thus $C = j$ and

$r(t) = \frac{1}{3}t^3\,i + t^4\,j - \frac{1}{3}t^3\,k + j = \frac{1}{3}t^3\,i + (t^4 + 1)j - \frac{1}{3}t^3\,k$.

For Exercises 41–43, let $\mathbf{u}(t) = \langle u_1(t), u_2(t), u_3(t)\rangle$ and $\mathbf{v}(t) = \langle v_1(t), v_2(t), v_3(t)\rangle$. In each of these exercises, the procedure is to apply Theorem 2 so that the corresponding properties of derivatives of real-valued functions can be used.

41. $\dfrac{d}{dt}\left[\mathbf{u}(t) + \mathbf{v}(t)\right] = \dfrac{d}{dt}\langle u_1(t) + v_1(t), u_2(t) + v_2(t), u_3(t) + v_3(t)\rangle$

$$= \left\langle \frac{d}{dt}\left[u_1(t) + v_1(t)\right], \frac{d}{dt}\left[u_2(t) + v_2(t)\right], \frac{d}{dt}\left[u_3(t) + v_3(t)\right]\right\rangle$$

$$= \langle u_1'(t) + v_1'(t), u_2'(t) + v_2'(t), u_3'(t) + v_3'(t)\rangle$$

$$= \langle u_1'(t), u_2'(t), u_3'(t)\rangle + \langle v_1'(t), v_2'(t), v_3'(t)\rangle = \mathbf{u}'(t) + \mathbf{v}'(t).$$

43. $\dfrac{d}{dt}\left[\mathbf{u}(t) \times \mathbf{v}(t)\right]$

$$= \frac{d}{dt}\langle u_2(t)v_3(t) - u_3(t)v_2(t), u_3(t)v_1(t) - u_1(t)v_3(t), u_1(t)v_2(t) - u_2(t)v_1(t)\rangle$$

$$= \langle u_2'v_3(t) + u_2(t)v_3'(t) - u_3'(t)v_2(t) - u_3(t)v_2'(t),$$

$$u_3'(t)v_1(t) + u_3(t)v_1'(t) - u_1'(t)v_3(t) - u_1(t)v_3'(t),$$

$$u_1'(t)v_2(t) + u_1(t)v_2'(t) - u_2'(t)v_1(t) - u_2(t)v_1'(t)\rangle$$

$$= \langle u_2'(t)v_3(t) - u_3'(t)v_2(t), u_3'(t)v_1(t) - u_1'(t)v_3(t), u_1'(t)v_2(t) - u_2'(t)v_1(t)\rangle$$

$$+ \langle u_2(t)v_3'(t) - u_3(t)v_2'(t), u_3(t)v_1'(t) - u_1(t)v_3'(t), u_1(t)v_2'(t) - u_2(t)v_1'(t)\rangle$$

$$= \mathbf{u}'(t) \times \mathbf{v}(t) + \mathbf{u}(t) \times \mathbf{v}'(t)$$

Alternate Solution: Let $\mathbf{r}(t) = \mathbf{u}(t) \times \mathbf{v}(t)$. Then

$$\mathbf{r}(t + h) - \mathbf{r}(t) = \left[\mathbf{u}(t + h) \times \mathbf{v}(t + h)\right] - \left[\mathbf{u}(t) \times \mathbf{v}(t)\right]$$

$$= \left[\mathbf{u}(t + h) \times \mathbf{v}(t + h)\right] - \left[\mathbf{u}(t) \times \mathbf{v}(t)\right] + \left[\mathbf{u}(t + h) \times \mathbf{v}(t)\right] - \left[\mathbf{u}(t + h) \times \mathbf{v}(t)\right]$$

$$= \mathbf{u}(t + h) \times \left[\mathbf{v}(t + h) - \mathbf{v}(t)\right] + \left[\mathbf{u}(t + h) - \mathbf{u}(t)\right] \times \mathbf{v}(t)$$

(Be careful of the order of the cross product.)

Dividing through by h and taking the limit as $h \to 0$ we have

$$\mathbf{r}'(t) = \lim_{h\to 0}\frac{\mathbf{u}(t + h) \times \left[\mathbf{v}(t + h) - \mathbf{v}(t)\right]}{h} + \lim_{h\to 0}\frac{\left[\mathbf{u}(t + h) - \mathbf{u}(t)\right] \times \mathbf{v}(t)}{h}$$

$$= \mathbf{u}(t) \times \mathbf{v}'(t) + \mathbf{u}'(t) \times \mathbf{v}(t)$$

by Exercise 14.1.35(a) [ET 13.1.35(a)] and Definition 1.

45. $D_t\left[\mathbf{u}(t) \cdot \mathbf{v}(t)\right] = \mathbf{u}'(t) \cdot \mathbf{v}(t) + \mathbf{u}(t) \cdot \mathbf{v}'(t)$ by Formula 4 of Theorem 3

$$= \left(-4t\,\mathbf{j} + 9t^2\,\mathbf{k}\right) \cdot \left(t\,\mathbf{i} + \cos t\,\mathbf{j} + \sin t\,\mathbf{k}\right) + \left(\mathbf{i} - 2t^2\,\mathbf{j} + 3t^3\,\mathbf{k}\right) \cdot \left(\mathbf{i} - \sin t\,\mathbf{j} + \cos t\,\mathbf{k}\right)$$

$$= -4t\cos t + 9t^2\sin t + 1 + 2t^2\sin t + 3t^3\cos t$$

$$= 1 - 4t\cos t + 11t^2\sin t + 3t^3\cos t$$

47. $\dfrac{d}{dt}\left[\mathbf{r}\left(t\right)\times\mathbf{r}'\left(t\right)\right]=\mathbf{r}'\left(t\right)\times\mathbf{r}'\left(t\right)+\mathbf{r}\left(t\right)\times\mathbf{r}''\left(t\right)$ by Formula 5 of Theorem 3. But $\mathbf{r}'\left(t\right)\times\mathbf{r}'\left(t\right)=\mathbf{0}$ (see

Example 13.4.2 [ET 12.4.2]). Thus, $\dfrac{d}{dt}\left[\mathbf{r}\left(t\right)\times\mathbf{r}'\left(t\right)\right]=\mathbf{r}\left(t\right)\times\mathbf{r}''\left(t\right)$.

49. $\dfrac{d}{dt}\left|\mathbf{r}\left(t\right)\right|=\dfrac{d}{dt}\left[\mathbf{r}\left(t\right)\cdot\mathbf{r}\left(t\right)\right]^{1/2}=\tfrac{1}{2}\left[\mathbf{r}\left(t\right)\cdot\mathbf{r}\left(t\right)\right]^{-1/2}\left[2\mathbf{r}\left(t\right)\cdot\mathbf{r}'\left(t\right)\right]=\dfrac{\mathbf{r}\left(t\right)\cdot\mathbf{r}'\left(t\right)}{\left|\mathbf{r}\left(t\right)\right|}$

51. Since $\mathbf{u}\left(t\right)=\mathbf{r}\left(t\right)\cdot\left[\mathbf{r}'\left(t\right)\times\mathbf{r}''\left(t\right)\right]$,

$$\mathbf{u}'\left(t\right)=\mathbf{r}'\left(t\right)\cdot\left[\mathbf{r}'\left(t\right)\times\mathbf{r}''\left(t\right)\right]+\mathbf{r}\left(t\right)\cdot\dfrac{d}{dt}\left[\mathbf{r}'\left(t\right)\times\mathbf{r}''\left(t\right)\right]$$

$$=0+\mathbf{r}\left(t\right)\cdot\left[\mathbf{r}''\left(t\right)\times\mathbf{r}''\left(t\right)+\mathbf{r}'\left(t\right)\times\mathbf{r}'''\left(t\right)\right]\qquad\left[\text{since }\mathbf{r}'\left(t\right)\perp\mathbf{r}'\left(t\right)\times\mathbf{r}''\left(t\right)\right]$$

$$=\mathbf{r}\left(t\right)\cdot\left[\mathbf{r}'\left(t\right)\times\mathbf{r}'''\left(t\right)\right]\qquad\left[\text{since }\mathbf{r}''\left(t\right)\times\mathbf{r}''\left(t\right)=\mathbf{0}\right]$$

14.3 Arc Length and Curvature   ET 13.3

1. $\mathbf{r}'\left(t\right)=\langle 2\cos t,5,-2\sin t\rangle\ \Rightarrow\ \left|\mathbf{r}'\left(t\right)\right|=\sqrt{\left(2\cos t\right)^{2}+5^{2}+\left(-2\sin t\right)^{2}}=\sqrt{29}$. Then using Formula 3, we

have $L=\int_{-10}^{10}\left|\mathbf{r}'\left(t\right)\right|\,dt=\int_{-10}^{10}\sqrt{29}\,dt=\sqrt{29}\,t\,\Big]_{-10}^{10}=20\sqrt{29}$.

3. $\mathbf{r}'\left(t\right)=\sqrt{2}\mathbf{i}+e^{t}\mathbf{j}-e^{-t}\mathbf{k}\ \Rightarrow$

$\left|\mathbf{r}'\left(t\right)\right|=\sqrt{\left(\sqrt{2}\right)^{2}+\left(e^{t}\right)^{2}+\left(-e^{-t}\right)^{2}}=\sqrt{2+e^{2t}+e^{-2t}}=\sqrt{\left(e^{t}+e^{-t}\right)^{2}}=e^{t}+e^{-t}$ (since $e^{t}+e^{-t}>0$).

Then $L=\int_{0}^{1}\left|\mathbf{r}'\left(t\right)\right|\,dt=\int_{0}^{1}\left(e^{t}+e^{-t}\right)\,dt=\left[e^{t}-e^{-t}\right]_{0}^{1}=e-e^{-1}$.

5. The point $\left(2,4,8\right)$ corresponds to $t=2$, so by Equation 2, $L=\int_{0}^{2}\sqrt{\left(1\right)^{2}+\left(2t\right)^{2}+\left(3t^{2}\right)^{2}}\,dt$.

If $f\left(t\right)=\sqrt{1+4t^{2}+9t^{4}}$, then Simpson's Rule gives

$L\approx\dfrac{2-0}{10\cdot3}\left[f\left(0\right)+4f\left(0.2\right)+2f\left(0.4\right)+\cdots+4f\left(1.8\right)+f\left(2\right)\right]\approx9.5706$.

7. $\mathbf{r}'\left(t\right)=e^{t}\left(\cos t+\sin t\right)\mathbf{i}+e^{t}\left(\cos t-\sin t\right)\mathbf{j}$,

$ds/dt=\left|\mathbf{r}'\left(t\right)\right|=e^{t}\sqrt{\left(\cos t+\sin t\right)^{2}+\left(\cos t-\sin t\right)^{2}}=e^{t}\sqrt{2\cos^{2}t+2\sin^{2}t}=\sqrt{2}\,e^{t}$.

$s\left(t\right)=\int_{0}^{t}\left|\mathbf{r}'\left(u\right)\right|\,du=\int_{0}^{t}\sqrt{2}\,e^{u}\,du=\sqrt{2}\left(e^{t}-1\right)\ \Rightarrow\ \tfrac{1}{\sqrt{2}}s+1=e^{t}\ \Rightarrow\ t\left(s\right)=\ln\left(\tfrac{1}{\sqrt{2}}s+1\right)$.

Therefore, $\mathbf{r}\left(t\left(s\right)\right)=\left(\tfrac{1}{\sqrt{2}}s+1\right)\left[\sin\left(\ln\left(\tfrac{1}{\sqrt{2}}s+1\right)\right)\mathbf{i}+\cos\left(\ln\left(\tfrac{1}{\sqrt{2}}s+1\right)\right)\mathbf{j}\right]$.

9. $\left|\mathbf{r}'\left(t\right)\right|=\sqrt{\left(3\cos t\right)^{2}+16+\left(-3\sin t\right)^{2}}=\sqrt{9+16}=5$ and $s\left(t\right)=\int_{0}^{t}\left|\mathbf{r}'\left(u\right)\right|\,du=\int_{0}^{t}5\,du=5t\ \Rightarrow$

$t\left(s\right)=\tfrac{1}{5}s$. Therefore, $\mathbf{r}\left(t\left(s\right)\right)=3\sin\left(\tfrac{1}{5}s\right)\mathbf{i}+\tfrac{4}{5}s\mathbf{j}+3\cos\left(\tfrac{1}{5}s\right)\mathbf{k}$.

11. (a) $\mathbf{r}'\left(t\right)=\langle 2\cos t,5,-2\sin t\rangle\ \Rightarrow\ \left|\mathbf{r}'\left(t\right)\right|=\sqrt{4\cos^{2}t+25+4\sin^{2}t}=\sqrt{29}$. Then

$\mathbf{T}\left(t\right)=\dfrac{\mathbf{r}'\left(t\right)}{\left|\mathbf{r}'\left(t\right)\right|}=\tfrac{1}{\sqrt{29}}\langle 2\cos t,5,-2\sin t\rangle$ or $\left\langle\tfrac{2}{\sqrt{29}}\cos t,\tfrac{5}{\sqrt{29}},-\tfrac{2}{\sqrt{29}}\sin t\right\rangle$.

$\mathbf{T}'\left(t\right)=\tfrac{1}{\sqrt{29}}\langle -2\sin t,0,-2\cos t\rangle\ \Rightarrow\ \left|\mathbf{T}'\left(t\right)\right|=\tfrac{1}{\sqrt{29}}\sqrt{4\sin^{2}t+0+4\cos^{2}t}=\tfrac{2}{\sqrt{29}}$. Thus

$\mathbf{N}\left(t\right)=\dfrac{\mathbf{T}'\left(t\right)}{\left|\mathbf{T}'\left(t\right)\right|}=\dfrac{1/\sqrt{29}}{2/\sqrt{29}}\langle -2\sin t,0,-2\cos t\rangle=\langle -\sin t,0,-\cos t\rangle$.

(b) $\kappa\left(t\right)=\dfrac{\left|\mathbf{T}'\left(t\right)\right|}{\left|\mathbf{r}'\left(t\right)\right|}=\dfrac{2/\sqrt{29}}{\sqrt{29}}=\dfrac{2}{29}$.

13. (a) $\mathbf{r}'(t) = \langle t^2, 2t, 2 \rangle$ $\Rightarrow$ $|\mathbf{r}'(t)| = \sqrt{t^4 + 4t^2 + 4} = \sqrt{(t^2+2)^2} = t^2 + 2$. Then

$$\mathbf{T}(t) = \frac{\mathbf{r}'(t)}{|\mathbf{r}'(t)|} = \frac{1}{t^2+2} \langle t^2, 2t, 2 \rangle.$$

$$\mathbf{T}'(t) = \frac{-2t}{(t^2+2)^2} \langle t^2, 2t, 2 \rangle + \frac{1}{t^2+2} \langle 2t, 2, 0 \rangle \qquad \text{(by Theorem 14.2.3 [ET 13.2.3] \#3)}$$

$$= \frac{1}{(t^2+2)^2} \langle -2t^3, -4t^2, -4t \rangle + \frac{1}{(t^2+2)^2} \langle 2t^3 + 4t, 2t^2 + 4, 0 \rangle = \frac{1}{(t^2+2)^2} \langle 4t, 4 - 2t^2, -4t \rangle$$

$$|\mathbf{T}'(t)| = \frac{1}{(t^2+2)^2} \sqrt{16t^2 + (16 - 16t^2 + 4t^4) + 16t^2} = \frac{1}{(t^2+2)^2} \sqrt{4t^4 + 16t^2 + 16}$$

$$= \frac{1}{(t^2+2)^2} \sqrt{4(t^2+2)^2} = \frac{2(t^2+2)}{(t^2+2)^2} = \frac{2}{t^2+2}$$

Thus $\mathbf{N}(t) = \dfrac{\mathbf{T}'(t)}{|\mathbf{T}'(t)|} = \dfrac{1/(t^2+2)^2}{2/(t^2+2)} \langle 4t, 4 - 2t^2, -4t \rangle = \dfrac{1}{t^2+2} \langle 2t, 2 - t^2, -2t \rangle$.

(b) $\kappa(t) = \dfrac{|\mathbf{T}'(t)|}{|\mathbf{r}'(t)|} = \dfrac{2/(t^2+2)}{t^2+2} = \dfrac{2}{(t^2+2)^2}$

15. $\mathbf{r}'(t) = 2t\,\mathbf{i} + \mathbf{k}$, $\mathbf{r}''(t) = 2\mathbf{i}$, $|\mathbf{r}'(t)| = \sqrt{(2t)^2 + 0^2 + 1^2} = \sqrt{4t^2 + 1}$, $\mathbf{r}'(t) \times \mathbf{r}''(t) = 2\mathbf{j}$,

$|\mathbf{r}'(t) \times \mathbf{r}''(t)| = 2$. Then $\kappa(t) = \dfrac{|\mathbf{r}'(t) \times \mathbf{r}''(t)|}{|\mathbf{r}'(t)|^3} = \dfrac{2}{\left(\sqrt{4t^2+1}\right)^3} = \dfrac{2}{(4t^2+1)^{3/2}}$.

17. $\mathbf{r}'(t) = \langle \cos t, -\sin t, \cos t \rangle$, $\mathbf{r}''(t) = \langle -\sin t, -\cos t, -\sin t \rangle$, $|\mathbf{r}'(t)|^3 = \left(\sqrt{\cos^2 t + 1}\right)^3$,

$|\mathbf{r}'(t) \times \mathbf{r}''(t)| = |\langle 1, 0, -1 \rangle| = \sqrt{2}$, $\kappa(t) = \dfrac{|\mathbf{r}'(t) \times \mathbf{r}''(t)|}{|\mathbf{r}'(t)|^3} = \dfrac{\sqrt{2}}{(1 + \cos^2 t)^{3/2}}$

19. $\mathbf{r}'(t) = \langle \sqrt{2}, e^t, -e^{-t} \rangle$. The point $(0, 1, 1)$ corresponds to $t = 0$, and

$\mathbf{r}'(0) = \langle \sqrt{2}, 1, -1 \rangle$ $\Rightarrow$ $|\mathbf{r}'(0)| = \sqrt{(\sqrt{2})^2 + 1^2 + (-1)^2} = 2$.

$\mathbf{r}''(t) = \langle 0, e^t, e^{-t} \rangle$ $\Rightarrow$ $\mathbf{r}''(0) = \langle 0, 1, 1 \rangle$. $\mathbf{r}'(0) \times \mathbf{r}''(0) = \langle 2, -\sqrt{2}, \sqrt{2} \rangle$,

$|\mathbf{r}'(0) \times \mathbf{r}''(0)| = \sqrt{2^2 + (-\sqrt{2})^2 + (\sqrt{2})^2} = \sqrt{8} = 2\sqrt{2}$. Then $\kappa(0) = \dfrac{|\mathbf{r}'(0) \times \mathbf{r}''(0)|}{|\mathbf{r}'(0)|^3} = \dfrac{2\sqrt{2}}{2^3} = \dfrac{\sqrt{2}}{4}$.

21. $f(x) = x^3$, $f'(x) = 3x^2$, $f''(x) = 6x$, $\kappa(x) = \dfrac{|f''(x)|}{\left[1 + (f'(x))^2\right]^{3/2}} = \dfrac{6|x|}{(1 + 9x^4)^{3/2}}$

23. $y' = \cos x$, $y'' = -\sin x$, $\kappa(x) = \dfrac{|y''(x)|}{\left[1 + (y'(x))^2\right]^{3/2}} = \dfrac{|\sin x|}{(1 + \cos^2 x)^{3/2}}$

25. Since $y' = y'' = e^x$, the curvature is $\kappa(x) = \dfrac{|y''(x)|}{\left[1 + (y'(x))^2\right]^{3/2}} = \dfrac{e^x}{(1 + e^{2x})^{3/2}} = e^x \left(1 + e^{2x}\right)^{-3/2}$. To find

the maximum curvature, we first find the critical numbers of $\kappa(x)$:

$$\kappa'(x) = e^x \left(1 + e^{2x}\right)^{-3/2} + e^x \left(-\tfrac{3}{2}\right) \left(1 + e^{2x}\right)^{-5/2} \left(2e^{2x}\right) = e^x \dfrac{1 + e^{2x} - 3e^{2x}}{(1 + e^{2x})^{5/2}} = e^x \dfrac{1 - 2e^{2x}}{(1 + e^{2x})^{5/2}}.$$

$\kappa'(x) = 0$ when $1 - 2e^{2x} = 0$, so $e^{2x} = \tfrac{1}{2}$ or $x = -\tfrac{1}{2}\ln 2$. And since $1 - 2e^{2x} > 0$ for $x < -\tfrac{1}{2}\ln 2$ and

$1 - 2e^{2x} < 0$ for $x > -\tfrac{1}{2}\ln 2$, the maximum curvature is attained at the point

$\left(-\tfrac{1}{2}\ln 2, e^{(-\ln 2)/2}\right) = \left(-\tfrac{1}{2}\ln 2, \tfrac{1}{\sqrt{2}}\right)$. Since $\lim\limits_{x \to \infty} e^x \left(1 + e^{2x}\right)^{-3/2} = 0$, $\kappa(x)$ approaches 0 as $x \to \infty$.

27. (a) C appears to be changing direction more quickly at P than Q, so we would expect the curvature to be greater
at P.

(b) First we sketch approximate osculating circles at P and Q.
Using the axes scale as a guide, we measure the radius of the
osculating circle at P to be approximately 0.8 units, thus

$$\rho = \dfrac{1}{\kappa} \quad \Rightarrow \quad \kappa = \dfrac{1}{\rho} \approx \dfrac{1}{0.8} \approx 1.3. \text{ Similarly, we estimate}$$

the radius of the osculating circle at Q to be 1.4 units, so

$$\kappa = \dfrac{1}{\rho} \approx \dfrac{1}{1.4} \approx 0.7.$$

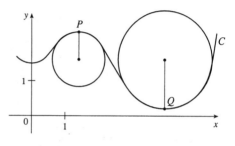

29. $y = x^4 \quad \Rightarrow \quad y' = 4x^3$, $y'' = 12x^2$, and $\kappa(x) = \dfrac{|y''|}{\left[1 + (y')^2\right]^{3/2}} = \dfrac{12x^2}{(1 + 16x^6)^{3/2}}$. The appearance of the two

humps in this graph is perhaps a little surprising, but it is explained by the fact that $y = x^4$ is very flat around the
origin, and so here the curvature is zero.

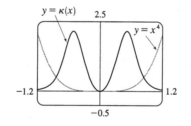

31. $\kappa(t) = \dfrac{|\dot{x}\ddot{y} - \ddot{x}\dot{y}|}{(\dot{x}^2 + \dot{y}^2)^{3/2}} = \dfrac{\left|(3t^2)(2) - (6t)(2t)\right|}{(9t^4 + 4t^2)^{3/2}} = \dfrac{6t^2}{(t^2)^{3/2}(9t^2 + 4)^{3/2}} = \dfrac{6t^2}{|t|^3(9t^2 + 4)^{3/2}}$

$$= \dfrac{6}{|t|(9t^2 + 4)^{3/2}}$$

33. $\left(1, \tfrac{2}{3}, 1\right)$ corresponds to $t = 1$. $\mathbf{T}(t) = \dfrac{\mathbf{r}'(t)}{|\mathbf{r}'(t)|} = \dfrac{\langle 2t, 2t^2, 1 \rangle}{\sqrt{4t^2 + 4t^4 + 1}} = \dfrac{\langle 2t, 2t^2, 1 \rangle}{2t^2 + 1}$, so $\mathbf{T}(1) = \langle \tfrac{2}{3}, \tfrac{2}{3}, \tfrac{1}{3} \rangle$.

$\mathbf{T}'(t) = -4t \left(2t^2 + 1\right)^{-2} \langle 2t, 2t^2, 1 \rangle + \left(2t^2 + 1\right)^{-1} \langle 2, 4t, 0 \rangle$ (By Theorem 14.2.3 [ET 13.2.3] #3)

$\quad = \left(2t^2 + 1\right)^{-2} \langle -8t^2 + 4t^2 + 2, -8t^3 + 8t^3 + 4t, -4t \rangle = 2 \left(2t^2 + 1\right)^{-2} \langle 1 - 2t^2, 2t, -2t \rangle$

$$\mathbf{N}(t) = \frac{\mathbf{T}'(t)}{|\mathbf{T}'(t)|} = \frac{2\left(2t^2+1\right)^{-2}\left\langle 1-2t^2, 2t, -2t\right\rangle}{2\left(2t^2+1\right)^{-2}\sqrt{\left(1-2t^2\right)^2+(2t)^2+(-2t)^2}} = \frac{\left\langle 1-2t^2, 2t, -2t\right\rangle}{\sqrt{1-4t^2+4t^4+8t^2}}$$

$$= \frac{\left\langle 1-2t^2, 2t, -2t\right\rangle}{1+2t^2}$$

$\mathbf{N}(1) = \left\langle -\frac{1}{3}, \frac{2}{3}, -\frac{2}{3}\right\rangle$ and $\mathbf{B}(1) = \mathbf{T}(1) \times \mathbf{N}(1) = \left\langle -\frac{4}{9}-\frac{2}{9}, -\left(-\frac{4}{9}+\frac{1}{9}\right), \frac{4}{9}+\frac{2}{9}\right\rangle = \left\langle -\frac{2}{3}, \frac{1}{3}, \frac{2}{3}\right\rangle.$

35. $(0, \pi, -2)$ corresponds to $t = \pi$. $\mathbf{r}(t) = \langle 2\sin 3t, t, 2\cos 3t\rangle \quad \Rightarrow$

$$\mathbf{T}(t) = \frac{\mathbf{r}'(t)}{|\mathbf{r}'(t)|} = \frac{\langle 6\cos 3t, 1, -6\sin 3t\rangle}{\sqrt{36\cos^2 3t + 1 + 36\sin^2 3t}} = \frac{1}{\sqrt{37}}\langle 6\cos 3t, 1, -6\sin 3t\rangle. \ \mathbf{T}(\pi) = \frac{1}{\sqrt{37}}\langle -6, 1, 0\rangle$$

is a normal vector for the normal plane, and so $\langle -6, 1, 0\rangle$ is also normal. Thus an equation for the plane is

$-6(x-0) + 1(y-\pi) + 0(z+2) = 0$ or $y - 6x = \pi$. $\mathbf{T}'(t) = \frac{1}{\sqrt{37}}\langle -18\sin 3t, 0, -18\cos 3t\rangle \quad \Rightarrow$

$|\mathbf{T}'(t)| = \frac{\sqrt{18^2\sin^2 3t + 18^2\cos^2 3t}}{\sqrt{37}} = \frac{18}{\sqrt{37}} \quad \Rightarrow \quad \mathbf{N}(t) = \frac{\mathbf{T}'(t)}{|\mathbf{T}'(t)|} = \langle -\sin 3t, 0, -\cos 3t\rangle.$ So

$\mathbf{N}(\pi) = \langle 0, 0, 1\rangle$ and $\mathbf{B}(\pi) = \frac{1}{\sqrt{37}}\langle -6, 1, 0\rangle \times \langle 0, 0, 1\rangle = \frac{1}{\sqrt{37}}\langle 1, 6, 0\rangle.$ Since $\mathbf{B}(\pi)$ is a normal to the

osculating plane, so is $\langle 1, 6, 0\rangle$ and an equation for the plane is $1(x-0) + 6(y-\pi) + 0(z+2) = 0$ or

$x + 6y = 6\pi.$

37. The ellipse is given by the parametric equations $x = 2\cos t$, $y = 3\sin t$, so using the result from Exercise 30,

$$\kappa(t) = \frac{|\dot{x}\ddot{y} - \ddot{x}\dot{y}|}{(\dot{x}^2 + \dot{y}^2)^{3/2}} = \frac{|(-2\sin t)(-3\sin t) - (3\cos t)(-2\cos t)|}{\left(4\sin^2 t + 9\cos^2 t\right)^{3/2}} = \frac{6}{\left(4\sin^2 t + 9\cos^2 t\right)^{3/2}}.$$

At $(2, 0)$, $t = 0$. Now $\kappa(0) = \frac{6}{27} = \frac{2}{9}$, so the radius of the osculating

circle is $1/\kappa(0) = \frac{9}{2}$ and its center is $\left(-\frac{5}{2}, 0\right)$. Its equation is

therefore $\left(x + \frac{5}{2}\right)^2 + y^2 = \frac{81}{4}$. At $(0, 3)$, $t = \frac{\pi}{2}$, and

$\kappa\left(\frac{\pi}{2}\right) = \frac{6}{8} = \frac{3}{4}$. So the radius of the osculating circle is $\frac{4}{3}$ and its

center is $\left(0, \frac{5}{3}\right)$. Hence its equation is $x^2 + \left(y - \frac{5}{3}\right)^2 = \frac{16}{9}$.

39. The tangent vector is normal to the normal plane, and the vector $\langle 6, 6, -8\rangle$ is normal to the given plane. But

$\mathbf{T}(t) \parallel \mathbf{r}'(t)$ and $\langle 6, 6, -8\rangle \parallel \langle 3, 3, -4\rangle$, so we need to find t such that $\mathbf{r}'(t) \parallel \langle 3, 3, -4\rangle$. $\mathbf{r}(t) = \langle t^3, 3t, t^4\rangle \quad \Rightarrow$

$\mathbf{r}'(t) = \langle 3t^2, 3, 4t^3\rangle \parallel \langle 3, 3, -4\rangle$ when $t = -1$. So the planes are parallel at the point $\mathbf{r}(-1) = (-1, -3, 1)$.

41. $\kappa = \left|\dfrac{d\mathbf{T}}{ds}\right| = \left|\dfrac{d\mathbf{T}/dt}{ds/dt}\right| = \dfrac{|d\mathbf{T}/dt|}{ds/dt}$ and $\mathbf{N} = \dfrac{d\mathbf{T}/dt}{|d\mathbf{T}/dt|}$, so $\kappa\mathbf{N} = \dfrac{\dfrac{d\mathbf{T}}{dt}\cdot\dfrac{d\mathbf{T}}{dt}}{\dfrac{d\mathbf{T}}{dt}\cdot\dfrac{ds}{dt}} = \dfrac{d\mathbf{T}/dt}{ds/dt} = \dfrac{d\mathbf{T}}{ds}$ by the Chain Rule.

43. (a) $|\mathbf{B}| = 1 \quad \Rightarrow \quad \mathbf{B} \cdot \mathbf{B} = 1 \quad \Rightarrow \quad \dfrac{d}{ds}(\mathbf{B} \cdot \mathbf{B}) = 0 \quad \Rightarrow \quad 2\dfrac{d\mathbf{B}}{ds} \cdot \mathbf{B} = 0 \quad \Rightarrow \quad \dfrac{d\mathbf{B}}{ds} \perp \mathbf{B}$

(b) $\mathbf{B} = \mathbf{T} \times \mathbf{N}$ $\Rightarrow$

$$\frac{d\mathbf{B}}{ds} = \frac{d}{ds}(\mathbf{T} \times \mathbf{N}) = \frac{d}{dt}(\mathbf{T} \times \mathbf{N})\frac{1}{ds/dt} = \frac{d}{dt}(\mathbf{T} \times \mathbf{N})\frac{1}{|\mathbf{r}'(t)|}$$

$$= [(\mathbf{T}' \times \mathbf{N}) + (\mathbf{T} \times \mathbf{N}')]\frac{1}{|\mathbf{r}'(t)|} = \left[\left(\mathbf{T}' \times \frac{\mathbf{T}'}{|\mathbf{T}'|}\right) + (\mathbf{T} \times \mathbf{N}')\right]\frac{1}{|\mathbf{r}'(t)|} = \frac{\mathbf{T} \times \mathbf{N}'}{|\mathbf{r}'(t)|}$$

$$\Rightarrow \quad \frac{d\mathbf{B}}{ds} \perp \mathbf{T}$$

(c) $\mathbf{B} = \mathbf{T} \times \mathbf{N}$ $\Rightarrow$ $\mathbf{T} \perp \mathbf{N}, \mathbf{B} \perp \mathbf{T}$ and $\mathbf{B} \perp \mathbf{N}$. So $\mathbf{B}, \mathbf{T}$ and $\mathbf{N}$ form an orthogonal set of vectors in the three-dimensional space $\mathbb{R}^3$. From parts (a) and (b), $d\mathbf{B}/ds$ is perpendicular to both $\mathbf{B}$ and $\mathbf{T}$, so $d\mathbf{B}/ds$ is parallel to $\mathbf{N}$. Therefore, $d\mathbf{B}/ds = -\tau(s)\mathbf{N}$, where $\tau(s)$ is a scalar.

(d) Since $\mathbf{B} = \mathbf{T} \times \mathbf{N}$, $\mathbf{T} \perp \mathbf{N}$ and both $\mathbf{T}$ and $\mathbf{N}$ are unit vectors, $\mathbf{B}$ is a unit vector mutually perpendicular to both $\mathbf{T}$ and $\mathbf{N}$. For a plane curve, $\mathbf{T}$ and $\mathbf{N}$ always lie in the plane of the curve, so that $\mathbf{B}$ is a constant unit vector always perpendicular to the plane. Thus $d\mathbf{B}/ds = \mathbf{0}$, but $d\mathbf{B}/ds = -\tau(s)\mathbf{N}$ and $\mathbf{N} \neq \mathbf{0}$, so $\tau(s) = 0$.

45. (a) $\mathbf{r}' = s'\mathbf{T}$ $\Rightarrow$ $\mathbf{r}'' = s''\mathbf{T} + s'\mathbf{T}' = s''\mathbf{T} + s'\frac{d\mathbf{T}}{ds}s' = s''\mathbf{T} + \kappa(s')^2\mathbf{N}$ by the first Serret-Frenet formula.

(b) Using part (a), we have

$$\mathbf{r}' \times \mathbf{r}'' = (s'\mathbf{T}) \times \left[s''\mathbf{T} + \kappa(s')^2\mathbf{N}\right]$$

$$\doteq [(s'\mathbf{T}) \times (s''\mathbf{T})] + \left[(s'\mathbf{T}) \times \left(\kappa(s')^2\mathbf{N}\right)\right] \quad \left(\text{By Theorem 13.4.8 [ET 12.4.8] \#3}\right)$$

$$= (s's'')(\mathbf{T} \times \mathbf{T}) + \kappa(s')^3(\mathbf{T} \times \mathbf{N}) = \mathbf{0} + \kappa(s')^3\mathbf{B} = \kappa(s')^3\mathbf{B}$$

(c) Using part (a), we have

$$\mathbf{r}''' = \left[s''\mathbf{T} + \kappa(s')^2\mathbf{N}\right]' = s'''\mathbf{T} + s''\mathbf{T}' + \kappa'(s')^2\mathbf{N} + 2\kappa s's''\mathbf{N} + \kappa(s')^2\mathbf{N}'$$

$$= s'''\mathbf{T} + s''\frac{d\mathbf{T}}{ds}s' + \kappa'(s')^2\mathbf{N} + 2\kappa s's''\mathbf{N} + \kappa(s')^2\frac{d\mathbf{N}}{ds}s'$$

$$= s'''\mathbf{T} + s''s'\kappa\mathbf{N} + \kappa'(s')^2\mathbf{N} + 2\kappa s's''\mathbf{N} + \kappa(s')^3(-\kappa\mathbf{T} + \tau\mathbf{B}) \text{ (by the second formula)}$$

$$= \left[s''' - \kappa^2(s')^3\right]\mathbf{T} + \left[3\kappa s's'' + \kappa'(s')^2\right]\mathbf{N} + \kappa\tau(s')^3\mathbf{B}$$

(d) Using parts (b) and (c) and the facts that $\mathbf{B} \cdot \mathbf{T} = 0$, $\mathbf{B} \cdot \mathbf{N} = 0$, and $\mathbf{B} \cdot \mathbf{B} = 1$, we get

$$\frac{(\mathbf{r}' \times \mathbf{r}'') \cdot \mathbf{r}'''}{|\mathbf{r}' \times \mathbf{r}''|^2} = \frac{\kappa(s')^3\mathbf{B} \cdot \left\{\left[s''' - \kappa^2(s')^3\right]\mathbf{T} + \left[3\kappa s's'' + \kappa'(s')^2\right]\mathbf{N} + \kappa\tau(s')^3\mathbf{B}\right\}}{|\kappa(s')^3\mathbf{B}|^2}$$

$$= \frac{\kappa(s')^3\,\kappa\tau(s')^3}{\left[\kappa(s')^3\right]^2} = \tau$$

47. $\mathbf{r} = \langle t, \frac{1}{2}t^2, \frac{1}{3}t^3 \rangle$ $\Rightarrow$ $\mathbf{r}' = \langle 1, t, t^2 \rangle$, $\mathbf{r}'' = \langle 0, 1, 2t \rangle$, $\mathbf{r}''' = \langle 0, 0, 2 \rangle$ $\Rightarrow$ $\mathbf{r}' \times \mathbf{r}'' = \langle t^2, -2t, 1 \rangle$ $\Rightarrow$

$$\tau = \frac{(\mathbf{r}' \times \mathbf{r}'') \cdot \mathbf{r}'''}{|\mathbf{r}' \times \mathbf{r}''|^2} = \frac{\langle t^2, -2t, 1 \rangle \cdot \langle 0, 0, 2 \rangle}{t^4 + 4t^2 + 1} = \frac{2}{t^4 + 4t^2 + 1}$$

49. For one helix, the vector equation is $\mathbf{r}(t) = \langle 10\cos t, 10\sin t, 34t/(2\pi) \rangle$ (measuring in angstroms), because the radius of each helix is 10 angstroms, and z increases by 34 angstroms for each increase of 2π in t. Using the arc length formula, letting t go from 0 to $2.9 \times 10^8 \times 2\pi$, we find the approximate length of each helix to be

$$L = \int_0^{2.9 \times 10^8 \times 2\pi} |\mathbf{r}'(t)|\, dt$$

$$= \int_0^{2.9 \times 10^8 \times 2\pi} \sqrt{(-10\sin t)^2 + (10\cos t)^2 + \left(\tfrac{34}{2\pi}\right)^2}\, dt$$

$$= \sqrt{100 + \left(\tfrac{34}{2\pi}\right)^2}\, t \,\Big]_0^{2.9 \times 10^8 \times 2\pi}$$

$$= 2.9 \times 10^8 \times 2\pi \sqrt{100 + \left(\tfrac{34}{2\pi}\right)^2}$$

$$\approx 2.07 \times 10^{10}\ \text{Å} \text{ — more than two meters!}$$

14.4 Motion in Space: Velocity and Acceleration ET 13.4

1. (a) If $\mathbf{r}(t) = x(t)\,\mathbf{i} + y(t)\,\mathbf{j} + z(t)\,\mathbf{k}$ is the position vector of the particle at time t, then the average velocity over the time interval $[0, 1]$ is

$$\mathbf{v}_{\text{ave}} = \frac{\mathbf{r}(1) - \mathbf{r}(0)}{1 - 0} = \frac{(4.5\,\mathbf{i} + 6.0\,\mathbf{j} + 3.0\,\mathbf{k}) - (2.7\,\mathbf{i} + 9.8\,\mathbf{j} + 3.7\,\mathbf{k})}{1} = 1.8\,\mathbf{i} - 3.8\,\mathbf{j} - 0.7\,\mathbf{k}. \text{ Similarly,}$$

over the other intervals we have

$$[0.5, 1]: \quad \mathbf{v}_{\text{ave}} = \frac{\mathbf{r}(1) - \mathbf{r}(0.5)}{1 - 0.5} = \frac{(4.5\,\mathbf{i} + 6.0\,\mathbf{j} + 3.0\,\mathbf{k}) - (3.5\,\mathbf{i} + 7.2\,\mathbf{j} + 3.3\,\mathbf{k})}{0.5}$$

$$= 2.0\,\mathbf{i} - 2.4\,\mathbf{j} - 0.6\,\mathbf{k}$$

$$[1, 2]: \quad \mathbf{v}_{\text{ave}} = \frac{\mathbf{r}(2) - \mathbf{r}(1)}{2 - 1} = \frac{(7.3\,\mathbf{i} + 7.8\,\mathbf{j} + 2.7\,\mathbf{k}) - (4.5\,\mathbf{i} + 6.0\,\mathbf{j} + 3.0\,\mathbf{k})}{1}$$

$$= 2.8\,\mathbf{i} + 1.8\,\mathbf{j} - 0.3\,\mathbf{k}$$

$$[1, 1.5]: \quad \mathbf{v}_{\text{ave}} = \frac{\mathbf{r}(1.5) - \mathbf{r}(1)}{1.5 - 1} = \frac{(5.9\,\mathbf{i} + 6.4\,\mathbf{j} + 2.8\,\mathbf{k}) - (4.5\,\mathbf{i} + 6.0\,\mathbf{j} + 3.0\,\mathbf{k})}{0.5}$$

$$= 2.8\,\mathbf{i} + 0.8\,\mathbf{j} - 0.4\,\mathbf{k}$$

(b) We can estimate the velocity at $t = 1$ by averaging the average velocities over the time intervals $[0.5, 1]$ and $[1, 1.5]$: $\mathbf{v}(1) \approx \frac{1}{2}\left[(2\,\mathbf{i} - 2.4\,\mathbf{j} - 0.6\,\mathbf{k}) + (2.8\,\mathbf{i} + 0.8\,\mathbf{j} - 0.4\,\mathbf{k})\right] = 2.4\,\mathbf{i} - 0.8\,\mathbf{j} - 0.5\,\mathbf{k}$. Then the speed is $|\mathbf{v}(1)| \approx \sqrt{(2.4)^2 + (-0.8)^2 + (-0.5)^2} \approx 2.58$.

3. $\mathbf{r}(t) = \langle t^2 - 1, t \rangle \quad \Rightarrow$ At $t = 1$:

$\mathbf{v}(t) = \mathbf{r}'(t) = \langle 2t, 1 \rangle,$ $\mathbf{v}(1) = \langle 2, 1 \rangle$

$\mathbf{a}(t) = \mathbf{r}''(t) = \langle 2, 0 \rangle,$ $\mathbf{a}(1) = \langle 2, 0 \rangle$

$|\mathbf{v}(t)| = \sqrt{4t^2 + 1}$

5. $\mathbf{r}(t) = e^t \mathbf{i} + e^{-t} \mathbf{j}$ ⇒

$\mathbf{v}(t) = e^t \mathbf{i} - e^{-t} \mathbf{j}$,

$\mathbf{a}(t) = e^t \mathbf{i} + e^{-t} \mathbf{j}$

At $t = 0$:

$\mathbf{v}(0) = \mathbf{i} - \mathbf{j}$,

$\mathbf{a}(0) = \mathbf{i} + \mathbf{j}$

$|\mathbf{v}(t)| = \sqrt{e^{2t} + e^{-2t}} = e^{-t}\sqrt{e^{4t} + 1}$

Since $x = e^t$, $t = \ln x$ and $y = e^{-t} = e^{-\ln x} = 1/x$, and

$x > 0, y > 0$.

7. $\mathbf{r}(t) = \langle \sin t, t, \cos t \rangle$ ⇒

$\mathbf{v}(t) = \langle \cos t, 1, -\sin t \rangle$, $\mathbf{v}(0) = \langle 1, 1, 0 \rangle$

$\mathbf{a}(t) = \langle -\sin t, 0, -\cos t \rangle$, $\mathbf{a}(0) = \langle 0, 0, -1 \rangle$

$|\mathbf{v}(t)| = \sqrt{\cos^2 t + 1 + \sin^2 t} = \sqrt{2}$

Since $x^2 + z^2 = 1$, $y = t$, the path of the particle is a helix about the

y-axis.

9. $\mathbf{r}(t) = \langle t, t^2, t^3 \rangle$ ⇒ $\mathbf{v}(t) = \mathbf{r}'(t) = \langle 1, 2t, 3t^2 \rangle$, $\mathbf{a}(t) = \mathbf{v}'(t) = \langle 0, 2, 6t \rangle$,

$|\mathbf{v}(t)| = \sqrt{1^2 + (2t)^2 + (3t^2)^2} = \sqrt{1 + 4t^2 + 9t^4}$.

11. $\mathbf{r}(t) = \sqrt{2}t \mathbf{i} + e^t \mathbf{j} + e^{-t} \mathbf{k}$ ⇒ $\mathbf{v}(t) = \mathbf{r}'(t) = \sqrt{2}\mathbf{i} + e^t \mathbf{j} - e^{-t} \mathbf{k}$, $\mathbf{a}(t) = \mathbf{v}'(t) = e^t \mathbf{j} + e^{-t} \mathbf{k}$,

$|\mathbf{v}(t)| = \sqrt{2 + e^{2t} + e^{-2t}} = \sqrt{(e^t + e^{-t})^2} = e^t + e^{-t}$.

13. $\mathbf{r}(t) = e^t \langle \cos t, \sin t, t \rangle$ ⇒

$\mathbf{v}(t) = \mathbf{r}'(t) = e^t \langle \cos t, \sin t, t \rangle + e^t \langle -\sin t, \cos t, 1 \rangle = e^t \langle \cos t - \sin t, \sin t + \cos t, t + 1 \rangle$

$\mathbf{a}(t) = \mathbf{v}'(t) = e^t \langle \cos t - \sin t - \sin t - \cos t, \sin t + \cos t + \cos t - \sin t, t + 1 + 1 \rangle$

$= e^t \langle -2 \sin t, 2 \cos t, t + 2 \rangle$

$|\mathbf{v}(t)| = e^t \sqrt{\cos^2 t + \sin^2 t - 2 \cos t \sin t + \sin^2 t + \cos^2 t + 2 \sin t \cos t + t^2 + 2t + 1}$

$= e^t \sqrt{t^2 + 2t + 3}$

15. $\mathbf{a}(t) = \mathbf{k}$ ⇒ $\mathbf{v}(t) = \int \mathbf{a}(t)\,dt = \int \mathbf{k}\,dt = t\mathbf{k} + \mathbf{c}_1$ and $\mathbf{i} - \mathbf{j} = \mathbf{v}(0) = 0\mathbf{k} + \mathbf{c}_1$, so $\mathbf{c}_1 = \mathbf{i} - \mathbf{j}$ and

$\mathbf{v}(t) = \mathbf{i} - \mathbf{j} + t\mathbf{k}$. $\mathbf{r}(t) = \int \mathbf{v}(t)\,dt = \int (\mathbf{i} - \mathbf{j} + t\mathbf{k})\,dt = t\mathbf{i} - t\mathbf{j} + \frac{1}{2}t^2 \mathbf{k} + \mathbf{c}_2$. But $\mathbf{0} = \mathbf{r}(0) = 0 + \mathbf{c}_2$, so

$\mathbf{c}_2 = \mathbf{0}$ and $\mathbf{r}(t) = t\mathbf{i} - t\mathbf{j} + \frac{1}{2}t^2 \mathbf{k}$.

17. (a) $\mathbf{a}(t) = \mathbf{i} + 2\mathbf{j} + 2t\mathbf{k}$ ⇒

(b)

$\mathbf{v}(t) = \int (\mathbf{i} + 2\mathbf{j} + 2t\mathbf{k})\,dt = t\mathbf{i} + 2t\mathbf{j} + t^2\mathbf{k} + \mathbf{c}_1$, and

$\mathbf{0} = \mathbf{v}(0) = 0 + \mathbf{c}_1$, so $\mathbf{c}_1 = \mathbf{0}$ and $\mathbf{v}(t) = \mathbf{i} + 2t\mathbf{j} + t^2\mathbf{k}$.

$\mathbf{r}(t) = \int (t\mathbf{i} + 2t\mathbf{j} + t^2\mathbf{k})\,dt = \frac{1}{2}t^2\mathbf{i} + t^2\mathbf{j} + \frac{1}{3}t^3\mathbf{k} + \mathbf{c}_2$.

But $\mathbf{i} + \mathbf{k} = \mathbf{r}(0) = 0 + \mathbf{c}_2$, so $\mathbf{c}_2 = \mathbf{i} + \mathbf{k}$ and

$\mathbf{r}(t) = \left(1 + \frac{1}{2}t^2\right)\mathbf{i} + t^2\mathbf{j} + \left(1 + \frac{1}{3}t^3\right)\mathbf{k}$.

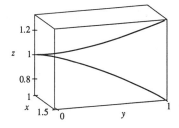

19. $\mathbf{r}(t) = \langle t^2, 5t, t^2 - 16t \rangle \quad \Rightarrow \quad \mathbf{v}(t) = \langle 2t, 5, 2t - 16 \rangle$,

$|\mathbf{v}(t)| = \sqrt{4t^2 + 25 + 4t^2 - 64t + 256} = \sqrt{8t^2 - 64t + 281}$ and

$\dfrac{d}{dt}|\mathbf{v}(t)| = \frac{1}{2}\left(8t^2 - 64t + 281\right)^{-1/2}(16t - 64)$. This is zero if and only if the numerator is zero, that is,

$16t - 64 = 0$ or $t = 4$. Since $\dfrac{d}{dt}|\mathbf{v}(t)| < 0$ for $t < 4$ and $\dfrac{d}{dt}|\mathbf{v}(t)| > 0$ for $t > 4$, the minimum speed of $\sqrt{153}$ is attained at $t = 4$ units of time.

21. $|\mathbf{F}(t)| = 20$ N in the direction of the positive z-axis, so $\mathbf{F}(t) = 20\,\mathbf{k}$. Also $m = 4$ kg, $\mathbf{r}(0) = \mathbf{0}$ and $\mathbf{v}(0) = \mathbf{i} - \mathbf{j}$. Since $20\mathbf{k} = \mathbf{F}(t) = 4\,\mathbf{a}(t)$, $\mathbf{a}(t) = 5\,\mathbf{k}$. Then $\mathbf{v}(t) = 5t\,\mathbf{k} + \mathbf{c}_1$ where $\mathbf{c}_1 = \mathbf{i} - \mathbf{j}$ so $\mathbf{v}(t) = \mathbf{i} - \mathbf{j} + 5t\,\mathbf{k}$ and the speed is $|\mathbf{v}(t)| = \sqrt{1 + 1 + 25t^2} = \sqrt{25t^2 + 2}$. Also $\mathbf{r}(t) = t\,\mathbf{i} - t\,\mathbf{j} + \frac{5}{2}t^2\,\mathbf{k} + \mathbf{c}_2$ and $\mathbf{0} = \mathbf{r}(0)$, so $\mathbf{c}_2 = \mathbf{0}$ and $\mathbf{r}(t) = t\,\mathbf{i} - t\,\mathbf{j} + \frac{5}{2}t^2\,\mathbf{k}$.

23. $|\mathbf{v}(0)| = 500$ m/s and since the angle of elevation is $30°$, the direction of the velocity is $\frac{1}{2}\left(\sqrt{3}\,\mathbf{i} + \mathbf{j}\right)$. Thus $\mathbf{v}(0) = 250\left(\sqrt{3}\,\mathbf{i} + \mathbf{j}\right)$ and if we set up the axes so the projectile starts at the origin, then $\mathbf{r}(0) = \mathbf{0}$. Ignoring air resistance, the only force is that due to gravity, so $\mathbf{F}(t) = -mg\,\mathbf{j}$ where $g \approx 9.8$ m/s^2. Thus $\mathbf{a}(t) = -g\,\mathbf{j}$ and $\mathbf{v}(t) = -gt\,\mathbf{j} + \mathbf{c}_1$. But $250\left(\sqrt{3}\,\mathbf{i} + \mathbf{j}\right) = \mathbf{v}(0) = \mathbf{c}_1$, so $\mathbf{v}(t) = 250\sqrt{3}\,\mathbf{i} + (250 - gt)\,\mathbf{j}$ and $\mathbf{r}(t) = 250\sqrt{3}t\,\mathbf{i} + \left(250t - \frac{1}{2}gt^2\right)\mathbf{j} + \mathbf{c}_2$ where $\mathbf{0} = \mathbf{r}(0) = \mathbf{c}_2$. Thus $\mathbf{r}(t) = 250\sqrt{3}t\,\mathbf{i} + \left(250t - \frac{1}{2}gt^2\right)\mathbf{j}$.

(a) Setting $250t - \frac{1}{2}gt^2 = 0$ gives $t = 0$ or $t = \frac{500}{g} \approx 51.0$ s. So the range is $250\sqrt{3} \cdot \frac{500}{g} \approx 22$ km.

(b) $0 = \dfrac{d}{dt}\left(250t - \frac{1}{2}gt^2\right) = 250 - gt$ implies that the maximum height is attained when $t = 250/g \approx 25.5$ s. Thus, the maximum height is $(250)(250/g) - g(250/g)^2\,\frac{1}{2} = (250)^2/(2g) \approx 3.2$ km.

(c) From part (a), impact occurs at $t = 500/g \approx 51.0$. Thus, the velocity at impact is

$\mathbf{v}(500/g) = 250\sqrt{3}\,\mathbf{i} + [250 - g(500/g)]\,\mathbf{j} = 250\sqrt{3}\,\mathbf{i} - 250\,\mathbf{j}$ and the speed is

$|\mathbf{v}(500/g)| = 250\sqrt{3 + 1} = 500$ m/s.

25. As in Example 5, $\mathbf{r}(t) = (v_0 \cos 45°)\,t\,\mathbf{i} + \left[(v_0 \sin 45°)\,t - \frac{1}{2}gt^2\right]\mathbf{j} = \frac{1}{2}\left[v_0\sqrt{2}t\,\mathbf{i} + \left(v_0\sqrt{2}t - gt^2\right)\mathbf{j}\right]$. Then the

ball lands at $t = \dfrac{v_0\sqrt{2}}{g}$ s. Now since it lands 90 m away, $90 = \frac{1}{2}v_0\sqrt{2}\dfrac{v_0\sqrt{2}}{g}$ or $v_0^2 = 90g$ and the initial velocity is

$v_0 = \sqrt{90g} \approx 30$ m/s.

27. From (4), $x = (v_0 \cos\alpha)\,t$ or $t = \dfrac{x}{v_0\cos\alpha}$. Thus

$y = (v_0\sin\alpha)\dfrac{x}{v_0\cos\alpha} - \dfrac{g}{2}\left(\dfrac{x}{v_0\cos\alpha}\right)^2 = (\tan\alpha)\,x - \dfrac{g}{2v_0^2\cos^2\alpha}x^2$. Thus

the trajectory is a parabola. Continuing by completing the square, we see that

$y - \dfrac{(\tan^2\alpha)\,v_0^2\cos^2\alpha}{2g} = -\dfrac{g}{2v_0^2\cos^2\alpha}\left[x - \dfrac{(\tan\alpha)\,v_0^2\left(\cos^2\alpha\right)}{g}\right]^2$ or

$y - \dfrac{v_0^2\sin^2\alpha}{2g} = -\dfrac{g}{2v_0^2\cos^2\alpha}\left(x - \dfrac{v_0^2\sin\alpha\cos\alpha}{g}\right)^2$. Thus the vertex of the parabola lies at

$\left(\dfrac{v_0^2\sin\alpha\cos\alpha}{g}, \dfrac{v_0^2\sin^2\alpha}{2g}\right)$, so the maximum height is $y = \dfrac{v_0^2\sin^2\alpha}{2g}$.

29. $r(t) = (3t - t^3)\,\mathbf{i} + 3t^2\,\mathbf{j} \quad \Rightarrow \quad r'(t) = (3 - 3t^2)\,\mathbf{i} + 6t\,\mathbf{j}$,

$|r'(t)| = \sqrt{(3 - 3t^2)^2 + (6t)^2} = \sqrt{9 + 18t^2 + 9t^4} = \sqrt{(3 + 3t^2)^2} = 3 + 3t^2$,

$r''(t) = -6t\,\mathbf{i} + 6\mathbf{j}$, $r'(t) \times r''(t) = (18 + 18t^2)\,\mathbf{k}$. Then Equation 9 gives

$a_T = \dfrac{r'(t) \cdot r''(t)}{|r'(t)|} = \dfrac{(3 - 3t^2)(-6t) + (6t)(6)}{3 + 3t^2} = \dfrac{18t + 18t^3}{3 + 3t^2} = \dfrac{18t(1 + t^2)}{3(1 + t^2)} = 6t$ (or by Equation 8,

$a_T = v' = \dfrac{d}{dt}[3 + 3t^2] = 6t$) and Equation 10 gives $a_N = \dfrac{|r'(t) \times r''(t)|}{|r'(t)|} = \dfrac{18 + 18t^2}{3 + 3t^2} = \dfrac{18(1 + t^2)}{3(1 + t^2)} = 6.$

31. $r(t) = \cos t\,\mathbf{i} + \sin t\,\mathbf{j} + t\,\mathbf{k} \quad \Rightarrow \quad r'(t) = -\sin t\,\mathbf{i} + \cos t\,\mathbf{j} + \mathbf{k}$, $|r'(t)| = \sqrt{\sin^2 t + \cos^2 t + 1} = \sqrt{2}$,

$r''(t) = -\cos t\,\mathbf{i} - \sin t\,\mathbf{j}$, $r'(t) \times r''(t) = \sin t\,\mathbf{i} - \cos t\,\mathbf{j} + \mathbf{k}$.

Then $a_T = \dfrac{r'(t) \cdot r''(t)}{|r'(t)|} = \dfrac{\sin t \cos t - \sin t \cos t}{\sqrt{2}} = 0$ and

$a_N = \dfrac{|r'(t) \times r''(t)|}{|r'(t)|} = \dfrac{\sqrt{\sin^2 t + \cos^2 t + 1}}{\sqrt{2}} = \dfrac{\sqrt{2}}{\sqrt{2}} = 1.$

33. $r(t) = e^t\,\mathbf{i} + \sqrt{2}\,t\,\mathbf{j} + e^{-t}\,\mathbf{k} \quad \Rightarrow \quad r'(t) = e^t\,\mathbf{i} + \sqrt{2}\,\mathbf{j} - e^{-t}\,\mathbf{k}$,

$|r(t)| = \sqrt{e^{2t} + 2 + e^{-2t}} = \sqrt{(e^t + e^{-t})^2} = e^t + e^{-t}$, $r''(t) = e^t\,\mathbf{i} + e^{-t}\,\mathbf{k}$.

Then $a_T = \dfrac{e^{2t} - e^{-2t}}{e^t + e^{-t}} = \dfrac{(e^t + e^{-t})(e^t - e^{-t})}{e^t + e^{-t}} = e^t - e^{-t} = 2\sinh t$ and

$a_N = \dfrac{|\sqrt{2}\,e^{-t}\mathbf{i} - 2\mathbf{j} - \sqrt{2}\,e^t\mathbf{k}|}{e^t + e^{-t}} = \dfrac{\sqrt{2(e^{-2t} + 2 + e^{2t})}}{e^t + e^{-t}} = \sqrt{2}\,\dfrac{e^t + e^{-t}}{e^t + e^{-t}} = \sqrt{2}.$

35. If the engines are turned off at time t, then the spacecraft will continue to travel in the direction of $v(t)$, so we need

a t such that for some scalar $s > 0$, $r(t) + s\,v(t) = \langle 6, 4, 9 \rangle$. $v(t) = r'(t) = \mathbf{i} + \dfrac{1}{t}\mathbf{j} + \dfrac{8t}{(t^2 + 1)^2}\mathbf{k} \quad \Rightarrow$

$r(t) + s\,v(t) = \left\langle 3 + t + s, 2 + \ln t + \dfrac{s}{t}, 7 - \dfrac{4}{t^2 + 1} + \dfrac{8st}{(t^2 + 1)^2} \right\rangle \quad \Rightarrow \quad 3 + t + s = 6 \quad \Rightarrow \quad s = 3 - t$, so

$7 - \dfrac{4}{t^2 + 1} + \dfrac{8(3 - t)t}{(t^2 + 1)^2} = 9 \quad \Leftrightarrow \quad \dfrac{24t - 12t^2 - 4}{(t^2 + 1)^2} = 2 \quad \Leftrightarrow \quad t^4 + 8t^2 - 12t + 3 = 0$. It is easily seen that

$t = 1$ is a root of this polynomial. Also $2 + \ln 1 + \dfrac{3 - 1}{1} = 4$, so $t = 1$ is the desired solution.

▬14 Review   ET 13

CONCEPT CHECK

1. A vector function is a function whose domain is a set of real numbers and whose range is a set of vectors. To find the derivative or integral, we can differentiate or integrate each component of the vector function.

2. The tip of the moving vector $r(t)$ of a continuous vector function traces out a space curve.

3. (a) A curve represented by the vector function $r(t)$ is smooth if $r'(t)$ is continuous and $r'(t) \neq 0$ on its parametric domain (except possibly at the endpoints).

(b) The tangent vector to a smooth curve at a point P with position vector $r(t)$ is the vector $r'(t)$. The tangent line at P is the line through P parallel to the tangent vector $r'(t)$. The unit tangent vector is $T(t) = \dfrac{r'(t)}{|r'(t)|}$.

4. (a)–(f) See Theorem 14.2.3 [ET 13.2.3].

5. Use Formula 14.3.2 [ET 13.3.2], or equivalently 14.3.3 [ET 13.3.3].

6. (a) The curvature of a curve is $\kappa = \left|\dfrac{d\mathbf{T}}{ds}\right|$ where $\mathbf{T}$ is the unit tangent vector.

(b) $\kappa(t) = \left|\dfrac{\mathbf{T}'(t)}{\mathbf{r}'(t)}\right|$ (c) $\kappa(t) = \dfrac{|\mathbf{r}'(t) \times \mathbf{r}''(t)|}{|\mathbf{r}'(t)|^3}$ (d) $\kappa(x) = \dfrac{|f''(x)|}{\left[1 + (f'(x))^2\right]^{3/2}}$

7. (a) The unit normal vector: $\mathbf{N}(t) = \dfrac{\mathbf{T}'(t)}{|\mathbf{T}'(t)|}$. The binormal vector: $\mathbf{B}(t) = \mathbf{T}(t) \times \mathbf{N}(t)$.

(b) See the discussion preceding Example 7 in Section 14.3 [ET 13.3].

8. (a) If $\mathbf{r}(t)$ is the position vector of the particle on the space curve, the velocity $\mathbf{v}(t) = \mathbf{r}'(t)$, the speed is given by $|\mathbf{v}(t)|$, and the acceleration $\mathbf{a}(t) = \mathbf{v}'(t) = \mathbf{r}''(t)$.

(b) $\mathbf{a} = a_T\mathbf{T} + a_N\mathbf{N}$ where $a_T = v'$ and $a_N = \kappa v^2$.

9. See the statement of Kepler's Laws on page 897 [ET 863].

TRUE-FALSE QUIZ

1. True. If we reparametrize the curve by replacing $u = t^3$, we have $\mathbf{r}(u) = u\mathbf{i} + 2u\mathbf{j} + 3u\mathbf{k}$, which is a line through the origin with direction vector $\mathbf{i} + 2\mathbf{j} + 3\mathbf{k}$.

3. False. $\mathbf{r}'(t) = \langle -\sin t, 2t, 4t^3 \rangle$, and since $\mathbf{r}'(0) = \langle 0, 0, 0 \rangle = \mathbf{0}$, the curve is not smooth.

5. False. By Formula 5 of Theorem 14.2.3 [ET 13.2.3], $\dfrac{d}{dt}[\mathbf{u}(t) \times \mathbf{v}(t)] = \mathbf{u}'(t) \times \mathbf{v}(t) + \mathbf{u}(t) \times \mathbf{v}'(t)$.

7. False. κ is the magnitude of the rate of change of the unit tangent vector $\mathbf{T}$ with respect to arc length s, not with respect to t.

9. True. See the discussion preceding Example 7 in Section 14.3 [ET 13.3].

EXERCISES

1. (a) Since $x = 2$ and $y^2 + z^2 = 1$, the curve is a circle in the plane $x = 2$ with center $(2, 0, 0)$ and radius 1.

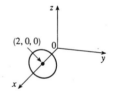

(b) $\mathbf{r}'(t) = \cos t\,\mathbf{j} - \sin t\,\mathbf{k} \;\Rightarrow\; \mathbf{r}''(t) = -\sin t\,\mathbf{j} - \cos t\,\mathbf{k}$

3. The projection of the curve C of intersection onto the xy-plane is the circle $x^2 + y^2 = 16$, $z = 0$. So we can write $x = 4\cos t$, $y = 4\sin t$, $0 \le t \le 2\pi$. From the equation of the plane, we have $z = 5 - x = 5 - 4\cos t$, so parametric equations for C are $x = 4\cos t$, $y = 4\sin t$, $z = 5 - 4\cos t$, $0 \le t \le 2\pi$, and the corresponding vector function is $\mathbf{r}(t) = 4\cos t\,\mathbf{i} + 4\sin t\,\mathbf{j} + (5 - 4\cos t)\,\mathbf{k}$, $0 \le t \le 2\pi$.

5. $\int_0^1 \left[(t + t^2)\,\mathbf{i} + (2 + t^3)\,\mathbf{j} + t^4\,\mathbf{k}\right] dt = \left[\left(\tfrac{1}{2}t^2 + \tfrac{1}{3}t^3\right)\mathbf{i} + \left(2t + \tfrac{1}{4}t^4\right)\mathbf{j} + \left(\tfrac{1}{5}t^5\right)\mathbf{k}\right]_0^1 = \tfrac{5}{6}\mathbf{i} + \tfrac{9}{4}\mathbf{j} + \tfrac{1}{5}\mathbf{k}$

7. $t = 1$ at $(1, 4, 2)$ and $t = 4$ at $(2, 1, 17)$, so

$$L = \int_1^4 \sqrt{\frac{1}{4t} + \frac{16}{t^4} + 4t^2}\; dt$$

$$\approx \frac{4 - 1}{3 \cdot 4}\left[\sqrt{\tfrac{1}{4} + 16 + 4} + 4 \cdot \sqrt{\frac{1}{4 \cdot \frac{7}{4}} + \frac{16}{\left(\frac{7}{4}\right)^4} + 4\left(\tfrac{7}{4}\right)^2} + 2 \cdot \sqrt{\frac{1}{4 \cdot \frac{10}{4}} + \frac{16}{\left(\frac{10}{4}\right)^4} + 4\left(\tfrac{10}{4}\right)^2}\right.$$

$$\left. + 4 \cdot \sqrt{\frac{1}{4 \cdot \frac{13}{4}} + \frac{16}{\left(\frac{13}{4}\right)^4} + 4\left(\tfrac{13}{4}\right)^2} + \sqrt{\frac{1}{4 \cdot 4} + \frac{16}{4^4} + 4 \cdot 4^2}\right]$$

$$\approx 15.9241$$

9. The angle of intersection of the two curves, θ, is the angle between their respective tangents at the point of intersection. For both curves the point $(1, 0, 0)$ occurs when $t = 0$. $\mathbf{r}_1'(t) = -\sin t\,\mathbf{i} + \cos t\,\mathbf{j} + \mathbf{k}$ $\Rightarrow$ $\mathbf{r}_1'(0) = \mathbf{j} + \mathbf{k}$ and $\mathbf{r}_2'(t) = \mathbf{i} + 2t\,\mathbf{j} + 3t^2\,\mathbf{k}$ $\Rightarrow$ $\mathbf{r}_2'(0) = \mathbf{i}$. $\mathbf{r}_1'(0) \cdot \mathbf{r}_2'(0) = (\mathbf{j} + \mathbf{k}) \cdot \mathbf{i} = 0$. Therefore, the curves intersect in a right angle, that is, $\theta = \frac{\pi}{2}$.

11. (a) $\mathbf{T}(t) = \dfrac{\mathbf{r}'(t)}{|\mathbf{r}'(t)|} = \dfrac{\langle t^2, t, 1 \rangle}{|\langle t^2, t, 1 \rangle|} = \dfrac{\langle t^2, t, 1 \rangle}{\sqrt{t^4 + t^2 + 1}}$

(b) $\mathbf{T}'(t) = -\frac{1}{2}\left(t^4 + t^2 + 1\right)^{-3/2}\left(4t^3 + 2t\right)\langle t^2, t, 1\rangle + \left(t^4 + t^2 + 1\right)^{-1/2}\langle 2t, 1, 0\rangle$

$= \dfrac{-2t^3 - t}{(t^4 + t^2 + 1)^{3/2}}\langle t^2, t, 1\rangle + \dfrac{1}{(t^4 + t^2 + 1)^{1/2}}\langle 2t, 1, 0\rangle$

$= \dfrac{\langle -2t^5 - t^3, -2t^4 - t^2, -2t^3 - t\rangle + \langle 2t^5 + 2t^3 + 2t, t^4 + t^2 + 1, 0\rangle}{(t^4 + t^2 + 1)^{3/2}}$

$= \dfrac{\langle 2t, -t^4 + 1, -2t^3 - t\rangle}{(t^4 + t^2 + 1)^{3/2}}$

$|\mathbf{T}'(t)| = \dfrac{\sqrt{4t^2 + t^8 - 2t^4 + 1 + 4t^6 + 4t^4 + t^2}}{(t^4 + t^2 + 1)^{3/2}} = \dfrac{\sqrt{t^8 + 4t^6 + 2t^4 + 5t^2}}{(t^4 + t^2 + 1)^{3/2}}$, and

$\mathbf{N}(t) = \dfrac{\langle 2t, 1 - t^4, -2t^3 - t\rangle}{\sqrt{t^8 + 4t^6 + 2t^4 + 5t^2}}$.

(c) $\kappa(t) = \dfrac{|\mathbf{T}'(t)|}{|\mathbf{r}'(t)|} = \dfrac{\sqrt{t^8 + 4t^6 + 2t^4 + 5t^2}}{(t^4 + t^2 + 1)^2}$

13. $y' = 4x^3$, $y'' = 12x^2$ and $\kappa(x) = \dfrac{|y''|}{[1 + (y')^2]^{3/2}} = \dfrac{|12x^2|}{(1 + 16x^6)^{3/2}}$, so $\kappa(1) = \dfrac{12}{(17)^{3/2}}$.

15. $\mathbf{r}(t) = \langle \sin 2t, t, \cos 2t \rangle$ $\Rightarrow$ $\mathbf{r}'(t) = \langle 2\cos 2t, 1, -2\sin 2t \rangle$ $\Rightarrow$ $\mathbf{T}(t) = \frac{1}{\sqrt{5}}\langle 2\cos 2t, 1, -2\sin 2t \rangle$ $\Rightarrow$ $\mathbf{T}'(t) = \frac{1}{\sqrt{5}}\langle -4\sin 2t, 0, -4\cos 2t \rangle$ $\Rightarrow$ $\mathbf{N}(t) = \langle -\sin 2t, 0, -\cos 2t \rangle$. So $\mathbf{N} = \mathbf{N}(\pi) = \langle 0, 0, -1 \rangle$ and $\mathbf{B} = \mathbf{T} \times \mathbf{N} = \frac{1}{\sqrt{5}}\langle -1, 2, 0 \rangle$. So a normal to the osculating plane is $\langle -1, 2, 0 \rangle$ and an equation is $-1(x - 0) + 2(y - \pi) + 0(z - 1) = 0$ or $x - 2y + 2\pi = 0$.

17. $\mathbf{v}(t) = 2\sqrt{2}\,\mathbf{i} + 2e^{2t}\,\mathbf{j} - 2e^{-2t}\,\mathbf{k}$, $|\mathbf{v}(t)| = \sqrt{8 + 4e^{4t} + 4e^{-4t}} = 2\left(e^{2t} + e^{-2t}\right)$, $\mathbf{a}(t) = 4e^{2t}\,\mathbf{j} + 4e^{-2t}\,\mathbf{k}$

19. We set up the axes so that the shot leaves the athlete's hand 7 ft above the origin. Then we are given $\mathbf{r}(0) = 7\mathbf{j}$, $|\mathbf{v}(0)| = 43$ ft/s, and $\mathbf{v}(0)$ has direction given by a $45°$ angle of elevation. Then a unit vector in the direction of $\mathbf{v}(0)$ is $\frac{1}{\sqrt{2}}(\mathbf{i} + \mathbf{j})$ $\Rightarrow$ $\mathbf{v}(0) = \frac{43}{\sqrt{2}}(\mathbf{i} + \mathbf{j})$. Assuming air resistance is negligible, the only external force is due to gravity, so as in Example 14.4.5 [ET 13.4.5] we have $\mathbf{a} = -g\,\mathbf{j}$ where here $g \approx 32$ ft/s^2. Since $\mathbf{v}'(t) = \mathbf{a}(t)$, we integrate, giving $\mathbf{v}(t) = -gt\,\mathbf{j} + \mathbf{C}$ where $\mathbf{C} = \mathbf{v}(0) = \frac{43}{\sqrt{2}}(\mathbf{i} + \mathbf{j})$ $\Rightarrow$ $\mathbf{v}(t) = \frac{43}{\sqrt{2}}\mathbf{i} + \left(\frac{43}{\sqrt{2}} - gt\right)\mathbf{j}$. Since $\mathbf{r}'(t) = \mathbf{v}(t)$ we integrate again, so $\mathbf{r}(t) = \frac{43}{\sqrt{2}}t\,\mathbf{i} + \left(\frac{43}{\sqrt{2}}t - \frac{1}{2}gt^2\right)\mathbf{j} + \mathbf{D}$. But

$\mathbf{D} = \mathbf{r}(0) = 7\mathbf{j}$ $\Rightarrow$ $\mathbf{r}(t) = \frac{43}{\sqrt{2}}t\,\mathbf{i} + \left(\frac{43}{\sqrt{2}}t - \frac{1}{2}gt^2 + 7\right)\mathbf{j}$.

(a) At 2 seconds, the shot is at $\mathbf{r}(2) = \frac{43}{\sqrt{2}}(2)\,\mathbf{i} + \left(\frac{43}{\sqrt{2}}(2) - \frac{1}{2}g(2)^2 + 7\right)\mathbf{j} \approx 60.8\,\mathbf{i} + 3.8\,\mathbf{j}$, so the shot is about 3.8 ft above the ground, at a horizontal distance of 60.8 ft from the athlete.

(b) The shot reaches its maximum height when the vertical component of velocity is 0: $\frac{43}{\sqrt{2}} - gt = 0 \Rightarrow$

$t = \dfrac{43}{\sqrt{2}g} \approx 0.95$ s. Then $\mathbf{r}(0.95) \approx 28.9\,\mathbf{i} + 21.4\,\mathbf{j}$, so the maximum height is approximately 21.4 ft.

(c) The shot hits the ground when the vertical component of $\mathbf{r}(t)$ is 0, so $\frac{43}{\sqrt{2}}t - \frac{1}{2}gt^2 + 7 = 0 \Rightarrow$

$-16t^2 + \frac{43}{\sqrt{2}}t + 7 = 0 \Rightarrow t \approx 2.11$ s. $\mathbf{r}(2.11) \approx 64.2\,\mathbf{i} - 0.08\,\mathbf{j}$, thus the shot lands approximately 64.2 ft

from the athlete.

21. (a) Instead of proceeding directly, we use Formula 3 of Theorem 14.2.3 [ET 13.2.3]: $\mathbf{r}(t) = t\,\mathbf{R}(t) \Rightarrow$

$\mathbf{v} = \mathbf{r}'(t) = \mathbf{R}(t) + t\,\mathbf{R}'(t) = \cos\omega t\,\mathbf{i} + \sin\omega t\,\mathbf{j} + t\,\mathbf{v}_d$.

(b) Using the same method as in part (a) and starting with $\mathbf{v} = \mathbf{R}(t) + t\,\mathbf{R}'(t)$, we have

$\mathbf{a} = \mathbf{v}' = \mathbf{R}'(t) + \mathbf{R}'(t) + t\,\mathbf{R}''(t) = 2\,\mathbf{R}'(t) + t\,\mathbf{R}''(t) = 2\,\mathbf{v}_d + t\,\mathbf{a}_d$.

(c) Here we have $\mathbf{r}(t) = e^{-t}\cos\omega t\,\mathbf{i} + e^{-t}\sin\omega t\,\mathbf{j} = e^{-t}\,\mathbf{R}(t)$. So, as in parts (a) and (b),

$\mathbf{v} = \mathbf{r}'(t) = e^{-t}\,\mathbf{R}'(t) - e^{-t}\,\mathbf{R}(t) = e^{-t}\,[\mathbf{R}'(t) - \mathbf{R}(t)] \Rightarrow$

$\mathbf{a} = \mathbf{v}' = e^{-t}\,[\mathbf{R}''(t) - \mathbf{R}'(t)] - e^{-t}\,[\mathbf{R}'(t) - \mathbf{R}(t)] = e^{-t}\,[\mathbf{R}''(t) - 2\,\mathbf{R}'(t) + \mathbf{R}(t)]$

$= e^{-t}\,\mathbf{a}_d - 2e^{-t}\,\mathbf{v}_d + e^{-t}\,\mathbf{R}$

Thus, the Coriolis acceleration (the sum of the "extra" terms not involving $\mathbf{a}_d$) is $-2e^{-t}\,\mathbf{v}_d + e^{-t}\,\mathbf{R}$.

Problems Plus

1. (a) $\mathbf{r}(t) = R\cos\omega t\,\mathbf{i} + R\sin\omega t\,\mathbf{j} \implies \mathbf{v} = \mathbf{r}'(t) = -\omega R\sin\omega t\,\mathbf{i} + \omega R\cos\omega t\,\mathbf{j}$, so $\mathbf{r} = R(\cos\omega t\,\mathbf{i} + \sin\omega t\,\mathbf{j})$ and $\mathbf{v} = \omega R(-\sin\omega t\,\mathbf{i} + \cos\omega t\,\mathbf{j})$. $\mathbf{v}\cdot\mathbf{r} = \omega R^2(-\cos\omega t\sin\omega t + \sin\omega t\cos\omega t) = 0$, so $\mathbf{v}\perp\mathbf{r}$. Since $\mathbf{r}$ points along a radius of the circle, and $\mathbf{v}\perp\mathbf{r}$, $\mathbf{v}$ is tangent to the circle. Because it is a velocity vector, $\mathbf{v}$ points in the direction of motion.

(b) In (a), we wrote $\mathbf{v}$ in the form $\omega R\,\mathbf{u}$, where $\mathbf{u}$ is the unit vector $-\sin\omega t\,\mathbf{i} + \cos\omega t\,\mathbf{j}$. Clearly $|\mathbf{v}| = \omega R\,|\mathbf{u}| = \omega R$. At speed ωR, the particle completes one revolution, a distance $2\pi R$, in time
$$T = \frac{2\pi R}{\omega R} = \frac{2\pi}{\omega}.$$

(c) $\mathbf{a} = \dfrac{d\mathbf{v}}{dt} = -\omega^2 R\cos\omega t\,\mathbf{i} - \omega^2 R\sin\omega t\,\mathbf{j} = -\omega^2 R(\cos\omega t\,\mathbf{i} + \sin\omega t\,\mathbf{j})$, so $\mathbf{a} = -\omega^2\mathbf{r}$. This shows that $\mathbf{a}$ is proportional to $\mathbf{r}$ and points in the opposite direction (toward the origin). Also, $|\mathbf{a}| = \omega^2\,|\mathbf{r}| = \omega^2 R$.

(d) By Newton's Second Law (see Section 14.4 [ET 13.4]), $\mathbf{F} = m\mathbf{a}$, so
$$|\mathbf{F}| = m\,|\mathbf{a}| = mR\omega^2 = \frac{m\,(\omega R)^2}{R} = \frac{m\,|\mathbf{v}|^2}{R}.$$

3. (a) The projectile reaches maximum height when $0 = \dfrac{dy}{dt} = \dfrac{d}{dt}\left[(v_0\sin\alpha)\,t - \tfrac{1}{2}gt^2\right] = v_0\sin\alpha - gt$; that is,

when $t = \dfrac{v_0\sin\alpha}{g}$ and $y = (v_0\sin\alpha)\left(\dfrac{v_0\sin\alpha}{g}\right) - \dfrac{1}{2}g\left(\dfrac{v_0\sin\alpha}{g}\right)^2 = \dfrac{v_0^2\sin^2\alpha}{2g}$. This is the maximum height attained when the projectile is fired with an angle of elevation α. This maximum height is largest when $\alpha = \frac{\pi}{2}$. In that case, $\sin\alpha = 1$ and the maximum height is $\dfrac{v_0^2}{2g}$.

(b) Let $R = v_0^2/g$. We are asked to consider the parabola $x^2 + 2Ry - R^2 = 0$ which can be rewritten as $y = -\dfrac{1}{2R}x^2 + \dfrac{R}{2}$. The points on or inside this parabola are those for which $-R\le x\le R$ and $0 \le y \le \dfrac{-1}{2R}x^2 + \dfrac{R}{2}$. When the projectile is fired at angle of elevation α, the points (x,y) along its path satisfy the relations $x = (v_0\cos\alpha)\,t$ and $y = (v_0\sin\alpha)\,t - \tfrac{1}{2}gt^2$, where $0\le t\le (2v_0\sin\alpha)/g$ (as in Example 5 in Section 14.4 [ET 13.4]). Thus $|x| \le \left|v_0\cos\alpha\left(\dfrac{2v_0\sin\alpha}{g}\right)\right| = \left|\dfrac{v_0^2}{g}\sin 2\alpha\right| \le \left|\dfrac{v_0^2}{g}\right| = |R|$. This shows that $-R\le x\le R$.

For t in the specified range, we also have $y = t\left(v_0\sin\alpha - \tfrac{1}{2}gt\right) = \tfrac{1}{2}gt\left(\dfrac{2v_0\sin\alpha}{g} - t\right) \ge 0$ and
$$y = (v_0\sin\alpha)\frac{x}{v_0\cos\alpha} - \frac{g}{2}\left(\frac{x}{v_0\cos\alpha}\right)^2 = (\tan\alpha)\,x - \frac{g}{2v_0^2\cos^2\alpha}x^2 = -\frac{1}{2R\cos^2\alpha}x^2 + (\tan\alpha)\,x.\ \text{Thus}$$

$$y - \left(\frac{-1}{2R}x^2 + \frac{R}{2}\right) = \frac{-1}{2R\cos^2\alpha}x^2 + \frac{1}{2R}x^2 + (\tan\alpha)\,x - \frac{R}{2}$$

$$= \frac{x^2}{2R}\left(1 - \frac{1}{\cos^2\alpha}\right) + (\tan\alpha)\,x - \frac{R}{2} = \frac{x^2(1 - \sec^2\alpha) + 2R(\tan\alpha)\,x - R^2}{2R}$$

$$= \frac{-(\tan^2\alpha)\,x^2 + 2R(\tan\alpha)\,x - R^2}{2R} = \frac{-[(\tan\alpha)\,x - R]^2}{2R} \le 0$$

We have shown that every target that can be hit by the projectile lies on or inside the parabola

$y = -\dfrac{1}{2R}x^2 + \dfrac{R}{2}$. Now let (a, b) be any point on or inside the parabola $y = -\dfrac{1}{2R}x^2 + \dfrac{R}{2}$. Then

$-R \le a \le R$ and $0 \le b \le -\dfrac{1}{2R}a^2 + \dfrac{R}{2}$. We seek an angle α such that (a, b) lies in the path of the projectile;

that is, we wish to find an angle α such that $b = -\dfrac{1}{2R\cos^2\alpha}a^2 + (\tan\alpha)\,a$ or

equivalently $b = \dfrac{-1}{2R}\left(\tan^2\alpha + 1\right)a^2 + (\tan\alpha)\,a$. Rearranging this equation we get

$\dfrac{a^2}{2R}\tan^2\alpha - a\tan\alpha + \left(\dfrac{a^2}{2R} + b\right) = 0$ or $a^2\,(\tan\alpha)^2 - 2aR\,(\tan\alpha) + \left(a^2 + 2bR\right) = 0$ (★). This

quadratic equation for $\tan\alpha$ has real solutions exactly when the discriminant is nonnegative. Now

$B^2 - 4AC \ge 0 \;\Leftrightarrow\; (-2aR)^2 - 4a^2\left(a^2 + 2bR\right) \ge 0 \;\Leftrightarrow\; 4a^2\left(R^2 - a^2 - 2bR\right) \ge 0 \;\Leftrightarrow$

$-a^2 - 2bR + R^2 \ge 0 \;\Leftrightarrow\; b \le \dfrac{1}{2R}\left(R^2 - a^2\right) \;\Leftrightarrow\; b \le \dfrac{-1}{2R}a^2 + \dfrac{R}{2}$. This condition is satisfied since

(a, b) is on or inside the parabola $y = -\dfrac{1}{2R}x^2 + \dfrac{R}{2}$. It follows that (a, b) lies in the path of the projectile when

$\tan\alpha$ satisfies (★), that is, when $\tan\alpha = \dfrac{2aR \pm \sqrt{4a^2\left(R^2 - a^2 - 2bR\right)}}{2a^2} = \dfrac{R \pm \sqrt{R^2 - 2bR - a^2}}{a}$.

(c)

If the gun is pointed at a target with height h at a distance D downrange, then

$\tan\alpha = h/D$. When the projectile reaches a distance D downrange

(remember we are assuming that it doesn't hit the ground first), we have

$D = x = (v_0\cos\alpha)\,t$, so $t = \dfrac{D}{v_0\cos\alpha}$ and

$y = (v_0\sin\alpha)\,t - \tfrac{1}{2}gt^2 = D\tan\alpha - \dfrac{gD^2}{2v_0^2\cos^2\alpha}$. Meanwhile, the target, whose x-coordinate is also D, has

fallen from height h to height $h - \tfrac{1}{2}gt^2 = D\tan\alpha - \dfrac{gD^2}{2v_0^2\cos^2\alpha}$. Thus the projectile hits the target.

5. (a) $m\dfrac{d^2\mathbf{R}}{dt^2} = -mg\,\mathbf{j} - k\dfrac{d\mathbf{R}}{dt} \;\Rightarrow\; \dfrac{d}{dt}\left(m\dfrac{d\mathbf{R}}{dt} + k\,\mathbf{R} + mgt\,\mathbf{j}\right) = 0 \;\Rightarrow\; m\dfrac{d\mathbf{R}}{dt} + k\mathbf{R} + mgt\,\mathbf{j} = \mathbf{c}$ ($\mathbf{c}$ is a

constant vector in the xy-plane). At $t = 0$, this says that $m\mathbf{v}\,(0) + k\mathbf{R}\,(0) = \mathbf{c}$. Since $\mathbf{v}\,(0) = \mathbf{v}_0$ and

$\mathbf{R}\,(0) = \mathbf{0}$, we have $\mathbf{c} = m\mathbf{v}_0$. Therefore $\dfrac{d\mathbf{R}}{dt} + \dfrac{k}{m}\mathbf{R} + gt\,\mathbf{j} = \mathbf{v}_0$, or $\dfrac{d\mathbf{R}}{dt} + \dfrac{k}{m}\mathbf{R} = \mathbf{v}_0 - gt\,\mathbf{j}$.

(b) Multiplying by $e^{(k/m)t}$ gives $e^{(k/m)t}\dfrac{d\mathbf{R}}{dt} + \dfrac{k}{m}e^{(k/m)t}\mathbf{R} = e^{(k/m)t}\mathbf{v}_0 - gte^{(k/m)t}\,\mathbf{j}$ or

$\dfrac{d}{dt}\left(e^{(k/m)t}\mathbf{R}\right) = e^{(k/m)t}\mathbf{v}_0 - gte^{(k/m)t}\,\mathbf{j}$. Integrating gives

$e^{(k/m)t}\mathbf{R} = \dfrac{m}{k}e^{(k/m)t}\mathbf{v}_0 - \left[\dfrac{mg}{k}te^{(k/m)t} - \dfrac{m^2g}{k^2}e^{(k/m)t}\right]\mathbf{j} + \mathbf{b}$ for some constant vector $\mathbf{b}$. Setting $t = 0$

yields the relation $\mathbf{R}\,(0) = \dfrac{m}{k}\mathbf{v}_0 + \dfrac{m^2g}{k^2}\,\mathbf{j} + \mathbf{b}$, so $\mathbf{b} = -\dfrac{m}{k}\mathbf{v}_0 - \dfrac{m^2g}{k^2}\,\mathbf{j}$. Thus

$e^{(k/m)t}\mathbf{R} = \dfrac{m}{k}\left[e^{(k/m)t} - 1\right]\mathbf{v}_0 - \left[\dfrac{mg}{k}te^{(k/m)t} - \dfrac{m^2g}{k^2}\left(e^{(k/m)t} - 1\right)\right]\mathbf{j}$ and

$\mathbf{R}\,(t) = \dfrac{m}{k}\left[1 - e^{-kt/m}\right]\mathbf{v}_0 + \dfrac{mg}{k}\left[\dfrac{m}{k}\left(1 - e^{-kt/m}\right) - t\right]\mathbf{j}$.

7. (a) $\mathbf{a} = -g\mathbf{j} \implies \mathbf{v} = \mathbf{v}_0 - gt\mathbf{j} = 2\mathbf{i} - gt\mathbf{j} \implies \mathbf{s} = \mathbf{s}_0 + 2t\mathbf{i} - \frac{1}{2}gt^2\mathbf{j} = 3.5\mathbf{j} + 2t\mathbf{i} - \frac{1}{2}gt^2\mathbf{j} \implies$

$\mathbf{s} = 2t\mathbf{i} + \left(3.5 - \frac{1}{2}gt^2\right)\mathbf{j}$. Therefore $y = 0$ when $t = \sqrt{7/g}$ seconds. At that instant, the ball is

$2\sqrt{7/g} \approx 0.94$ ft to the right of the table top. Its coordinates (relative to an origin on the floor directly under the

table's edge) are $(0.94, 0)$. At impact, the velocity is $\mathbf{v} = 2\mathbf{i} - \sqrt{7g}\,\mathbf{j}$, so the speed is

$|\mathbf{v}| = \sqrt{4 + 7g} \approx 15$ ft/s.

(b) The slope of the curve when $t = \sqrt{\dfrac{7}{g}}$ is $\dfrac{dy}{dx} = \dfrac{dy/dt}{dx/dt} = \dfrac{-gt}{2} = \dfrac{-g\sqrt{7/g}}{2} = \dfrac{-\sqrt{7g}}{2}$. Thus $\cot\theta = \sqrt{7g}/2$

and $\theta \approx 7.6°$.

(c) From (a), $|\mathbf{v}| = \sqrt{4 + 7g}$. So the ball rebounds with speed $0.8\sqrt{4 + 7g} \approx 12.08$ ft/s at angle of inclination

$90° - \theta \approx 82.3886°$. By Example 14.4.5 [ET 13.4.5], the horizontal distance traveled between bounces is

$d = \dfrac{v_0^2 \sin 2\alpha}{g}$, where $v_0 \approx 12.08$ ft/s and $\alpha \approx 82.3886°$. Therefore, $d \approx 1.197$ ft. So the ball strikes the floor

at about $2\sqrt{7/g} + 1.197 \approx 2.13$ ft to the right of the table's edge.

Partial Derivatives

15.1 Functions of Several Variables

1. (a) From Table 1, $f(8, 60) = -7$, which means that if the temperature is $8\,°C$ and the wind speed is 60 km/h, then the air would feel equivalent to approximately $-7\,°C$ without wind.

(b) The question is asking: when the temperature is $-12\,°C$, what wind speed gives a wind-chill index of $-26\,°C$? From Table 1, the speed is 20 km/h.

(c) The question is asking: when the wind speed is 80 km/h, what temperature gives a wind-chill index of $-14\,°C$? From Table 1, the temperature is $4\,°C$.

(d) The function $I = f(-4, v)$ means that we fix T at -4 and allow v to vary, resulting in a function of one variable. In other words, the function gives wind-chill index values for different wind speeds when the temperature is $-4\,°C$. From Table 1 (look at the row corresponding to $T = -4$), the function decreases and appears to approach a constant value as v increases.

(e) The function $I = f(T, 50)$ means that we fix v at 50 and allow T to vary, again giving a function of one variable. In other words, the function gives wind-chill index values for different temperatures when the wind speed is 50 km/h. From Table 1 (look at the column corresponding to $v = 50$), the function increases almost linearly as T increases.

3. If the amounts of labor and capital are both doubled, we replace L, K in the function with $2L, 2K$, giving

$$P(2L, 2K) = 1.01(2L)^{0.75}(2K)^{0.25} = 1.01\left(2^{0.75}\right)\left(2^{0.25}\right)L^{0.75}K^{0.25} = \left(2^{1}\right)1.01L^{0.75}K^{0.25}$$
$$= 2P(L, K)$$

Thus, the production is doubled. It is also true for the general case $P(L, K) = bL^{\alpha}K^{1-\alpha}$:

$$P(2L, 2K) = b(2L)^{\alpha}(2K)^{1-\alpha} = b(2^{\alpha})\left(2^{1-\alpha}\right)L^{\alpha}K^{1-\alpha} = \left(2^{\alpha+1-\alpha}\right)bL^{\alpha}K^{1-\alpha} = 2P(L, K).$$

5. (a) According to the table, $f(40, 15) = 25$, which means that if a 40-knot wind has been blowing in the open sea for 15 hours, it will create waves with estimated heights of 25 feet.

(b) $h = f(30, t)$ means we fix v at 30 and allow t to vary, resulting in a function of one variable. Thus here, $h = f(30, t)$ gives the wave heights produced by 30-knot winds blowing for t hours. From the table (look at the row corresponding to $v = 30$), the function increases but at a declining rate as t increases. In fact, the function values appear to be approaching a limiting value of approximately 19, which suggests that 30-knot winds cannot produce waves higher than about 19 feet.

(c) $h = f(v, 30)$ means we fix t at 30, again giving a function of one variable. So, $h = f(v, 30)$ gives the wave heights produced by winds of speed v blowing for 30 hours. From the table (look at the column corresponding to $t = 30$), the function appears to increase at an increasing rate, with no apparent limiting value. This suggests that faster winds (lasting 30 hours) always create higher waves.

7. (a) $f(2,4) = e^{2^2-4} = e^0 = 1$.

(b) The exponential function is defined everywhere, so no matter what values of x and y we use, e^{x^2-y} is defined. So the domain of f is $\mathbb{R}^2$.

(c) Because the range of $g(x,y) = x^2 - y$ is $\mathbb{R}$, and the range of e^x is $(0,\infty)$, the range of $e^{g(x,y)} = e^{x^2-y}$ is $\{z \mid z > 0\}$.

9. (a) $f(3,6,4) = 3^2 \ln(3 - 6 + 4) = 9 \ln 1 = 0$.

(b) For the logarithmic function to be defined, we need $x - y + z > 0$. Thus the domain of f is $\{(x,y,z) \mid x + z > y\}$.

(c) Since $x^2 \ln(x - y + z)$ can be any real number, the range of f is $\mathbb{R}$.

11. $\sqrt{x+y}$ is defined only when $x + y \geq 0$, or $y \geq -x$. So the domain of f is $\{(x,y) \mid y \geq -x\}$.

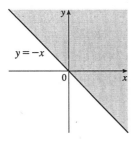

13. $\ln(9 - x^2 - 9y^2)$ is defined only when $9 - x^2 - 9y^2 > 0$, or $\frac{1}{9}x^2 + y^2 < 1$. So the domain of f is $\{(x,y) \mid \frac{1}{9}x^2 + y^2 < 1\}$, the interior of an ellipse.

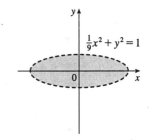

15. $\dfrac{3x + 5y}{x^2 + y^2 - 4}$ is defined only when $x^2 + y^2 - 4 \neq 0$, or $x^2 + y^2 \neq 4$. So the domain of f is $\{(x,y) \mid x^2 + y^2 \neq 4\}$.

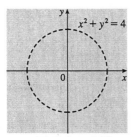

17. $D = \{(x,y) \mid x^2 + y \geq 0\} = \{(x,y) \mid y \geq -x^2\}$

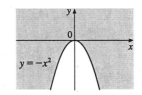

19. We need $1 - x^2 - y^2 - z^2 \geq 0$ or
$x^2 + y^2 + z^2 \leq 1$, so
$D = \{(x, y, z) \mid x^2 + y^2 + z^2 \leq 1\}$ (the points
inside or on the sphere of radius 1, center the
origin).

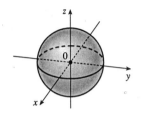

21. $z = 3$, a horizontal plane through the point
$(0, 0, 3)$.

23. $z = 1 - x - y$ or $x + y + z = 1$, a plane with
intercepts 1, 1, and 1.

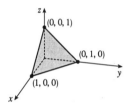

25. $z = 1 - x^2$, a parabolic cylinder.

27. $z = x^2 + 9y^2$, an elliptic paraboloid with vertex
the origin.

29. $z = \sqrt{x^2 + y^2}$ so $x^2 + y^2 = z^2$ and $z \geq 0$, the
top half of a right circular cone.

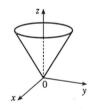

31. The point $(-3, 3)$ lies between the level curves with z-values 50 and 60. Since the point is a little closer to the level curve with $z = 60$, we estimate that $f(-3, 3) \approx 56$. The point $(3, -2)$ appears to be just about halfway between the level curves with z-values 30 and 40, so we estimate $f(3, -2) \approx 35$. The graph rises as we approach the origin, gradually from above, steeply from below.

33. Near A, the level curves are very close together, indicating that the terrain is quite steep. At B, the level curves are much farther apart, so we would expect the terrain to be much less steep than near A, perhaps almost flat.

35. The level curves are $xy = k$. For $k = 0$ the curves are the coordinate axis; if $k > 0$, they are hyperbolas in the first and third quadrants; if $k < 0$, they are hyperbolas in the second and fourth quadrants.

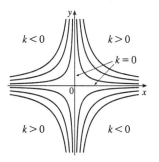

37. $k = x/y$ or $x = ky$ is a family of lines without the point $(0, 0)$.

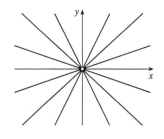

39. $k = \sqrt{x + y}$ or for $x + y \geq 0$, $k^2 = x + y$, or $y = -x + k^2$. *Note:* $k \geq 0$ since $k = \sqrt{x + y}$.

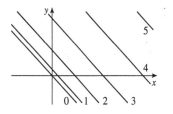

41. $k = x - y^2$, or $x - k = y^2$, a family of parabolas with vertex $(k, 0)$.

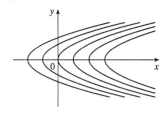

43. The contour map consists of the level curves $k = x^2 + 9y^2$, a family of ellipses with major axis the x-axis. (Or, if $k = 0$, the origin.)

The graph of $f(x, y)$ is the surface $z = x^2 + 9y^2$, an elliptic paraboloid.

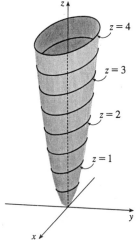

If we visualize lifting each ellipse $k = x^2 + 9y^2$ of the contour map to the plane $z = k$, we have horizontal traces that indicate the shape of the graph of f.

45. The isothermals are given by $k = 100/\left(1 + x^2 + 2y^2\right)$ or $x^2 + 2y^2 = (100 - k)/k$ $(0 < k \leq 100)$, a family of ellipses.

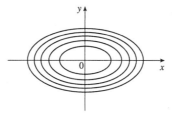

47. $f(x, y) = x^3 + y^3$

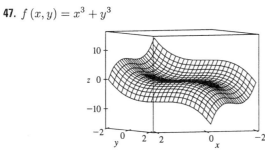

Note that the function is 0 along the line $y = -x$.

49. $f(x, y) = xy^2 - x^3$

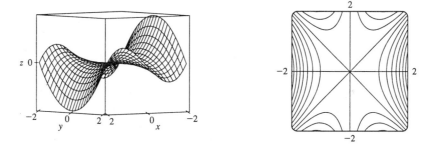

The cross-sections parallel to the yz-plane (such as the left-front trace in the graph above) are parabolas; those parallel to the xz-plane (such as the right-front trace) are cubic curves. The surface is called a monkey saddle because a monkey sitting on the surface near the origin has places for both legs and tail to rest.

51. (a) B   *Reasons:* This function is constant on any circle centered at the origin, a description which matches

   (b) III   only B and III.

53. (a) F   *Reasons:* z increases without bound as we use points closer to the origin, a condition satisfied only

   (b) V   by F and V.

55. (a) D   *Reasons:* This function is periodic in both x and y, with period 2π in each variable.

   (b) IV

57. $k = x + 3y + 5z$ is a family of parallel planes with normal vector $\langle 1, 3, 5 \rangle$.

59. $k = x^2 - y^2 + z^2$ are the equations of the level surfaces. For $k = 0$, the surface is a right circular cone with vertex the origin and axis the y-axis. For $k > 0$, we have a family of hyperboloids of one sheet with axis the y-axis. For $k < 0$, we have a family of hyperboloids of two sheets with axis the y-axis.

61. $f(x, y) = 3x - x^4 - 4y^2 - 10xy$

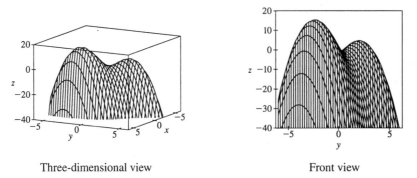

| Three-dimensional view | Front view |

It does appear that the function has a maximum value, at the higher of the two "hilltops." From the front view graph, the maximum value appears to be approximately 15. Both hilltops could be considered local maximum points, as the values of f there are larger than at the neighboring points. There does not appear to be any local minimum point; although the valley shape between the two peaks looks like a minimum of some kind, some neighboring points have lower function values.

63.

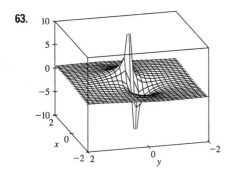

$f(x, y) = \dfrac{x + y}{x^2 + y^2}$. As both x and y become large, the function values appear to approach 0, regardless of which direction is considered. As (x, y) approaches the origin, the graph exhibits asymptotic behavior. From some directions, $f(x, y) \to \infty$, while in others $f(x, y) \to -\infty$. (These are the vertical spikes visible in the graph.) If the graph is examined carefully, however, one can see that $f(x, y)$ approaches 0 along the line $y = -x$.

65. $f(x, y) = e^{cx^2 + y^2}$. First, if $c = 0$, the graph is the cylindrical surface $z = e^{y^2}$ (whose level curves are parallel lines). When $c > 0$, the vertical trace above the y-axis remains fixed while the sides of the surface in the x-direction "curl" upward, giving the graph a shape resembling an elliptic paraboloid. The level curves of the surface are ellipses centered at the origin.

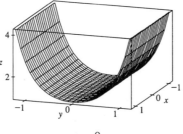

$c = 0$

For $0 < c < 1$, the ellipses have major axis the x-axis and the eccentricity increases as $c \to 0$.

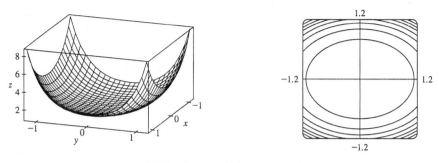

$c = 0.5$ (level curves in increments of 1)

For $c = 1$ the level curves are circles centered at the origin.

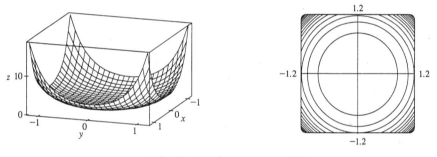

$c = 1$ (level curves in increments of 1)

When $c > 1$, the level curves are ellipses with major axis the y-axis, and the eccentricity increases as c increases.

$c = 2$ (level curves in increments of 4)

For values of $c < 0$, the sides of the surface in the x-direction curl downward and approach the xy-plane (while the vertical trace $x = 0$ remains fixed), giving a saddle-shaped appearance to the graph near the point $(0, 0, 1)$. The level curves consist of a family of hyperbolas. As c decreases, the surface becomes flatter in the x-direction and the surface's approach to the curve in the trace $x = 0$ becomes steeper, as the graphs demonstrate.

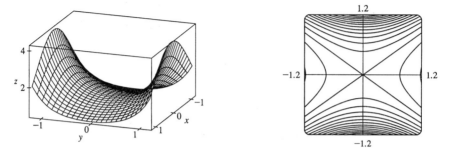

$c = -0.5$ (level curves in increments of 0.25)

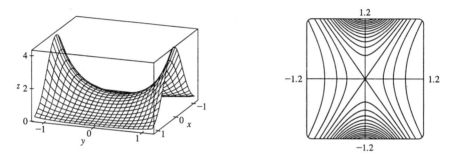

$c = -2$ (level curves in increments of 0.25)

67. (a) $P = bL^\alpha K^{1-\alpha} \implies \dfrac{P}{K} = bL^\alpha K^{-\alpha} \implies \dfrac{P}{K} = b\left(\dfrac{L}{K}\right)^\alpha \implies \ln\dfrac{P}{K} = \ln\left(b\left(\dfrac{L}{K}\right)^\alpha\right) \implies$

$\ln\dfrac{P}{K} = \ln b + \alpha \ln\left(\dfrac{L}{K}\right)$

(b) We list the values for $\ln\left(L/K\right)$ and $\ln\left(P/K\right)$ for the years 1899-1922. (Historically, these values were rounded to 2 decimal places.)

Year	$x = \ln\left(L/K\right)$	$y = \ln\left(P/K\right)$	Year	$x = \ln\left(L/K\right)$	$y = \ln\left(P/K\right)$
1899	0	0	1911	−0.38	−0.34
1900	−0.02	−0.06	1912	−0.38	−0.24
1901	−0.04	−0.02	1913	−0.41	−0.25
1902	−0.04	0	1914	−0.47	−0.37
1903	−0.07	−0.05	1915	−0.53	−0.34
1904	−0.13	−0.12	1916	−0.49	−0.28
1905	−0.18	−0.04	1917	−0.53	−0.39
1906	−0.20	−0.07	1918	−0.60	−0.50
1907	−0.23	−0.15	1919	−0.68	−0.57
1908	−0.41	−0.38	1920	−0.74	−0.57
1909	−0.33	−0.24	1921	−1.05	−0.85
1910	−0.35	−0.27	1922	−0.98	−0.59

After entering the (x, y) pairs into a calculator or CAS, the resulting least squares regression line through the points is approximately $y = 0.75136x + 0.01053$, which we round to $y = 0.75x + 0.01$.

(c) Comparing the regression line from part (b) to the equation $y = \ln b + \alpha x$ with $x = \ln(L/K)$ and $y = \ln(P/K)$, we have $\alpha = 0.75$ and $\ln b = 0.01 \Rightarrow b = e^{0.01} \approx 1.01$. Thus, the Cobb-Douglas production function is $P = bL^\alpha K^{1-\alpha} = 1.01L^{0.75}K^{0.25}$.

15.2 Limits and Continuity   ET 14.2

1. In general, we can't say anything about $f(3, 1)$! $\lim\limits_{(x,y) \to (3,1)} f(x, y) = 6$ means that the values of $f(x, y)$ approach 6 as (x, y) approaches, but is not equal to, $(3, 1)$. If f is continuous, we know that $\lim\limits_{(x,y) \to (a,b)} f(x, y) = f(a, b)$, so $\lim\limits_{(x,y) \to (3,1)} f(x, y) = f(3, 1) = 6$.

3. We make a table of values of $f(x, y) = \dfrac{x^2y^3 + x^3y^2 - 5}{2 - xy}$ for a set of (x, y) points near the origin.

$\dfrac{}{x}\ \dfrac{y}{}$	−0.2	−0.1	−0.05	0	0.05	0.1	0.2
−0.2	−2.551	−2.525	−2.513	−2.500	−2.488	−2.475	−2.451
−0.1	−2.525	−2.513	−2.506	−2.500	−2.494	−2.488	−2.475
−0.05	−2.513	−2.506	−2.503	−2.500	−2.497	−2.494	−2.488
0	−2.500	−2.500	−2.500		−2.500	−2.500	−2.500
0.05	−2.488	−2.494	−2.497	−2.500	−2.503	−2.506	−2.513
0.1	−2.475	−2.488	−2.494	−2.500	−2.506	−2.513	−2.525
0.2	−2.451	−2.475	−2.488	−2.500	−2.513	−2.525	−2.551

As the table shows, the values of $f(x, y)$ seem to approach -2.5 as (x, y) approaches the origin from a variety of different directions. This suggests that $\lim\limits_{(x,y) \to (0,0)} f(x, y) = -2.5$.

Since f is a rational function, it is continuous on its domain. f is defined at $(0, 0)$, so we can use direct substitution to establish that $\lim\limits_{(x,y) \to (0,0)} f(x, y) = \dfrac{0^2 0^3 + 0^3 0^2 - 5}{2 - 0 \cdot 0} = -\dfrac{5}{2}$, verifying our guess.

5. $f(x, y) = x^5 + 4x^3y - 5xy^2$ is a polynomial, and hence continuous, so
$$\lim\limits_{(x,y) \to (5,-2)} f(x, y) = f(5, -2) = 5^5 + 4(5)^3(-2) - 5(5)(-2)^2 = 2025.$$

7. $f(x, y) = x^2/(x^2 + y^2)$. First approach $(0, 0)$ along the x-axis. Then $f(x, 0) = x^2/x^2 = 1$ for $x \neq 0$, so $f(x, y) \to 1$. Now approach $(0, 0)$ along the y-axis. Then for $y \neq 0$, $f(0, y) = 0$, so $f(x, y) \to 0$. Since f has two different limits along two different lines, the limit does not exist.

9. $f(x, y) = 8x^2y^2/(x^4 + y^4)$. Approaching $(0, 0)$ along the x-axis gives $f(x, y) \to 0$. Approaching $(0, 0)$ along the line $y = x$, $f(x, x) = 8x^4/2x^4 = 4$ for $x \neq 0$, so along this line $f(x, y) \to 4$ as $(x, y) \to (0, 0)$. Thus the limit doesn't exist.

11. $f(x,y) = \dfrac{xy}{\sqrt{x^2+y^2}}$. We can see that the limit along any line through $(0,0)$ is 0, as well as along other paths

through $(0,0)$ such as $x = y^2$ and $y = x^2$. So we suspect that the limit exists and equals 0; we use the Squeeze

Theorem to prove our assertion. $0 \le \left| \dfrac{xy}{\sqrt{x^2+y^2}} \right| \le |x|$ since $|y| \le \sqrt{x^2+y^2}$, and $|x| \to 0$ as $(x,y) \to (0,0)$.

So $\lim\limits_{(x,y)\to(0,0)} f(x,y) = 0$.

13. Let $f(x,y) = \dfrac{2x^2 y}{x^4+y^2}$. Then $f(x,0) = 0$ for $x \ne 0$, so $f(x,y) \to 0$ as $(x,y) \to (0,0)$ along the x-axis. But

$f(x,x^2) = \dfrac{2x^4}{2x^4} = 1$ for $x \ne 0$, so $f(x,y) \to 1$ as $(x,y) \to (0,0)$ along the parabola $y = x^2$. Thus the limit

doesn't exist.

15. $\lim\limits_{(x,y)\to(0,0)} \dfrac{x^2+y^2}{\sqrt{x^2+y^2+1}-1} = \lim\limits_{(x,y)\to(0,0)} \dfrac{x^2+y^2}{\sqrt{x^2+y^2+1}-1} \cdot \dfrac{\sqrt{x^2+y^2+1}+1}{\sqrt{x^2+y^2+1}+1}$

$$= \lim\limits_{(x,y)\to(0,0)} \dfrac{\left(x^2+y^2\right)\left(\sqrt{x^2+y^2+1}+1\right)}{x^2+y^2}$$

$$= \lim\limits_{(x,y)\to(0,0)} \left(\sqrt{x^2+y^2+1}+1\right) = 2$$

17. e^{-xy} and $\sin(\pi z/2)$ are each compositions of continuous functions, and hence continuous, so their product

$f(x,y,z) = e^{-xy}\sin(\pi z/2)$ is a continuous function. Then

$\lim\limits_{(x,y,z)\to(3,0,1)} f(x,y,z) = f(3,0,1) = e^{-(3)(0)}\sin(\pi \cdot 1/2) = 1$.

19. $f(x,y,z) = \dfrac{xy+yz^2+xz^2}{x^2+y^2+z^4}$. Then $f(x,0,0) = 0/x^2 = 0$ for $x \ne 0$, so as $(x,y,z) \to (0,0,0)$ along the

x-axis, $f(x,y,z) \to 0$. But $f(x,x,0) = x^2/(2x^2) = \frac{1}{2}$ for $x \ne 0$, so as $(x,y,z) \to (0,0,0)$ along the line

$y = x$, $z = 0$, $f(x,y,z) \to \frac{1}{2}$. Thus the limit doesn't exist.

21.

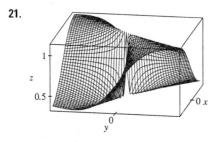

From the ridges on the graph, we see that as $(x,y) \to (0,0)$ along

the lines under the two ridges, $f(x,y)$ approaches different

values. So the limit does not exist.

23. $h(x,y) = g(f(x,y)) = (2x+3y-6)^2 + \sqrt{2x+3y-6}$. Since f is a polynomial, it is continuous on $\mathbb{R}^2$ and

g is continuous on its domain $\{t \mid t \ge 0\}$. Thus h is continuous on its domain

$D = \{(x,y) \mid 2x+3y-6 \ge 0\} = \{(x,y) \mid y \ge -\frac{2}{3}x+2\}$, which consists of all points on and above the line

$y = -\frac{2}{3}x+2$.

25.

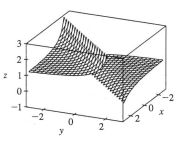

From the graph, it appears that f is discontinuous along the line $y = x$. If we consider $f(x, y) = e^{1/(x-y)}$ as a composition of functions, $g(x, y) = 1/(x - y)$ is a rational function and therefore continuous except where $x - y = 0 \implies y = x$. Since the function $h(t) = e^t$ is continuous everywhere, the composition $h(g(x, y)) = e^{1/(x-y)} = f(x, y)$ is continuous except along the line $y = x$, as we suspected.

27. $F(x, y) = \dfrac{1}{x^2 - y}$ is a rational function and thus is continuous on its domain

$\{(x, y) \mid x^2 - y \neq 0\} = \{(x, y) \mid y \neq x^2\}$, so F is continuous on $\mathbb{R}^2$ except the parabola $y = x^2$.

29. $F(x, y) = \arctan(x + \sqrt{y}) = g(f(x, y))$ where $f(x, y) = x + \sqrt{y}$, continuous on its domain $\{(x, y) \mid y \geq 0\}$, and $g(t) = \arctan t$ is continuous everywhere. Thus F is continuous on its domain $\{(x, y) \mid y \geq 0\}$.

31. $G(x, y) = g_1(f_1(x, y)) - g_2(f_2(x, y))$ where $f_1(x, y) = x + y$ and $f_2(x, y) = x - y$, both of which are polynomials so continuous on $\mathbb{R}^2$, and $g_1(t) = \sqrt{t}$, $g_2(s) = \sqrt{s}$, both of which are continuous on their respective domains $\{t \mid t \geq 0\}$ and $\{s \mid s \geq 0\}$. Thus $g_1 \circ f_1$ is continuous on its domain $D_1 = \{(x, y) \mid x + y \geq 0\} = \{(x, y) \mid y \geq -x\}$ and $g_2 \circ f_2$ is continuous on its domain $D_2 = \{(x, y) \mid x - y \geq 0\} = \{(x, y) \mid y \leq x\}$. Then G, being the difference of these two composite functions, is continuous on its domain $D = D_1 \cap D_2 = \{(x, y) \mid -x \leq y \leq x\} = \{(x, y) \mid |y| \leq x\}$.

33. $f(x, y, z) = \dfrac{xyz}{x^2 + y^2 - z}$ is a rational function and thus is continuous on its domain

$\{(x, y, z) \mid x^2 + y^2 - z \neq 0\} = \{(x, y, z) \mid z \neq x^2 + y^2\}$, so f is continuous on $\mathbb{R}^3$ except on the circular paraboloid $z = x^2 + y^2$.

35. $f(x, y) = \begin{cases} \dfrac{x^2 y^3}{2x^2 + y^2} & \text{if } (x, y) \neq (0, 0) \\ 1 & \text{if } (x, y) = (0, 0) \end{cases}$ The first piece of f is a rational function defined everywhere except

at the origin, so f is continuous on $\mathbb{R}^2$ except possibly at the origin. Since $x^2 \leq 2x^2 + y^2$, we have $|x^2 y^3 / (2x^2 + y^2)| \leq |y^3|$. We know that $|y^3| \to 0$ as $(x, y) \to (0, 0)$. So, by the Squeeze Theorem,

$\displaystyle \lim_{(x,y)\to(0,0)} f(x, y) = \lim_{(x,y)\to(0,0)} \frac{x^2 y^3}{2x^2 + y^2} = 0$. But $f(0, 0) = 1$, so f is discontinuous at $(0, 0)$. Therefore, f is continuous on the set $\{(x, y) \mid (x, y) \neq (0, 0)\}$.

37. $\displaystyle \lim_{(x,y)\to(0,0)} \frac{x^3 + y^3}{x^2 + y^2} = \lim_{r\to0^+} \frac{(r\cos\theta)^3 + (r\sin\theta)^3}{r^2} = \lim_{r\to0^+} (r\cos^3\theta + r\sin^3\theta) = 0$

39. $\displaystyle \lim_{(x,y,z)\to(0,0,0)} \frac{xyz}{x^2 + y^2 + z^2} = \lim_{\rho\to0^+} \frac{(\rho\sin\phi\cos\theta)(\rho\sin\phi\sin\theta)(\rho\cos\phi)}{\rho^2}$

$\displaystyle = \lim_{\rho\to0^+} (\rho\sin^2\phi\cos\phi\sin\theta\cos\theta) = 0$

41. Since $|\mathbf{x} - \mathbf{a}|^2 = |\mathbf{x}|^2 + |\mathbf{a}|^2 - 2|\mathbf{x}||\mathbf{a}|\cos\theta \geq |\mathbf{x}|^2 + |\mathbf{a}|^2 - 2|\mathbf{x}||\mathbf{a}| = (|\mathbf{x}| - |\mathbf{a}|)^2$, we have $||\mathbf{x}| - |\mathbf{a}|| \leq |\mathbf{x} - \mathbf{a}|$. Let $\epsilon > 0$ be given and set $\delta = \epsilon$. Then whenever $0 < |\mathbf{x} - \mathbf{a}| < \delta$, $||\mathbf{x}| - |\mathbf{a}|| \leq |\mathbf{x} - \mathbf{a}| < \delta = \epsilon$. Hence $\lim_{\mathbf{x}\to\mathbf{a}} |\mathbf{x}| = |\mathbf{a}|$ and $f(\mathbf{x}) = |\mathbf{x}|$ is continuous on $\mathbb{R}^n$.

15.3 Partial Derivatives

1. (a) $\partial T/\partial x$ represents the rate of change of T when we fix y and t and consider T as a function of the single variable x, which describes how quickly the temperature changes when longitude changes but latitude and time are constant. $\partial T/\partial y$ represents the rate of change of T when we fix x and t and consider T as a function of y, which describes how quickly the temperature changes when latitude changes but longitude and time are constant. $\partial T/\partial t$ represents the rate of change of T when we fix x and y and consider T as a function of t, which describes how quickly the temperature changes over time for a constant longitude and latitude.

(b) $f_x(158, 21, 9)$ represents the rate of change of temperature at longitude $158\,°$ W, latitude $21\,°$ N at 9:00 A.M. when only longitude varies. Since the air is warmer to the west than to the east, increasing longitude results in an increased air temperature, so we would expect $f_x(158, 21, 9)$ to be positive. $f_y(158, 21, 9)$ represents the rate of change of temperature at the same time and location when only latitude varies. Since the air is warmer to the south and cooler to the north, increasing latitude results in a decreased air temperature, so we would expect $f_y(158, 21, 9)$ to be negative. $f_t(158, 21, 9)$ represents the rate of change of temperature at the same time and location when only time varies. Since typically air temperature increases from the morning to the afternoon as the sun warms it, we would expect $f_t(158, 21, 9)$ to be positive.

3. (a) By Definition 4, $f_T(12, 20) = \lim\limits_{h \to 0} \dfrac{f(12 + h, 20) - f(12, 20)}{h}$, which we can approximate by considering $h = 4$ and $h = -4$ and using the values given in the table:

$f_T(12, 20) \approx \dfrac{f(16, 20) - f(12, 20)}{4} = \dfrac{11 - 5}{4} = 1.5$,

$f_T(12, 20) \approx \dfrac{f(8, 20) - f(12, 20)}{-4} = \dfrac{0 - 5}{-4} = 1.25$. Averaging these values, we estimate $f_T(12, 20)$ to be approximately 1.375. Thus, when the actual temperature is $12\,°$C and the wind speed is 20 km/h, the apparent temperature rises by about $1.375\,°$C for every degree that the actual temperature rises.

Similarly, $f_v(12, 20) = \lim\limits_{h \to 0} \dfrac{f(12, 20 + h) - f(12, 20)}{h}$ which we can approximate by considering

$h = 10$ and $h = -10$: $f_v(12, 20) \approx \dfrac{f(12, 30) - f(12, 20)}{10} = \dfrac{3 - 5}{10} = -0.2$,

$f_v(12, 20) \approx \dfrac{f(12, 10) - f(12, 20)}{-10} = \dfrac{9 - 5}{-10} = -0.4$. Averaging these values, we estimate $f_v(12, 20)$ to be approximately -0.3. Thus, when the actual temperature is $12\,°$C and the wind speed is 20 km/h, the apparent temperature decreases by about $0.3\,°$C for every km/h that the wind speed increases.

(b) For a fixed wind speed v, the values of the wind-chill index I increase as temperature T increases (look at a column of the table), so $\dfrac{\partial I}{\partial T}$ is positive. For a fixed temperature T, the values of I decrease (or remain constant) as v increases (look at a row of the table), so $\dfrac{\partial I}{\partial v}$ is negative (or perhaps 0).

(c) For fixed values of T, the function values $f(T, v)$ appear to become constant (or nearly constant) as v increases, so the corresponding rate of change is 0 or near 0 as v increases. This suggests that $\lim\limits_{v \to \infty} \left(\dfrac{\partial I}{\partial v} \right) = 0$.

5. First of all, if we start at the point $(3, -3)$ and move in the positive y-direction, we see that both b and c decrease, while a increases. Both b and c have a low point at about $(3, -1.5)$, while a is 0 at this point. So a is definitely the graph of f_y, and one of b and c is the graph of f. To see which is which, we start at the point $(-3, -1.5)$ and move in the positive x-direction. b traces out a line with negative slope, while c traces out a parabola opening downward. This tells us that b is the x-derivative of c. So c is the graph of f, b is the graph of f_x, and a is the graph of f_y.

7. $f(x, y) = 16 - 4x^2 - y^2 \Rightarrow f_x(x, y) = -8x$ and $f_y(x, y) = -2y \Rightarrow f_x(1, 2) = -8$ and $f_y(1, 2) = -4$. The graph of f is the paraboloid $z = 16 - 4x^2 - y^2$ and the vertical plane $y = 2$ intersects it in the parabola $z = 12 - 4x^2$, $y = 2$ (the curve C_1 in the first figure). The slope of the tangent line to this parabola at $(1, 2, 8)$ is $f_x(1, 2) = -8$. Similarly the plane $x = 1$ intersects the paraboloid in the parabola $z = 12 - y^2$, $x = 1$ (the curve C_2 in the second figure) and the slope of the tangent line at $(1, 2, 8)$ is $f_y(1, 2) = -4$.

9. $f(x, y) = x^2 + y^2 + x^2 y \Rightarrow f_x = 2x + 2xy$, $f_y = 2y + x^2$

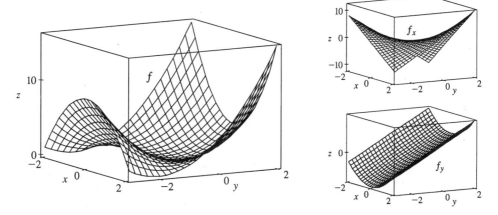

Note that the traces of f in planes parallel to the xz-plane are parabolas which open downward for $y < -1$ and upward for $y > -1$, and the traces of f_x in these planes are straight lines, which have negative slopes for $y < -1$ and positive slopes for $y > -1$. The traces of f in planes parallel to the yz-plane are parabolas which always open upward, and the traces of f_y in these planes are straight lines with positive slopes.

11. $f(x, y) = 3x - 2y^4 \Rightarrow f_x(x, y) = 3 - 0 = 3$, $f_y(x, y) = 0 - 8y^3 = -8y^3$

13. $z = xe^{3y} \Rightarrow \dfrac{\partial z}{\partial x} = e^{3y}$, $\dfrac{\partial z}{\partial y} = 3xe^{3y}$

15. $f(x, y) = \dfrac{x - y}{x + y} \Rightarrow f_x(x, y) = \dfrac{(1)(x + y) - (x - y)(1)}{(x + y)^2} = \dfrac{2y}{(x + y)^2}$,

$f_y(x, y) = \dfrac{(-1)(x + y) - (x - y)(1)}{(x + y)^2} = -\dfrac{2x}{(x + y)^2}$

17. $w = \sin \alpha \cos \beta \quad \Rightarrow \quad \dfrac{\partial w}{\partial \alpha} = \cos \alpha \cos \beta, \; \dfrac{\partial w}{\partial \beta} = -\sin \alpha \sin \beta$

19. $f(u,v) = \tan^{-1}\left(\dfrac{u}{v}\right) \quad \Rightarrow \quad f_u(u,v) = \dfrac{1}{1+(u/v)^2}\left(\dfrac{1}{v}\right) = \dfrac{1}{v}\left(\dfrac{v^2}{u^2+v^2}\right) = \dfrac{v}{u^2+v^2},$

$f_v(u,v) = \dfrac{1}{1+(u/v)^2}\left(-\dfrac{u}{v^2}\right) = -\dfrac{u}{v^2}\left(\dfrac{v^2}{u^2+v^2}\right) = -\dfrac{u}{u^2+v^2}$

21. $z = \ln\left(x + \sqrt{x^2+y^2}\right) \quad \Rightarrow$

$\dfrac{\partial z}{\partial x} = \dfrac{1}{x+\sqrt{x^2+y^2}}\left[1 + \tfrac{1}{2}\left(x^2+y^2\right)^{-1/2}(2x)\right] = \dfrac{\left(\sqrt{x^2+y^2}+x\right)\Big/\sqrt{x^2+y^2}}{\left(x+\sqrt{x^2+y^2}\right)} = \dfrac{1}{\sqrt{x^2+y^2}}$

$\dfrac{\partial z}{\partial y} = \dfrac{1}{x+\sqrt{x^2+y^2}}\left(\dfrac{1}{2}\right)\left(x^2+y^2\right)^{-1/2}(2y) = \dfrac{y}{x\sqrt{x^2+y^2}+x^2+y^2}$

23. $f(x,y,z) = xy^2z^3 + 3yz \quad \Rightarrow \quad f_x(x,y,z) = y^2z^3, \; f_y(x,y,z) = 2xyz^3 + 3z, \; f_z(x,y,z) = 3xy^2z^2 + 3y$

25. $w = \ln(x+2y+3z) \quad \Rightarrow \quad \dfrac{\partial w}{\partial x} = \dfrac{1}{x+2y+3z}, \; \dfrac{\partial w}{\partial y} = \dfrac{2}{x+2y+3z}, \; \dfrac{\partial w}{\partial z} = \dfrac{3}{x+2y+3z}$

27. $u = xe^{-t}\sin\theta \quad \Rightarrow \quad \dfrac{\partial u}{\partial x} = e^{-t}\sin\theta, \; \dfrac{\partial u}{\partial t} = -xe^{-t}\sin\theta, \; \dfrac{\partial u}{\partial \theta} = xe^{-t}\cos\theta$

29. $f(x,y,z,t) = \dfrac{x-y}{z-t} \quad \Rightarrow \quad f_x(x,y,z,t) = \dfrac{1}{z-t}, \; f_y(x,y,z,t) = -\dfrac{1}{z-t},$

$f_z(x,y,z,t) = (x-y)(-1)(z-t)^{-2} = \dfrac{y-x}{(z-t)^2}, \text{ and}$

$f_t(x,y,z,t) = (x-y)(-1)(z-t)^{-2}(-1) = \dfrac{x-y}{(z-t)^2}.$

31. $u = \sqrt{x_1^2 + x_2^2 + \cdots + x_n^2}.$ For each $i = 1, \ldots, n,$

$u_{x_i} = \tfrac{1}{2}\left(x_1^2 + x_2^2 + \cdots + x_n^2\right)^{-1/2}(2x_i) = \dfrac{x_i}{\sqrt{x_1^2 + x_2^2 + \cdots + x_n^2}}.$

33. $f(x,y) = \sqrt{x^2+y^2} \quad \Rightarrow \quad f_x(x,y) = \tfrac{1}{2}\left(x^2+y^2\right)^{-1/2}(2x) = \dfrac{x}{\sqrt{x^2+y^2}}, \text{ so } f_x(3,4) = \dfrac{3}{\sqrt{3^2+4^2}} = \dfrac{3}{5}.$

35. $f(x,y,z) = \dfrac{x}{y+z} = x(y+z)^{-1} \quad \Rightarrow \quad f_z(x,y,z) = x(-1)(y+z)^{-2} = -\dfrac{x}{(y+z)^2}, \text{ so}$

$f_z(3,2,1) = -\dfrac{3}{(2+1)^2} = -\dfrac{1}{3}.$

37. $f(x,y) = x^2 - xy + 2y^2 \quad \Rightarrow$

$f_x(x,y) = \lim_{h\to 0}\dfrac{f(x+h,y)-f(x,y)}{h} = \lim_{h\to 0}\dfrac{(x+h)^2 - (x+h)y + 2y^2 - \left(x^2 - xy + 2y^2\right)}{h}$

$= \lim_{h\to 0}\dfrac{h(2x-y+h)}{h} = \lim_{h\to 0}(2x-y+h) = 2x-y$

$f_y(x,y) = \lim_{h\to 0}\dfrac{f(x,y+h)-f(x,y)}{h} = \lim_{h\to 0}\dfrac{x^2 - x(y+h) + 2(y+h)^2 - \left(x^2 - xy + 2y^2\right)}{h}$

$= \lim_{h\to 0}\dfrac{h(4y-x+2h)}{h} = \lim_{h\to 0}(4y-x+2h) = 4y-x$

39. $xy + yz = xz \Rightarrow \frac{\partial}{\partial x}(xy + yz) = \frac{\partial}{\partial x}(xz) \Leftrightarrow y + y\frac{\partial z}{\partial x} = z + x\frac{\partial z}{\partial x} \Leftrightarrow (y - x)\frac{\partial z}{\partial x} = z - y$, so

$\frac{\partial z}{\partial x} = \frac{z - y}{y - x}$. $\frac{\partial}{\partial y}(xy + yz) = \frac{\partial}{\partial y}(xz) \Leftrightarrow x + z + y\frac{\partial z}{\partial y} = x\frac{\partial z}{\partial y} \Leftrightarrow (y - x)\frac{\partial z}{\partial y} = -(x + z)$, so

$\frac{\partial z}{\partial y} = \frac{x + z}{x - y}$.

41. $x^2 + y^2 - z^2 = 2x(y + z) \Leftrightarrow \frac{\partial}{\partial x}(x^2 + y^2 - z^2) = \frac{\partial}{\partial x}[2x(y + z)] \Leftrightarrow$

$2x - 2z\frac{\partial z}{\partial x} = 2(y + z) + 2x\frac{\partial z}{\partial x} \Leftrightarrow 2(x + z)\frac{\partial z}{\partial x} = 2(x - y - z)$, so $\frac{\partial z}{\partial x} = \frac{x - y - z}{x + z}$.

$\frac{\partial}{\partial y}(x^2 + y^2 - z^2) = \frac{\partial}{\partial y}[2x(y + z)] \Leftrightarrow 2y - 2z\frac{\partial z}{\partial y} = 2x\left(1 + \frac{\partial z}{\partial y}\right) \Leftrightarrow 2(x + z)\frac{\partial z}{\partial y} = 2(y - x)$, so

$\frac{\partial z}{\partial y} = \frac{y - x}{x + z}$.

43. (a) $z = f(x) + g(y) \Rightarrow \frac{\partial z}{\partial x} = f'(x), \frac{\partial z}{\partial y} = g'(y)$

(b) $z = f(x + y)$. Let $u = x + y$. Then $\frac{\partial z}{\partial x} = \frac{df}{du}\frac{\partial u}{\partial x} = \frac{df}{du}(1) = f'(u) = f'(x + y)$,

$\frac{\partial z}{\partial y} = \frac{df}{du}\frac{\partial u}{\partial y} = \frac{df}{du}(1) = f'(u) = f'(x + y)$.

45. $f(x, y) = x^4 - 3x^2y^3 \Rightarrow f_x(x, y) = 4x^3 - 6xy^3, f_y(x, y) = -9x^2y^2$. Then $f_{xx}(x, y) = 12x^2 - 6y^3$,

$f_{xy}(x, y) = -18xy^2, f_{yx}(x, y) = -18xy^2$, and $f_{yy}(x, y) = -18x^2y$.

47. $z = \frac{x}{x + y} = x(x + y)^{-1} \Rightarrow z_x = \frac{1(x + y) - 1(x)}{(x + y)^2} = \frac{y}{(x + y)^2}, z_y = x(-1)(x + y)^{-2} = -\frac{x}{(x + y)^2}$.

Then $z_{xx} = y(-2)(x + y)^{-3} = -\frac{2y}{(x + y)^3}, z_{xy} = \frac{1(x + y)^2 - y(2)(x + y)}{[(x + y)^2]^2} = \frac{x + y - 2y}{(x + y)^3} = \frac{x - y}{(x + y)^3}$,

$z_{yx} = -\frac{1(x + y)^2 - x(2)(x + y)}{[(x + y)^2]^2} = -\frac{-x^2 + xy + y^2}{(x + y)^2} = \frac{(x + y)(x - y)}{(x + y)^2} = \frac{x - y}{(x + y)^3}$, and

$z_{yy} = -x(-2)(x + y)^{-3} = \frac{2x}{(x + y)^3}$.

49. $u = e^{-s}\sin t \Rightarrow u_s = -e^{-s}\sin t, u_t = e^{-s}\cos t$. Then $u_{ss} = e^{-s}\sin t, u_{st} = -e^{-s}\cos t$,

$u_{ts} = -e^{-s}\cos t$, and $u_{tt} = -e^{-s}\sin t$.

51. $u = x^5y^4 - 3x^2y^3 + 2x^2 \Rightarrow u_x = 5x^4y^4 - 6xy^3 + 4x, u_{xy} = 20x^4y^3 - 18xy^2$ and

$u_y = 4x^5y^3 - 9x^2y^2, u_{yx} = 20x^4y^3 - 18xy^2$. Thus $u_{xy} = u_{yx}$.

53. $u = \ln\sqrt{x^2 + y^2} = \ln(x^2 + y^2)^{1/2} = \frac{1}{2}\ln(x^2 + y^2) \Rightarrow u_x = \frac{1}{2}\frac{1}{x^2 + y^2}\cdot 2x = \frac{x}{x^2 + y^2}$,

$u_{xy} = x(-1)(x^2 + y^2)^{-2}(2y) = -\frac{2xy}{(x^2 + y^2)^2}$ and $u_y = \frac{1}{2}\frac{1}{x^2 + y^2}\cdot 2y = \frac{y}{x^2 + y^2}$,

$u_{yx} = y(-1)(x^2 + y^2)^{-2}(2x) = -\frac{2xy}{(x^2 + y^2)^2}$. Thus $u_{xy} = u_{yx}$.

55. $f(x, y) = x^2y^3 - 2x^4y \Rightarrow f_x = 2xy^3 - 8x^3y, f_{xx} = 2y^3 - 24x^2y, f_{xxx} = -48xy$

57. $f(x, y, z) = x^5 + x^4y^4z^3 + yz^2 \Rightarrow f_x = 5x^4 + 4x^3y^4z^3, f_{xy} = 16x^3y^3z^3$, and $f_{xyz} = 48x^3y^3z^2$

59. $z = x \sin y \quad \Rightarrow \quad \dfrac{\partial z}{\partial x} = \sin y, \; \dfrac{\partial^2 z}{\partial y \partial x} = \cos y,$ and $\dfrac{\partial^3 z}{\partial y^2 \partial x} = -\sin y.$

61. $u = \ln\left(x + 2y^2 + 3z^3\right) \quad \Rightarrow \quad \dfrac{\partial u}{\partial z} = \dfrac{1}{x + 2y^2 + 3z^3}\left(9z^2\right) = \dfrac{9z^2}{x + 2y^2 + 3z^3},$

$\dfrac{\partial^2 u}{\partial y\, \partial z} = -9z^2 \left(x + 2y^2 + 3z^3\right)^{-2}(4y) = -\dfrac{36yz^2}{\left(x + 2y^2 + 3z^3\right)^2},$ and $\dfrac{\partial^3 u}{\partial x\, \partial y\, \partial z} = \dfrac{72yz^2}{\left(x + 2y^2 + 3z^3\right)^3}.$

63. By Definition 4, $f_x\,(3, 2) = \lim\limits_{h \to 0} \dfrac{f\,(3 + h, 2) - f\,(3, 2)}{h}$ which we can approximate by considering $h = 0.5$

and $h = -0.5$: $f_x\,(3, 2) \approx \dfrac{f\,(3.5, 2) - f\,(3, 2)}{0.5} = \dfrac{22.4 - 17.5}{0.5} = 9.8,$

$f_x\,(3, 2) \approx \dfrac{f\,(2.5, 2) - f\,(3, 2)}{-0.5} = \dfrac{10.2 - 17.5}{-0.5} = 14.6.$ Averaging these values, we estimate $f_x\,(3, 2)$ to be

approximately 12.2. Similarly, $f_x\,(3, 2.2) = \lim\limits_{h \to 0} \dfrac{f\,(3 + h, 2.2) - f\,(3, 2.2)}{h}$ which we can approximate by

considering $h = 0.5$ and $h = -0.5$: $f_x\,(3, 2.2) \approx \dfrac{f\,(3.5, 2.2) - f\,(3, 2.2)}{0.5} = \dfrac{26.1 - 15.9}{0.5} = 20.4,$

$f_x\,(3, 2.2) \approx \dfrac{f\,(2.5, 2.2) - f\,(3, 2.2)}{-0.5} = \dfrac{9.3 - 15.9}{-0.5} = 13.2.$ Averaging these values, we have $f_x\,(3, 2.2) \approx 16.8.$

To estimate $f_{xy}\,(3, 2)$, we first need an estimate for $f_x\,(3, 1.8)$:

$f_x\,(3, 1.8) \approx \dfrac{f\,(3.5, 1.8) - f\,(3, 1.8)}{0.5} = \dfrac{20.0 - 18.1}{0.5} = 3.8,$

$f_x\,(3, 1.8) \approx \dfrac{f\,(2.5, 1.8) - f\,(3, 1.8)}{-0.5} = \dfrac{12.5 - 18.1}{-0.5} = 11.2.$ Averaging these values, we get $f_x\,(3, 1.8) \approx 7.5.$

Now $f_{xy}\,(x, y) = \dfrac{\partial}{\partial y}\left[f_x\,(x, y)\right]$ and $f_x\,(x, y)$ is itself a function of 2 variables, so

Definition 4 says that $f_{xy}\,(x, y) = \dfrac{\partial}{\partial y}\left[f_x\,(x, y)\right] = \lim\limits_{h \to 0} \dfrac{f_x\,(x, y + h) - f_x\,(x, y)}{h} \quad \Rightarrow$

$f_{xy}\,(3, 2) = \lim\limits_{h \to 0} \dfrac{f_x\,(3, 2 + h) - f_x\,(3, 2)}{h}.$ We can estimate this value using our previous work with

$h = 0.2$ and $h = -0.2$: $f_{xy}\,(3, 2) \approx \dfrac{f_x\,(3, 2.2) - f_x\,(3, 2)}{0.2} = \dfrac{16.8 - 12.2}{0.2} = 23,$

$f_{xy}\,(3, 2) \approx \dfrac{f_x\,(3, 1.8) - f_x\,(3, 2)}{-0.2} = \dfrac{7.5 - 12.2}{-0.2} = 23.5.$ Averaging these values, we estimate $f_{xy}\,(3, 2)$ to be

approximately 23.25.

65. $u = e^{-\alpha^2 k^2 t} \sin kx \quad \Rightarrow \quad u_x = ke^{-\alpha^2 k^2 t} \cos kx, \; u_{xx} = -k^2 e^{-\alpha^2 k^2 t} \sin kx,$ and $u_t = -\alpha^2 k^2 e^{-\alpha^2 k^2 t} \sin kx.$

Thus $\alpha^2 u_{xx} = u_t.$

67. $u = \dfrac{1}{\sqrt{x^2 + y^2 + z^2}} \quad \Rightarrow \quad u_x = \left(-\tfrac{1}{2}\right)\left(x^2 + y^2 + z^2\right)^{-3/2}(2x) = -x\left(x^2 + y^2 + z^2\right)^{-3/2}$ and

$u_{xx} = -\left(x^2 + y^2 + z^2\right)^{-3/2} - x\left(-\tfrac{3}{2}\right)\left(x^2 + y^2 + z^2\right)^{-5/2}(2x) = \dfrac{2x^2 - y^2 - z^2}{\left(x^2 + y^2 + z^2\right)^{5/2}}.$

By symmetry, $u_{yy} = \dfrac{2y^2 - x^2 - z^2}{\left(x^2 + y^2 + z^2\right)^{5/2}}$ and $u_{zz} = \dfrac{2z^2 - x^2 - y^2}{\left(x^2 + y^2 + z^2\right)^{5/2}}.$ Thus

$u_{xx} + u_{yy} + u_{zz} = \dfrac{2x^2 - y^2 - z^2 + 2y^2 - x^2 - z^2 + 2z^2 - x^2 - y^2}{\left(x^2 + y^2 + z^2\right)^{5/2}} = 0.$

69. Let $v = x + at$, $w = x - at$. Then $u_t = \dfrac{\partial\,[f\,(v) + g\,(w)]}{\partial t} = \dfrac{df\,(v)}{dv}\dfrac{\partial v}{\partial t} + \dfrac{dg\,(w)}{dw}\dfrac{\partial w}{\partial t} = af'\,(v) - ag'\,(w)$ and

$u_{tt} = \dfrac{\partial\,[af'\,(v) - ag'\,(w)]}{\partial t} = a\,[af''\,(v) + ag''\,(w)] = a^2\,[f''\,(v) + g''\,(w)]$. Similarly, by using the Chain Rule

we have $u_x = f'\,(v) + g'\,(w)$ and $u_{xx} = f''\,(v) + g''\,(w)$. Thus $u_{tt} = a^2 u_{xx}$.

71. $z_x = e^y + ye^x$, $z_{xx} = ye^x$, $\partial^3 z/\partial x^3 = ye^x$. By symmetry $z_y = xe^y + e^x$, $z_{yy} = xe^y$ and $\partial^3 z/\partial y^3 = xe^y$.
Then $\partial^3 z/\partial x \partial y^2 = e^y$ and $\partial^3 z/\partial x^2 \partial y = e^x$. Thus $z = xe^y + ye^x$ satisfies the given partial differential
equation.

73. If we fix $K = K_0$, $P\,(L, K_0)$ is a function of a single variable L, and $\dfrac{dP}{dL} = \alpha\dfrac{P}{L}$ is a separable differential

equation. Then $\dfrac{dP}{P} = \alpha\dfrac{dL}{L}$ $\Rightarrow$ $\displaystyle\int\dfrac{dP}{P} = \int\alpha\dfrac{dL}{L}$ $\Rightarrow$ $\ln|P| = \alpha\ln|L| + C\,(K_0)$, where $C\,(K_0)$ can

depend on K_0. Then $|P| = e^{\alpha\ln|L| + C(K_0)}$, and since $P > 0$ and $L > 0$, we have
$P = e^{\alpha\ln L}e^{C(K_0)} = e^{C(K_0)}e^{\ln L^\alpha} = C_1\,(K_0)\,L^\alpha$ where $C_1\,(K_0) = e^{C(K_0)}$.

75. By the Chain Rule, taking the partial derivative of both sides with respect to R_1 gives

$$\dfrac{\partial R^{-1}}{\partial R}\dfrac{\partial R}{\partial R_1} = \dfrac{\partial\,[(1/R_1) + (1/R_2) + (1/R_3)]}{\partial R_1} \text{ or } -R^{-2}\dfrac{\partial R}{\partial R_1} = -R_1^{-2}. \text{ Thus } \dfrac{\partial R}{\partial R_1} = \dfrac{R^2}{R_1^2}.$$

77. $\dfrac{\partial K}{\partial m} = \tfrac{1}{2}V^2$, $\dfrac{\partial K}{\partial V} = mV$, $\dfrac{\partial^2 K}{\partial V^2} = m$. Thus $\dfrac{\partial K}{\partial m} \cdot \dfrac{\partial^2 K}{\partial V^2} = \tfrac{1}{2}V^2 m = K$.

79. $f_x\,(x, y) = x + 4y$ $\Rightarrow$ $f_{xy}\,(x, y) = 4$ and $f_y\,(x, y) = 3x - y$ $\Rightarrow$ $f_{yx}\,(x, y) = 3$. Since f_{xy} and f_{yx} are
continuous everywhere but $f_{xy}\,(x, y) \ne f_{yx}\,(x, y)$, Clairaut's Theorem implies that such a function $f\,(x, y)$ does
not exist.

81. By the geometry of partial derivatives, the slope of the tangent line is $f_x\,(1, 2)$. By implicit differentiation of
$4x^2 + 2y^2 + z^2 = 16$, we get $8x + 2z\,(\partial z/\partial x) = 0$ $\Rightarrow$ $\partial z/\partial x = -4x/z$, so when $x = 1$ and $z = 2$ we have
$\partial z/\partial x = -2$. So the slope is $f_x\,(1, 2) = -2$. Thus the tangent line is given by $z - 2 = -2\,(x - 1)$, $y = 2$. Taking
the parameter to be $t = x - 1$, we can write parametric equations for this line: $x = 1 + t$, $y = 2$, $z = 2 - 2t$.

83. By Clairaut's Theorem, $f_{xyy} = (f_{xy})_y = (f_{yx})_y = f_{yxy} = (f_y)_{xy} = (f_y)_{yx} = f_{yyx}$.

85. Let $g\,(x) = f\,(x, 0) = x\,(x^2)^{-3/2}\,e^0 = x\,|x|^{-3}$. But we are using the point $(1, 0)$, so near $(1, 0)$, $g\,(x) = x^{-2}$.
Then $g'\,(x) = -2x^{-3}$ and $g'\,(1) = -2$, so using (1) we have $f_x\,(1, 0) = g'\,(1) = -2$.

87. (a)

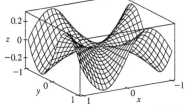

(b) For $(x, y) \ne (0, 0)$, $f_x\,(x, y) = \dfrac{(3x^2 y - y^3)\,(x^2 + y^2) - (x^3 y - xy^3)\,(2x)}{(x^2 + y^2)^2} = \dfrac{x^4 y + 4x^2 y^3 - y^5}{(x^2 + y^2)^2}$, and by

symmetry $f_y\,(x, y) = \dfrac{x^5 - 4x^3 y^2 - xy^4}{(x^2 + y^2)^2}$.

(c) $f_x\,(0, 0) = \displaystyle\lim_{h \to 0}\dfrac{f\,(h, 0) - f\,(0, 0)}{h} = \lim_{h \to 0}\dfrac{(0/h^2) - 0}{h} = 0$ and $f_y\,(0, 0) = \displaystyle\lim_{h \to 0}\dfrac{f\,(0, h) - f\,(0, 0)}{h} = 0$.

(d) By (3), $f_{xy}(0,0) = \dfrac{\partial f_x}{\partial y} = \lim_{h \to 0} \dfrac{f_x(0,h) - f_x(0,0)}{h} = \lim_{h \to 0} \dfrac{(-h^5 - 0)/h^4}{h} = -1$ while by (2),

$f_{yx}(0,0) = \dfrac{\partial f_y}{\partial x} = \lim_{h \to 0} \dfrac{f_y(h,0) - f_y(0,0)}{h} = \lim_{h \to 0} \dfrac{h^5/h^4}{h} = 1.$

(e) For $(x,y) \neq (0,0)$, we use a CAS to compute $f_{xy}(x,y) = \dfrac{x^6 + 9x^4 y^2 - 4x^2 y^4 + 4y^6}{(x^2 + y^2)^3}$. Now as

$(x,y) \to (0,0)$ along the x-axis, $f_{xy}(x,y) \to 1$ while as $(x,y) \to (0,0)$ along the y-axis, $f_{xy}(x,y) \to 4$. Thus f_{xy} isn't continuous at $(0,0)$ and Clairaut's Theorem doesn't apply, so there is no contradiction. The graphs of f_{xy} and f_{yx} are identical except at the origin, where we observe the discontinuity.

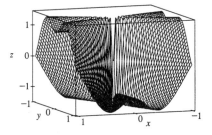

15.4 Tangent Planes and Linear Approximations

1. $z = f(x,y) = y^2 - x^2 \Rightarrow f_x(x,y) = -2x$, $f_y(x,y) = 2y$, so $f_x(-4,5) = 8$, $f_y(-4,5) = 10$. By Equation 2, an equation of the tangent plane is $z - 9 = f_x(-4,5)[x - (-4)] + f_y(-4,5)(y - 5) \Rightarrow$ $z - 9 = 8(x + 4) + 10(y - 5)$ or $z = 8x + 10y - 9$.

3. $z = f(x,y) = \sqrt{4 - x^2 - 2y^2} \Rightarrow f_x(x,y) = \tfrac{1}{2}(4 - x^2 - 2y^2)^{-1/2}(-2x) = -\dfrac{x}{\sqrt{4 - x^2 - 2y^2}}$,

$f_y(x,y) = \tfrac{1}{2}(4 - x^2 - 2y^2)^{-1/2}(-4y) = -\dfrac{2y}{\sqrt{4 - x^2 - 2y^2}}$, so $f_x(1,-1) = -1$

and $f_y(1,-1) = 2$. Thus, an equation of the tangent plane is

$z - 1 = f_x(1,-1)(x - 1) + f_y(1,-1)[y - (-1)] \Rightarrow z - 1 = -1(x - 1) + 2(y + 1)$ or $x - 2y + z = 4$.

5. $z = f(x,y) = \ln(2x + y) \Rightarrow f_x(x,y) = \dfrac{2}{2x + y}$, $f_y(x,y) = \dfrac{1}{2x + y}$, $f_x(-1,3) = 2$, $f_y(-1,3) = 1$.

Thus an equation of the tangent plane is $z = 2(x + 1) + (y - 3)$ or $z = 2x + y - 1$.

7. $z = f(x,y) = xy$, so $f_x(x,y) = y \Rightarrow f_x(-1,2) = 2$, $f_y(x,y) = x \Rightarrow f_y(-1,2) = -1$ and an equation of the tangent plane is $z + 2 = 2(x + 1) + (-1)(y - 2)$ or $z = 2x - y + 2$. After zooming in, the surface and the tangent plane become almost indistinguishable. (Here, the tangent plane is shown with fewer traces than the surface.) If we zoom in farther, the surface and the tangent plane will appear to coincide.

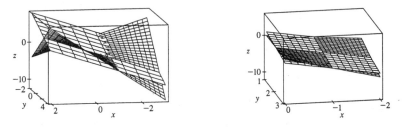

9. $f(x,y) = e^{-(x^2+y^2)/15} (\sin^2 x + \cos^2 y)$. A CAS gives

$f_x = -\frac{2}{15} e^{-(x^2+y^2)/15} (x \sin^2 x + x \cos^2 y - 15 \sin x \cos x)$ and

$f_y = -\frac{2}{15} e^{-(x^2+y^2)/15} (y \sin^2 x + y \cos^2 y + 15 \sin y \cos y)$. We use the CAS to evaluate these at $(2, 3)$, and

then substitute the results into Equation 2 in order to plot the tangent plane. After zooming in, the surface and the

tangent plane become almost indistinguishable. (Here, the tangent plane is above the surface.) If we zoom in farther,

the surface and the tangent plane will appear to coincide.

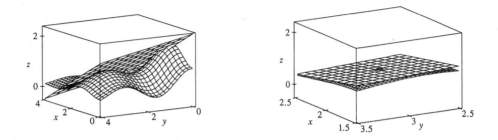

11. $f(x,y) = x\sqrt{y}$. The partial derivatives are $f_x(x,y) = \sqrt{y}$ and $f_y(x,y) = \dfrac{x}{2\sqrt{y}}$, so $f_x(1, 4) = 2$ and

$f_y(1, 4) = \frac{1}{4}$. Both f_x and f_y are continuous functions for $y > 0$, so by Theorem 8, f is

differentiable at $(1, 4)$. By Equation 3, the linearization of f at $(1, 4)$ is given by

$L(x,y) = f(1, 4) + f_x(1, 4)(x - 1) + f_y(1, 4)(y - 4) = 2 + 2(x - 1) + \frac{1}{4}(y - 4) = 2x + \frac{1}{4}y - 1$.

13. $f(x,y) = e^x \cos xy$. The partial derivatives are $f_x(x,y) = e^x (\cos xy - y \sin xy)$ and $f_y(x,y) = -xe^x \sin xy$,

so $f_x(0, 0) = 1$ and $f_y(0, 0) = 0$. Both f_x and f_y are continuous functions, so f is

differentiable at $(0, 0)$ by Theorem 8. The linearization of f at $(0, 0)$ is given by

$L(x,y) = f(0, 0) + f_x(0, 0)(x - 0) + f_y(0, 0)(y - 0) = 1 + 1(x - 0) + 0(y - 0) = x + 1$.

15. $f(x,y) = \tan^{-1}(x + 2y)$. The partial derivatives are $f_x(x,y) = \dfrac{1}{1 + (x + 2y)^2}$ and

$f_y(x,y) = \dfrac{2}{1 + (x + 2y)^2}$, so $f_x(1, 0) = \frac{1}{2}$ and $f_y(1, 0) = 1$. Both f_x and f_y are continuous

functions, so f is differentiable at $(1, 0)$, and the linearization of f at $(1, 0)$ is

$L(x,y) = f(1, 0) + f_x(1, 0)(x - 1) + f_y(1, 0)(y - 0) = \frac{\pi}{4} + \frac{1}{2}(x - 1) + 1(y) = \frac{1}{2}x + y + \frac{\pi}{4} - \frac{1}{2}$.

17. $f(x,y) = \sqrt{20 - x^2 - 7y^2}$ $\Rightarrow$ $f_x(x,y) = -\dfrac{x}{\sqrt{20 - x^2 - 7y^2}}$ and $f_y(x,y) = -\dfrac{7y}{\sqrt{20 - x^2 - 7y^2}}$, so

$f_x(2, 1) = -\frac{2}{3}$ and $f_y(2, 1) = -\frac{7}{3}$. Then the linear approximation of f at $(2, 1)$ is given by

$$f(x,y) \approx f(2, 1) + f_x(2, 1)(x - 2) + f_y(2, 1)(y - 1) = 3 - \frac{2}{3}(x - 2) - \frac{7}{3}(y - 1)$$

$$= -\frac{2}{3}x - \frac{7}{3}y + \frac{20}{3}$$

Thus $f(1.95, 1.08) \approx -\frac{2}{3}(1.95) - \frac{7}{3}(1.08) + \frac{20}{3} = 2.84\overline{6}$.

19. $f(x, y, z) = \sqrt{x^2 + y^2 + z^2} \Rightarrow f_x(x, y, z) = \dfrac{x}{\sqrt{x^2 + y^2 + z^2}}, f_y(x, y, z) = \dfrac{y}{\sqrt{x^2 + y^2 + z^2}}$, and

$f_z(x, y, z) = \dfrac{z}{\sqrt{x^2 + y^2 + z^2}}$, so $f_x(3, 2, 6) = \frac{3}{7}, f_y(3, 2, 6) = \frac{2}{7}$, and $f_z(3, 2, 6) = \frac{6}{7}$. Then the linear

approximation of f at $(3, 2, 6)$ is given by

$$f(x, y, z) \approx f(3, 2, 6) + f_x(3, 2, 6)(x - 3) + f_y(3, 2, 6)(y - 2) + f_z(3, 2, 6)(z - 6)$$

$$= 7 + \tfrac{3}{7}(x - 3) + \tfrac{2}{7}(y - 2) + \tfrac{6}{7}(z - 6) = \tfrac{3}{7}x + \tfrac{2}{7}y + \tfrac{6}{7}z$$

Thus $\sqrt{(3.02)^2 + (1.97)^2 + (5.99)^2} = f(3.02, 1.97, 5.99) \approx \tfrac{3}{7}(3.02) + \tfrac{2}{7}(1.97) + \tfrac{6}{7}(5.99) \approx 6.9914$.

21. From the table, $f(94, 80) = 127$. To estimate $f_T(94, 80)$ and $f_H(94, 80)$ we follow the procedure used in

Section 15.3 [ET 14.3]. Since $f_T(94, 80) = \lim\limits_{h \to 0} \dfrac{f(94 + h, 80) - f(94, 80)}{h}$, we approximate this quantity with

$h = \pm 2$ and use the values given in the table: $f_T(94, 80) \approx \dfrac{f(96, 80) - f(94, 80)}{2} = \dfrac{135 - 127}{2} = 4$,

$f_T(94, 80) \approx \dfrac{f(92, 80) - f(94, 80)}{-2} = \dfrac{119 - 127}{-2} = 4$.

Averaging these values gives $f_T(94, 80) \approx 4$. Similarly, $f_H(94, 80) = \lim\limits_{h \to 0} \dfrac{f(94, 80 + h) - f(94, 80)}{h}$,

so we use $h = \pm 5$: $f_H(94, 80) \approx \dfrac{f(94, 85) - f(94, 80)}{5} = \dfrac{132 - 127}{5} = 1$,

$f_H(94, 80) \approx \dfrac{f(94, 75) - f(94, 80)}{-5} = \dfrac{122 - 127}{-5} = 1$. Averaging these values gives $f_H(94, 80) \approx 1$. The

linear approximation, then, is

$$f(T, H) \approx f(94, 80) + f_T(94, 80)(T - 94) + f_H(94, 80)(H - 80)$$

$$\approx 127 + 4(T - 94) + 1(H - 80)$$

Thus when $T = 95$ and $H = 78$, $f(95, 78) \approx 127 + 4(95 - 94) + 1(78 - 80) = 129$, so we estimate the heat
index to be approximately $129\,°F$.

23. $z = x^2 y^3 \Rightarrow dz = \dfrac{\partial z}{\partial x}dx + \dfrac{\partial z}{\partial y}dy = 2xy^3\,dx + 3x^2 y^2\,dy$

25. $u = e^t \sin\theta \Rightarrow du = \dfrac{\partial u}{\partial t}dt + \dfrac{\partial u}{\partial \theta}d\theta = e^t \sin\theta\,dt + e^t \cos\theta\,d\theta$

27. $w = \ln\sqrt{x^2 + y^2 + z^2} \Rightarrow$

$$dw = \frac{\partial w}{\partial x}dx + \frac{\partial w}{\partial y}dy + \frac{\partial w}{\partial z}dz$$

$$= \left(\frac{1}{2}\right)\frac{2x\left(x^2 + y^2 + z^2\right)^{-1/2}dx + 2y\left(x^2 + y^2 + z^2\right)^{-1/2}dy + 2z\left(x^2 + y^2 + z^2\right)^{-1/2}dz}{\left(x^2 + y^2 + z^2\right)^{1/2}}$$

$$= \frac{x\,dx + y\,dy + z\,dz}{x^2 + y^2 + z^2}$$

29. $dx = \Delta x = 0.05$, $dy = \Delta y = 0.1$, $z = 5x^2 + y^2$, $z_x = 10x$, $z_y = 2y$. Thus when $x = 1$ and
$y = 2$, $dz = z_x(1, 2)\,dx + z_y(1, 2)\,dy = (10)(0.05) + (4)(0.1) = 0.9$ while
$\Delta z = f(1.05, 2.1) - f(1, 2) = 5(1.05)^2 + (2.1)^2 - 5 - 4 = 0.9225$.

31. $dA = \dfrac{\partial A}{\partial x}\, dx + \dfrac{\partial A}{\partial y}\, dy = y\, dx + x\, dy$ and $|\Delta x| \le 0.1$, $|\Delta y| \le 0.1$. We use $dx = 0.1$, $dy = 0.1$ with $x = 30$,

$y = 24$; then the maximum error in the area is about $dA = 24\,(0.1) + 30\,(0.1) = 5.4 \text{ cm}^2$.

33. The volume of a can is $V = \pi r^2 h$ and $\Delta V \approx dV$ is an estimate of the amount of tin. Here

$dV = 2\pi r h\, dr + \pi r^2\, dh$, so put $dr = 0.04$, $dh = 0.08$ (0.04 on top, 0.04 on bottom) and then

$\Delta V \approx dV = 2\pi\,(48)\,(0.04) + \pi\,(16)\,(0.08) \approx 16.08 \text{ cm}^3$. Thus the amount of tin is about 16 cm^3.

35. The area of the rectangle is $A = xy$, and $\Delta A \approx dA$ is an estimate of the area of paint in the stripe. Here

$dA = y\, dx + x\, dy$, so with $dx = dy = \frac{3+3}{12} = \frac{1}{2}$, $\Delta A \approx dA = (100)\left(\frac{1}{2}\right) + (200)\left(\frac{1}{2}\right) = 150 \text{ ft}^2$. Thus there are

approximately 150 ft^2 of paint in the stripe.

37. First we find $\dfrac{\partial R}{\partial R_1}$ implicitly by taking partial derivatives of both sides with respect to R_1:

$$\frac{\partial}{\partial R_1}\left[\frac{1}{R}\right] = \frac{\partial\,[(1/R_1) + (1/R_2) + (1/R_3)]}{\partial R_1} \;\Rightarrow\; -R^{-2}\frac{\partial R}{\partial R_1} = -R_1^{-2} \;\Rightarrow\; \frac{\partial R}{\partial R_1} = \frac{R^2}{R_1^2}.\; \text{Then by}$$

symmetry, $\dfrac{\partial R}{\partial R_2} = \dfrac{R^2}{R_2^2}$, $\dfrac{\partial R}{\partial R_3} = \dfrac{R^2}{R_3^2}$. When $R_1 = 25$, $R_2 = 40$ and $R_3 = 50$, $\dfrac{1}{R} = \dfrac{17}{200} \;\Leftrightarrow\; R = \frac{200}{17}$ ohms.

Since the possible error for each R_i is 0.5%, the maximum error of R is attained by setting $\Delta R_i = 0.005 R_i$. So

$$\Delta R \approx dR = \frac{\partial R}{\partial R_1}\,\Delta R_1 + \frac{\partial R}{\partial R_2}\,\Delta R_2 + \frac{\partial R}{\partial R_3}\,\Delta R_3 = (0.005)\,R^2\left[\frac{1}{R_1} + \frac{1}{R_2} + \frac{1}{R_3}\right]$$

$$= (0.005)\,R = \frac{1}{17} \approx 0.059 \text{ ohms}$$

39. $\Delta z = f\,(a + \Delta x, b + \Delta y) - f\,(a, b) = (a + \Delta x)^2 + (b + \Delta y)^2 - (a^2 + b^2)$

$= a^2 + 2a\,\Delta x + (\Delta x)^2 + b^2 + 2b\,\Delta y + (\Delta y)^2 - a^2 - b^2 = 2a\,\Delta x + (\Delta x)^2 + 2b\,\Delta y + (\Delta y)^2$

But $f_x\,(a, b) = 2a$ and $f_y\,(a, b) = 2b$ and so $\Delta z = f_x\,(a, b)\,\Delta x + f_y\,(a, b)\,\Delta y + \Delta x \Delta x + \Delta y \Delta y$, which is

Definition 7 with $\epsilon_1 = \Delta x$ and $\epsilon_2 = \Delta y$. Hence f is differentiable.

41. To show that f is continuous at (a, b) we need to show that $\displaystyle\lim_{(x,y)\to(a,b)} f\,(x, y) = f\,(a, b)$ or equivalently

$\displaystyle\lim_{(\Delta x, \Delta y)\to(0,0)} f\,(a + \Delta x, b + \Delta y) = f\,(a, b)$. Since f is differentiable at (a, b),

$f\,(a + \Delta x, b + \Delta y) - f\,(a, b) = \Delta z = f_x\,(a, b)\,\Delta x + f_y\,(a, b)\,\Delta y + \epsilon_1\,\Delta x + \epsilon_2\,\Delta y$, where ϵ_1 and $\epsilon_2 \to 0$ as

$(\Delta x, \Delta y) \to (0, 0)$. Thus $f\,(a + \Delta x, b + \Delta y) = f\,(a, b) + f_x\,(a, b)\,\Delta x + f_y\,(a, b)\,\Delta y + \epsilon_1\,\Delta x + \epsilon_2\,\Delta y$.

Taking the limit of both sides as $(\Delta x, \Delta y) \to (0, 0)$ gives $\displaystyle\lim_{(\Delta x, \Delta y)\to(0,0)} f\,(a + \Delta x, b + \Delta y) = f\,(a, b)$. Thus f is

continuous at (a, b).

▅▅ 15.5 The Chain Rule   ET 14.5

1. $z = x^2 y + xy^2$, $x = 2 + t^4$, $y = 1 - t^3$ $\Rightarrow$

$\dfrac{dz}{dt} = \dfrac{\partial z}{\partial x}\dfrac{dx}{dt} + \dfrac{\partial z}{\partial y}\dfrac{dy}{dt} = \left(2xy + y^2\right)\left(4t^3\right) + \left(x^2 + 2xy\right)\left(-3t^2\right) = 4\left(2xy + y^2\right)t^3 - 3\left(x^2 + 2xy\right)t^2$

3. $z = \sin x \cos y$, $x = \pi t$, $y = \sqrt{t}$ $\Rightarrow$

$\dfrac{dz}{dt} = \dfrac{\partial z}{\partial x}\dfrac{dx}{dt} + \dfrac{\partial z}{\partial y}\dfrac{dy}{dt} = \cos x \cos y \cdot \pi + \sin x\,(-\sin y) \cdot \tfrac{1}{2}t^{-1/2} = \pi \cos x \cos y - \dfrac{1}{2\sqrt{t}}\sin x \sin y$

5. $w = xe^{y/z}$, $x = t^2$, $y = 1 - t$, $z = 1 + 2t$ $\Rightarrow$

$$\frac{dw}{dt} = \frac{\partial w}{\partial x}\frac{dx}{dt} + \frac{\partial w}{\partial y}\frac{dy}{dt} + \frac{\partial w}{\partial z}\frac{dz}{dt} = e^{y/z} \cdot 2t + xe^{y/z}\left(\frac{1}{z}\right) \cdot (-1) + xe^{y/z}\left(-\frac{y}{z^2}\right) \cdot 2 = e^{y/z}\left(2t - \frac{x}{z} - \frac{2xy}{z^2}\right)$$

7. $z = x^2 + xy + y^2$, $x = s + t$, $y = st$ $\Rightarrow$

$$\frac{\partial z}{\partial s} = \frac{\partial z}{\partial x}\frac{\partial x}{\partial s} + \frac{\partial z}{\partial y}\frac{\partial y}{\partial s} = (2x + y)(1) + (x + 2y)(t) = 2x + y + xt + 2yt$$

$$\frac{\partial z}{\partial t} = \frac{\partial z}{\partial x}\frac{\partial x}{\partial t} + \frac{\partial z}{\partial y}\frac{\partial y}{\partial t} = (2x + y)(1) + (x + 2y)(s) = 2x + y + xs + 2ys$$

9. $z = \arctan(2x + y)$, $x = s^2 t$, $y = s \ln t$ $\Rightarrow$

$$\frac{\partial z}{\partial s} = \frac{\partial z}{\partial x}\frac{\partial x}{\partial s} + \frac{\partial z}{\partial y}\frac{\partial y}{\partial s} = \frac{2}{1 + (2x + y)^2} \cdot 2st + \frac{1}{1 + (2x + y)^2} \cdot \ln t = \frac{4st + \ln t}{1 + (2x + y)^2}$$

$$\frac{\partial z}{\partial t} = \frac{\partial z}{\partial x}\frac{\partial x}{\partial t} + \frac{\partial z}{\partial y}\frac{\partial y}{\partial t} = \frac{2}{1 + (2x + y)^2} \cdot s^2 + \frac{1}{1 + (2x + y)^2} \cdot \frac{s}{t} = \frac{2s^2 + s/t}{1 + (2x + y)^2}$$

11. $z = e^r \cos\theta$, $r = st$, $\theta = \sqrt{s^2 + t^2}$ $\Rightarrow$

$$\frac{\partial z}{\partial s} = \frac{\partial z}{\partial r}\frac{\partial r}{\partial s} + \frac{\partial z}{\partial \theta}\frac{\partial \theta}{\partial s} = e^r \cos\theta \cdot t + e^r (-\sin\theta) \cdot \tfrac{1}{2}(s^2 + t^2)^{-1/2}(2s)$$

$$= te^r \cos\theta - e^r \sin\theta \cdot \frac{s}{\sqrt{s^2 + t^2}} = e^r \left(t\cos\theta - \frac{s}{\sqrt{s^2 + t^2}}\sin\theta\right)$$

$$\frac{\partial z}{\partial t} = \frac{\partial z}{\partial r}\frac{\partial r}{\partial t} + \frac{\partial z}{\partial \theta}\frac{\partial \theta}{\partial t} = e^r \cos\theta \cdot s + e^r (-\sin\theta) \cdot \tfrac{1}{2}(s^2 + t^2)^{-1/2}(2t)$$

$$= se^r \cos\theta - e^r \sin\theta \cdot \frac{t}{\sqrt{s^2 + t^2}} = e^r \left(s\cos\theta - \frac{t}{\sqrt{s^2 + t^2}}\sin\theta\right)$$

13. When $t = 3$, $x = g(3) = 2$ and $y = h(3) = 7$. By the Chain Rule (2),

$$\frac{dz}{dt} = \frac{\partial f}{\partial x}\frac{dx}{dt} + \frac{\partial f}{\partial y}\frac{dy}{dt} = f_x(2, 7)\,g'(3) + f_y(2, 7)\,h'(3) = (6)(5) + (-8)(-4) = 62.$$

15.

$u = f(x, y)$, $x = x(r, s, t)$, $y = y(r, s, t)$ $\Rightarrow$

$$\frac{\partial u}{\partial r} = \frac{\partial u}{\partial x}\frac{\partial x}{\partial r} + \frac{\partial u}{\partial y}\frac{\partial y}{\partial r}, \quad \frac{\partial u}{\partial s} = \frac{\partial u}{\partial x}\frac{\partial x}{\partial s} + \frac{\partial u}{\partial y}\frac{\partial y}{\partial s},$$

$$\frac{\partial u}{\partial t} = \frac{\partial u}{\partial x}\frac{\partial x}{\partial t} + \frac{\partial u}{\partial y}\frac{\partial y}{\partial t}$$

17.

$v = f(p, q, r)$, $p = p(x, y, z)$, $q = q(x, y, z)$, $r = r(x, y, z)$ $\Rightarrow$

$$\frac{\partial v}{\partial x} = \frac{\partial v}{\partial p}\frac{\partial p}{\partial x} + \frac{\partial v}{\partial q}\frac{\partial q}{\partial x} + \frac{\partial v}{\partial r}\frac{\partial r}{\partial x}, \quad \frac{\partial v}{\partial y} = \frac{\partial v}{\partial p}\frac{\partial p}{\partial y} + \frac{\partial v}{\partial q}\frac{\partial q}{\partial y} + \frac{\partial v}{\partial r}\frac{\partial r}{\partial y},$$

$$\frac{\partial v}{\partial z} = \frac{\partial v}{\partial p}\frac{\partial p}{\partial z} + \frac{\partial v}{\partial q}\frac{\partial q}{\partial z} + \frac{\partial v}{\partial r}\frac{\partial r}{\partial z}$$

19. $w = x^2 + y^2 + z^2$, $x = st$, $y = s\cos t$, $z = s\sin t$ $\Rightarrow$

$$\frac{\partial w}{\partial s} = \frac{\partial w}{\partial x}\frac{\partial x}{\partial s} + \frac{\partial w}{\partial y}\frac{\partial y}{\partial s} + \frac{\partial w}{\partial z}\frac{\partial z}{\partial s} = 2xt + 2y\cos t + 2z\sin t.$$ When $s = 1$, $t = 0$, we have $x = 0$, $y = 1$ and

$z = 0$, so $\partial w/\partial s = 2\cos 0 = 2$. Similarly $\partial w/\partial t = 2xs + 2y(-s\sin t) + 2z(s\cos t) = 0 + (-2)\sin 0 + 0 = 0$

when $s = 1$ and $t = 0$.

21. $z = y^2 \tan x,\ x = t^2 uv,\ y = u + tv^2 \quad \Rightarrow$

$\partial z / \partial t = \left(y^2 \sec^2 x\right) 2tuv + (2y \tan x) v^2,\ \partial z / \partial u = \left(y^2 \sec^2 x\right) t^2 v + 2y \tan x,$

$\partial z / \partial v = \left(y^2 \sec^2 x\right) t^2 u + (2y \tan x) 2tv.$ When $t = 2,\ u = 1$ and $v = 0$, we have $x = 0,\ y = 1$, so $\partial z / \partial t = 0,$

$\partial z / \partial u = 0,\ \partial z / \partial v = 4.$

23. $u = \dfrac{x + y}{y + z},\ x = p + r + t,\ y = p - r + t,\ z = p + r - t \quad \Rightarrow$

$\dfrac{\partial u}{\partial p} = \dfrac{1}{y + z} + \dfrac{(y + z) - (x + y)}{(y + z)^2} - \dfrac{x + y}{(y + z)^2} = \dfrac{(y + z) + (z - x) - (x + y)}{(y + z)^2} = 2\dfrac{z - x}{(y + z)^2}$

$\qquad = 2\dfrac{-2t}{4p^2} = -\dfrac{t}{p^2},$

$\dfrac{\partial u}{\partial r} = \dfrac{1}{y + z} + \dfrac{z - x}{(y + z)^2}(-1) - \dfrac{x + y}{(y + z)^2} = 0,\ \text{and}$

$\dfrac{\partial u}{\partial t} = \dfrac{1}{y + z} + \dfrac{z - x}{(y + z)^2} + \dfrac{x + y}{(y + z)^2} = 2\dfrac{y + z}{(y + z)^2} = \dfrac{2}{2p} = \dfrac{1}{p}.$

25. $x^2 - xy + y^3 = 8$, so let $F(x, y) = x^2 - xy + y^3 - 8 = 0$. Then $\dfrac{dy}{dx} = -\dfrac{F_x}{F_y} = -\dfrac{-(2x - y)}{-x + 3y^2} = \dfrac{y - 2x}{3y^2 - x}.$

27. $\cos(x - y) = xe^y$, so let $F(x, y) = \cos(x - y) - xe^y = 0.$

Then $\dfrac{dy}{dx} = -\dfrac{F_x}{F_y} = -\dfrac{-\sin(x - y) - e^y}{-\sin(x - y)(-1) - xe^y} = \dfrac{\sin(x - y) + e^y}{\sin(x - y) - xe^y}.$

29. $xy^2 + yz^2 + zx^2 = 3$, so let $F(x, y) = xy^2 + yz^2 + zx^2 - 3 = 0.$

Then $\dfrac{\partial z}{\partial x} = -\dfrac{F_x}{F_z} = -\dfrac{y^2 + 2zx}{2yz + x^2}$ and $\dfrac{\partial z}{\partial y} = -\dfrac{F_y}{F_z} = -\dfrac{2xy + z^2}{2yz + x^2}.$

31. Let $F(x, y, z) = xe^y + yz + ze^x = 0$. Then $\dfrac{\partial z}{\partial x} = -\dfrac{F_x}{F_z} = -\dfrac{e^y + ze^x}{y + e^x},\ \dfrac{\partial z}{\partial y} = -\dfrac{F_y}{F_z} = -\dfrac{xe^y + z}{y + e^x}.$

33. Since x and y are each functions of t, $T(x, y)$ is a function of t, so by the Chain Rule, $\dfrac{dT}{dt} = \dfrac{\partial T}{\partial x}\dfrac{dx}{dt} + \dfrac{\partial T}{\partial y}\dfrac{dy}{dt}.$

After 3 seconds, $x = \sqrt{1 + t} = \sqrt{1 + 3} = 2,\ y = 2 + \tfrac{1}{3}t = 2 + \tfrac{1}{3}(3) = 3,\ \dfrac{dx}{dt} = \dfrac{1}{2\sqrt{1 + t}} = \dfrac{1}{2\sqrt{1 + 3}} = \dfrac{1}{4},$

and $\dfrac{dy}{dt} = \dfrac{1}{3}$. Then $\dfrac{dT}{dt} = T_x(2, 3)\dfrac{dx}{dt} + T_y(2, 3)\dfrac{dy}{dt} = 4\left(\tfrac{1}{4}\right) + 3\left(\tfrac{1}{3}\right) = 2$. Thus the temperature is rising at a

rate of 2 degrees Celsius per second.

35. $C = 1449.2 + 4.6T - 0.055T^2 + 0.00029T^3 + 0.016D$, so $\dfrac{\partial C}{\partial T} = 4.6 - 0.11T + 0.00087T^2$ and $\dfrac{\partial C}{\partial D} = 0.016.$

According to the graph, the diver is experiencing a temperature of approximately $12.5\,^\circ$C at $t = 20$ minutes, so

$\dfrac{\partial C}{\partial T} = 4.6 - 0.11(12.5) + 0.00087(12.5)^2 \approx 3.36$. By sketching tangent lines at $t = 20$ to the graphs given, we

estimate $\dfrac{dD}{dt} \approx \dfrac{1}{2}$ and $\dfrac{dT}{dt} \approx -\dfrac{1}{10}$. Then, by the Chain Rule,

$\dfrac{dC}{dt} = \dfrac{\partial C}{\partial T}\dfrac{dT}{dt} + \dfrac{\partial C}{\partial D}\dfrac{dD}{dt} \approx (3.36)\left(-\tfrac{1}{10}\right) + (0.016)\left(\tfrac{1}{2}\right) \approx -0.33$. Thus the speed of sound experienced by the

diver is decreasing at a rate of approximately 0.33 m/s per minute.

37. (a) $V = \ell w h$, so by the Chain Rule,

$$\frac{dV}{dt} = \frac{\partial V}{\partial \ell}\frac{d\ell}{dt} + \frac{\partial V}{\partial w}\frac{dw}{dt} + \frac{\partial V}{\partial h}\frac{dh}{dt} = wh\frac{d\ell}{dt} + \ell h\frac{dw}{dt} + \ell w\frac{dh}{dt}$$

$$= 2\cdot 2\cdot 2 + 1\cdot 2\cdot 2 + 1\cdot 2\cdot(-3) = 6\text{ m}^3/\text{s}$$

(b) $S = 2(\ell w + \ell h + wh)$, so by the Chain Rule,

$$\frac{dS}{dt} = \frac{\partial S}{\partial \ell}\frac{d\ell}{dt} + \frac{\partial S}{\partial w}\frac{dw}{dt} + \frac{\partial S}{\partial h}\frac{dh}{dt} = 2(w+h)\frac{d\ell}{dt} + 2(\ell+h)\frac{dw}{dt} + 2(\ell+w)\frac{dh}{dt}$$

$$= 2(2+2)2 + 2(1+2)2 + 2(1+2)(-3) = 10\text{ m}^2/\text{s}$$

(c) $L^2 = \ell^2 + w^2 + h^2 \;\Rightarrow\; 2L\dfrac{dL}{dt} = 2\ell\dfrac{d\ell}{dt} + 2w\dfrac{dw}{dt} + 2h\dfrac{dh}{dt} = 2(1)(2) + 2(2)(2) + 2(2)(-3) = 0 \;\Rightarrow$

$dL/dt = 0$ m/s.

39. $\dfrac{dP}{dt} = 0.05$, $\dfrac{dT}{dt} = 0.15$, $V = 8.31\dfrac{T}{P}$ and $\dfrac{dV}{dt} = \dfrac{8.31}{P}\dfrac{dT}{dt} - 8.31\dfrac{T}{P^2}\dfrac{dP}{dt}$. Thus when $P = 20$ and $T = 320$,

$\dfrac{dV}{dt} = 8.31\left[\dfrac{0.15}{20} - \dfrac{(0.05)(320)}{400}\right] \approx -0.27$ L/s.

41. (a) By the Chain Rule, $\dfrac{\partial z}{\partial r} = \dfrac{\partial z}{\partial x}\cos\theta + \dfrac{\partial z}{\partial y}\sin\theta$, $\dfrac{\partial z}{\partial\theta} = \dfrac{\partial z}{\partial x}(-r\sin\theta) + \dfrac{\partial z}{\partial y}r\cos\theta$.

(b) $\left(\dfrac{\partial z}{\partial r}\right)^2 = \left(\dfrac{\partial z}{\partial x}\right)^2\cos^2\theta + 2\dfrac{\partial z}{\partial x}\dfrac{\partial z}{\partial y}\cos\theta\sin\theta + \left(\dfrac{\partial z}{\partial y}\right)^2\sin^2\theta$,

$\left(\dfrac{\partial z}{\partial\theta}\right)^2 = \left(\dfrac{\partial z}{\partial x}\right)^2 r^2\sin^2\theta - 2\dfrac{\partial z}{\partial x}\dfrac{\partial z}{\partial y}r^2\cos\theta\sin\theta + \left(\dfrac{\partial z}{\partial y}\right)^2 r^2\cos^2\theta$. Thus

$\left(\dfrac{\partial z}{\partial r}\right)^2 + \dfrac{1}{r^2}\left(\dfrac{\partial z}{\partial\theta}\right)^2 = \left[\left(\dfrac{\partial z}{\partial x}\right)^2 + \left(\dfrac{\partial z}{\partial y}\right)^2\right](\cos^2\theta + \sin^2\theta) = \left(\dfrac{\partial z}{\partial x}\right)^2 + \left(\dfrac{\partial z}{\partial y}\right)^2$.

43. Let $u = x - y$. Then $\dfrac{\partial z}{\partial x} = \dfrac{dz}{du}\dfrac{\partial u}{\partial x} = \dfrac{dz}{du}$ and $\dfrac{\partial z}{\partial y} = \dfrac{dz}{du}(-1)$. Thus $\dfrac{\partial z}{\partial x} + \dfrac{\partial z}{\partial y} = 0$.

45. Let $u = x + at$, $v = x - at$. Then $z = f(u) + g(v)$, so $\partial z/\partial u = f'(u)$ and $\partial z/\partial v = g'(v)$.

Thus $\dfrac{\partial z}{\partial t} = \dfrac{\partial z}{\partial u}\dfrac{\partial u}{\partial t} + \dfrac{\partial z}{\partial v}\dfrac{\partial v}{\partial t} = af'(u) - ag'(v)$ and

$\dfrac{\partial^2 z}{\partial t^2} = a\dfrac{\partial}{\partial t}[f'(u) - g'(v)] = a\left(\dfrac{df'(u)}{du}\dfrac{\partial u}{\partial t} - \dfrac{dg'(v)}{dv}\dfrac{\partial v}{\partial t}\right) = a^2 f''(u) + a^2 g''(v)$. Similarly

$\dfrac{\partial z}{\partial x} = f'(u) + g'(v)$ and $\dfrac{\partial^2 z}{\partial x^2} = f''(u) + g''(v)$. Thus $\dfrac{\partial^2 z}{\partial t^2} = a^2\dfrac{\partial^2 z}{\partial x^2}$.

47. $\dfrac{\partial z}{\partial s} = \dfrac{\partial z}{\partial x}2s + \dfrac{\partial z}{\partial y}2r$. Then

$$\frac{\partial^2 z}{\partial r\partial s} = \frac{\partial}{\partial r}\left(\frac{\partial z}{\partial x}2s\right) + \frac{\partial}{\partial r}\left(\frac{\partial z}{\partial y}2r\right)$$

$$= \frac{\partial^2 z}{\partial x^2}\frac{\partial x}{\partial r}2s + \frac{\partial}{\partial y}\left(\frac{\partial z}{\partial x}\right)\frac{\partial y}{\partial r}2s + \frac{\partial z}{\partial x}\frac{\partial}{\partial r}(2s) + \frac{\partial^2 z}{\partial y^2}\frac{\partial y}{\partial r}2r + \frac{\partial}{\partial x}\left(\frac{\partial z}{\partial y}\right)\frac{\partial x}{\partial r}2r + \frac{\partial z}{\partial y}2$$

$$= 4rs\frac{\partial^2 z}{\partial x^2} + \frac{\partial^2 z}{\partial y\partial x}4s^2 + 0 + 4rs\frac{\partial^2 z}{\partial y^2} + \frac{\partial^2 z}{\partial x\partial y}4r^2 + 2\frac{\partial z}{\partial y}$$

By the continuity of the partials, $\dfrac{\partial^2 z}{\partial r\partial s} = 4rs\dfrac{\partial^2 z}{\partial x^2} + 4rs\dfrac{\partial^2 z}{\partial y^2} + (4r^2 + 4s^2)\dfrac{\partial^2 z}{\partial x\partial y} + 2\dfrac{\partial z}{\partial y}$.

49. $\dfrac{\partial z}{\partial r} = \dfrac{\partial z}{\partial x}\cos\theta + \dfrac{\partial z}{\partial y}\sin\theta$ and $\dfrac{\partial z}{\partial \theta} = -\dfrac{\partial z}{\partial x}r\sin\theta + \dfrac{\partial z}{\partial y}r\cos\theta$. Then

$$\frac{\partial^2 z}{\partial r^2} = \cos\theta\left(\frac{\partial^2 z}{\partial x^2}\cos\theta + \frac{\partial^2 z}{\partial y\,\partial x}\sin\theta\right) + \sin\theta\left(\frac{\partial^2 z}{\partial y^2}\sin\theta + \frac{\partial^2 z}{\partial x\,\partial y}\cos\theta\right)$$

$$= \cos^2\theta\frac{\partial^2 z}{\partial x^2} + 2\cos\theta\sin\theta\frac{\partial^2 z}{\partial x\,\partial y} + \sin^2\theta\frac{\partial^2 z}{\partial y^2}$$

and

$$\frac{\partial^2 z}{\partial\theta^2} = -r\cos\theta\frac{\partial z}{\partial x} + (-r\sin\theta)\left(\frac{\partial^2 z}{\partial x^2}(-r\sin\theta) + \frac{\partial^2 z}{\partial y\,\partial x}r\cos\theta\right)$$

$$-r\sin\theta\frac{\partial z}{\partial y} + r\cos\theta\left(\frac{\partial^2 z}{\partial y^2}r\cos\theta + \frac{\partial^2 z}{\partial x\,\partial y}(-r\sin\theta)\right)$$

$$= -r\cos\theta\frac{\partial z}{\partial x} - r\sin\theta\frac{\partial z}{\partial y} + r^2\sin^2\theta\frac{\partial^2 z}{\partial x^2} - 2r^2\cos\theta\sin\theta\frac{\partial^2 z}{\partial x\,\partial y} + r^2\cos^2\theta\frac{\partial^2 z}{\partial y^2}$$

Thus

$$\frac{\partial^2 z}{\partial r^2} + \frac{1}{r^2}\frac{\partial^2 z}{\partial\theta^2} + \frac{1}{r}\frac{\partial z}{\partial r} = (\cos^2\theta + \sin^2\theta)\frac{\partial^2 z}{\partial x^2} + (\sin^2\theta + \cos^2\theta)\frac{\partial^2 z}{\partial y^2} - \frac{1}{r}\cos\theta\frac{\partial z}{\partial x}$$

$$-\frac{1}{r}\sin\theta\frac{\partial z}{\partial y} + \frac{1}{r}\left(\cos\theta\frac{\partial z}{\partial x} + \sin\theta\frac{\partial z}{\partial y}\right)$$

$$= \frac{\partial^2 z}{\partial x^2} + \frac{\partial^2 z}{\partial y^2} \text{ as desired.}$$

51. (a) Since f is a polynomial, it has continuous second-order partial derivatives, and

$$f(tx, ty) = (tx)^2(ty) + 2(tx)(ty)^2 + 5(ty)^3 = t^3 x^2 y + 2t^3 xy^2 + 5t^3 y^3$$

$$= t^3(x^2 y + 2xy^2 + 5y^3) = t^3 f(x, y)$$

Thus, f is homogeneous of degree 3.

(b) Differentiating both sides of $f(tx, ty) = t^n f(x, y)$ with respect to t using the Chain Rule, we get

$$\frac{\partial}{\partial t}f(tx, ty) = \frac{\partial}{\partial t}[t^n f(x, y)] \quad\Leftrightarrow$$

$$\frac{\partial}{\partial(tx)}f(tx, ty)\cdot\frac{\partial(tx)}{\partial t} + \frac{\partial}{\partial(ty)}f(tx, ty)\cdot\frac{\partial(ty)}{\partial t} = x\frac{\partial}{\partial(tx)}f(tx, ty) + y\frac{\partial}{\partial(ty)}f(tx, ty) = nt^{n-1}f(x, y).$$

Setting $t = 1$: $x\dfrac{\partial}{\partial x}f(x, y) + y\dfrac{\partial}{\partial y}f(x, y) = nf(x, y)$.

53. Differentiating both sides of $f(tx, ty) = t^n f(x, y)$ with respect to x using the Chain Rule, we get

$$\frac{\partial}{\partial x}f(tx, ty) = \frac{\partial}{\partial x}[t^n f(x, y)] \quad\Leftrightarrow$$

$$\frac{\partial}{\partial(tx)}f(tx, ty)\cdot\frac{\partial(tx)}{\partial x} + \frac{\partial}{\partial(ty)}f(tx, ty)\cdot\frac{\partial(ty)}{\partial x} = t^n\frac{\partial}{\partial x}f(x, y) \quad\Leftrightarrow\quad tf_x(tx, ty) = t^n f_x(x, y).$$

Thus $f_x(tx, ty) = t^{n-1}f_x(x, y)$.

▰15.6 Directional Derivatives and the Gradient Vector ET 14.6

1. First we draw a line passing through Raleigh and the eye of the hurricane. We can approximate the directional derivative at Raleigh in the direction of the eye of the hurricane by the average rate of change of pressure between the points where this line intersects the contour lines closest to Raleigh. In the direction of the eye of the hurricane, the pressure changes from 996 millibars to 992 millibars. We estimate the distance between these two points to be approximately 40 miles, so the rate of change of pressure in the direction given is approximately $\frac{992-996}{40} = -0.1$ millibars/mi.

3. $f(x,y) = x^2 y^3 + 2x^4 y \Rightarrow f_x(x,y) = 2xy^3 + 8x^3 y$ and $f_y(x,y) = 3x^2 y^2 + 2x^4$. If **u** is a unit vector in the direction of $\theta = \frac{\pi}{3}$, then from Equation 6,

$$D_{\mathbf{u}}f(1,-2) = f_x(1,-2)\cos\tfrac{\pi}{3} + f_y(1,-2)\sin\tfrac{\pi}{3} = (-32)\left(\tfrac{1}{2}\right) + (14)\left(\tfrac{\sqrt{3}}{2}\right) = 7\sqrt{3} - 16.$$

5. $f(x,y) = \sqrt{5x-4y} \Rightarrow f_x(x,y) = \frac{1}{2}(5x-4y)^{-1/2}(5) = \dfrac{5}{2\sqrt{5x-4y}}$ and

$f_y(x,y) = \frac{1}{2}(5x-4y)^{-1/2}(-4) = -\dfrac{2}{\sqrt{5x-4y}}$. If **u** is a unit vector in the direction of $\theta = -\frac{\pi}{6}$, then from

Equation 6, $D_{\mathbf{u}}f(4,1) = f_x(4,1)\cos\left(-\tfrac{\pi}{6}\right) + f_y(4,1)\sin\left(-\tfrac{\pi}{6}\right) = \tfrac{5}{8}\cdot\tfrac{\sqrt{3}}{2} + \left(-\tfrac{1}{2}\right)\left(-\tfrac{1}{2}\right) = \tfrac{5\sqrt{3}}{16} + \tfrac{1}{4}$.

7. $f(x,y) = 5xy^2 - 4x^3 y$

  (a) $\nabla f(x,y) = \langle f_x(x,y), f_y(x,y)\rangle = \langle 5y^2 - 12x^2 y, 10xy - 4x^3 \rangle$

  (b) $\nabla f(1,2) = \langle 5(2)^2 - 12(1)^2(2), 10(1)(2) - 4(1)^3 \rangle = \langle -4, 16 \rangle$

  (c) By Equation 9, $D_{\mathbf{u}}f(1,2) = \nabla f(1,2)\cdot\mathbf{u} = \langle -4,16\rangle\cdot\langle\tfrac{5}{13},\tfrac{12}{13}\rangle = (-4)\left(\tfrac{5}{13}\right) + (16)\left(\tfrac{12}{13}\right) = \tfrac{172}{13}$.

9. $f(x,y,z) = xy^2 z^3$

  (a) $\nabla f(x,y,z) = \langle f_x(x,y,z), f_y(x,y,z), f_z(x,y,z)\rangle = \langle y^2 z^3, 2xyz^3, 3xy^2 z^2 \rangle$

  (b) $\nabla f(1,-2,1) = \langle 4, -4, 12 \rangle$

  (c) $\nabla f(1,-2,1)\cdot\mathbf{u} = \tfrac{4}{\sqrt{3}} + \tfrac{4}{\sqrt{3}} + \tfrac{12}{\sqrt{3}} = \tfrac{20}{\sqrt{3}}$

11. $f(x,y) = 1 + 2x\sqrt{y} \Rightarrow \nabla f(x,y) = \left\langle 2\sqrt{y}, 2x\cdot\tfrac{1}{2}y^{-1/2}\right\rangle = \langle 2\sqrt{y}, x/\sqrt{y}\rangle$, $\nabla f(3,4) = \langle 4, \tfrac{3}{2}\rangle$,

  and a unit vector in the direction of **v** is $\mathbf{u} = \dfrac{1}{\sqrt{4^2+(-3)^2}}\langle 4,-3\rangle = \langle\tfrac{4}{5},-\tfrac{3}{5}\rangle$, so

  $D_{\mathbf{u}}f(3,4) = \nabla f(3,4)\cdot\mathbf{u} = \langle 4,\tfrac{3}{2}\rangle\cdot\langle\tfrac{4}{5},-\tfrac{3}{5}\rangle = \tfrac{23}{10}$.

13. $g(s,t) = s^2 e^t \Rightarrow \nabla g(s,t) = 2se^t\,\mathbf{i} + s^2 e^t\,\mathbf{j}$, $\nabla g(2,0) = 4\,\mathbf{i} + 4\,\mathbf{j}$, and a unit vector in the direction of **v** is $\mathbf{u} = \tfrac{1}{\sqrt{2}}(\mathbf{i}+\mathbf{j})$, so $D_{\mathbf{u}}g(2,0) = \nabla g(2,0)\cdot\mathbf{u} = (4\,\mathbf{i}+4\,\mathbf{j})\cdot\tfrac{1}{\sqrt{2}}(\mathbf{i}+\mathbf{j}) = \tfrac{8}{\sqrt{2}} = 4\sqrt{2}$.

15. $f(x,y,z) = \sqrt{x^2+y^2+z^2} \Rightarrow \nabla f(x,y,z) = \left\langle \dfrac{x}{\sqrt{x^2+y^2+z^2}}, \dfrac{y}{\sqrt{x^2+y^2+z^2}}, \dfrac{z}{\sqrt{x^2+y^2+z^2}} \right\rangle$,

  $\nabla f(1,2,-2) = \langle\tfrac{1}{3},\tfrac{2}{3},-\tfrac{2}{3}\rangle$, and a unit vector in the direction of **v** is $\mathbf{u} = \tfrac{1}{9}\langle -6,6,-3\rangle = \langle -\tfrac{2}{3},\tfrac{2}{3},-\tfrac{1}{3}\rangle$, so

  $D_{\mathbf{u}}f(1,2,-2) = \nabla f(1,2,-2)\cdot\mathbf{u} = \langle\tfrac{1}{3},\tfrac{2}{3},-\tfrac{2}{3}\rangle\cdot\langle-\tfrac{2}{3},\tfrac{2}{3},-\tfrac{1}{3}\rangle = \tfrac{4}{9}$.

17. $g(x, y, z) = x\tan^{-1}(y/z)$ $\Rightarrow$ $\nabla g(x, y, z) = \left\langle \tan^{-1}(y/z), xz/(y^2 + z^2), -xy/(y^2 + z^2)\right\rangle$,
$\nabla g(1, 2, -2) = \left\langle -\frac{\pi}{4}, -\frac{1}{4}, -\frac{1}{4}\right\rangle$, $\mathbf{u} = \frac{1}{\sqrt{3}}\langle 1, 1, -1\rangle$ and

$$D_{\mathbf{u}}g(1, 2, -2) = \frac{(-\pi)(1)}{4\sqrt{3}} + \frac{(-1)(1)}{4\sqrt{3}} + \frac{(-1)(-1)}{4\sqrt{3}} = -\frac{\pi}{4\sqrt{3}}.$$

19. $f(x, y) = \sqrt{xy}$ $\Rightarrow$ $\nabla f(x, y) = \left\langle \frac{1}{2}(xy)^{-1/2}(y), \frac{1}{2}(xy)^{-1/2}(x)\right\rangle = \left\langle \frac{y}{2\sqrt{xy}}, \frac{x}{2\sqrt{xy}}\right\rangle$, so

$\nabla f(2, 8) = \langle 1, \frac{1}{4}\rangle$. The unit vector in the direction of $\overrightarrow{PQ} = \langle 5 - 2, 4 - 8\rangle = \langle 3, -4\rangle$ is $\mathbf{u} = \langle \frac{3}{5}, -\frac{4}{5}\rangle$, so
$D_{\mathbf{u}}f(2, 8) = \nabla f(2, 8) \cdot \mathbf{u} = \langle 1, \frac{1}{4}\rangle \cdot \langle \frac{3}{5}, -\frac{4}{5}\rangle = \frac{2}{5}$.

21. $f(x, y) = xe^{-y} + 3y$ $\Rightarrow$ $\nabla f(x, y) = \langle e^{-y}, 3 - xe^{-y}\rangle$, $\nabla f(1, 0) = \langle 1, 2\rangle$ is the direction of maximum rate
of change and the maximum rate is $|\nabla f(1, 0)| = \sqrt{5}$.

23. $f(x, y) = \sin(xy)$ $\Rightarrow$ $\nabla f(x, y) = \langle y\cos(xy), x\cos(xy)\rangle$, $\nabla f(1, 0) = \langle 0, 1\rangle$. Thus the maximum rate of
change is $|\nabla f(1, 0)| = 1$ in the direction $\langle 0, 1\rangle$.

25. $f(x, y, z) = x + y/z$ $\Rightarrow$ $\nabla f(x, y, z) = \left\langle 1, \frac{1}{z}, -\frac{y}{z^2}\right\rangle$, so the maximum rate of change is

$|\nabla f(4, 3, -1)| = \sqrt{11}$ in the direction $\langle 1, -1, -3\rangle$.

27. (a) As in the proof of Theorem 15, $D_{\mathbf{u}}f = |\nabla f|\cos\theta$. Since the minimum value of $\cos\theta$ is -1 occurring when
$\theta = \pi$, the minimum value of $D_{\mathbf{u}}f$ is $-|\nabla f|$ occurring when $\theta = \pi$, that is when $\mathbf{u}$ is in the opposite direction
of ∇f (assuming $\nabla f \neq \mathbf{0}$).

(b) $f(x, y) = x^4y - x^2y^3$ $\Rightarrow$ $\nabla f(x, y) = \langle 4x^3y - 2xy^3, x^4 - 3x^2y^2\rangle$, so f decreases fastest at the point
$(2, -3)$ in the direction $-\nabla f(2, -3) = -\langle 12, -92\rangle = \langle -12, 92\rangle$.

29. $T = \dfrac{k}{\sqrt{x^2 + y^2 + z^2}}$ and $120 = T(1, 2, 2) = \dfrac{k}{3}$ so $k = 360$.

(a) $\mathbf{u} = \dfrac{\langle 1, -1, 1\rangle}{\sqrt{3}}$,

$$D_{\mathbf{u}}T(1, 2, 2) = \nabla T(1, 2, 2) \cdot \mathbf{u} = \left[-360(x^2 + y^2 + z^2)^{-3/2}\langle x, y, z\rangle\right]_{(1,2,2)} \cdot \mathbf{u}$$

$$= -\frac{40}{3}\langle 1, 2, 2\rangle \cdot \frac{1}{\sqrt{3}}\langle 1, -1, 1\rangle = -\frac{40}{3\sqrt{3}}$$

(b) From (a), $\nabla T = -360(x^2 + y^2 + z^2)^{-3/2}\langle x, y, z\rangle$, and since $\langle x, y, z\rangle$ is the position vector of the point
(x, y, z), the vector $-\langle x, y, z\rangle$, and thus ∇T, always points toward the origin.

31. $\nabla V(x, y, z) = \langle 10x - 3y + yz, xz - 3x, xy\rangle$, $\nabla V(3, 4, 5) = \langle 38, 6, 12\rangle$

(a) $D_{\mathbf{u}}V(3, 4, 5) = \langle 38, 6, 12\rangle \cdot \frac{1}{\sqrt{3}}\langle 1, 1, -1\rangle = \frac{32}{\sqrt{3}}$

(b) $\nabla V(3, 4, 5) = \langle 38, 6, 12\rangle$ or equivalently $\langle 19, 3, 6\rangle$.

(c) $|\nabla V(3, 4, 5)| = \sqrt{38^2 + 6^2 + 12^2} = \sqrt{1624} = 2\sqrt{406}$

33. A unit vector in the direction of $\overrightarrow{AB}$ is $\mathbf{i}$ and a unit vector in the direction of $\overrightarrow{AC}$ is $\mathbf{j}$. Thus
$D_{\overrightarrow{AB}}f(1, 3) = f_x(1, 3) = 3$ and $D_{\overrightarrow{AC}}f(1, 3) = f_y(1, 3) = 26$. Therefore
$\nabla f(1, 3) = \langle f_x(1, 3), f_y(1, 3)\rangle = \langle 3, 26\rangle$, and by definition, $D_{\overrightarrow{AD}}f(1, 3) = \nabla f \cdot \mathbf{u}$ where $\mathbf{u}$ is a unit vector in
the direction of $\overrightarrow{AD}$, which is $\langle \frac{5}{13}, \frac{12}{13}\rangle$. Therefore, $D_{\overrightarrow{AB}}f(1, 3) = \langle 3, 26\rangle \cdot \langle \frac{5}{13}, \frac{12}{13}\rangle = 3 \cdot \frac{5}{13} + 26 \cdot \frac{12}{13} = \frac{327}{13}$.

35. (a) $\nabla (au + bv) = \left\langle \dfrac{\partial (au + bv)}{\partial x}, \dfrac{\partial (au + bv)}{\partial y} \right\rangle = \left\langle a\dfrac{\partial u}{\partial x} + b\dfrac{\partial v}{\partial x}, a\dfrac{\partial u}{\partial y} + b\dfrac{\partial v}{\partial y} \right\rangle$

$= a \left\langle \dfrac{\partial u}{\partial x}, \dfrac{\partial u}{\partial y} \right\rangle + b \left\langle \dfrac{\partial v}{\partial x}, \dfrac{\partial v}{\partial y} \right\rangle = a\nabla u + b\nabla v$

(b) $\nabla (uv) = \left\langle v\dfrac{\partial u}{\partial x} + u\dfrac{\partial v}{\partial x}, v\dfrac{\partial u}{\partial y} + u\dfrac{\partial v}{\partial y} \right\rangle = v\left\langle \dfrac{\partial u}{\partial x}, \dfrac{\partial u}{\partial y} \right\rangle + u\left\langle \dfrac{\partial v}{\partial x}, \dfrac{\partial v}{\partial y} \right\rangle = v\nabla u + u\nabla v$

(c) $\nabla \left(\dfrac{u}{v} \right) = \left\langle \dfrac{v\dfrac{\partial u}{\partial x} - u\dfrac{\partial v}{\partial x}}{v^2}, \dfrac{v\dfrac{\partial u}{\partial y} - u\dfrac{\partial v}{\partial y}}{v^2} \right\rangle = \dfrac{v\left\langle \dfrac{\partial u}{\partial x}, \dfrac{\partial u}{\partial y} \right\rangle - u\left\langle \dfrac{\partial v}{\partial x}, \dfrac{\partial v}{\partial y} \right\rangle}{v^2} = \dfrac{v\nabla u - u\nabla v}{v^2}$

(d) $\nabla u^n = \left\langle \dfrac{\partial (u^n)}{\partial x}, \dfrac{\partial (u^n)}{\partial y} \right\rangle = \left\langle nu^{n-1}\dfrac{\partial u}{\partial x}, nu^{n-1}\dfrac{\partial u}{\partial y} \right\rangle = nu^{n-1}\nabla u.$

37. Let $F (x, y, z) = x^2 + 2y^2 + 3z^2$. Then $x^2 + 2y^2 + 3z^2 = 21$ is a level surface of F.
$F_x (x, y, z) = 2x \Rightarrow F_x (4, -1, 1) = 8$, $F_y (x, y, z) = 4y \Rightarrow F_y (4, -1, 1) = -4$, and
$F_z (x, y, z) = 6z \Rightarrow F_z (4, -1, 1) = 6$.

(a) Equation 19 gives an equation of the tangent plane at $(4, -1, 1)$ as $8 (x - 4) - 4 [y - (-1)] + 6 (z - 1) = 0$ or
$4x - 2y + 3z = 21$.

(b) By Equation 20, the normal line has symmetric equations $\dfrac{x - 4}{8} = \dfrac{y + 1}{-4} = \dfrac{z - 1}{6}$ or

$\dfrac{x - 4}{4} = \dfrac{y + 1}{-2} = \dfrac{z - 1}{3}$.

39. $F (x, y, z) = x^2 + y^2 - z^2 - 2xy + 4xz \Rightarrow \nabla F (x, y, z) = \langle 2x - 2y + 4z, 2y - 2x, -2z + 4x \rangle$,
$\nabla F (1, 0, 1) = \langle 6, -2, 2 \rangle$

(a) $6 (x - 1) - 2 (y - 0) + 2 (z - 1) = 0$ or $3x - y + z = 4$ (b) $\dfrac{x - 1}{3} = -y = z - 1$

41. $F (x, y, z) = -z + xe^y \cos z \Rightarrow \nabla F (x, y, z) = \langle e^y \cos z, xe^y \cos z, -1 - xe^y \sin z \rangle$,
$\nabla F (1, 0, 0) = \langle 1, 1, -1 \rangle$
(a) $1 (x - 1) + 1 (y - 0) - 1 (z - 0) = 0$ or $x + y - z = 1$ (b) $x - 1 = y = -z$

43. $F (x, y, z) = xy + yz + zx$, $\nabla F (x, y, z) = \langle y + z, x + z, y + x \rangle$, $\nabla F (1, 1, 1) = \langle 2, 2, 2 \rangle$, so an equation of the tangent plane is $2x + 2y + 2z = 6$ or $x + y + z = 3$, and the normal line is given by $x - 1 = y - 1 = z - 1$ or
$x = y = z$.

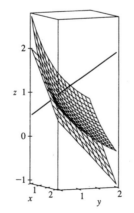

45. $\nabla f(x, y) = \langle 2x, 8y \rangle$, $\nabla f(2, 1) = \langle 4, 8 \rangle$. The tangent line has

equation $\nabla f(2, 1) \cdot \langle x - 2, y - 1 \rangle = 0 \Rightarrow$

$4(x - 2) + 8(y - 1) = 0$, which simplifies to $x + 2y = 4$.

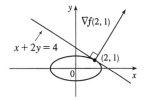

47. $\nabla F(x_0, y_0, z_0) = \left\langle \dfrac{2x_0}{a^2}, \dfrac{2y_0}{b^2}, \dfrac{2z_0}{c^2} \right\rangle$. Thus an equation of the tangent plane at (x_0, y_0, z_0) is

$\dfrac{2x_0}{a^2} x + \dfrac{2y_0}{b^2} y + \dfrac{2z_0}{c^2} z = 2 \left(\dfrac{x_0^2}{a^2} + \dfrac{y_0^2}{b^2} + \dfrac{z_0^2}{c^2} \right) = 2(1) = 2$ since (x_0, y_0, z_0) is a point on the ellipsoid. Hence

$\dfrac{x_0}{a^2} x + \dfrac{y_0}{b^2} y + \dfrac{z_0}{c^2} z = 1$ is an equation of the tangent plane.

49. $\nabla F(x_0, y_0, z_0) = \left\langle \dfrac{2x_0}{a^2}, \dfrac{2y_0}{b^2}, \dfrac{-1}{c} \right\rangle$, so an equation of the tangent plane is

$\dfrac{2x_0}{a^2} x + \dfrac{2y_0}{b^2} y - \dfrac{1}{c} z = \dfrac{2x_0^2}{a^2} + \dfrac{2y_0^2}{b^2} - \dfrac{z_0}{c}$ or $\dfrac{2x_0}{a^2} x + \dfrac{2y_0}{b^2} y = \dfrac{z}{c} + 2 \left(\dfrac{x_0^2}{a^2} + \dfrac{y_0^2}{b^2} \right) - \dfrac{z_0}{c}$. But $\dfrac{z_0}{c} = \dfrac{x_0^2}{a^2} + \dfrac{y_0^2}{b^2}$, so

the equation can be written as $\dfrac{2x_0}{a^2} x + \dfrac{2y_0}{b^2} y = \dfrac{z + z_0}{c}$.

51. $\nabla f(x_0, y_0, z_0) = \langle 2x_0, -2y_0, 4z_0 \rangle$ and the given line has direction numbers $2, 4, 6$, so

$\langle 2x_0, -2y_0, 4z_0 \rangle = k \langle 2, 4, 6 \rangle$ or $x_0 = k$, $y_0 = -2k$ and $z_0 = \frac{3}{2} k$. But $x_0^2 - y_0^2 + 2z_0^2 = 1$ or

$\left(1 - 4 + \frac{9}{2} \right) k^2 = 1$, so $k = \pm \sqrt{\frac{2}{3}} = \pm \frac{\sqrt{6}}{3}$ and there are two such points: $\left(\pm \frac{\sqrt{6}}{3}, \mp \frac{2\sqrt{6}}{3}, \pm \frac{\sqrt{6}}{2} \right)$.

53. Let (x_0, y_0, z_0) be a point on the cone [other than $(0, 0, 0)$]. Then an equation of the tangent plane to the cone at this

point is $2x_0 x + 2y_0 y - 2z_0 z = 2 \left(x_0^2 + y_0^2 - z_0^2 \right)$. But $x_0^2 + y_0^2 = z_0^2$ so the tangent plane is given by

$x_0 x + y_0 y - z_0 z = 0$, a plane which always contains the origin.

55. Let (x_0, y_0, z_0) be a point on the surface. Then an equation of the tangent plane at the point is

$\dfrac{x}{2\sqrt{x_0}} + \dfrac{y}{2\sqrt{y_0}} + \dfrac{z}{2\sqrt{z_0}} = \dfrac{\sqrt{x_0} + \sqrt{y_0} + \sqrt{z_0}}{2}$. But $\sqrt{x_0} + \sqrt{y_0} + \sqrt{z_0} = \sqrt{c}$, so the equation is

$\dfrac{x}{\sqrt{x_0}} + \dfrac{y}{\sqrt{y_0}} + \dfrac{z}{\sqrt{z_0}} = \sqrt{c}$. The x-, y-, and z-intercepts are $\sqrt{cx_0}$, $\sqrt{cy_0}$ and $\sqrt{cz_0}$ respectively. (The x-intercept

is found by setting $y = z = 0$ and solving the resulting equation for x, and the y- and z-intercepts are found

similarly.) So the sum of the intercepts is $\sqrt{c} \left(\sqrt{x_0} + \sqrt{y_0} + \sqrt{z_0} \right) = c$, a constant.

57. If $f(x, y, z) = z - x^2 - y^2$ and $g(x, y, z) = 4x^2 + y^2 + z^2$, then the tangent line is perpendicular to both ∇f and

∇g at $(-1, 1, 2)$. The vector $\mathbf{v} = \nabla f \times \nabla g$ will therefore be parallel to the tangent line. We have:

$\nabla f(x, y, z) = \langle -2x, -2y, 1 \rangle \Rightarrow \nabla f(-1, 1, 2) = \langle 2, -2, 1 \rangle$, and $\nabla g(x, y, z) = \langle 8x, 2y, 2z \rangle \Rightarrow$

$\nabla g(-1, 1, 2) = \langle -8, 2, 4 \rangle$. Hence $\mathbf{v} = \nabla f \times \nabla g = \begin{vmatrix} \mathbf{i} & \mathbf{j} & \mathbf{k} \\ 2 & -2 & 1 \\ -8 & 2 & 4 \end{vmatrix} = -10\mathbf{i} - 16\mathbf{j} - 12\mathbf{k}$. Parametric equations

are: $x = -1 - 10t$, $y = 1 - 16t$, $z = 2 - 12t$.

59. (a) The direction of the normal line of F is given by ∇F, and that of G by ∇G. Assuming that
$\nabla F \neq 0 \neq \nabla G$, the two normal lines are perpendicular at P if $\nabla F \cdot \nabla G = 0$ at $P \quad \Leftrightarrow$
$\langle \partial F/\partial x, \partial F/\partial y, \partial F/\partial z \rangle \cdot \langle \partial G/\partial x, \partial G/\partial y, \partial G/\partial z \rangle = 0$ at $P \quad \Leftrightarrow \quad F_x G_x + F_y G_y + F_z G_z = 0$ at P.

(b) Here $F = x^2 + y^2 - z^2$ and $G = x^2 + y^2 + z^2 - r^2$, so

$\nabla F \cdot \nabla G = \langle 2x, 2y, -2z \rangle \cdot \langle 2x, 2y, 2z \rangle = 4x^2 + 4y^2 - 4z^2 = 4F = 0$, since the point $\langle x, y, z \rangle$ lies on the

graph of $F = 0$. To see that this is true without using calculus, note that $G = 0$ is the equation of a sphere

centered at the origin and $F = 0$ is the equation of a right circular cone with vertex at the origin (which is

generated by lines through the origin). At any point of intersection, the sphere's normal line (which passes

through the origin) lies on the cone, and thus is perpendicular to the cone's normal line. So the surfaces with

equations $F = 0$ and $G = 0$ are everywhere orthogonal.

61. Let $\mathbf{u} = \langle a, b \rangle$ and $\mathbf{v} = \langle c, d \rangle$. Then we know that at the given point, $D_{\mathbf{u}} f = \nabla f \cdot \mathbf{u} = a f_x + b f_y$ and
$D_{\mathbf{v}} f = \nabla f \cdot \mathbf{v} = c f_x + d f_y$. But these are just two linear equations in the two unknowns f_x and f_y, and since $\mathbf{u}$
and $\mathbf{v}$ are not parallel, we can solve the equations to find $\nabla f = \langle f_x, f_y \rangle$ at the given point. In fact,

$$\nabla f = \left\langle \frac{d D_{\mathbf{u}} f - b D_{\mathbf{v}} f}{ad - bc}, \frac{a D_{\mathbf{v}} f - c D_{\mathbf{u}} f}{ad - bc} \right\rangle.$$

▰▰15.7 Maximum and Minimum Values ET 14.7

1. (a) First we compute $D(1, 1) = f_{xx}(1, 1) f_{yy}(1, 1) - [f_{xy}(1, 1)]^2 = (4)(2) - (1)^2 = 7$. Since $D(1, 1) > 0$ and
$f_{xx}(1, 1) > 0$, f has a local minimum at $(1, 1)$ by the Second Derivatives Test.

(b) $D(1, 1) = f_{xx}(1, 1) f_{yy}(1, 1) - [f_{xy}(1, 1)]^2 = (4)(2) - (3)^2 = -1$. Since $D(1, 1) < 0$, f has a saddle
point at $(1, 1)$ by the Second Derivatives Test.

3. In the figure, a point at approximately $(1, 1)$ is enclosed by level curves which are oval in shape and indicate that as
we move away from the point in any direction the values of f are increasing. Hence we would expect a local
minimum at or near $(1, 1)$. The level curves near $(0, 0)$ resemble hyperbolas, and as we move away from the origin,
the values of f increase in some directions and decrease in others, so we would expect to find a saddle point there.
To verify our predictions, we have $f(x, y) = 4 + x^3 + y^3 - 3xy \quad \Rightarrow \quad f_x(x, y) = 3x^2 - 3y$,
$f_y(x, y) = 3y^2 - 3x$. We have critical points where these partial derivatives are equal to 0: $3x^2 - 3y = 0$,
$3y^2 - 3x = 0$. Substituting $y = x^2$ from the first equation into the second equation gives $3(x^2)^2 - 3x = 0 \quad \Rightarrow$
$3x(x^3 - 1) = 0 \quad \Rightarrow \quad x = 0$ or $x = 1$. Then we have two critical points, $(0, 0)$ and $(1, 1)$. The second
partial derivatives are $f_{xx}(x, y) = 6x$, $f_{xy}(x, y) = -3$, and $f_{yy}(x, y) = 6y$, so
$D(x, y) = f_{xx}(x, y) f_{yy}(x, y) - [f_{xy}(x, y)]^2 = (6x)(6y) - (-3)^2 = 36xy - 9$. Then
$D(0, 0) = 36(0)(0) - 9 = -9$, and $D(1, 1) = 36(1)(1) - 9 = 27$. Since $D(0, 0) < 0$, f has a saddle point at
$(0, 0)$ by the Second Derivatives Test. Since $D(1, 1) > 0$ and $f_{xx}(1, 1) > 0$, f has a local minimum at $(1, 1)$.

5. $f(x, y) = 9 - 2x + 4y - x^2 - 4y^2$ $\Rightarrow$ $f_x = -2 - 2x$, $f_y = 4 - 8y$, $f_{xx} = -2$, $f_{xy} = 0$, $f_{yy} = -8$. Then $f_x = 0$ and $f_y = 0$ imply $x = -1$ and $y = \frac{1}{2}$, so the only critical point is $\left(-1, \frac{1}{2}\right)$.

$D(x, y) = f_{xx} f_{yy} - (f_{xy})^2 = (-2)(-8) - 0^2 = 16$, so since $D\left(-1, \frac{1}{2}\right) = 16 > 0$ and $f_{xx}\left(-1, \frac{1}{2}\right) = -2 < 0$, $f\left(-1, \frac{1}{2}\right) = 11$ is a local maximum by the Second Derivatives Test.

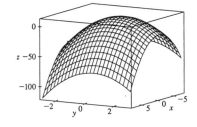

7. $f(x, y) = x^2 + y^2 + x^2 y + 4$ $\Rightarrow$ $f_x = 2x + 2xy$, $f_y = 2y + x^2$, $f_{xx} = 2 + 2y$, $f_{yy} = 2$, $f_{xy} = 2x$. Then $f_y = 0$ implies $y = -\frac{1}{2}x^2$, substituting into $f_x = 0$ gives $2x - x^3 = 0$ so $x = 0$ or $x = \pm\sqrt{2}$. Thus the critical points are $(0, 0)$, $(\sqrt{2}, -1)$ and $(-\sqrt{2}, -1)$. Now $D(0, 0) = 4$, $D(\sqrt{2}, -1) = -8 = D(-\sqrt{2}, -1)$, $f_{xx}(0, 0) = 2$, $f_{xx}(\pm\sqrt{2}, -1) = 0$.

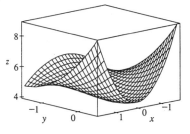

Thus $f(0, 0) = 4$ is a local minimum and $(\pm\sqrt{2}, -1)$ are saddle points.

9. $f(x, y) = 1 + 2xy - x^2 - y^2$ $\Rightarrow$ $f_x = 2y - 2x$, $f_y = 2x - 2y$, $f_{xx} = f_{yy} = -2$, $f_{xy} = 2$. Then $f_x = 0$ and $f_y = 0$ implies $x = y$ so the critical points are all points of the form (x_0, x_0). But $D(x_0, x_0) = 4 - 4 = 0$ so the Second Derivatives Test gives no information. However $1 + 2xy - x^2 - y^2 = 1 - (x - y)^2$ and $1 - (x - y)^2 \leq 1$ for all (x, y), with equality if and only if $x = y$. Thus $f(x_0, x_0) = 1$ are local maxima.

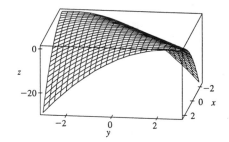

11. $f(x, y) = xy - 2x - y$ $\Rightarrow$ $f_x = y - 2$, $f_y = x - 1$, $f_{xx} = f_{yy} = 0$, $f_{xy} = 1$ and the only critical point is $(1, 2)$. Now $D(1, 2) = -1$, so $(1, 2)$ is a saddle point and f has no local maximum or minimum.

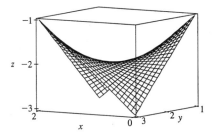

13. $f(x,y) = \dfrac{x^2y^2 - 8x + y}{xy}$ $\Rightarrow$ $f_x = y - x^{-2}$, $f_y = x + 8y^{-2}$, $f_{xx} = 2x^{-3}$, $f_{yy} = -16y^{-3}$ and $f_{xy} = 1$.

Then $f_x = 0$ implies $y = x^{-2}$, substituting into $f_y = 0$ gives $x + 8x^4 = 0$ so $x = 0$ or $x = -\frac{1}{2}$ but $(0,y)$ is not in

the domain of f. Thus the only critical point is $\left(-\frac{1}{2}, 4\right)$. Then $f_{xx}\left(-\frac{1}{2}, 4\right) = -16$ and $D\left(-\frac{1}{2}, 4\right) = 4 - 1 > 0$ so

$f\left(-\frac{1}{2}, 4\right) = -6$ is a local maximum.

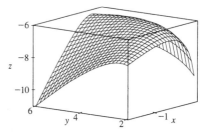

15. $f(x,y) = e^x \cos y$ $\Rightarrow$ $f_x = e^x \cos y$, $f_y = -e^x \sin y$. Now
$f_x = 0$ implies $\cos y = 0$ or $y = \frac{\pi}{2} + n\pi$ for n an integer. But
$\sin\left(\frac{\pi}{2} + n\pi\right) \neq 0$, so there are no critical points.

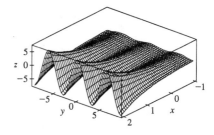

17. $f(x,y) = x \sin y$ $\Rightarrow$ $f_x = \sin y$, $f_y = x \cos y$, $f_{xx} = 0$, $f_{yy} = -x \sin y$ and $f_{xy} = \cos y$. Then $f_x = 0$ if and

only if $y = n\pi$, n an integer, and substituting into $f_y = 0$ requires $x = 0$ for each of these y-values. Thus the

critical points are $(0, n\pi)$, n an integer. But $D(0, n\pi) = -\cos^2(n\pi) < 0$ so each critical point is a saddle point.

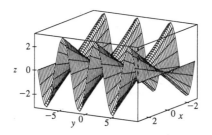

19. $f(x, y) = 3x^2 y + y^3 - 3x^2 - 3y^2 + 2$

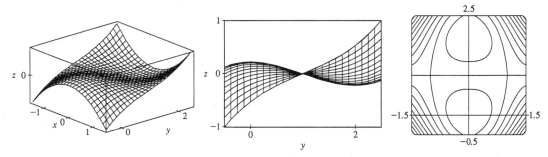

From the graphs, it appears that f has a local maximum $f(0, 0) \approx 2$ and a local minimum $f(0, 2) \approx -2$. There appear to be saddle points near $(\pm 1, 1)$.

$f_x = 6xy - 6x$, $f_y = 3x^2 + 3y^2 - 6y$. Then $f_x = 0$ implies $x = 0$ or $y = 1$ and when $x = 0$, $f_y = 0$ implies $y = 0$ or $y = 2$; when $y = 1$, $f_y = 0$ implies $x^2 = 1$ or $x = \pm 1$. Thus the critical points are $(0, 0)$, $(0, 2)$, $(\pm 1, 1)$. Now $f_{xx} = 6y - 6$, $f_{yy} = 6y - 6$ and $f_{xy} = 6x$, so $D(0, 0) = D(0, 2) = 36 > 0$ while $D(\pm 1, 1) = -36 < 0$ and $f_{xx}(0, 0) = -6$, $f_{xx}(0, 2) = 6$. Hence $(\pm 1, 1)$ are saddle points while $f(0, 0) = 2$ is a local maximum and $f(0, 2) = -2$ is a local minimum.

21. $f(x, y) = \sin x + \sin y + \sin(x + y)$, $0 \le x \le 2\pi$, $0 \le y \le 2\pi$

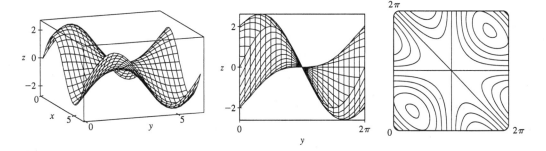

From the graphs it appears that f has a local maximum at about $(1, 1)$ with value approximately 2.6, a local minimum at about $(5, 5)$ with value approximately -2.6, and a saddle point at about $(3, 3)$.

$f_x = \cos x + \cos(x + y)$, $f_y = \cos y + \cos(x + y)$, $f_{xx} = -\sin x - \sin(x + y)$, $f_{yy} = -\sin y - \sin(x + y)$, $f_{xy} = -\sin(x + y)$. Setting $f_x = 0$ and $f_y = 0$ and subtracting gives $\cos x - \cos y = 0$ or $\cos x = \cos y$. Thus $x = y$ or $x = 2\pi - y$. If $x = y$, $f_x = 0$ becomes $\cos x + \cos 2x = 0$ or $2\cos^2 x + \cos x - 1 = 0$, a quadratic in $\cos x$. Thus $\cos x = -1$ or $\frac{1}{2}$ and $x = \pi$, $\frac{\pi}{3}$, or $\frac{5\pi}{3}$, yielding the critical points (π, π), $\left(\frac{\pi}{3}, \frac{\pi}{3}\right)$ and $\left(\frac{5\pi}{3}, \frac{5\pi}{3}\right)$. Similarly if $x = 2\pi - y$, $f_x = 0$ becomes $(\cos x) + 1 = 0$ and the resulting critical point is (π, π). Now $D(x, y) = \sin x \sin y + \sin x \sin(x + y) + \sin y \sin(x + y)$. So $D(\pi, \pi) = 0$ and the Second Derivatives Test doesn't apply. $D\left(\frac{\pi}{3}, \frac{\pi}{3}\right) = \frac{9}{4} > 0$ and $f_{xx}\left(\frac{\pi}{3}, \frac{\pi}{3}\right) < 0$ so $f\left(\frac{\pi}{3}, \frac{\pi}{3}\right) = \frac{3\sqrt{3}}{2}$ is a local maximum while $D\left(\frac{5\pi}{3}, \frac{5\pi}{3}\right) = \frac{9}{4} > 0$ and $f_{xx}\left(\frac{5\pi}{3}, \frac{5\pi}{3}\right) > 0$, so $f\left(\frac{5\pi}{3}, \frac{5\pi}{3}\right) = -\frac{3\sqrt{3}}{2}$ is a local minimum.

23. $f(x, y) = x^4 - 5x^2 + y^2 + 3x + 2 \quad \Rightarrow \quad f_x(x, y) = 4x^3 - 10x + 3$ and $f_y(x, y) = 2y$. $f_y = 0 \quad \Rightarrow \quad y = 0$,

and the graph of f_x shows that the roots of $f_x = 0$ are approximately $x = -1.714, 0.312$ and 1.402 .

(Alternatively, we could have used a calculator or a CAS to find these roots.) So to three decimal places, the critical

points are $(-1.714, 0)$, $(1.402, 0)$, and $(0.312, 0)$. Now since $f_{xx} = 12x^2 - 10$, $f_{xy} = 0$, $f_{yy} = 2$, and

$D = 24x^2 - 20$, we have $D(-1.714, 0) > 0$, $f_{xx}(-1.714, 0) > 0$, $D(1.402, 0) > 0$, $f_{xx}(1.402, 0) > 0$, and

$D(0.312, 0) < 0$. Therefore $f(-1.714, 0) \approx -9.200$ and $f(1.402, 0) \approx 0.242$ are local minima, and $(0.312, 0)$

is a saddle point. The lowest point on the graph is approximately $(-1.714, 0, -9.200)$.

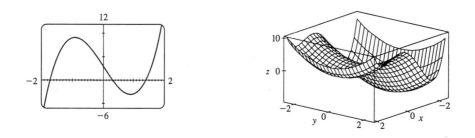

25. $f(x, y) = 2x + 4x^2 - y^2 + 2xy^2 - x^4 - y^4 \quad \Rightarrow \quad f_x(x, y) = 2 + 8x + 2y^2 - 4x^3$,

$f_y(x, y) = -2y + 4xy - 4y^3$. Now $f_y = 0 \quad \Leftrightarrow \quad 2y(2y^2 - 2x + 1) = 0 \quad \Leftrightarrow \quad y = 0$ or $y^2 = x - \frac{1}{2}$.

The first of these implies that $f_x = -4x^3 + 8x + 2$, and the second implies that

$f_x = 2 + 8x + 2\left(x - \frac{1}{2}\right) - 4x^3 = -4x^3 + 10x + 1$. From the graphs, we see that the first possibility for f_x has

roots at approximately -1.267, -0.259, and 1.526, and the second has a root at approximately 1.629 (the negative

roots do not give critical points, since $y^2 = x - \frac{1}{2}$ must be positive). So to three decimal places, f has critical points

at $(-1.267, 0)$, $(-0.259, 0)$, $(1.526, 0)$, and $(1.629, \pm 1.063)$. Now since $f_{xx} = 8 - 12x^2$, $f_{xy} = 4y$,

$f_{yy} = 4x - 12y^2$, and $D = (8 - 12x^2)(4x - 12y^2) - 16y^2$, we have $D(-1.267, 0) > 0$, $f_{xx}(-1.267, 0) > 0$,

$D(-0.259, 0) < 0$, $D(1.526, 0) < 0$, $D(1.629, \pm 1.063) > 0$, and $f_{xx}(1.629, \pm 1.063) < 0$. Therefore, to three

decimal places, $f(-1.267, 0) \approx 1.310$ and $f(1.629, \pm 1.063) \approx 8.105$ are local maxima, and $(-0.259, 0)$ and

$(1.526, 0)$ are saddle points. The highest points on the graph are approximately $(1.629, \pm 1.063, 8.105)$.

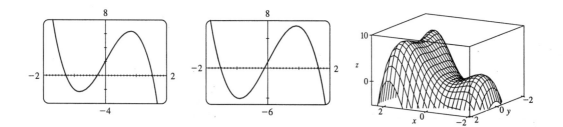

27. Since f is a polynomial it is continuous on D, so an absolute maximum and minimum exist. Here $f_x = -3$, $f_y = 4$ so there are no critical points inside D. Thus the absolute extrema must both occur on the boundary. Along L_1, $y = 0$ and $f(x, 0) = 5 - 3x$, a decreasing function in x, so the maximum value is $f(0, 0) = 5$ and the minimum value is $f(4, 0) = -7$. Along L_2, $x = 4$ and $f(4, y) = -7 + 4y$, an increasing function in y, so the minimum value is

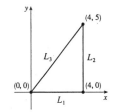

$f(4, 0) = -7$ and the maximum value is $f(4, 5) = 13$. Along L_3, $y = \frac{5}{4}x$ and $f\left(x, \frac{5}{4}x\right) = 5 + 2x$, an increasing function in x, so the minimum value is $f(0, 0) = 5$ and the maximum value is $f(4, 5) = 13$. Thus the absolute minimum of f on D is $f(4, 0) = -7$ and the absolute maximum is $f(4, 5) = 13$.

29. In Exercise 7, we found the critical points of f; only $(0, 0)$ with $f(0, 0) = 4$ is in D. On L_1: $y = -1$, $f(x, -1) = 5$, a constant. On L_2: $x = 1$, $f(1, y) = y^2 + y + 5$, a quadratic in y which attains its maximum at $(1, 1)$, $f(1, 1) = 7$ and its minimum at $\left(1, -\frac{1}{2}\right)$, $f\left(1, -\frac{1}{2}\right) = \frac{17}{4}$. On L_3: $f(x, 1) = 2x^2 + 5$ which attains its maximum at $(-1, 1)$ and $(1, 1)$ with $f(\pm 1, 1) = 7$ and its minimum at $(0, 1)$, $f(0, 1) = 5$. On L_4: $f(-1, y) = y^2 + y + 5$ with maximum at $(-1, 1)$, $f(-1, 1) = 7$ and minimum at $\left(-1, -\frac{1}{2}\right)$, $f\left(-1, -\frac{1}{2}\right) = \frac{17}{4}$. Thus the absolute maximum is attained at both $(\pm 1, 1)$ with $f(\pm 1, 1) = 7$ and the absolute minimum on D is attained at $(0, 0)$ with $f(0, 0) = 4$.

31. $f_x(x, y) = y - 1$ and $f_y(x, y) = x - 1$ and so the critical point is $(1, 1)$ (in D), where $f(1, 1) = 0$. Along L_1: $y = 4$, so $f(x, 4) = 1 + 4x - x - 4 = 3x - 3$, $-2 \leq x \leq 2$, which is an increasing function and has a maximum value when $x = 2$ where $f(2, 4) = 3$ and a minimum of $f(-2, 4) = -9$. Along L_2: $y = x^2$, so let $g(x) = f(x, x^2) = x^3 - x^2 - x + 1$. Then $g'(x) = 3x^2 - 2x - 1 = 0 \iff x = -\frac{1}{3}$ or $x = 1$. $f\left(-\frac{1}{3}, \frac{1}{9}\right) = \frac{32}{27}$ and $f(1, 1) = 0$. As a result, the absolute maximum and minimum values of f on D are $f(2, 4) = 3$ and $f(-2, 4) = -9$.

33. $f_x(x, y) = 6x^2$ and $f_y(x, y) = 4y^3$. And so $f_x = 0$ and $f_y = 0$ only occur when $x = y = 0$. Hence, the only critical point inside the disk is at $x = y = 0$ where $f(0, 0) = 0$. Now on the circle $x^2 + y^2 = 1$, $y^2 = 1 - x^2$ so let $g(x) = f(x, y) = 2x^3 + \left(1 - x^2\right)^2 = x^4 + 2x^3 - 2x^2 + 1$, $-1 \leq x \leq 1$. Then $g'(x) = 4x^3 + 6x^2 - 4x = 0 \Rightarrow x = 0, -2$, or $\frac{1}{2}$. $f(0, \pm 1) = g(0) = 1$, $f\left(\frac{1}{2}, \pm\frac{\sqrt{3}}{2}\right) = g\left(\frac{1}{2}\right) = \frac{13}{16}$, and $(-2, -3)$ is not in D. Checking the endpoints, we get $f(-1, 0) = g(-1) = -2$ and $f(1, 0) = g(1) = 2$. Thus the absolute maximum and minimum of f on D are $f(1, 0) = 2$ and $f(-1, 0) = -2$.

Another method: On the boundary $x^2 + y^2 = 1$ we can write $x = \cos\theta$, $y = \sin\theta$, so $f(\cos\theta, \sin\theta) = 2\cos^3\theta + \sin^4\theta$, $0 \leq \theta \leq 2\pi$.

35. $f(x, y) = -(x^2 - 1)^2 - (x^2 y - x - 1)^2$ $\Rightarrow$ $f_x(x, y) = -2(x^2 - 1)(2x) - 2(x^2 y - x - 1)(2xy - 1)$

and $f_y(x, y) = -2(x^2 y - x - 1)x^2$. Setting $f_y(x, y) = 0$ gives either $x = 0$ or $x^2 y - x - 1 = 0$. There are no

critical points for $x = 0$, since $f_x(0, y) = -2$, so we set $x^2 y - x - 1 = 0$ $\Leftrightarrow$ $y = \dfrac{x + 1}{x^2}$ $(x \neq 0)$, so

$f_x\left(x, \dfrac{x+1}{x^2}\right) = -2(x^2 - 1)(2x) - 2\left(x^2 \dfrac{x+1}{x^2} - x - 1\right)\left(2x\dfrac{x+1}{x^2} - 1\right) = -4x(x^2 - 1)$. Therefore

$f_x(x, y) = f_y(x, y) = 0$ at the points $(1, 2)$ and $(-1, 0)$.

To classify these critical points, we calculate

$f_{xx}(x, y) = -12x^2 - 12x^2 y^2 + 12xy + 4y + 2$,

$f_{yy}(x, y) = -2x^4$, and $f_{xy}(x, y) = -8x^3 y + 6x^2 + 4x$. In

order to use the Second Derivatives Test we calculate

$D(-1, 0) = f_{xx}(-1, 0) f_{yy}(-1, 0) - [f_{xy}(-1, 0)]^2$

$\qquad = 16 > 0$,

$f_{xx}(-1, 0) = -10 < 0$, $D(1, 2) = 16 > 0$, and

$f_{xx}(1, 2) = -26 < 0$, so both $(-1, 0)$ and $(1, 2)$ give local

maxima.

37. $d = \sqrt{(x-2)^2 + (y+2)^2 + (z-3)^2}$, where $z = \frac{1}{3}(6x + 4y - 2)$, so we minimize

$d^2 = f(x, y) = (x-2)^2 + (y+2)^2 + \left(2x + \frac{4}{3}y - \frac{11}{3}\right)^2$. Then $f_x = 10x + \frac{16}{3}y - \frac{56}{3}$ and

$f_y = \frac{50}{9}y + \frac{16}{3}x - \frac{52}{9}$. Solving $50y + 48x = 52$ and $16y + 30x = 56$ simultaneously gives $x = \frac{164}{61}$, $y = -\frac{94}{61}$.

The absolute minimum must occur at a critical point. Thus $d^2 = \left(\frac{42}{61}\right)^2 + \left(\frac{28}{61}\right)^2 + \left(-\frac{21}{61}\right)^2$ or $d = \frac{7}{\sqrt{61}}$.

39. Minimize $d^2 = x^2 + y^2 + z^2 = x^2 + y^2 + xy + 1$. Then $f_x = 2x + y$, $f_y = 2y + x$ so the critical point is $(0, 0)$

and $D(0, 0) = 4 - 1 > 0$ with $f_{xx}(0, 0) = 2$ so this is a minimum. Thus $z^2 = 1$ or $z = \pm 1$ and the points on the

surface are $(0, 0, \pm 1)$.

41. $x + y + z = 100$, so maximize $f(x, y) = xy(100 - x - y)$. $f_x = 100y - 2xy - y^2$, $f_y = 100x - x^2 - 2xy$,

$f_{xx} = -2y$, $f_{yy} = -2x$, $f_{xy} = 100 - 2x - 2y$. Then $f_x = 0$ implies $y = 0$ or $y = 100 - 2x$. Substituting $y = 0$

into $f_y = 0$ gives $x = 0$ or $x = 100$ and substituting $y = 100 - 2x$ into $f_y = 0$ gives $3x^2 - 100x = 0$ so

$x = 0$ or $\frac{100}{3}$. Thus the critical points are $(0, 0)$, $(100, 0)$, $(0, 100)$ and $\left(\frac{100}{3}, \frac{100}{3}\right)$.

$D(0, 0) = D(100, 0) = D(0, 100) = -10{,}000$ while $D\left(\frac{100}{3}, \frac{100}{3}\right) = \frac{10{,}000}{3}$ and $f_{xx}\left(\frac{100}{3}, \frac{100}{3}\right) = -\frac{200}{3} < 0$.

Thus $(0, 0)$, $(100, 0)$ and $(0, 100)$ are saddle points whereas $f\left(\frac{100}{3}, \frac{100}{3}\right)$ is a local maximum. Thus the numbers are

$x = y = z = \frac{100}{3}$.

43. Maximize $f(x, y) = xy \left(36 - 9x^2 - 36y^2\right)^{1/2}/2$ with (x, y, z) in first octant. Then

$$f_x = \frac{y\left(36 - 9x^2 - 36y^2\right)^{1/2}}{2} + \frac{-9x^2 y\left(36 - 9x^2 - 36y^2\right)^{-1/2}}{2} = \frac{\left(36y - 18x^2 y - 36y^3\right)}{2\left(36 - 9x^2 - 36y^2\right)^{1/2}} \text{ and}$$

$$f_y = \frac{36x - 9x^3 - 72xy^2}{2\left(36 - 9x^2 - 36y^2\right)^{1/2}}. \text{ Setting } f_x = 0 \text{ gives } y = 0 \text{ or } y^2 = \frac{2 - x^2}{2} \text{ but } y > 0, \text{ so only the latter solution}$$

applies. Substituting this y into $f_y = 0$ gives $x^2 = \frac{4}{3}$ or $x = \frac{2}{\sqrt{3}}$, $y = \frac{1}{\sqrt{3}}$ and then $z^2 = (36 - 12 - 12)/4 = 3$.
The fact that this gives a maximum volume follows from the geometry. This maximum volume is

$$V = (2x)(2y)(2z) = 8\left(\frac{2}{\sqrt{3}}\right)\left(\frac{1}{\sqrt{3}}\right)\left(\sqrt{3}\right) = \frac{16}{\sqrt{3}}.$$

45. Maximize $f(x, y) = \frac{xy}{3}(6 - x - 2y)$, then the maximum volume is $V = xyz$.

$f_x = \frac{1}{3}\left(6y - 2xy - y^2\right) = \frac{1}{3}y(6 - 2x - 2y)$ and $f_y = \frac{1}{3}x(6 - x - 4y)$. Setting $f_x = 0$ and $f_y = 0$ gives the
critical point $(2, 1)$ which geometrically must yield a maximum. Thus the volume of the largest such box is
$V = (2)(1)\left(\frac{2}{3}\right) = \frac{4}{3}$.

47. Let the dimensions be x, y, and z; then $4x + 4y + 4z = c$ and the volume is
$V = xyz = xy\left(\frac{1}{4}c - x - y\right) = \frac{1}{4}cxy - x^2 y - xy^2$, $x > 0$, $y > 0$. Then $V_x = \frac{1}{4}cy - 2xy - y^2$ and
$V_y = \frac{1}{4}cx - x^2 - 2xy$, so $V_x = 0 = V_y$ when $2x + y = \frac{1}{4}c$ and $x + 2y = \frac{1}{4}c$. Solving, we get $x = \frac{1}{12}c$, $y = \frac{1}{12}c$
and $z = \frac{1}{4}c - x - y = \frac{1}{12}c$. From the geometrical nature of the problem, this critical point must give an absolute
maximum. Thus the box is a cube with edge length $\frac{1}{12}c$.

49. Let the dimensions be x, y and z, then minimize $xy + 2(xz + yz)$ if $xyz = 32{,}000$ m^3. Then
$f(x, y) = xy + [64{,}000(x + y)/xy] = xy + 64{,}000\left(x^{-1} + y^{-1}\right)$, $f_x = y - 64{,}000x^{-2}$, $f_y = x - 64{,}000y^{-2}$.
And $f_x = 0$ implies $y = 64{,}000/x^2$; substituting into $f_y = 0$ implies $x^3 = 64{,}000$ or $x = 40$ and then $y = 40$.
Now $D(x, y) = [(2)(64{,}000)]^2 x^{-3}y^{-3} - 1 > 0$ for $(40, 40)$ and $f_{xx}(40, 40) > 0$ so this is indeed a minimum.
Thus the dimensions of the box are $x = y = 40$ cm, $z = 20$ cm.

51. Note that here the variables are m and b, and $f(m, b) = \sum\limits_{i=1}^{n} [y_i - (mx_i + b)]^2$. Then

$$f_m = \sum_{i=1}^{n} -2x_i [y_i - (mx_i + b)] = 0 \text{ implies } \sum_{i=1}^{n} \left(x_i y_i - mx_i^2 - bx_i\right) = 0 \text{ or } \sum_{i=1}^{n} x_i y_i = m\sum_{i=1}^{n} x_i^2 + b\sum_{i=1}^{n} x_i$$

and $f_b = \sum\limits_{i=1}^{n} -2[y_i - (mx_i + b)] = 0$ implies $\sum\limits_{i=1}^{n} y_i = m\sum\limits_{i=1}^{n} x_i + \sum\limits_{i=1}^{n} b = m\left(\sum\limits_{i=1}^{n} x_i\right) + nb$. Thus we have

the two desired equations. Now $f_{mm} = \sum\limits_{i=1}^{n} 2x_i^2$, $f_{bb} = \sum\limits_{i=1}^{n} 2 = 2n$ and $f_{mb} = \sum\limits_{i=1}^{n} 2x_i$. And $f_{mm}(m, b) > 0$

always and $D(m, b) = 4n\left(\sum\limits_{i=1}^{n} x_i^2\right) - 4\left(\sum\limits_{i=1}^{n} x_i\right)^2 = 4\left[n\left(\sum\limits_{i=1}^{n} x_i^2\right) - \left(\sum\limits_{i=1}^{n} x_i\right)^2\right] > 0$ always so the

solutions of these two equations do indeed minimize $\sum\limits_{i=1}^{n} d_i^2$.

15.8 Lagrange Multipliers

1. At the extreme values of f, the level curves of f just touch the curve $g\,(x,y) = 8$ with a common tangent line. (See Figure 1 and the accompanying discussion.) We can observe several such occurrences on the contour map, but the level curve $f\,(x,y) = c$ with the largest value of c which still intersects the curve $g\,(x,y) = 8$ is approximately $c = 59$, and the smallest value of c corresponding to a level curve which intersects $g\,(x,y) = 8$ appears to be $c = 30$. Thus we estimate the maximum value of f subject to the constraint $g\,(x,y) = 8$ to be about 59 and the minimum to be 30.

3. $f\,(x,y) = x^2 - y^2$, $g\,(x,y) = x^2 + y^2 = 1$ $\Rightarrow$ $\nabla f = \langle 2x, -2y \rangle$, $\lambda \nabla g = \langle 2\lambda x, 2\lambda y \rangle$. Then $2x = 2\lambda x$ implies $x = 0$ or $\lambda = 1$. If $x = 0$, then $x^2 + y^2 = 1$ implies $y = \pm 1$ and if $\lambda = 1$, then $-2y = 2\lambda y$ implies $y = 0$ and thus $x = \pm 1$. Thus the possible points for the extreme values of f are $(\pm 1, 0)$, $(0, \pm 1)$. But $f\,(\pm 1, 0) = 1$ while $f\,(0, \pm 1) = -1$ so the maximum value of f on $x^2 + y^2 = 1$ is $f\,(\pm 1, 0) = 1$ and the minimum value is $f\,(0, \pm 1) = -1$.

5. $f\,(x,y) = x^2 y$, $g\,(x,y) = x^2 + 2y^2 = 6$ $\Rightarrow$ $\nabla f = \langle 2xy, x^2 \rangle$, $\lambda \nabla g = \langle 2\lambda x, 4\lambda y \rangle$. Then $2xy = 2\lambda x$ implies $x = 0$ or $\lambda = y$. If $x = 0$, then $x^2 = 4\lambda y$ implies $\lambda = 0$ or $y = 0$. However, if $y = 0$ then $g\,(x,y) = 0$, a contradiction. So $\lambda = 0$ and then $g\,(x,y) = 6$ $\Rightarrow$ $y = \pm\sqrt{3}$. If $\lambda = y$, then $x^2 = 4\lambda y$ implies $x^2 = 4y^2$, and so $g\,(x,y) = 6$ $\Rightarrow$ $4y^2 + 2y^2 = 6$ $\Rightarrow$ $y^2 = 1$ $\Rightarrow$ $y = \pm 1$. Thus f has possible extreme values at the points $(0, \pm\sqrt{3})$, $(\pm 2, 1)$, and $(\pm 2, -1)$. After evaluating f at these points, we find the maximum value to be $f\,(\pm 2, 1) = 4$ and the minimum to be $f\,(\pm 2, -1) = -4$.

7. $f\,(x,y,z) = 2x + 6y + 10z$, $g\,(x,y,z) = x^2 + y^2 + z^2 = 35$ $\Rightarrow$ $\nabla f = \langle 2, 6, 10 \rangle$, $\lambda \nabla g = \langle 2\lambda x, 2\lambda y, 2\lambda z \rangle$. Then $2\lambda x = 2$, $2\lambda y = 6$, $2\lambda z = 10$ imply $x = \dfrac{1}{\lambda}$, $y = \dfrac{3}{\lambda}$, and $z = \dfrac{5}{\lambda}$. But
$$35 = x^2 + y^2 + z^2 = \left(\frac{1}{\lambda}\right)^2 + \left(\frac{3}{\lambda}\right)^2 + \left(\frac{5}{\lambda}\right)^2 \Rightarrow 35 = \frac{35}{\lambda^2} \Rightarrow \lambda = \pm 1, \text{ so } f \text{ has possible extreme}$$
values at the points $(1, 3, 5)$, $(-1, -3, -5)$. The maximum value of f on $x^2 + y^2 + z^2 = 35$ is $f\,(1, 3, 5) = 70$, and the minimum is $f\,(-1, -3, -5) = -70$.

9. $f\,(x,y,z) = xyz$, $g\,(x,y,z) = x^2 + 2y^2 + 3z^2 = 6$ $\Rightarrow$ $\nabla f = \langle yz, xz, xy \rangle$, $\lambda \nabla g = \langle 2\lambda x, 4\lambda y, 6\lambda z \rangle$. Then $\nabla f = \lambda \nabla g$ implies $\lambda = (yz)/(2x) = (xz)/(4y) = (xy)/(6z)$ or $x^2 = 2y^2$ and $z^2 = \frac{2}{3}y^2$. Thus $x^2 + 2y^2 + 3z^2 = 6$ implies $6y^2 = 6$ or $y = \pm 1$. Then the possible points are $\left(\sqrt{2}, \pm 1, \sqrt{\frac{2}{3}}\right)$, $\left(\sqrt{2}, \pm 1, -\sqrt{\frac{2}{3}}\right)$, $\left(-\sqrt{2}, \pm 1, \sqrt{\frac{2}{3}}\right)$, $\left(-\sqrt{2}, \pm 1, -\sqrt{\frac{2}{3}}\right)$. And the maximum value of f on the ellipsoid is $\dfrac{2}{\sqrt{3}}$, occurring when all coordinates are positive or exactly two are negative and the minimum is $-\dfrac{2}{\sqrt{3}}$ occurring when 1 or 3 of the coordinates are negative.

11. $f\,(x,y,z) = x^2 + y^2 + z^2$, $g\,(x,y,z) = x^4 + y^4 + z^4 = 1$ $\Rightarrow$ $\nabla f = \langle 2x, 2y, 2z \rangle$, $\lambda \nabla g = \langle 4\lambda x^3, 4\lambda y^3, 4\lambda z^3 \rangle$.
Case 1: If $x \neq 0$, $y \neq 0$ and $z \neq 0$, then $\nabla f = \lambda \nabla g$ implies $\lambda = 1/\left(2x^2\right) = 1/\left(2y^2\right) = 1/\left(2z^2\right)$ or $x^2 = y^2 = z^2$ and $3x^4 = 1$ or $x = \pm\dfrac{1}{\sqrt[4]{3}}$ giving the points $\left(\pm\dfrac{1}{\sqrt[4]{3}}, \dfrac{1}{\sqrt[4]{3}}, \dfrac{1}{\sqrt[4]{3}}\right)$, $\left(\pm\dfrac{1}{\sqrt[4]{3}}, -\dfrac{1}{\sqrt[4]{3}}, \dfrac{1}{\sqrt[4]{3}}\right)$, $\left(\pm\dfrac{1}{\sqrt[4]{3}}, \dfrac{1}{\sqrt[4]{3}}, -\dfrac{1}{\sqrt[4]{3}}\right)$, $\left(\pm\dfrac{1}{\sqrt[4]{3}}, -\dfrac{1}{\sqrt[4]{3}}, -\dfrac{1}{\sqrt[4]{3}}\right)$ all with an f-value of $\sqrt{3}$.

Case 2: If one of the variables equals zero and the other two are not zero, then the squares of the two nonzero coordinates are equal with common value $\frac{1}{\sqrt{2}}$ and corresponding f value of $\sqrt{2}$.

Case 3: If exactly two of the variables are zero, then the third variable has value ± 1 with the corresponding f value of 1. Thus on $x^4 + y^4 + z^4 = 1$, the maximum value of f is $\sqrt{3}$ and the minimum value is 1.

13. $f(x, y, z, t) = x + y + z + t$, $g(x, y, z, t) = x^2 + y^2 + z^2 + t^2 = 1$ $\Rightarrow$ $\langle 1, 1, 1, 1 \rangle = \langle 2\lambda x, 2\lambda y, 2\lambda z, 2\lambda t \rangle$, so $\lambda = 1/(2x) = 1/(2y) = 1/(2z) = 1/(2t)$ and $x = y = z = t$. But $x^2 + y^2 + z^2 + t^2 = 1$, so the possible points are $\left(\pm \frac{1}{2}, \pm \frac{1}{2}, \pm \frac{1}{2}, \pm \frac{1}{2} \right)$. Thus the maximum value of f is $f \left(\frac{1}{2}, \frac{1}{2}, \frac{1}{2}, \frac{1}{2} \right) = 2$ and the minimum value is $f \left(-\frac{1}{2}, -\frac{1}{2}, -\frac{1}{2}, -\frac{1}{2} \right) = -2$.

15. $f(x, y, z) = x + 2y$, $g(x, y, z) = x + y + z = 1$, $h(x, y, z) = y^2 + z^2 = 4$ $\Rightarrow$ $\nabla f = \langle 1, 2, 0 \rangle$, $\lambda \nabla g = \langle \lambda, \lambda, \lambda \rangle$ and $\mu \nabla h = \langle 0, 2\mu y, 2\mu z \rangle$. Then $1 = \lambda$, $2 = \lambda + 2\mu y$ and $0 = \lambda + 2\mu z$ so $\mu y = \frac{1}{2} = -\mu z$ or $y = 1/(2\mu)$, $z = -1/(2\mu)$. Thus $x + y + z = 1$ implies $x = 1$ and $y^2 + z^2 = 4$ implies $\mu = \pm \frac{1}{2\sqrt{2}}$. Then the possible points are $\left(1, \pm \sqrt{2}, \mp \sqrt{2} \right)$ and the maximum value is $f \left(1, \sqrt{2}, -\sqrt{2} \right) = 1 + 2\sqrt{2}$ and the minimum value is $f \left(1, -\sqrt{2}, \sqrt{2} \right) = 1 - 2\sqrt{2}$.

17. $f(x, y, z) = yz + xy$, $g(x, y, z) = xy = 1$, $h(x, y, z) = y^2 + z^2 = 1$ $\Rightarrow$ $\nabla f = \langle y, x + z, y \rangle$, $\lambda \nabla g = \langle \lambda y, \lambda x, 0 \rangle$, $\mu \nabla h = \langle 0, 2\mu y, 2\mu z \rangle$. Then $y = \lambda y$ implies $\lambda = 1$ [$y \neq 0$ since $g(x, y, z) = 1$], $x + z = \lambda x + 2\mu y$ and $y = 2\mu z$. Thus $\mu = z/(2y) = y/(2z)$ or $y^2 = z^2$, and so $y^2 + z^2 = 1$ implies $y = \pm \frac{1}{\sqrt{2}}$, $z = \pm \frac{1}{\sqrt{2}}$. Then $xy = 1$ implies $x = \pm \sqrt{2}$ and the possible points are $\left(\pm \sqrt{2}, \pm \frac{1}{\sqrt{2}}, \frac{1}{\sqrt{2}} \right)$, $\left(\pm \sqrt{2}, \pm \frac{1}{\sqrt{2}}, -\frac{1}{\sqrt{2}} \right)$. Hence the maximum of f subject to the constraints is $f \left(\pm \sqrt{2}, \pm \frac{1}{\sqrt{2}}, \pm \frac{1}{\sqrt{2}} \right) = \frac{3}{2}$ and the minimum is $f \left(\pm \sqrt{2}, \pm \frac{1}{\sqrt{2}}, \mp \frac{1}{\sqrt{2}} \right) = \frac{1}{2}$.

Note: Since $xy = 1$ is one of the constraints we could have solved the problem by solving $f(y, z) = yz + 1$ subject to $y^2 + z^2 = 1$.

19. $f(x, y) = e^{-xy}$. For the interior of the region, we find the critical points: $f_x = -ye^{-xy}$, $f_y = -xe^{-xy}$, so the only critical point is $(0, 0)$, and $f(0, 0) = 1$. For the boundary, we use Lagrange multipliers.
$g(x, y) = x^2 + 4y^2 = 1$ $\Rightarrow$ $\lambda \nabla g = \langle 2\lambda x, 8\lambda y \rangle$, so setting $\nabla f = \lambda \nabla g$ we get $-ye^{-xy} = 2\lambda x$ and $-xe^{-xy} = 8\lambda y$. The first of these gives $e^{-xy} = -2\lambda x/y$, and then the second gives $-x(-2\lambda x/y) = 8\lambda y$ $\Rightarrow$ $x^2 = 4y^2$. Solving this last equation with the constraint $x^2 + 4y^2 = 1$ gives $x = \pm \frac{1}{\sqrt{2}}$ and $y = \pm \frac{1}{2\sqrt{2}}$. Now $f \left(\pm \frac{1}{\sqrt{2}}, \mp \frac{1}{2\sqrt{2}} \right) = e^{1/4} \approx 1.284$ and $f \left(\pm \frac{1}{\sqrt{2}}, \pm \frac{1}{2\sqrt{2}} \right) = e^{-1/4} \approx 0.779$. The former are the maxima on the region and the latter are the minima.

21. $P(L, K) = bL^\alpha K^{1-\alpha}$, $g(L, K) = mL + nK = p$ $\Rightarrow$ $\nabla P = \langle \alpha b L^{\alpha-1} K^{1-\alpha}, (1-\alpha) bL^\alpha K^{-\alpha} \rangle$, $\lambda \nabla g = \langle \lambda m, \lambda n \rangle$. Then $\alpha b (K/L)^{1-\alpha} = \lambda m$ and $(1-\alpha) b (L/K)^\alpha = \lambda n$ and $mL + nK = p$, so $\alpha b (K/L)^{1-\alpha}/m = (1-\alpha) b (L/K)^\alpha/n$ or $n\alpha/[m(1-\alpha)] = (L/K)^\alpha (L/K)^{1-\alpha}$ or $L = Kn\alpha/[m(1-\alpha)]$. Substituting into $mL + nK = p$ gives $K = (1-\alpha) p/n$ and $L = \alpha p/m$ for the maximum production.

23. Let the sides of the rectangle be x and y. Then $f(x, y) = xy$, $g(x, y) = 2x + 2y = p$ $\Rightarrow$ $\nabla f(x, y) = \langle y, x \rangle$, $\lambda \nabla g = \langle 2\lambda, 2\lambda \rangle$. Then $\lambda = \frac{1}{2} y = \frac{1}{2} x$ implies $x = y$ and the rectangle with maximum area is a square with side length $\frac{1}{4} p$.

25. $f(x, y, z) = (x - 2)^2 + (y + 2)^2 + (z - 3)^2$, $g(x, y, z) = 6x + 4y - 3z = 2$ ⇒
$\nabla f = \langle 2(x - 2), 2(y + 2), 2(z - 3) \rangle = \lambda \nabla g = \langle 6\lambda, 4\lambda, -3\lambda \rangle$, so $x = 3\lambda + 2$, $y = 2\lambda - 2$, $z = -\frac{3}{2}\lambda + 3$ and
$(18\lambda + 12) + (8\lambda - 8) + \frac{9}{2}\lambda - 9 = 2$ implies $\lambda = \frac{14}{61}$. Thus the shortest distance is
$$\sqrt{\left(\frac{42}{61}\right)^2 + \left(\frac{28}{61}\right)^2 + \left(-\frac{21}{61}\right)^2} = \frac{7}{\sqrt{61}}.$$

27. $f(x, y, z) = x^2 + y^2 + z^2$, $g(x, y, z) = z^2 - xy - 1 = 0$ ⇒
$\nabla f = \langle 2x, 2y, 2z \rangle = \lambda \nabla g = \langle -\lambda y, -\lambda x, 2\lambda z \rangle$. Then $2z = 2\lambda z$ implies $z = 0$ or $\lambda = 1$. If $z = 0$ then
$g(x, y, z) = 1$ implies $xy = -1$ or $x = -1/y$. Thus $2x = -\lambda y$ and $2y = -\lambda x$ imply $\lambda = 2/y^2 = 2y^2$ or
$y = \pm 1$, $x = \pm 1$. If $\lambda = 1$, then $2x = -y$ and $2y = -x$ imply $x = y = 0$, so $z = \pm 1$. Hence the possible points
are $(\pm 1, \mp 1, 0)$, $(0, 0, \pm 1)$ and the minimum value of f is $f(0, 0, \pm 1) = 1$, so the points closest to the origin are
$(0, 0, \pm 1)$.

29. $f(x, y, z) = xyz$, $g(x, y, z) = x + y + z = 100$ ⇒ $\nabla f = \langle yz, xz, xy \rangle = \lambda \nabla g = \langle \lambda, \lambda, \lambda \rangle$. Then
$\lambda = yz = xz = xy$ implies $x = y = z = \frac{100}{3}$.

31. If the dimensions are $2x$, $2y$ and $2z$, then $f(x, y, z) = 8xyz$ and $g(x, y, z) = 9x^2 + 36y^2 + 4z^2 = 36$ ⇒
$\nabla f = \langle 8yz, 8xz, 8xy \rangle = \lambda \nabla g = \langle 18\lambda x, 72\lambda y, 8\lambda z \rangle$. Thus $18\lambda x = 8yz$, $72\lambda y = 8xz$, $8\lambda z = 8xy$ so $x^2 = 4y^2$,
$z^2 = 9y^2$ and $36y^2 + 36y^2 + 36y^2 = 36$ or $y = \frac{1}{\sqrt{3}}$ $(y > 0)$. Thus the volume of the largest such rectangle is
$$8\left(\frac{1}{\sqrt{3}}\right)\left(\frac{2}{\sqrt{3}}\right)\left(\frac{3}{\sqrt{3}}\right) = 16\sqrt{3}.$$

33. $f(x, y, z) = xyz$, $g(x, y, z) = x + 2y + 3z = 6$ ⇒ $\nabla f = \langle yz, xz, xy \rangle = \lambda \nabla g = \langle \lambda, 2\lambda, 3\lambda \rangle$.
Then $\lambda = yz = \frac{1}{2}xz = \frac{1}{3}xy$ implies $x = 2y$, $z = \frac{2}{3}y$. But $2y + 2y + 2y = 6$ so $y = 1$, $x = 2$, $z = \frac{2}{3}$ and the
volume is $V = \frac{4}{3}$.

35. $f(x, y, z) = xyz$, $g(x, y, z) = 4(x + y + z) = c$ ⇒ $\nabla f = \langle yz, xz, xy \rangle$, $\lambda \nabla g = \langle 4\lambda, 4\lambda, 4\lambda \rangle$. Thus
$4\lambda = yz = xz = xy$ or $x = y = z = \frac{1}{12}c$ are the dimensions giving the maximum volume.

37. $f(x, y, z) = xy + 2xz + 2yz$, $g(x, y, z) = xyz = 32{,}000 \text{ cm}^3$ ⇒
$\nabla f = \langle 2z + y, 2z + x, 2(x + y) \rangle = \lambda \nabla g = \langle \lambda yz, \lambda xz, \lambda xy \rangle$. Then (1) $\lambda yz = 2z + y$, (2) $\lambda xz = 2z + x$, and
(3) $\lambda xy = 2(x + y)$. Now (1)–(2) implies $\lambda z(y - x) = y - x$, so $x = y$ or $\lambda = 1/z$. If $\lambda = 1/z$ then (1) implies
$z = 0$ which can't be, so $x = y$. But twice (2) minus (3) together with $x = y$ implies
$\lambda y(2x - y) = (4z + 2y) - 4y$ or $\lambda y(2z - y) = 2(2z - y)$ so $z = y/2$ or $\lambda = 2/y$. If $\lambda = 2/y$ then (3) implies
$y = 0$ which can't be. Thus $x = y = 2z$ and $\frac{1}{2}y^3 = 32{,}000$ or $y = 40$ and the dimensions which minimize the
volume are $x = y = 40\text{cm}$, $z = 20\text{cm}$.

39. We need to find the extreme values of $f(x, y, z) = x^2 + y^2 + z^2$ subject to the two constraints
$g(x, y, z) = x + y + 2z = 2$ and $h(x, y, z) = x^2 + y^2 - z = 0$. $\nabla f = \langle 2x, 2y, 2z \rangle$, $\lambda \nabla g = \langle \lambda, \lambda, 2\lambda \rangle$ and
$\mu \nabla h = \langle 2\mu x, 2\mu y, -\mu \rangle$. Thus we need (1) $2x = \lambda + 2\mu x$, (2) $2y = \lambda + 2\mu y$, (3) $2z = 2\lambda - \mu$,
(4) $x + y + 2z = 2$, and (5) $x^2 + y^2 - z = 0$. From (1) and (2), $2(x - y) = 2\mu(x - y)$, so if $x \neq y$, $\mu = 1$.
Putting this in (3) gives $2z = 2\lambda - 1$ or $\lambda = z + \frac{1}{2}$, but putting $\mu = 1$ into (1) says $\lambda = 0$. Hence $z + \frac{1}{2} = 0$ or
$z = -\frac{1}{2}$. Then (4) and (5) become $x + y - 3 = 0$ and $x^2 + y^2 + \frac{1}{2} = 0$. The last equation cannot be true, so this
case gives no solution. So we must have $x = y$. Then (4) and (5) become $2x + 2z = 2$ and $2x^2 - z = 0$ which
imply $z = 1 - x$ and $z = 2x^2$. Thus $2x^2 = 1 - x$ or $2x^2 + x - 1 = (2x - 1)(x + 1) = 0$ so $x = \frac{1}{2}$ or $x = -1$.
The two points to check are $\left(\frac{1}{2}, \frac{1}{2}, \frac{1}{2}\right)$ and $(-1, -1, 2)$: $f\left(\frac{1}{2}, \frac{1}{2}, \frac{1}{2}\right) = \frac{3}{4}$ and $f(-1, -1, 2) = 6$. Thus $\left(\frac{1}{2}, \frac{1}{2}, \frac{1}{2}\right)$ is
the point on the ellipse nearest the origin and $(-1, -1, 2)$ is the one farthest from the origin.

41. (a) We wish to maximize $f(x_1, x_2, \ldots, x_n) = \sqrt[n]{x_1 x_2 \cdots x_n}$ subject to

$g(x_1, x_2, \ldots, x_n) = x_1 + x_2 + \cdots + x_n = c$ and $x_i > 0$.

$$\nabla f = \left\langle \frac{1}{n}(x_1 x_2 \cdots x_n)^{\frac{1}{n}-1}(x_2 \cdots x_n), \frac{1}{n}(x_1 x_2 \cdots x_n)^{\frac{1}{n}-1}(x_1 x_3 \cdots x_n), \ldots, \right.$$

$$\left. \frac{1}{n}(x_1 x_2 \cdots x_n)^{\frac{1}{n}-1}(x_1 \cdots x_{n-1}) \right\rangle$$

and $\lambda \nabla g = \langle \lambda, \lambda, \ldots, \lambda \rangle$, so we need to solve the system of equations

$$\frac{1}{n}(x_1 x_2 \cdots x_n)^{\frac{1}{n}-1}(x_2 \cdots x_n) = \lambda \quad \Rightarrow \quad x_1^{1/n} x_2^{1/n} \cdots x_n^{1/n} = n\lambda x_1$$

$$\frac{1}{n}(x_1 x_2 \cdots x_n)^{\frac{1}{n}-1}(x_1 x_3 \cdots x_n) = \lambda \quad \Rightarrow \quad x_1^{1/n} x_2^{1/n} \cdots x_n^{1/n} = n\lambda x_2$$

$$\vdots$$

$$\frac{1}{n}(x_1 x_2 \cdots x_n)^{\frac{1}{n}-1}(x_1 \cdots x_{n-1}) = \lambda \quad \Rightarrow \quad x_1^{1/n} x_2^{1/n} \cdots x_n^{1/n} = n\lambda x_n$$

This implies $n\lambda x_1 = n\lambda x_2 = \cdots = n\lambda x_n$. Note $\lambda \neq 0$, otherwise we can't have all $x_i > 0$. Thus

$x_1 = x_2 = \cdots = x_n$. But $x_1 + x_2 + \cdots + x_n = c \quad \Rightarrow \quad nx_1 = c \quad \Rightarrow \quad x_1 = \dfrac{c}{n} = x_2 = x_3 = \cdots = x_n$.

Then the only point where f can have an extreme value is $\left(\dfrac{c}{n}, \dfrac{c}{n}, \ldots, \dfrac{c}{n}\right)$. Since we can choose values for

$(x_1, x_2, \ldots, x_n)$ that make f as close to zero (but not equal) as we like, f has no minimum value. Thus the

maximum value is $f\left(\dfrac{c}{n}, \dfrac{c}{n}, \ldots, \dfrac{c}{n}\right) = \sqrt[n]{\dfrac{c}{n} \cdot \dfrac{c}{n} \cdots \dfrac{c}{n}} = \dfrac{c}{n}$.

(b) From part (a), $\dfrac{c}{n}$ is the maximum value of f. Thus $f(x_1, x_2, \ldots, x_n) = \sqrt[n]{x_1 x_2 \cdots x_n} \leq \dfrac{c}{n}$. But

$x_1 + x_2 + \cdots + x_n = c$, so $\sqrt[n]{x_1 x_2 \cdots x_n} \leq \dfrac{x_1 + x_2 + \cdots + x_n}{n}$. These two means are equal when f attains

its maximum value $\dfrac{c}{n}$, but this can occur only at the point $\left(\dfrac{c}{n}, \dfrac{c}{n}, \ldots, \dfrac{c}{n}\right)$ we found in part (a). So the means

are equal only when $x_1 = x_2 = x_3 = \cdots = x_n = \dfrac{c}{n}$.

▤15 Review   ET 14

CONCEPT CHECK

1. (a) A function f of two variables is a rule that assigns to each ordered pair (x, y) of real numbers in its domain a unique real number denoted by $f(x, y)$.

(b) One way to visualize a function of two variables is by graphing it, resulting in the surface $z = f(x, y)$. Another method for visualizing a function of two variables is a contour map. The contour map consists of level curves of the function which are horizontal traces of the graph of the function projected onto the xy-plane. Also, we can use an arrow diagram such as Figure 1 in Section 15.1 [ET 14.1].

2. A function f of three variables is a rule that assigns to each ordered triple (x, y, z) in its domain a unique real number $f(x, y, z)$. We can visualize a function of three variables by examining its level surfaces $f(x, y, z) = k$, where k is a constant.

3. $\displaystyle\lim_{(x,y)\to(a,b)} f(x, y) = L$ means the values of $f(x, y)$ approach the number L as the point (x, y) approaches the

point (a, b) along any path that is within the domain of f. We can show that a limit at a point does not exist by finding two different paths approaching the point along which $f(x, y)$ has different limits.

4. (a) See Definition 15.2.4 [ET 14.2.4].

(b) If f is continuous on $\mathbb{R}^2$, its graph will appear as a surface without holes or breaks.

5. (a) See (2) and (3) in Section 15.3 [ET 14.3].

(b) See the discussion preceding Example 2 in Section 15.3 [ET 14.3].

(c) To find f_x, regard y as a constant and differentiate $f(x, y)$ with respect to x. To find f_y, regard x as a constant and differentiate $f(x, y)$ with respect to y.

6. See the statement of Clairaut's Theorem on page 936 [ET 902].

7. (a) See (2) in Section 15.4 [ET 14.4].

(b) See (19) and the preceding discussion in Section 15.6 [ET 14.6].

8. See (3) and (4) and the accompanying discussion in Section 15.4 [ET 14.4]. We can interpret the linearization of f at (a, b) geometrically as the linear function whose graph is the tangent plane to the graph of f at (a, b). Thus it is the linear function which best approximates f near (a, b).

9. (a) See Definition 15.4.7 [ET 14.4.7].

(b) Use Theorem 15.4.8 [ET 14.4.8].

10. See (10) and the associated discussion in Section 15.4 [ET 14.4].

11. See (2) and (3) in Section 15.5 [ET 14.5].

12. See (7) and the preceding discussion in Section 15.5 [ET 14.5].

13. (a) See Definition 15.6.2 [ET 14.6.2]. We can interpret it as the rate of change of f at (x_0, y_0) in the direction of $\mathbf{u}$. Geometrically, if P is the point $(x_0, y_0, f(x_0, y_0))$ on the graph of f and C is the curve of intersection of the graph of f with the vertical plane that passes through P in the direction $\mathbf{u}$, the directional derivative of f at (x_0, y_0) in the direction of $\mathbf{u}$ is the slope of the tangent line to C at P. (See Figure 5 in Section 15.6 [ET 14.6].)

(b) See Theorem 15.6.3 [ET 14.6.3].

14. (a) See (8) and (13) in Section 15.6 [ET 14.6].

(b) $D_{\mathbf{u}} f(x, y) = \nabla f(x, y) \cdot \mathbf{u}$ or $D_{\mathbf{u}} f(x, y, z) = \nabla f(x, y, z) \cdot \mathbf{u}$

(c) The gradient vector of a function points in the direction of maximum rate of increase of the function. On a graph of the function, the gradient points in the direction of steepest ascent.

15. (a) f has a local maximum at (a, b) if $f(x, y) \leq f(a, b)$ when (x, y) is near (a, b).

(b) f has an absolute maximum at (a, b) if $f(x, y) \leq f(a, b)$ for all points (x, y) in the domain of f.

(c) f has a local minimum at (a, b) if $f(x, y) \geq f(a, b)$ when (x, y) is near (a, b).

(d) f has an absolute minimum at (a, b) if $f(x, y) \geq f(a, b)$ for all points (x, y) in the domain of f.

(e) f has a saddle point at (a, b) if $f(a, b)$ is a local maximum in one direction but a local minimum in another.

16. (a) By Theorem 15.7.2 [ET 14.7.2], if f has a local maximum at (a, b) and the first-order partial derivatives of f exist there, then $f_x(a, b) = 0$ and $f_y(a, b) = 0$.

(b) A critical point of f is a point (a, b) such that $f_x(a, b) = 0$ and $f_y(a, b) = 0$ or one of these partial derivatives does not exist.

17. See (3) in Section 15.7 [ET 14.7].

18. (a) See Figure 11 and the accompanying discussion in Section 15.7 [ET 14.7].

(b) See Theorem 15.7.8 [ET 14.7.8].

(c) See the procedure outlined in (9) in Section 15.7 [ET 14.7].

19. See the discussion beginning on page 985 [ET 951]; see the discussion preceding Example 5 in Section 15.8 [ET 14.8].

TRUE-FALSE QUIZ

1. True. $f_y(a, b) = \lim\limits_{h \to 0} \dfrac{f(a, b + h) - f(a, b)}{h}$ from Equation 15.3.3 [ET 14.3.3]. Let $h = y - b$. As $h \to 0$,

$y \to b$. Then by substituting, we get $f_y(a, b) = \lim\limits_{y \to b} \dfrac{f(a, y) - f(a, b)}{y - b}$.

3. False. $f_{xy} = \dfrac{\partial^2 f}{\partial y \partial x}$.

5. False. See Example 15.2.3 [ET 14.2.3].

7. True. If f has a local minimum and f is differentiable at (a, b) then by Theorem 15.7.2 [ET 14.7.2], $f_x(a, b) = 0$ and $f_y(a, b) = 0$, so $\nabla f(a, b) = \langle f_x(a, b), f_y(a, b) \rangle = \langle 0, 0 \rangle = \mathbf{0}$.

9. False. $\nabla f(x, y) = \langle 0, 1/y \rangle$.

11. True. $\nabla f = \langle \cos x, \cos y \rangle$, so $|\nabla f| = \sqrt{\cos^2 x + \cos^2 y}$. But $|\cos \theta| \le 1$, so $|\nabla f| \le \sqrt{2}$. Now $D_{\mathbf{u}} f(x, y) = \nabla f \cdot \mathbf{u} = |\nabla f| |\mathbf{u}| \cos \theta$, but $\mathbf{u}$ is a unit vector, so $|D_{\mathbf{u}} f(x, y)| \le \sqrt{2} \cdot 1 \cdot 1 = \sqrt{2}$.

EXERCISES

1. The domain of $\sin^{-1} x$ is $-1 \le x \le 1$ while the domain of $\tan^{-1} y$ is all real numbers, so the domain of

$f(x, y) = \sin^{-1} x + \tan^{-1} y$ is $\{(x, y) \mid -1 \le x \le 1\}$.

3. $z = f(x, y) = 1 - x^2 - y^2$, a paraboloid with

vertex $(0, 0, 1)$.

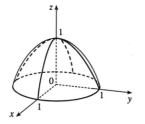

5. Let $k = e^{-c} = e^{-(x^2 + y^2)}$ be the level curves. Then $-\ln k = c = x^2 + y^2$, so we have a family of concentric circles.

7. f is a rational function, so it is continuous on its domain. Since f is defined at $(1, 1)$, we use direct substitution to

evaluate the limit: $\lim\limits_{(x,y) \to (1,1)} \dfrac{2xy}{x^2 + 2y^2} = \dfrac{2(1)(1)}{1^2 + 2(1)^2} = \dfrac{2}{3}$.

9. (a) $T_x(6,4) = \lim\limits_{h \to 0} \dfrac{T(6+h,4) - T(6,4)}{h}$, so we can approximate $T_x(6,4)$ by considering $h = \pm 2$ and using

the values given in the table: $T_x(6,4) \approx \dfrac{T(8,4) - T(6,4)}{2} = \dfrac{86 - 80}{2} = 3$,

$T_x(6,4) \approx \dfrac{T(4,4) - T(6,4)}{-2} = \dfrac{72 - 80}{-2} = 4$. Averaging these values, we estimate $T_x(6,4)$ to be

approximately $3.5\,°\mathrm{C}/\mathrm{m}$. Similarly, $T_y(6,4) = \lim\limits_{h \to 0} \dfrac{T(6,4+h) - T(6,4)}{h}$, which we can

approximate with $h = \pm 2$: $T_y(6,4) \approx \dfrac{T(6,6) - T(6,4)}{2} = \dfrac{75 - 80}{2} = -2.5$,

$T_y(6,4) \approx \dfrac{T(6,2) - T(6,4)}{-2} = \dfrac{87 - 80}{-2} = -3.5$. Averaging these values, we estimate $T_y(6,4)$ to be

approximately $-3.0\,°\mathrm{C}/\mathrm{m}$.

(b) Here $\mathbf{u} = \left\langle \frac{1}{\sqrt{2}}, \frac{1}{\sqrt{2}} \right\rangle$, so by Equation 15.6.9 [ET 14.6.9],

$D_{\mathbf{u}}T(6,4) = \nabla T(6,4) \cdot \mathbf{u} = T_x(6,4)\,\frac{1}{\sqrt{2}} + T_y(6,4)\,\frac{1}{\sqrt{2}}$. Using our estimates from part (a), we have

$D_{\mathbf{u}}T(6,4) \approx (3.5)\,\frac{1}{\sqrt{2}} + (-3.0)\,\frac{1}{\sqrt{2}} = \frac{1}{2\sqrt{2}} \approx 0.35$. This means that as we move through the point $(6,4)$ in

the direction of $\mathbf{u}$, the temperature increases at a rate of approximately $0.35\,°\mathrm{C}/\mathrm{m}$.

Alternatively, we can use Definition 15.6.2 [ET 14.6.2]:

$D_{\mathbf{u}}T(6,4) = \lim\limits_{h \to 0} \dfrac{T\left(6 + h\frac{1}{\sqrt{2}}, 4 + h\frac{1}{\sqrt{2}}\right) - T(6,4)}{h}$, which we can estimate with $h = \pm 2\sqrt{2}$. Then

$D_{\mathbf{u}}T(6,4) \approx \dfrac{T(8,6) - T(6,4)}{2\sqrt{2}} = \dfrac{80 - 80}{2\sqrt{2}} = 0$, $D_{\mathbf{u}}T(6,4) \approx \dfrac{T(4,2) - T(6,4)}{-2\sqrt{2}} = \dfrac{74 - 80}{-2\sqrt{2}} = \dfrac{3}{\sqrt{2}}$.

Averaging these values, we have $D_{\mathbf{u}}T(6,4) \approx \frac{3}{2\sqrt{2}} \approx 1.1\,°\mathrm{C}/\mathrm{m}$.

(c) $T_{xy}(x,y) = \dfrac{\partial}{\partial y}[T_x(x,y)] = \lim\limits_{h \to 0} \dfrac{T_x(x, y+h) - T_x(x,y)}{h}$, so $T_{xy}(6,4) = \lim\limits_{h \to 0} \dfrac{T_x(6, 4+h) - T_x(6,4)}{h}$

which we can estimate with $h = \pm 2$. We have $T_x(6,4) \approx 3.5$ from part (a), but we will also need values for

$T_x(6,6)$ and $T_x(6,2)$. If we use $h = \pm 2$ and the values given in the table, we have

$T_x(6,6) \approx \dfrac{T(8,6) - T(6,6)}{2} = \dfrac{80 - 75}{2} = 2.5$, $T_x(6,6) \approx \dfrac{T(4,6) - T(6,6)}{-2} = \dfrac{68 - 75}{-2} = 3.5$.

Averaging these values, we estimate $T_x(6,6) \approx 3.0$. Similarly,

$T_x(6,2) \approx \dfrac{T(8,2) - T(6,2)}{2} = \dfrac{90 - 87}{2} = 1.5$, $T_x(6,2) \approx \dfrac{T(4,2) - T(6,2)}{-2} = \dfrac{74 - 87}{-2} = 6.5$.

Averaging these values, we estimate $T_x(6,2) \approx 4.0$. Finally, we estimate $T_{xy}(6,4)$:

$T_{xy}(6,4) \approx \dfrac{T_x(6,6) - T_x(6,4)}{2} = \dfrac{3.0 - 3.5}{2} = -0.25$,

$T_{xy}(6,4) \approx \dfrac{T_x(6,2) - T_x(6,4)}{-2} = \dfrac{4.0 - 3.5}{-2} = -0.25$. Averaging these values, we have

$T_{xy}(6,4) \approx -0.25$.

11. $f(x,y) = 3x^4 - x\sqrt{y} \quad \Rightarrow \quad f_x = 12x^3 - \sqrt{y},\ f_y = -\frac{1}{2}xy^{-1/2}$

13. $f(s,t) = e^{2s}\cos \pi t \quad \Rightarrow \quad f_s = 2e^{2s}\cos \pi t,\ f_t = -\pi e^{2s}\sin \pi t$

15. $f(x,y,z) = xy^z \quad \Rightarrow \quad f_x = y^z,\ f_y = xzy^{z-1},\ f_z = xy^z \ln y$

17. $f(x,y) = x^2 y^3 - 2x^4 + y^2 \quad \Rightarrow \quad f_x = 2xy^3 - 8x^3,\ f_y = 3x^2 y^2 + 2y,\ f_{xx} = 2y^3 - 24x^2,\ f_{yy} = 6x^2 y + 2$,

and $f_{xy} = f_{yx} = 6xy^2$.

19. $f(x, y, z) = xy^2z^3 \Rightarrow f_x = y^2z^3, f_y = 2xyz^3, f_z = 3xy^2z^2, f_{xx} = 0, f_{yy} = 2xz^3, f_{zz} = 6xy^2z,$
$f_{xy} = f_{yx} = 2yz^3, f_{xz} = f_{zx} = 3y^2z^2,$ and $f_{yz} = f_{zy} = 6xyz^2.$

21. $u = x^y \Rightarrow u_x = yx^{y-1}, u_y = x^y \ln x$ and $(x/y) u_x + (\ln x)^{-1} u_y = x^y + x^y = 2u.$

23. $z_x(0, 1) = 0, z_y(0, 1) = 6$ and an equation of the tangent plane is $z - 5 = 6(y - 1)$ or $z - 6y = -1.$

25. $F(x, y, z) = xy^2z^3, \nabla F = \langle y^2z^3, 2xyz^3, 3xy^2z^2 \rangle$ and $\nabla F(3, 2, 1) = \langle 4, 12, 36 \rangle.$ Thus an equation of the
tangent plane is $4(x - 3) + 12(y - 2) + 36(z - 1) = 0$ or $x + 3y + 9z = 18.$

27. $F(x, y, z) = x^2 + 2y^2 - 3z^2, F_x = 2x, F_y = 4y, F_z = -6z; F_x(3, 2, -1) = 6, F_y(3, 2, -1) = 8,$
$F_z(3, 2, -1) = 6.$ So an equation of the tangent plane is $6(x - 3) + 8(y - 2) + 6(z + 1) = 0$ or
$3x + 4y + 3z = 14.$

29. $F(x, y, z) = x^2 + y^2 + z^2, \nabla F(x_0, y_0, z_0) = \langle 2x_0, 2y_0, 2z_0 \rangle = k\langle 2, 1, -3 \rangle$ or $x_0 = k, y_0 = \frac{1}{2}k$ and
$z_0 = -\frac{3}{2}k.$ But $x_0^2 + y_0^2 + z_0^2 = 1,$ so $\frac{7}{2}k^2 = 1$ and $k = \pm\sqrt{\frac{2}{7}}.$ Hence there are two such points:
$\left(\pm\sqrt{\frac{2}{7}}, \pm\frac{1}{\sqrt{14}}, \mp\frac{3}{\sqrt{14}} \right).$

31. $f(x, y, z) = x^3\sqrt{y^2 + z^2} \Rightarrow f_x(x, y, z) = 3x^2\sqrt{y^2 + z^2}, f_y(x, y, z) = \dfrac{yx^3}{\sqrt{y^2 + z^2}},$ and

$f_z(x, y, z) = \dfrac{zx^3}{\sqrt{y^2 + z^2}},$ so $f(2, 3, 4) = 8(5) = 40, f_x(2, 3, 4) = 3(4)\sqrt{25} = 60, f_y(2, 3, 4) = \dfrac{3(8)}{\sqrt{25}} = \dfrac{24}{5},$

and $f_z(2, 3, 4) = \dfrac{4(8)}{\sqrt{25}} = \dfrac{32}{5}.$ Then the linear approximation of f at $(2, 3, 4)$ is

$$f(x, y, z) \approx f(2, 3, 4) + f_x(2, 3, 4)(x - 2) + f_y(2, 3, 4)(y - 3) + f_z(2, 3, 4)(z - 4)$$
$$= 40 + 60(x - 2) + \tfrac{24}{5}(y - 3) + \tfrac{32}{5}(z - 4) = 60x + \tfrac{24}{5}y + \tfrac{32}{5}z - 120$$

Then

$$(1.98)^3\sqrt{(3.01)^2 + (3.97)^2} = f(1.98, 3.01, 3.97) \approx 60(1.98) + \tfrac{24}{5}(3.01) + \tfrac{32}{5}(3.97) - 120$$
$$= 38.656$$

33. $\dfrac{dw}{dt} = \dfrac{1}{2\sqrt{x}}(2e^{2t}) + \dfrac{2y}{z}(3t^2 + 4) + \dfrac{-y^2}{z^2}(2t) = e^t + \dfrac{2y}{z}(3t^2 + 4) - 2t\dfrac{y^2}{z^2}$

35. By the Chain Rule, $\dfrac{\partial z}{\partial s} = \dfrac{\partial z}{\partial x}\dfrac{\partial x}{\partial s} + \dfrac{\partial z}{\partial y}\dfrac{\partial y}{\partial s}.$ When $s = 1$ and $t = 2, x = g(1, 2) = 3$ and $y = h(1, 2) = 6,$ so

$\dfrac{\partial z}{\partial s} = f_x(3, 6) g_s(1, 2) + f_y(3, 6) h_s(1, 2) = (7)(-1) + (8)(-5) = -47.$ Similarly, $\dfrac{\partial z}{\partial t} = \dfrac{\partial z}{\partial x}\dfrac{\partial x}{\partial t} + \dfrac{\partial z}{\partial y}\dfrac{\partial y}{\partial t},$

so $\dfrac{\partial z}{\partial t} = f_x(3, 6) g_t(1, 2) + f_y(3, 6) h_t(1, 2) = (7)(4) + (8)(10) = 108.$

37. $\dfrac{\partial z}{\partial x} = 2xf'(x^2 - y^2), \dfrac{\partial z}{\partial y} = 1 - 2yf'(x^2 - y^2)\ \left[\text{where } f' = \dfrac{df}{d(x^2 - y^2)} \right].$ Then

$y\dfrac{\partial z}{\partial x} + x\dfrac{\partial z}{\partial y} = 2xyf'(x^2 - y^2) + x - 2xyf'(x^2 - y^2) = x.$

39. $\dfrac{\partial z}{\partial x} = \dfrac{\partial z}{\partial u}y + \dfrac{\partial z}{\partial v}\dfrac{-y}{x^2}$ and

$$\frac{\partial^2 z}{\partial x^2} = y\frac{\partial}{\partial x}\left(\frac{\partial z}{\partial u}\right) + \frac{2y}{x^3}\frac{\partial z}{\partial v} + \frac{-y}{x^2}\frac{\partial}{\partial x}\left(\frac{\partial z}{\partial v}\right)$$

$$= \frac{2y}{x^3}\frac{\partial z}{\partial v} + y\left(\frac{\partial^2 z}{\partial u^2}y + \frac{\partial^2 z}{\partial v\partial u}\frac{-y}{x^2}\right) + \frac{-y}{x^2}\left(\frac{\partial^2 z}{\partial v^2}\frac{-y}{x^2} + \frac{\partial^2 z}{\partial u\partial v}y\right)$$

$$= \frac{2y}{x^3}\frac{\partial z}{\partial v} + y^2\frac{\partial^2 z}{\partial u^2} - \frac{2y^2}{x^2}\frac{\partial^2 z}{\partial u\partial v} + \frac{y^2}{x^4}\frac{\partial^2 z}{\partial v^2}$$

Also $\dfrac{\partial z}{\partial y} = x\dfrac{\partial z}{\partial u} + \dfrac{1}{x}\dfrac{\partial z}{\partial v}$ and

$$\frac{\partial^2 z}{\partial y^2} = x\frac{\partial}{\partial y}\left(\frac{\partial z}{\partial u}\right) + \frac{1}{x}\frac{\partial}{\partial y}\left(\frac{\partial z}{\partial v}\right) = x\left(\frac{\partial^2 z}{\partial u^2}x + \frac{\partial^2 z}{\partial v\partial u}\frac{1}{x}\right) + \frac{1}{x}\left(\frac{\partial^2 z}{\partial v^2}\frac{1}{x} + \frac{\partial^2 z}{\partial u\partial v}x\right)$$

$$= x^2\frac{\partial^2 z}{\partial u^2} + 2\frac{\partial^2 z}{\partial u\partial v} + \frac{1}{x^2}\frac{\partial^2 z}{\partial v^2}$$

Thus

$$x^2\frac{\partial^2 z}{\partial x^2} - y^2\frac{\partial^2 z}{\partial y^2} = \frac{2y}{x}\frac{\partial z}{\partial v} + x^2y^2\frac{\partial^2 z}{\partial u^2} - 2y^2\frac{\partial^2 z}{\partial u\partial v} + \frac{y^2}{x^2}\frac{\partial^2 z}{\partial v^2} - x^2y^2\frac{\partial^2 z}{\partial u^2} - 2y^2\frac{\partial^2 z}{\partial u\partial v} - \frac{y^2}{x^2}\frac{\partial^2 z}{\partial v^2}$$

$$= \frac{2y}{x}\frac{\partial z}{\partial v} - 4y^2\frac{\partial^2 z}{\partial u\partial v} = 2v\frac{\partial z}{\partial v} - 4uv\frac{\partial^2 z}{\partial u\partial v}$$

since $y = xv = \dfrac{uv}{y}$ or $y^2 = uv$.

41. $\nabla f = \left\langle z^2\sqrt{y}e^{x\sqrt{y}}, \dfrac{xz^2 e^{x\sqrt{y}}}{2\sqrt{y}}, 2ze^{x\sqrt{y}}\right\rangle = ze^{x\sqrt{y}}\left\langle z\sqrt{y}, \dfrac{xz}{2\sqrt{y}}, 2\right\rangle$

43. $\nabla f = \langle 1/\sqrt{x}, -2y\rangle$, $\nabla f(1,5) = \langle 1, -10\rangle$, $\mathbf{u} = \frac{1}{5}\langle 3, -4\rangle$. Then $D_{\mathbf{u}}f(1,5) = \frac{43}{5}$.

45. $\nabla f = \langle 2xy, x^2 + 1/(2\sqrt{y})\rangle$, $|\nabla f(2,1)| = |\langle 4, \frac{9}{2}\rangle|$. Thus the maximum rate of change of f at $(2,1)$ is $\frac{\sqrt{145}}{2}$ in the direction $\langle 4, \frac{9}{2}\rangle$.

47. First we draw a line passing through Homestead and the eye of the hurricane. We can approximate the directional derivative at Homestead in the direction of the eye of the hurricane by the average rate of change of wind speed between the points where this line intersects the contour lines closest to Homestead. In the direction of the eye of the hurricane, the wind speed changes from 45 to 50 knots. We estimate the distance between these two points to be approximately 8 miles, so the rate of change of wind speed in the direction given is approximately $\frac{50-45}{8} = \frac{5}{8} = 0.625$ knots/mi.

49. $f(x,y) = x^2 - xy + y^2 + 9x - 6y + 10 \Rightarrow f_x = 2x - y + 9$, $f_y = -x + 2y - 6$, $f_{xx} = 2 = f_{yy}$, $f_{xy} = -1$. Then $f_x = 0$ and $f_y = 0$ imply $y = 1$, $x = -4$. Thus the only critical point is $(-4, 1)$ and $f_{xx}(-4,1) > 0$, $D(-4,1) = 3 > 0$, so $f(-4,1) = -11$ is a local minimum.

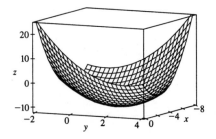

51. $f(x, y) = 3xy - x^2y - xy^2 \Rightarrow f_x = 3y - 2xy - y^2$,

$f_y = 3x - x^2 - 2xy$, $f_{xx} = -2y$, $f_{yy} = -2x$, $f_{xy} = 3 - 2x - 2y$. $\frac{1}{n}$ Then

$f_x = 0$ implies $y(3 - 2x - y) = 0$ so $y = 0$ or $y = 3 - 2x$. Substituting

into $f_y = 0$ implies $x(3 - x) = 0$ or $3x(-1 + x) = 0$. Hence the critical

points are $(0, 0)$, $(3, 0)$, $(0, 3)$ and $(1, 1)$.

$D(0, 0) = D(3, 0) = D(0, 3) = -9 < 0$ so $(0, 0)$, $(3, 0)$, and $(0, 3)$ are

saddle points. $D(1, 1) = 3 > 0$ and $f_{xx}(1, 1) = -2 < 0$, so $f(1, 1) = 1$ is

a local maximum.

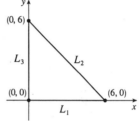

53. First solve inside D. Here $f_x = 4y^2 - 2xy^2 - y^3$,

$f_y = 8xy - 2x^2y - 3xy^2$. Then $f_x = 0$ implies $y = 0$ or $y = 4 - 2x$,

but $y = 0$ isn't inside D. Substituting $y = 4 - 2x$ into $f_y = 0$ implies

$x = 0$, $x = 2$ or $x = 1$, but $x = 0$ isn't inside D, and when $x = 2$,

$y = 0$ but $(2, 0)$ isn't inside D. Thus the only critical point inside D

is $(1, 2)$ and $f(1, 2) = 4$. Secondly we consider the boundary of D.

On L_1, $f(x, 0) = 0$ and so $f = 0$ on L_1. On L_2, $x = -y + 6$ and

$f(-y + 6, y) = y^2(6 - y)(-2) = -2(6y^2 - y^3)$ which has

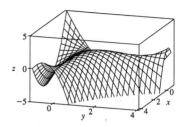

critical points at $y = 0$ and $y = 4$. Then $f(6, 0) = 0$ while $f(2, 4) = -64$. On L_3, $f(0, y) = 0$, so $f = 0$ on L_3.

Thus on D the absolute maximum of f is $f(1, 2) = 4$ while the absolute minimum is $f(2, 4) = -64$.

55. $f(x, y) = x^3 - 3x + y^4 - 2y^2$

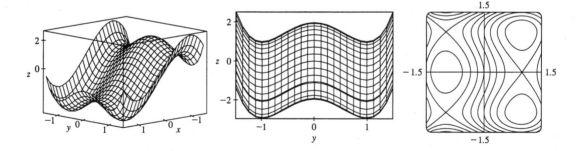

From the graphs, it appears that f has a local maximum $f(-1, 0) \approx 2$, local minima $f(1, \pm 1) \approx -3$, and saddle

points at $(-1, \pm 1)$ and $(1, 0)$.

To find the exact quantities, we calculate $f_x = 3x^2 - 3 = 0 \Leftrightarrow x = \pm 1$ and $f_y = 4y^3 - 4y = 0 \Leftrightarrow$

$y = 0, \pm 1$, giving the critical points estimated above. Also $f_{xx} = 6x$, $f_{xy} = 0$, $f_{yy} = 12y^2 - 4$, so using the

Second Derivatives Test, $D(-1, 0) = 24 > 0$ and $f_{xx}(-1, 0) = -6 < 0$ indicating a local maximum

$f(-1, 0) = 2$; $D(1, \pm 1) = 48 > 0$ and $f_{xx}(1, \pm 1) = 6 > 0$ indicating local minima $f(1, \pm 1) = -3$; and

$D(-1, \pm 1) = -48$ and $D(1, 0) = -24$, indicating saddle points.

57. $f(x,y) = x^2 y$, $g(x,y) = x^2 + y^2 = 1$ $\Rightarrow$ $\nabla f = \langle 2xy, x^2 \rangle = \lambda \nabla g = \langle 2\lambda x, 2\lambda y \rangle$. Then $2xy = 2\lambda x$ and $x^2 = 2\lambda y$ imply $\lambda = x^2/(2y)$ and $\lambda = y$ if $x \neq 0$ and $y \neq 0$. Hence $x^2 = 2y^2$. Then $x^2 + y^2 = 1$ implies $3y^2 = 1$ so $y = \pm\frac{1}{\sqrt{3}}$ and $x = \pm\sqrt{\frac{2}{3}}$. [Note if $x = 0$ then $x^2 = 2\lambda y$ implies $y = 0$ and $f(0,0) = 0$.] Thus the possible points are $\left(\pm\sqrt{\frac{2}{3}}, \pm\frac{1}{\sqrt{3}}\right)$ and the absolute maxima are $f\left(\pm\sqrt{\frac{2}{3}}, \frac{1}{\sqrt{3}}\right) = \frac{2}{3\sqrt{3}}$ while the absolute minima are $f\left(\pm\sqrt{\frac{2}{3}}, -\frac{1}{\sqrt{3}}\right) = -\frac{2}{3\sqrt{3}}$.

59. $f(x,y,z) = x + y + z$, $g(x,y,z) = 1/x + 1/y + 1/z = 1$ $\Rightarrow$ $\nabla f = \langle 1,1,1 \rangle = \lambda \nabla g = \langle -\lambda x^{-2}, -\lambda y^{-2}, -\lambda z^{-2} \rangle$. Thus $\lambda = -x^2 = -y^2 = -z^2$ or $y = \pm x$, $z = \pm x$. Substituting into $1/x + 1/y + 1/z = 1$ gives (1) $3/x = 1$ so $x = 3$, or (2) $1/x = 1$ so $x = 1$, or (3) $-1/x = 1$ so $x = -1$ with the associated points (1) $(3,3,3)$, (2) $(1,1,-1)$ or $(1,-1,1)$, (3) $(-1,1,1)$. Thus the absolute maximum is $f(3,3,3) = 9$ and the absolute minimum is $f(1,1,-1) = f(1,-1,1) = f(-1,1,1) = 1$.

61. $f(x,y,z) = x^2 + y^2 + z^2$, $g(x,y,z) = xy^2 z^3 = 2$ $\Rightarrow$ $\nabla f = \langle 2x, 2y, 2z \rangle = \lambda \nabla g = \langle \lambda y^2 z^3, 2\lambda xyz^3, 3\lambda xy^2 z^2 \rangle$. Since $xy^2 z^3 = 2$, $x \neq 0$, $y \neq 0$ and $z \neq 0$, so (1) $2x = \lambda y^2 z^3$, (2) $1 = \lambda xz^3$, (3) $2 = 3\lambda xy^2 z$. Then (2) and (3) imply $\frac{1}{xz^3} = \frac{2}{3xy^2 z}$ or $y^2 = \frac{2}{3}z^2$ so $y = \pm z\sqrt{\frac{2}{3}}$. Similarly (1) and (3) imply $\frac{2x}{y^2 z^3} = \frac{2}{3xy^2 z}$ or $3x^2 = z^2$ so $x = \pm\frac{1}{\sqrt{3}}z$. But $xy^2 z^3 = 2$ so x and z must have the same sign, that is, $x = \frac{1}{\sqrt{3}}z$. Thus $g(x,y,z) = 2$ implies $\frac{1}{\sqrt{3}}z\left(\frac{2}{3}z^2\right)z^3 = 2$ or $z = \pm 3^{1/4}$ and the possible points are $\left(\pm 3^{-1/4}, 3^{-1/4}\sqrt{2}, \pm 3^{1/4}\right)$, $\left(\pm 3^{-1/4}, -3^{-1/4}\sqrt{2}, \pm 3^{1/4}\right)$. However at each of these points f takes on the same value, $2\sqrt{3}$. But $(2,1,1)$ also satisfies $g(x,y,z) = 2$ and $f(2,1,1) = 6 > 2\sqrt{3}$. Thus f has an absolute minimum value of $2\sqrt{3}$ and no absolute maximum subject to the constraint $xy^2 z^3 = 2$.

Alternate Solution: $g(x,y,z) = xy^2 z^3 = 2$ implies $y^2 = \frac{2}{xz^3}$, so minimize $f(x,z) = x^2 + \frac{2}{xz^3} + z^2$. Then $f_x = 2x - \frac{2}{x^2 z^3}$, $f_z = -\frac{6}{xz^4} + 2z$, $f_{xx} = 2 + \frac{4}{x^3 z^3}$, $f_{zz} = \frac{24}{xz^5} + 2$ and $f_{xz} = \frac{6}{x^2 z^4}$. Now $f_x = 0$ implies $2x^3 z^3 - 2 = 0$ or $z = 1/x$. Substituting into $f_y = 0$ implies $-6x^3 + 2x^{-1} = 0$ or $x = \frac{1}{\sqrt[4]{3}}$, so the two critical points are $\left(\pm\frac{1}{\sqrt[4]{3}}, \pm\sqrt[4]{3}\right)$. Then $D\left(\pm\frac{1}{\sqrt[4]{3}}, \pm\sqrt[4]{3}\right) = (2+4)\left(2+\frac{24}{3}\right) - \left(\frac{6}{\sqrt{3}}\right)^2 > 0$ and $f_{xx}\left(\pm\frac{1}{\sqrt[4]{3}}, \pm\sqrt[4]{3}\right) = 6 > 0$, so each point is a minimum. Finally, $y^2 = \frac{2}{xz^3}$, so the four points closest to the origin are $\left(\pm\frac{1}{\sqrt[4]{3}}, \frac{\sqrt{2}}{\sqrt[4]{3}}, \pm\sqrt[4]{3}\right)$, $\left(\pm\frac{1}{\sqrt[4]{3}}, -\frac{\sqrt{2}}{\sqrt[4]{3}}, \pm\sqrt[4]{3}\right)$.

63.

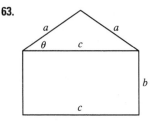

The area of the triangle is $\frac{1}{2}ca\sin\theta$ and the area of the rectangle is bc. Thus, the area of the whole object is $f(a,b,c) = \frac{1}{2}ca\sin\theta + bc$. The perimeter of the object is $g(a,b,c) = 2a + 2b + c = P$. To simplify $\sin\theta$ in terms of a, b, and c notice that $a^2\sin^2\theta + \left(\frac{1}{2}c\right)^2 = a^2$ $\Rightarrow$

$\sin\theta = \frac{1}{2a}\sqrt{4a^2 - c^2}$. Thus $f(a,b,c) = \frac{c}{4}\sqrt{4a^2 - c^2} + bc$.

(Instead of using θ, we could just have used the Pythagorean Theorem.) As a result, by Lagrange's method, we must find a, b, c, and λ by solving $\nabla f = \lambda \nabla g$ which gives the following equations: (1) $ca\left(4a^2 - c^2\right)^{-1/2} = 2\lambda$, (2) $c = 2\lambda$, (3) $\frac{1}{4}\left(4a^2 - c^2\right)^{1/2} - \frac{1}{4}c^2\left(4a^2 - c^2\right)^{-1/2} + b = \lambda$, and (4) $2a + 2b + c = P$. From (2), $\lambda = \frac{1}{2}c$ and so (1) produces $ca\left(4a^2 - c^2\right)^{-1/2} = c \;\Rightarrow\; \left(4a^2 - c^2\right)^{1/2} = a \;\Rightarrow\; 4a^2 - c^2 = a^2 \;\Rightarrow\;$ (5) $c = \sqrt{3}a$.

Similarly, since $\left(4a^2 - c^2\right)^{1/2} = a$ and $\lambda = \frac{1}{2}c$, (3) gives $\frac{a}{4} - \frac{c^2}{4a} + b = \frac{c}{2}$, so from (5), $\frac{a}{4} - \frac{3a}{4} + b = \frac{\sqrt{3}a}{2}$

$\Rightarrow\; -\frac{a}{2} - \frac{\sqrt{3}a}{2} = -b \;\Rightarrow\;$ (6) $b = \frac{a}{2}\left(1 + \sqrt{3}\right)$. Substituting (5) and (6) into (4) we get:

$2a + a\left(1 + \sqrt{3}\right) + \sqrt{3}a = P \;\Rightarrow\; 3a + 2\sqrt{3}a = P \;\Rightarrow\; a = \dfrac{P}{3 + 2\sqrt{3}} = \dfrac{2\sqrt{3} - 3}{3}P$ and thus

$b = \dfrac{\left(2\sqrt{3} - 3\right)\left(1 + \sqrt{3}\right)}{6}P = \dfrac{3 - \sqrt{3}}{6}P$ and $c = \left(2 - \sqrt{3}\right)P$.

Problems Plus

1. The areas of the smaller rectangles are $A_1 = xy$,

$A_2 = (L - x)y$, $A_3 = (L - x)(W - y)$, $A_4 = x(W - y)$. For

$0 \le x \le L$, $0 \le y \le W$, let

$$f(x, y) = A_1^2 + A_2^2 + A_3^2 + A_4^2$$

$$= x^2 y^2 + (L - x)^2 y^2 + (L - x)^2 (W - y)^2$$

$$+ x^2 (W - y)^2$$

$$= \left[x^2 + (L - x)^2 \right] \left[y^2 + (W - y)^2 \right]$$

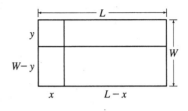

Then we need to find the maximum and minimum values of $f(x, y)$. Here

$f_x(x, y) = [2x - 2(L - x)] \left[y^2 + (W - y)^2 \right] = 0 \quad \Rightarrow \quad 4x - 2L = 0$ or $x = \frac{1}{2}L$, and

$f_y(x, y) = \left[x^2 + (L - x)^2 \right] [2y - 2(W - y)] = 0 \quad \Rightarrow \quad 4y - 2W = 0$ or $y = W/2$. Also

$f_{xx} = 4 \left[y^2 + (W - y)^2 \right]$, $f_{yy} = 4 \left[x^2 + (L - x)^2 \right]$, and $f_{xy} = (4x - 2L)(4y - 2W)$. Then

$D = 16 \left[y^2 + (W - y)^2 \right] \left[x^2 + (L - x)^2 \right] - (4x - 2L)^2 (4y - 2W)^2$. Thus when $x = \frac{1}{2}L$ and $y = \frac{1}{2}W$,

$D > 0$ and $f_{xx} = 2W^2 > 0$. Thus a minimum of f occurs at $\left(\frac{1}{2}L, \frac{1}{2}W \right)$ and this minimum value is

$f \left(\frac{1}{2}L, \frac{1}{2}W \right) = \frac{1}{4}L^2 W^2$. There are no other critical points, so the maximum must occur on the boundary. Now

along the width of the rectangle let $g(y) = f(0, y) = f(L, y) = L^2 \left[y^2 + (W - y)^2 \right]$, $0 \le y \le W$. Then

$g'(y) = L^2 [2y - 2(W - y)] = 0 \quad \Leftrightarrow \quad y = \frac{1}{2}W$. And $g \left(\frac{1}{2} \right) = \frac{1}{2}L^2 W^2$. Checking the

endpoints, we get $g(0) = g(W) = L^2 W^2$. Along the length of the rectangle let

$h(x) = f(x, 0) = f(x, W) = W^2 \left[x^2 + (L - x)^2 \right]$, $0 \le x \le L$. By symmetry $h'(x) = 0 \quad \Leftrightarrow \quad x = \frac{1}{2}L$ and

$h \left(\frac{1}{2}L \right) = \frac{1}{2}L^2 W^2$. At the endpoints we have $h(0) = h(L) = L^2 W^2$. Therefore $L^2 W^2$ is the maximum value of

f. This maximum value of f occurs when the "cutting" lines correspond to sides of the rectangle.

3. (a) The area of a trapezoid is $\frac{1}{2}h(b_1 + b_2)$, where h is the height (the distance between the two parallel sides) and

b_1, b_2 are the lengths of the bases (the parallel sides). From the figure in the text, we see that $h = x \sin \theta$,

$b_1 = w - 2x$, and $b_2 = w - 2x + 2x \cos \theta$. Therefore the cross-sectional area of the rain gutter is

$$A(x, \theta) = \frac{1}{2} x \sin \theta \left[(w - 2x) + (w - 2x + 2x \cos \theta) \right] = (x \sin \theta)(w - 2x + x \cos \theta)$$

$$= wx \sin \theta - 2x^2 \sin \theta + x^2 \sin \theta \cos \theta, \ 0 < x \le \frac{1}{2}w, \ 0 < \theta \le \frac{\pi}{2}$$

We look for the critical points of A: $\partial A/\partial x = w\sin\theta - 4x\sin\theta + 2x\sin\theta\cos\theta$ and

$\partial A/\partial\theta = wx\cos\theta - 2x^2\cos\theta + x^2\left(\cos^2\theta - \sin^2\theta\right)$, so $\partial A/\partial x = 0 \;\Leftrightarrow\; \sin\theta\,(w - 4x + 2x\cos\theta) = 0$

$\Leftrightarrow\; \cos\theta = \dfrac{4x-w}{2x} = 2 - \dfrac{w}{2x}\;(0 < \theta \le \frac{\pi}{2}) \;\Rightarrow\; \sin\theta > 0)$. If, in addition, $\partial A/\partial\theta = 0$, then

$0 = wx\cos\theta - 2x^2\cos\theta + x^2\left(2\cos^2\theta - 1\right)$

$= wx\left(2 - \dfrac{w}{2x}\right) - 2x^2\left(2 - \dfrac{w}{2x}\right) + x^2\left[2\left(2 - \dfrac{w}{2x}\right)^2 - 1\right]$

$= 2wx - \tfrac{1}{2}w^2 - 4x^2 + wx + x^2\left[8 - \dfrac{4w}{x} + \dfrac{w^2}{2x^2} - 1\right] = -wx + 3x^2 = x\,(3x - w)$

Since $x > 0$, we must have $x = \tfrac{1}{3}w$, in which case $\cos\theta = \tfrac{1}{2}$, so $\theta = \tfrac{\pi}{3}$, $\sin\theta = \tfrac{\sqrt{3}}{2}$, $k = \tfrac{\sqrt{3}}{6}w$, $b_1 = \tfrac{1}{3}w$, $b_2 = \tfrac{2}{3}w$, and $A = \tfrac{\sqrt{3}}{12}w^2$. As in Example 15.7.6 [ET 14.7.6], we can argue from the physical nature of this problem that we have found a relative maximum of A. Now checking the boundary of A, let

$g\,(\theta) = A\,(w/2, \theta) = \tfrac{1}{2}w^2\sin\theta - \tfrac{1}{2}w^2\sin\theta + \tfrac{1}{4}w^2\sin\theta\cos\theta = \tfrac{1}{8}w^2\sin 2\theta,\; 0 < \theta \le \tfrac{\pi}{2}$. Clearly g is

maximized when $\sin 2\theta = 1$ in which case $A = \tfrac{1}{8}w^2$. Also along the line $\theta = \tfrac{\pi}{2}$, let

$h\,(x) = A\left(x, \tfrac{\pi}{2}\right) = wx - 2x^2,\, 0 < x < \tfrac{1}{2}w \;\Rightarrow\; h'\,(x) = w - 4x = 0 \;\Leftrightarrow\; x = \tfrac{1}{4}w$, and

$h\left(\tfrac{1}{4}w\right) = w\left(\tfrac{1}{4}w\right) - 2\left(\tfrac{1}{4}w\right)^2 = \tfrac{1}{8}w^2$. Since $\tfrac{1}{8}w^2 < \tfrac{\sqrt{3}}{12}w^2$, we conclude that the relative maximum found earlier was an absolute maximum.

(b) If the metal were bent into a semi-circular gutter of radius r, we would have $w = \pi r$ and

$A = \tfrac{1}{2}\pi r^2 = \tfrac{1}{2}\pi\left(\dfrac{w}{\pi}\right)^2 = \dfrac{w^2}{2\pi}$. Since $\dfrac{w^2}{2\pi} > \dfrac{\sqrt{3}w^2}{12}$, it *would* be better to bend the metal into a gutter with a semicircular cross-section.

5. Let $g\,(x, y) = xf\left(\dfrac{y}{x}\right)$. Then $g_x\,(x, y) = f\left(\dfrac{y}{x}\right) + xf'\left(\dfrac{y}{x}\right)\left(-\dfrac{y}{x^2}\right) = f\left(\dfrac{y}{x}\right) - \dfrac{y}{x}f'\left(\dfrac{y}{x}\right)$ and

$g_y\,(x, y) = xf'\left(\dfrac{y}{x}\right)\left(\dfrac{1}{x}\right) = f'\left(\dfrac{y}{x}\right)$. Thus the tangent plane at (x_0, y_0, z_0) on the surface has equation

$z - x_0 f\left(\dfrac{y_0}{x_0}\right) = \left[f\left(\dfrac{y_0}{x_0}\right) - y_0 x_0^{-1}f'\left(\dfrac{y_0}{x_0}\right)\right](x - x_0) + f'\left(\dfrac{y_0}{x_0}\right)(y - y_0) \;\Rightarrow$

$\left[f\left(\dfrac{y_0}{x_0}\right) - y_0 x_0^{-1}f'\left(\dfrac{y_0}{x_0}\right)\right]x + \left[f'\left(\dfrac{y_0}{x_0}\right)\right]y - z = 0$. But any plane whose equation is of the form

$ax + by + cz = 0$ passes through the origin. Thus the origin is the common point of intersection.

7. (a) $x = r\cos\theta,\, y = r\sin\theta,\, z = z$. Then $\dfrac{\partial u}{\partial r} = \dfrac{\partial u}{\partial x}\dfrac{\partial x}{\partial r} + \dfrac{\partial u}{\partial y}\dfrac{\partial y}{\partial r} + \dfrac{\partial u}{\partial z}\dfrac{\partial z}{\partial r} = \dfrac{\partial u}{\partial x}\cos\theta + \dfrac{\partial u}{\partial y}\sin\theta$ and

$\dfrac{\partial^2 u}{\partial r^2} = \cos\theta\left[\dfrac{\partial^2 u}{\partial x^2}\dfrac{\partial x}{\partial r} + \dfrac{\partial^2 u}{\partial y\partial x}\dfrac{\partial y}{\partial r} + \dfrac{\partial^2 u}{\partial z\partial x}\dfrac{\partial z}{\partial r}\right] + \sin\theta\left[\dfrac{\partial^2 u}{\partial y^2}\dfrac{\partial y}{\partial r} + \dfrac{\partial^2 u}{\partial x\partial y}\dfrac{\partial x}{\partial r} + \dfrac{\partial^2 u}{\partial z\partial y}\dfrac{\partial z}{\partial r}\right]$

$= \dfrac{\partial^2 u}{\partial x^2}\cos^2\theta + \dfrac{\partial^2 u}{\partial y^2}\sin^2\theta + 2\dfrac{\partial^2 u}{\partial y\partial x}\cos\theta\sin\theta$

Similarly $\dfrac{\partial u}{\partial \theta} = -\dfrac{\partial u}{\partial x} r \sin \theta + \dfrac{\partial u}{\partial y} r \cos \theta$ and

$$\dfrac{\partial^2 u}{\partial \theta^2} = \dfrac{\partial^2 u}{\partial x^2} r^2 \sin^2 \theta + \dfrac{\partial^2 u}{\partial y^2} r^2 \cos^2 \theta - 2 \dfrac{\partial^2 u}{\partial y \partial x} r^2 \sin \theta \cos \theta - \dfrac{\partial u}{\partial x} r \cos \theta - \dfrac{\partial u}{\partial y} r \sin \theta. \text{ So}$$

$$\dfrac{\partial^2 u}{\partial r^2} + \dfrac{1}{r} \dfrac{\partial u}{\partial r} + \dfrac{1}{r^2} \dfrac{\partial^2 u}{\partial \theta^2} + \dfrac{\partial^2 u}{\partial z^2}$$

$$= \dfrac{\partial^2 u}{\partial x^2} \cos^2 \theta + \dfrac{\partial^2 u}{\partial y^2} \sin^2 \theta + 2 \dfrac{\partial^2 u}{\partial y \partial x} \cos \theta \sin \theta + \dfrac{\partial u}{\partial x} \dfrac{\cos \theta}{r} + \dfrac{\partial u}{\partial y} \dfrac{\sin \theta}{r}$$

$$+ \dfrac{\partial^2 u}{\partial x^2} \sin^2 \theta + \dfrac{\partial^2 u}{\partial y^2} \cos^2 \theta - 2 \dfrac{\partial^2 u}{\partial y \partial x} \sin \theta \cos \theta - \dfrac{\partial u}{\partial x} \dfrac{\cos \theta}{r} - \dfrac{\partial u}{\partial y} \dfrac{\sin \theta}{r} + \dfrac{\partial^2 u}{\partial z^2}$$

$$= \dfrac{\partial^2 u}{\partial x^2} + \dfrac{\partial^2 u}{\partial y^2} + \dfrac{\partial^2 u}{\partial z^2}$$

(b) $x = \rho \sin \phi \cos \theta$, $y = \rho \sin \phi \sin \theta$, $z = \rho \cos \phi$. Then

$$\dfrac{\partial u}{\partial \rho} = \dfrac{\partial u}{\partial x} \dfrac{\partial x}{\partial \rho} + \dfrac{\partial u}{\partial y} \dfrac{\partial y}{\partial \rho} + \dfrac{\partial u}{\partial z} \dfrac{\partial z}{\partial \rho} = \dfrac{\partial u}{\partial x} \sin \phi \cos \theta + \dfrac{\partial u}{\partial y} \sin \phi \sin \theta + \dfrac{\partial u}{\partial z} \cos \phi, \text{ and}$$

$$\dfrac{\partial^2 u}{\partial \rho^2} = \sin \phi \cos \theta \left[\dfrac{\partial^2 u}{\partial x^2} \dfrac{\partial x}{\partial \rho} + \dfrac{\partial^2 u}{\partial y \partial x} \dfrac{\partial y}{\partial \rho} + \dfrac{\partial^2 u}{\partial z \partial x} \dfrac{\partial z}{\partial \rho} \right]$$

$$+ \sin \phi \sin \theta \left[\dfrac{\partial^2 u}{\partial y^2} \dfrac{\partial y}{\partial \rho} + \dfrac{\partial^2 u}{\partial x \partial y} \dfrac{\partial x}{\partial \rho} + \dfrac{\partial^2 u}{\partial z \partial y} \dfrac{\partial z}{\partial \rho} \right]$$

$$+ \cos \phi \left[\dfrac{\partial^2 u}{\partial z^2} \dfrac{\partial z}{\partial \rho} + \dfrac{\partial^2 u}{\partial x \partial z} \dfrac{\partial x}{\partial \rho} + \dfrac{\partial^2 u}{\partial y \partial z} \dfrac{\partial y}{\partial \rho} \right]$$

$$= 2 \dfrac{\partial^2 u}{\partial y \partial x} \sin^2 \phi \sin \theta \cos \theta + 2 \dfrac{\partial^2 u}{\partial z \partial x} \sin \phi \cos \phi \cos \theta + 2 \dfrac{\partial^2 u}{\partial y \partial z} \sin \phi \cos \phi \sin \theta$$

$$+ \dfrac{\partial^2 u}{\partial x^2} \sin^2 \phi \cos^2 \theta + \dfrac{\partial^2 u}{\partial y^2} \sin^2 \phi \sin^2 \theta + \dfrac{\partial^2 u}{\partial z^2} \cos^2 \phi$$

Similarly $\dfrac{\partial u}{\partial \phi} = \dfrac{\partial u}{\partial x} \rho \cos \phi \cos \theta + \dfrac{\partial u}{\partial y} \rho \cos \phi \sin \theta - \dfrac{\partial u}{\partial z} \rho \sin \phi$, and

$$\dfrac{\partial^2 u}{\partial \phi^2} = 2 \dfrac{\partial^2 u}{\partial y \partial x} \rho^2 \cos^2 \phi \sin \theta \cos \theta - 2 \dfrac{\partial^2 u}{\partial x \partial z} \rho^2 \sin \phi \cos \phi \cos \theta$$

$$- 2 \dfrac{\partial^2 u}{\partial y \partial z} \rho^2 \sin \phi \cos \phi \sin \theta + \dfrac{\partial^2 u}{\partial x^2} \rho^2 \cos^2 \phi \cos^2 \theta + \dfrac{\partial^2 u}{\partial y^2} \rho^2 \cos^2 \phi \sin^2 \theta$$

$$+ \dfrac{\partial^2 u}{\partial z^2} \rho^2 \sin^2 \phi - \dfrac{\partial u}{\partial x} \rho \sin \phi \cos \theta - \dfrac{\partial u}{\partial y} \rho \sin \phi \sin \theta - \dfrac{\partial u}{\partial z} \rho \cos \phi$$

And $\dfrac{\partial u}{\partial \theta} = -\dfrac{\partial u}{\partial x} \rho \sin \phi \sin \theta + \dfrac{\partial u}{\partial y} \rho \sin \phi \cos \theta$, while

$$\dfrac{\partial^2 u}{\partial \theta^2} = -2 \dfrac{\partial^2 u}{\partial y \partial x} \rho^2 \sin^2 \phi \cos \theta \sin \theta + \dfrac{\partial^2 u}{\partial x^2} \rho^2 \sin^2 \phi \sin^2 \theta$$

$$+ \dfrac{\partial^2 u}{\partial y^2} \rho^2 \sin^2 \phi \cos^2 \theta - \dfrac{\partial u}{\partial x} \rho \sin \phi \cos \theta - \dfrac{\partial u}{\partial y} \rho \sin \phi \sin \theta$$

Therefore

$$\frac{\partial^2 u}{\partial \rho^2} + \frac{2}{\rho} \frac{\partial u}{\partial \rho} + \frac{\cot \phi}{\rho^2} \frac{\partial u}{\partial \phi} + \frac{1}{\rho^2} \frac{\partial^2 u}{\partial \phi^2} + \frac{1}{\rho^2 \sin^2 \phi} \frac{\partial^2 u}{\partial \theta^2}$$

$$= \frac{\partial^2 u}{\partial x^2} \left[\left(\sin^2 \phi \cos^2 \theta \right) + \left(\cos^2 \phi \cos^2 \theta \right) + \sin^2 \theta \right]$$

$$+ \frac{\partial^2 u}{\partial y^2} \left[\left(\sin^2 \phi \sin^2 \theta \right) + \left(\cos^2 \phi \sin^2 \theta \right) + \cos^2 \theta \right] + \frac{\partial^2 u}{\partial z^2} \left[\cos^2 \phi + \sin^2 \phi \right]$$

$$+ \frac{\partial u}{\partial x} \left[\frac{2 \sin^2 \phi \cos \theta + \cos^2 \phi \cos \theta - \sin^2 \phi \cos \theta - \cos \theta}{\rho \sin \phi} \right]$$

$$+ \frac{\partial u}{\partial y} \left[\frac{2 \sin^2 \phi \sin \theta + \cos^2 \phi \sin \theta - \sin^2 \phi \sin \theta - \sin \theta}{\rho \sin \phi} \right]$$

But $2 \sin^2 \phi \cos \theta + \cos^2 \phi \cos \theta - \sin^2 \phi \cos \theta - \cos \theta = \left(\sin^2 \phi + \cos^2 \phi - 1 \right) \cos \theta = 0$
and similarly the coefficient of $\partial u / \partial y$ is 0. Also
$\sin^2 \phi \cos^2 \theta + \cos^2 \phi \cos^2 \theta + \sin^2 \theta = \cos^2 \theta \left(\sin^2 \phi + \cos^2 \phi \right) + \sin^2 \theta = 1$, and similarly the coefficient of
$\partial^2 u / \partial y^2$ is 1. So Laplace's Equation in spherical coordinates is as stated.

9. Since we are minimizing the area of the ellipse, and the circle lies above
the x-axis, the ellipse will intersect the circle for only one value of y. This
y-value must satisfy both the equation of the circle and the equation of the
ellipse. Now $\frac{x^2}{a^2} + \frac{y^2}{b^2} = 1 \Rightarrow x^2 = \frac{a^2}{b^2} \left(b^2 - y^2 \right)$. Substituting into
the equation of the circle gives $\frac{a^2}{b^2} \left(b^2 - y^2 \right) + y^2 - 2y = 0 \Rightarrow$

$$\left(\frac{b^2 - a^2}{b^2} \right) y^2 - 2y + a^2 = 0.$$ In order for there to be only one solution to this quadratic equation, the discriminant

must be 0, so $4 - 4a^2 \frac{b^2 - a^2}{b^2} = 0 \Rightarrow b^2 - a^2 b^2 + a^4 = 0$. The area of the ellipse is $A(a, b) = \pi a b$, and we

minimize this function subject to the constraint $g(a, b) = b^2 - a^2 b^2 + a^4 = 0$.

Now $\nabla A = \lambda \nabla g \Leftrightarrow \pi b = \lambda \left(4a^3 - 2ab^2 \right), \pi a = \lambda \left(2b - 2ba^2 \right) \Rightarrow$ (1) $\lambda = \frac{\pi b}{2a \left(2a^2 - b^2 \right)}$,

(2) $\lambda = \frac{\pi a}{2b \left(1 - a^2 \right)}$, (3) $b^2 - a^2 b^2 + a^4 = 0$. Comparing (1) and (2) gives $\frac{\pi b}{2a \left(2a^2 - b^2 \right)} = \frac{\pi a}{2b \left(1 - a^2 \right)} \Rightarrow$

$2 \pi b^2 = 4 \pi a^4 \Leftrightarrow a^2 = \frac{1}{\sqrt{2}} b$. Substitute this into (3) to get $b = \frac{3}{\sqrt{2}} \Rightarrow a = \sqrt{\frac{3}{2}}$.

Multiple Integrals

1. (a) $\sum\limits_{i=1}^{2}\sum\limits_{j=1}^{2} f\left(x_{ij}^{*}, y_{ij}^{*}\right) \Delta A = f\left(0, \tfrac{3}{2}\right)\Delta A + f\left(0, 2\right)\Delta A + f\left(1, \tfrac{3}{2}\right)\Delta A + f\left(1, 2\right)\Delta A$

$$= \left(-\tfrac{27}{4}\right)\tfrac{1}{2} + (-12)\tfrac{1}{2} + \left(1 - \tfrac{27}{4}\right)\tfrac{1}{2} + (1 - 12)\tfrac{1}{2} = -17.75$$

(b) $\tfrac{1}{2}\left[f\left(1, \tfrac{3}{2}\right) + f\left(1, 2\right) + f\left(2, \tfrac{3}{2}\right) + f\left(2, 2\right)\right] = \tfrac{1}{2}\left[-\tfrac{23}{4} + (-11) + \left(-\tfrac{19}{4}\right) + (-10)\right]$

$$= \tfrac{1}{2}\left(-\tfrac{63}{2}\right) = -15.75$$

(c) $\tfrac{1}{2}\left[f\left(0, 1\right) + f\left(0, \tfrac{3}{2}\right) + f\left(1, 1\right) + f\left(1, \tfrac{3}{2}\right)\right] = \tfrac{1}{2}\left[-3 - \tfrac{27}{4} - 2 - \tfrac{23}{4}\right] = -8.75$

(d) $\tfrac{1}{2}\left[f\left(1, 1\right) + f\left(1, \tfrac{3}{2}\right) + f\left(2, 1\right) + f\left(2, \tfrac{3}{2}\right)\right] = \tfrac{1}{2}\left[-2 - \tfrac{23}{4} - 1 - \tfrac{19}{4}\right] = -6.75$

3. (a) The subrectangles are shown in the figure.

The surface is the graph of $f\left(x, y\right) = x^2 + 4y$ and $\Delta A = 1$, so we estimate

$$V \approx \sum\limits_{i=1}^{2}\sum\limits_{j=1}^{3} f\left(x_i, y_j\right) \Delta A$$

$$= f\left(1, 1\right)\Delta A + f\left(1, 2\right)\Delta A + f\left(1, 3\right)\Delta A + f\left(2, 1\right)\Delta A + f\left(2, 2\right)\Delta A + f\left(2, 3\right)\Delta A$$

$$= 5\left(1\right) + 9\left(1\right) + 13\left(1\right) + 8\left(1\right) + 12\left(1\right) + 16\left(1\right) = 63$$

(b) $V \approx \sum\limits_{i=1}^{2}\sum\limits_{j=1}^{3} f\left(\overline{x}_i, \overline{y}_j\right) \Delta A$

$$= f\left(\tfrac{1}{2}, \tfrac{1}{2}\right)\Delta A + f\left(\tfrac{1}{2}, \tfrac{3}{2}\right)\Delta A + f\left(\tfrac{1}{2}, \tfrac{5}{2}\right)\Delta A$$

$$+ f\left(\tfrac{3}{2}, \tfrac{1}{2}\right)\Delta A + f\left(\tfrac{3}{2}, \tfrac{3}{2}\right)\Delta A + f\left(\tfrac{3}{2}, \tfrac{5}{2}\right)\Delta A$$

$$= \tfrac{9}{4}\left(1\right) + \tfrac{25}{4}\left(1\right) + \tfrac{41}{4}\left(1\right) + \tfrac{17}{4}\left(1\right) + \tfrac{33}{4}\left(1\right) + \tfrac{49}{4}\left(1\right) = \tfrac{87}{2} = 43.5$$

5. (a) Each subrectangle and its midpoint are shown in the figure. The area of each subrectangle is $\Delta A = 2$, so we evaluate f at each midpoint and estimate

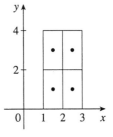

$$\iint_R f(x,y)\,dA \approx \sum_{i=1}^{2}\sum_{j=1}^{2} f(\bar{x}_i, \bar{y}_j)\,\Delta A$$

$$= f(1.5,1)\,\Delta A + f(1.5,3)\,\Delta A$$

$$+ f(2.5,1)\,\Delta A + f(2.5,3)\,\Delta A$$

$$= 1(2) + (-8)(2) + 5(2) + (-1)(2) = -6$$

(b) The subrectangles are shown in the figure. In each subrectangle, the sample point farthest from the origin is the upper right corner, and the area of each subrectangle is $\Delta A = \frac{1}{2}$. Thus we estimate

$$\iint_R f(x,y)\,dA \approx \sum_{i=1}^{4}\sum_{j=1}^{4} f(x_i, y_j)\,\Delta A$$

$$= f(1.5,1)\,\Delta A + f(1.5,2)\,\Delta A + f(1.5,3)\,\Delta A + f(1.5,4)\,\Delta A$$

$$+ f(2,1)\,\Delta A + f(2,2)\,\Delta A + f(2,3)\,\Delta A + f(2,4)\,\Delta A$$

$$+ f(2.5,1)\,\Delta A + f(2.5,2)\,\Delta A + f(2.5,3)\,\Delta A + f(2.5,4)\,\Delta A$$

$$+ f(3,1)\,\Delta A + f(3,2)\,\Delta A + f(3,3)\,\Delta A + f(3,4)\,\Delta A$$

$$= 1\left(\tfrac{1}{2}\right) + (-4)\left(\tfrac{1}{2}\right) + (-8)\left(\tfrac{1}{2}\right) + (-6)\left(\tfrac{1}{2}\right) + 3\left(\tfrac{1}{2}\right) + 0\left(\tfrac{1}{2}\right) + (-5)\left(\tfrac{1}{2}\right) + (-8)\left(\tfrac{1}{2}\right)$$

$$+ 5\left(\tfrac{1}{2}\right) + 3\left(\tfrac{1}{2}\right) + (-1)\left(\tfrac{1}{2}\right) + (-4)\left(\tfrac{1}{2}\right) + 8\left(\tfrac{1}{2}\right) + 6\left(\tfrac{1}{2}\right) + 3\left(\tfrac{1}{2}\right) + 0\left(\tfrac{1}{2}\right)$$

$$= -3.5$$

7. The values of $f(x,y) = \sqrt{52 - x^2 - y^2}$ get smaller as we move farther from the origin, so on any of the subrectangles in the problem, the function will have its largest value at the lower left corner of the subrectangle and its smallest value at the upper right corner, and any other value will lie between these two. So using these subrectangles we have $U < V < L$. (Note that this is true no matter how R is divided into subrectangles.)

9. With $m = n = 2$, we have $\Delta A = 4$. Using the contour map to estimate the value of f at the center of each subrectangle, we have

$$\iint_R f(x,y)\,dA \approx \sum_{i=1}^{2}\sum_{j=1}^{2} f(\bar{x}_i, \bar{y}_j)\,\Delta A = \Delta A\,[f(1,1) + f(1,3) + f(3,1) + f(3,3)]$$

$$\approx 4\,(27 + 4 + 14 + 17) = 248$$

11. $z = 3 > 0$, so we can interpret the integral as the volume of the solid S that lies below the plane $z = 3$ and above the rectangle $[-2,2] \times [1,6]$. S is a rectangular solid, thus $\iint_R 3\,dA = 4 \cdot 5 \cdot 3 = 60$.

13. $z = f(x, y) = 4 - 2y \geq 0$ for $0 \leq y \leq 1$. Thus the integral represents the volume of that part of the rectangular solid $[0, 1] \times [0, 1] \times [0, 4]$ which lies below the plane $z = 4 - 2$. So

$$\iint_R (4 - y)\, dA = (1)(1)(2) + \tfrac{1}{2}(1)(1)(2) = 3$$

15. To calculate the estimates using a programmable calculator, we can use an algorithm similar to that of Exercise 5.1.7 [ET 5.1.7]. In Maple, we can define the function $f(x, y) = e^{-x^2 - y^2}$ (calling it f), load the `student` package, and then use the command

$$\text{middlesum(middlesum(f,x=0..1,m),}$$
$$\text{y=0..1,m);}$$

to get the estimate with $n = m^2$ squares of equal size. Mathematica has no special Riemann sum command, but we can define f and then use nested `Sum` commands to calculate the estimates.

n	estimate
1	0.6065
4	0.5694
16	0.5606
64	0.5585
256	0.5579
1024	0.5578

17. If we divide R into mn subrectangles, $\iint_R k\, dA \approx \sum_{i=1}^{m} \sum_{j=1}^{n} f(x_{ij}^*, y_{ij}^*)\, \Delta A$ for any choice of sample points (x_{ij}^*, y_{ij}^*). But $f(x_{ij}^*, y_{ij}^*) = k$ always and $\sum_{i=1}^{m} \sum_{j=1}^{n} \Delta A =$ area of $R = (b - a)(d - c)$. Thus, no matter how we choose the sample points, $\sum_{i=1}^{m} \sum_{j=1}^{n} f(x_{ij}^*, y_{ij}^*)\, \Delta A = k \sum_{i=1}^{m} \sum_{j=1}^{n} \Delta A = k(b - a)(d - c)$ and so

$$\iint_R k\, dA = \lim_{m,n \to \infty} \sum_{i=1}^{m} \sum_{j=1}^{n} f(x_{ij}^*, y_{ij}^*)\, \Delta A = \lim_{m,n \to \infty} k \sum_{i=1}^{m} \sum_{j=1}^{n} \Delta A$$
$$= \lim_{m,n \to \infty} k(b - a)(d - c) = k(b - a)(d - c)$$

16.2 Iterated Integrals   ET 15.2

1. $\int_0^3 (2x + 3x^2 y)\, dx = \left[x^2 + x^3 y\right]_{x=0}^{x=3} = (9 + 27y) - (0 + 0) = 9 + 27y,$

$\int_0^4 (2x + 3x^2 y)\, dy = \left[2xy + 3x^2 \dfrac{y^2}{2}\right]_{y=0}^{y=4} = \left(8x + 3x^2 \cdot \dfrac{16}{2}\right) - (0 + 0) = 8x + 24x^2$

3. $\int_1^3 \int_0^1 (1 + 4xy)\, dx\, dy = \int_1^3 \left[x + 2x^2 y\right]_{x=0}^{x=1} dy = \int_1^3 (1 + 2y)\, dy$
$$= \left[y + y^2\right]_1^3 = (3 + 9) - (1 + 1) = 10$$

5. $\int_0^{\pi/2} \int_0^{\pi/2} \sin x \cos y\, dy\, dx = \int_0^{\pi/2} \sin x\, dx \int_0^{\pi/2} \cos y\, dy$ (as in Example 5)
$$= \left[-\cos x\right]_0^{\pi/2} \left[\sin y\right]_0^{\pi/2} = -(0 - 1)(1 - 0) = 1$$

7. $\int_0^3 \int_0^1 \sqrt{x+y}\,dx\,dy = \int_0^3 \left[\frac{2}{3}(x+y)^{3/2}\right]_{x=0}^{x=1} dy = \frac{2}{3}\int_0^3 \left[(1+y)^{3/2} - y^{3/2}\right] dy$

$$= \frac{2}{3}\left[\frac{2}{5}(1+y)^{5/2} - \frac{2}{5}y^{5/2}\right]_0^3 = \frac{4}{15}\left[32 - 3^{5/2} - 1\right] = \frac{4}{15}\left(31 - 9\sqrt{3}\right)$$

9. $\int_1^4 \int_1^2 \left(\frac{x}{y} + \frac{y}{x}\right) dy\,dx = \int_1^4 \left[x\ln|y| + \frac{1}{x}\cdot\frac{1}{2}y^2\right]_{y=1}^{y=2} dx = \int_1^4 \left(x\ln 2 + \frac{3}{2x}\right) dx$

$$= \left[\frac{1}{2}x^2 \ln 2 + \frac{3}{2}\ln|x|\right]_1^4 = 8\ln 2 + \frac{3}{2}\ln 4 - \frac{1}{2}\ln 2$$

$$= \frac{15}{2}\ln 2 + 3\ln 4^{1/2} = \frac{21}{2}\ln 2$$

11. $\int_0^{\ln 2} \int_0^{\ln 5} e^{2x-y}\,dx\,dy = \left(\int_0^{\ln 5} e^{2x}\,dx\right)\left(\int_0^{\ln 2} e^{-y}\,dy\right) = \left[\frac{1}{2}e^{2x}\right]_0^{\ln 5}\left[-e^{-y}\right]_0^{\ln 2}$

$$= \left(\frac{25}{2} - \frac{1}{2}\right)\left(-\frac{1}{2} + 1\right) = 6$$

13. $\iint_R \left(6x^2 y^3 - 5y^4\right) dA = \int_0^3 \int_0^1 \left(6x^2 y^3 - 5y^4\right) dy\,dx = \int_0^3 \left[\frac{3}{2}x^2 y^4 - y^5\right]_{y=0}^{y=1} dx$

$$= \int_0^3 \left(\frac{3}{2}x^2 - 1\right) dx = \left[\frac{1}{2}x^3 - x\right]_0^3 = \frac{27}{2} - 3 = \frac{21}{2}$$

15. $\iint_R \frac{xy^2}{x^2+1}\,dA = \int_0^1 \int_{-3}^3 \frac{xy^2}{x^2+1}\,dy\,dx = \int_0^1 \frac{x}{x^2+1}\,dx \int_{-3}^3 y^2\,dy$

$$= \left[\frac{1}{2}\ln\left(x^2+1\right)\right]_0^1 \left[\frac{1}{3}y^3\right]_{-3}^3 = \frac{1}{2}\left(\ln 2 - \ln 1\right)\cdot\frac{1}{3}(27+27) = 9\ln 2$$

17. $\int_0^{\pi/6} \int_0^{\pi/3} x\sin(x+y)\,dy\,dx$

$$= \int_0^{\pi/6} \left[-x\cos(x+y)\right]_{y=0}^{y=\pi/3} dx = \int_0^{\pi/6} \left[x\cos x - x\cos\left(x+\frac{\pi}{3}\right)\right] dx$$

$$= x\left[\sin x - \sin\left(x+\frac{\pi}{3}\right)\right]_0^{\pi/6} - \int_0^{\pi/6}\left[\sin x - \sin\left(x+\frac{\pi}{3}\right)\right] dx$$

$$\text{(by integrating by parts separately for each term)}$$

$$= \frac{\pi}{6}\left[\frac{1}{2} - 1\right] - \left[-\cos x + \cos\left(x+\frac{\pi}{3}\right)\right]_0^{\pi/6} = -\frac{\pi}{12} - \left[-\frac{\sqrt{3}}{2} + 0 - \left(-1 + \frac{1}{2}\right)\right]$$

$$= \frac{\sqrt{3}-1}{2} - \frac{\pi}{12}$$

19. $\int_0^1 \int_1^2 \frac{1}{x+y}\,dx\,dy = \int_0^1 \left[\ln(x+y)\right]_{x=1}^{x=2} dy = \int_0^1 \left[\ln(2+y) - \ln(1+y)\right] dy$

$$= \left[\left[(2+y)\ln(2+y) - (2+y)\right] - \left[(1+y)\ln(1+y) - (1+y)\right]\right]_0^1$$

$$\text{(by integrating by parts separately for each term or by the Table of Integrals)}$$

$$= (3\ln 3) - 3 - (2\ln 2) + 2 - \left[(2\ln 2 - 2) - (0 - 1)\right] = 3\ln 3 - 4\ln 2 = \ln\frac{27}{16}$$

21. $z = f(x,y) = 4 - x - 2y \geq 0$ for $0 \leq x \leq 1$ and $0 \leq y \leq 1$.
So the solid is the region in the first octant which lies below the
plane $z = 4 - x - 2y$ and above $[0,1] \times [0,1]$.

23. $V = \iint_R (2x + 5y + 1)\, dA = \int_1^4 \int_{-1}^0 (2x + 5y + 1)\, dx\, dy = \int_1^4 \left[x^2 + 5xy + x \right]_{x=-1}^{x=0} dy$

$= \int_1^4 5y\, dy = \frac{5}{2} y^2 \Big]_1^4 = \frac{75}{2}$

25. $V = \int_{-2}^2 \int_{-1}^1 \left(1 - \frac{1}{4}x^2 - \frac{1}{9}y^2 \right) dx\, dy = 4 \int_0^2 \int_0^1 \left(1 - \frac{1}{4}x^2 - \frac{1}{9}y^2 \right) dx\, dy$

$= 4 \int_0^2 \left[x - \frac{1}{12}x^3 - \frac{1}{9}y^2 x \right]_{x=0}^{x=1} dy = 4 \int_0^2 \left(\frac{11}{12} - \frac{1}{9}y^2 \right) dy = 4 \left[\frac{11}{12}y - \frac{1}{27}y^3 \right]_0^2 = 4 \cdot \frac{83}{54} = \frac{166}{27}$

27. Here we need the volume of the solid lying under the surface $z = x\sqrt{x^2 + y}$ and above the square $R = [0, 1] \times [0, 1]$ in the xy-plane.

$V = \int_0^1 \int_0^1 x\sqrt{x^2 + y}\, dx\, dy = \int_0^1 \frac{1}{3} \left[(x^2 + y)^{3/2} \right]_{x=0}^{x=1} dy = \frac{1}{3} \int_0^1 \left[(1 + y)^{3/2} - y^{3/2} \right] dy$

$= \frac{1}{3} \cdot \frac{2}{5} \left[(1 + y)^{5/2} - y^{5/2} \right]_0^1 = \frac{4}{15} \left(2\sqrt{2} - 1 \right)$

29. In the first octant, $z \geq 0 \implies y \leq 3$, so

$V = \int_0^3 \int_0^2 (9 - y^2)\, dx\, dy = \int_0^3 \left[9x - y^2 x \right]_{x=0}^{x=2} dy = \int_0^3 (18 - 2y^2)\, dy = \left[18y - \frac{2}{3}y^3 \right]_0^3 = 36$

31. In Maple, we can calculate the integral by defining the integrand as `f`
and then using the command `int(int(f,x=0..1),y=0..1);`.
In Mathematica, we can use the command
`Integrate[Integrate[f,{x,0,1}],{y,0,1}]`. We find
that $\iint_R x^5 y^3 e^{xy}\, dA = 21e - 57 \approx 0.0839$. We can use `plot3d` (in
Maple) or `Plot3d` (in Mathematica) to graph the function.

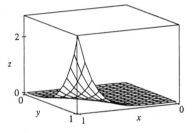

33. R is the rectangle $[-1, 1] \times [0, 5]$. Thus, $A(R) = 2 \cdot 5 = 10$ and

$f_{\text{ave}} = \frac{1}{A(R)} \iint_R f(x, y)\, dA = \frac{1}{10} \int_0^5 \int_{-1}^1 x^2 y\, dx\, dy = \frac{1}{10} \int_0^5 \left[\frac{1}{3}x^3 y \right]_{x=-1}^{x=1} dy = \frac{1}{10} \int_0^5 \frac{2}{3} y\, dy$

$= \frac{1}{10} \left[\frac{1}{3}y^2 \right]_0^5 = \frac{5}{6}$

35. Let $f(x, y) = \frac{x - y}{(x + y)^3}$. Then a CAS gives $\int_0^1 \int_0^1 f(x, y)\, dy\, dx = \frac{1}{2}$ and $\int_0^1 \int_0^1 f(x, y)\, dx\, dy = -\frac{1}{2}$.

To explain the seeming violation of Fubini's Theorem, note that f has an infinite discontinuity at $(0, 0)$ and thus does not satisfy the conditions of Fubini's Theorem. In fact, both iterated integrals involve improper integrals which diverge at their lower limits of integration.

16.3   Double Integrals over General Regions

ET 15.3

1. $\int_0^1 \int_0^{x^2} (x + 2y)\, dy\, dx = \int_0^1 \left[xy + y^2 \right]_{y=0}^{y=x^2} dx = \int_0^1 \left[x(x^2) + (x^2)^2 - 0 - 0 \right] dx$

$= \int_0^1 (x^3 + x^4)\, dx = \left[\frac{1}{4}x^4 + \frac{1}{5}x^5 \right]_0^1 = \frac{9}{20}$

3. $\int_0^1 \int_y^{e^y} \sqrt{x}\, dx\, dy = \int_0^1 \left[\frac{2}{3}x^{3/2} \right]_{x=y}^{x=e^y} dy = \frac{2}{3} \int_0^1 \left(e^{3y/2} - y^{3/2} \right) dy = \frac{2}{3} \left[\frac{2}{3}e^{3y/2} - \frac{2}{5}y^{5/2} \right]_0^1$

$= \frac{2}{3} \left(\frac{2}{3}e^{3/2} - \frac{2}{5} - \frac{2}{3}e^0 + 0 \right) = \frac{4}{9}e^{3/2} - \frac{32}{45}$

5. $\int_0^{\pi/2} \int_0^{\cos\theta} e^{\sin\theta}\, dr\, d\theta = \int_0^{\pi/2} \left[r e^{\sin\theta} \right]_{r=0}^{r=\cos\theta} d\theta = \int_0^{\pi/2} (\cos\theta)\, e^{\sin\theta}\, d\theta = e^{\sin\theta} \Big]_0^{\pi/2}$

$$= e^{\sin(\pi/2)} - e^0 = e - 1$$

7. $\iint_D x^3 y^2\, dA = \int_0^2 \int_{-x}^x x^3 y^2\, dy\, dx = \int_0^2 \left[\frac{1}{3} x^3 y^3 \right]_{y=-x}^{y=x} dx = \frac{1}{3} \int_0^2 2x^6\, dx$

$$= \frac{2}{3} \left[\frac{1}{7} x^7 \right]_0^2 = \frac{2}{21} \left[2^7 - 0 \right] = \frac{256}{21}$$

9. $\int_0^1 \int_0^{\sqrt{x}} \frac{2y}{x^2+1}\, dy\, dx = \int_0^1 \left[\frac{y^2}{x^2+1} \right]_{y=0}^{y=\sqrt{x}} dx = \int_0^1 \frac{x}{x^2+1}\, dx$

$$= \frac{1}{2} \ln|x^2+1| \Big|_0^1 = \frac{1}{2}(\ln 2 - \ln 1) = \frac{1}{2} \ln 2$$

11. $\int_1^2 \int_y^{y^3} e^{x/y}\, dx\, dy = \int_1^2 \left[y e^{x/y} \right]_{x=y}^{x=y^3} dy = \int_1^2 \left(y e^{y^2} - ey \right) dy = \left[\frac{1}{2} e^{y^2} - \frac{1}{2} e y^2 \right]_1^2 = \frac{1}{2}\left(e^4 - 4e \right)$

13. $\int_0^1 \int_0^{x^2} x \cos y\, dy\, dx = \int_0^1 \left[x \sin y \right]_{y=0}^{y=x^2} dx = \int_0^1 x \sin x^2\, dx = -\frac{1}{2} \cos x^2 \Big]_0^1 = \frac{1}{2}(1 - \cos 1)$

15.

$\int_1^2 \int_{2-y}^{2y-1} y^3\, dx\, dy = \int_1^2 \left[x y^3 \right]_{x=2-y}^{x=2y-1} dy = \int_1^2 \left[(2y-1) - (2-y) \right] y^3\, dy$

$$= \int_1^2 \left(3y^4 - 3y^3 \right) dy = \left[\frac{3}{5} y^5 - \frac{3}{4} y^4 \right]_1^2$$

$$= \frac{96}{5} - 12 - \frac{3}{5} + \frac{3}{4} = \frac{147}{20}$$

(0, 2)   (3, 2)

D

$x = 2 - y$ $x = 2y - 1$

(1, 1)

17.

$y = \sqrt{4-x^2}$

D

-2 0   2 x

$y = -\sqrt{4-x^2}$

$\int_{-2}^2 \int_{-\sqrt{4-x^2}}^{\sqrt{4-x^2}} (2x - y)\, dy\, dx = \int_{-2}^2 \left[2xy - \frac{1}{2} y^2 \right]_{y=-\sqrt{4-x^2}}^{y=\sqrt{4-x^2}} dx$

$$= \int_{-2}^2 \left[2x\sqrt{4-x^2} - \frac{1}{2}(4-x^2) + 2x\sqrt{4-x^2} + \frac{1}{2}(4-x^2) \right] dx$$

$$= \int_{-2}^2 4x\sqrt{4-x^2}\, dx = -\frac{4}{3}(4-x^2)^{3/2} \Big]_{-2}^2 = 0$$

(Or, note that $4x\sqrt{4-x^2}$ is an odd function, so $\int_{-2}^2 4x\sqrt{4-x^2}\, dx = 0$.)

19.

y

(1, 1)

D

0   x

$V = \int_0^1 \int_{x^2}^{\sqrt{x}} (x^2 + y^2)\, dy\, dx = \int_0^1 \left[\left(x^2 y + \frac{y^3}{3} \right) \right]_{y=x^2}^{y=\sqrt{x}} dx$

$$= \int_0^1 \left(x^{5/2} - x^4 + \frac{1}{3} x^{3/2} - \frac{1}{3} x^6 \right) dx$$

$$= \left[\frac{2}{7} x^{7/2} - \frac{1}{5} x^5 + \frac{2}{15} x^{5/2} - \frac{1}{21} x^7 \right]_0^1$$

$$= \frac{18}{105} = \frac{6}{35}$$

21.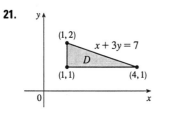

$$V = \int_1^2 \int_1^{7-3y} xy \, dx \, dy$$
$$= \int_1^2 \left[\tfrac{1}{2}x^2 y\right]_{x=1}^{x=7-3y} dy$$
$$= \tfrac{1}{2} \int_1^2 \left(48y - 42y^2 + 9y^3\right) dy$$
$$= \tfrac{1}{2} \left[24y^2 - 14y^3 + \tfrac{9}{4}y^4\right]_1^2 = \frac{31}{8}$$

23.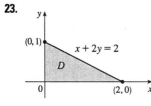

$$V = \int_0^2 \int_0^{1-x/2} \sqrt{9-x^2}\, dy\, dx = \int_0^2 \left[y\sqrt{9-x^2}\right]_{y=0}^{y=1-x/2} dx$$
$$= \int_0^2 \left(\sqrt{9-x^2} - \tfrac{1}{2}x\sqrt{9-x^2}\right) dx$$
$$= \int_0^2 \sqrt{9-x^2}\, dx + \tfrac{1}{4}\int_0^2 \left(-2x\sqrt{9-x^2}\right) dx$$
$$= \left[\tfrac{1}{2}x\sqrt{9-x^2} + \tfrac{9}{2}\sin^{-1}(x/3) + \tfrac{1}{6}\left(9-x^2\right)^{3/2}\right]_0^2$$
$$= \sqrt{5} + \tfrac{9}{2}\sin^{-1}\tfrac{2}{3} + \tfrac{5}{6}\sqrt{5} - \tfrac{1}{6}(27)$$
$$= \tfrac{1}{6}\left(11\sqrt{5} - 27\right) + \tfrac{9}{2}\sin^{-1}\tfrac{2}{3}$$

25.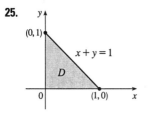

$$V = \int_0^1 \int_0^{1-x} (1-x-y)\, dy\, dx$$
$$= \int_0^1 \left[y - xy - \frac{y^2}{2}\right]_{y=0}^{y=1-x} dx$$
$$= \int_0^1 \left[(1-x)^2 - \tfrac{1}{2}(1-x)^2\right] dx$$
$$= \int_0^1 \tfrac{1}{2}(1-x)^2\, dx = \left[-\tfrac{1}{6}(1-x)^3\right]_0^1 = \tfrac{1}{6}$$

27.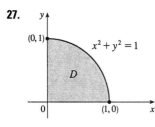

$$V = \int_0^1 \int_0^{\sqrt{1-x^2}} y\, dy\, dx$$
$$= \int_0^1 \left[\frac{y^2}{2}\right]_{y=0}^{y=\sqrt{1-x^2}} dx$$
$$= \int_0^1 \frac{1-x^2}{2}\, dx$$
$$= \tfrac{1}{2}\left[x - \tfrac{1}{3}x^3\right]_0^1 = \tfrac{1}{3}$$

29.

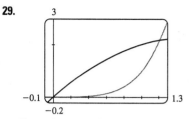

From the graph, it appears that the two curves intersect at $x = 0$ and at $x \approx 1.213$. Thus the desired integral is

$$\iint_D x \, dA \approx \int_0^{1.213} \int_{x^4}^{3x - x^2} x \, dy \, dx$$

$$= \int_0^{1.213} [xy]_{y = x^4}^{y = 3x - x^2} \, dx$$

$$= \int_0^{1.213} \left(3x^2 - x^3 - x^5\right) dx$$

$$= \left[x^3 - \tfrac{1}{4}x^4 - \tfrac{1}{6}x^6\right]_0^{1.213} \approx 0.713$$

31. The two bounding curves $y = x^3 - x$ and $y = x^2 + x$ intersect at the origin and at $x = 2$, with $x^2 + x > x^3 - x$ on $(0, 2)$. Using a CAS, we find that the volume is

$$V = \int_0^2 \int_{x^3 - x}^{x^2 + x} z \, dy \, dx = \int_0^2 \int_{x^3 - x}^{x^2 + x} \left(x^3 y^4 + xy^2\right) dy \, dx = \tfrac{13,984,735,616}{14,549,535}.$$

33.

Because the region of integration is

$$D = \{(x, y) \mid 0 \le y \le x, 0 \le x \le 1\}$$
$$= \{(x, y) \mid y \le x \le 1, 0 \le y \le 1\}$$

we have

$$\int_0^1 \int_0^x f(x, y) \, dy \, dx = \iint_D f(x, y) \, dA$$
$$= \int_0^1 \int_y^1 f(x, y) \, dx \, dy$$

35.

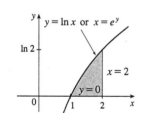

Because the region of integration is

$$D = \{(x, y) \mid 0 \le y \le \ln x, 1 \le x \le 2\}$$
$$= \{(x, y) \mid e^y \le x \le 2, 0 \le y \le \ln 2\}$$

we have

$$\int_1^2 \int_0^{\ln x} f(x, y) \, dy \, dx = \iint_D f(x, y) \, dA$$
$$= \int_0^{\ln 2} \int_{e^y}^2 f(x, y) \, dx \, dy$$

37.

Because the region of integration is

$$D = \{(x, y) \mid y/2 \le x \le 2, 0 \le y \le 4\}$$
$$= \{(x, y) \mid 0 \le y \le 2x, 0 \le x \le 2\}$$

we have

$$\int_0^4 \int_{y/2}^2 f(x, y) \, dx \, dy = \iint_D f(x, y) \, dA$$
$$= \int_0^2 \int_0^{2x} f(x, y) \, dy \, dx$$

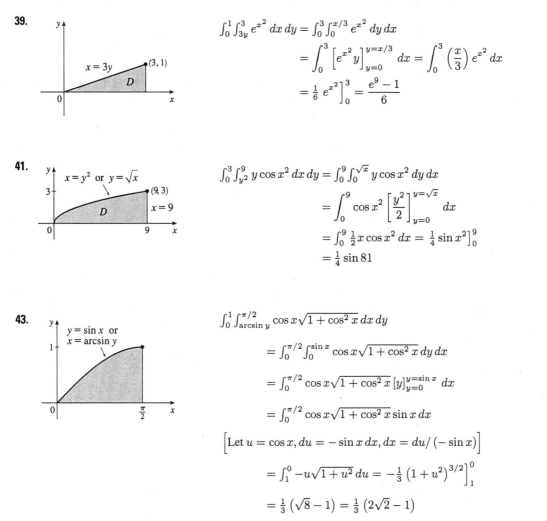

39.

$$\int_0^1 \int_{3y}^3 e^{x^2} \, dx \, dy = \int_0^3 \int_0^{x/3} e^{x^2} \, dy \, dx$$

$$= \int_0^3 \left[e^{x^2} y \right]_{y=0}^{y=x/3} dx = \int_0^3 \left(\frac{x}{3} \right) e^{x^2} \, dx$$

$$= \frac{1}{6} e^{x^2} \Big]_0^3 = \frac{e^9 - 1}{6}$$

41.

$$\int_0^3 \int_{y^2}^9 y \cos x^2 \, dx \, dy = \int_0^9 \int_0^{\sqrt{x}} y \cos x^2 \, dy \, dx$$

$$= \int_0^9 \cos x^2 \left[\frac{y^2}{2} \right]_{y=0}^{y=\sqrt{x}} dx$$

$$= \int_0^9 \frac{1}{2} x \cos x^2 \, dx = \frac{1}{4} \sin x^2 \Big]_0^9$$

$$= \frac{1}{4} \sin 81$$

43.

$$\int_0^1 \int_{\arcsin y}^{\pi/2} \cos x \sqrt{1 + \cos^2 x} \, dx \, dy$$

$$= \int_0^{\pi/2} \int_0^{\sin x} \cos x \sqrt{1 + \cos^2 x} \, dy \, dx$$

$$= \int_0^{\pi/2} \cos x \sqrt{1 + \cos^2 x} \, [y]_{y=0}^{y=\sin x} \, dx$$

$$= \int_0^{\pi/2} \cos x \sqrt{1 + \cos^2 x} \sin x \, dx$$

$$\left[\text{Let } u = \cos x, du = -\sin x \, dx, dx = du/(-\sin x) \right]$$

$$= \int_1^0 -u\sqrt{1 + u^2} \, du = -\frac{1}{3} \left(1 + u^2 \right)^{3/2} \Big]_1^0$$

$$= \frac{1}{3} \left(\sqrt{8} - 1 \right) = \frac{1}{3} \left(2\sqrt{2} - 1 \right)$$

45. $D = \{(x, y) \mid 0 \leq x \leq 1, -x + 1 \leq y \leq 1\} \cup \{(x, y) \mid -1 \leq x \leq 0, x + 1 \leq y \leq 1\}$

$\cup \{(x, y) \mid 0 \leq x \leq 1, -1 \leq y \leq x - 1\} \cup \{(x, y) \mid -1 \leq x \leq 0, -1 \leq y \leq -x - 1\}$,

all type I.

$$\iint_D x^2 \, dA = \int_0^1 \int_{1-x}^1 x^2 \, dy \, dx + \int_{-1}^0 \int_{x+1}^1 x^2 \, dy \, dx + \int_0^1 \int_{-1}^{x-1} x^2 \, dy \, dx + \int_{-1}^0 \int_{-1}^{-x-1} x^2 \, dy \, dx$$

$$= 4 \int_0^1 \int_{1-x}^1 x^2 \, dy \, dx \quad \text{[by symmetry of the regions and because } f(x, y) = x^2 \geq 0]$$

$$= 4 \int_0^1 x^3 \, dx = 4 \left[\frac{1}{4} x^4 \right]_0^1 = 1$$

47. For $D = [0, 1] \times [0, 1], 0 \leq \sqrt{x^3 + y^3} \leq \sqrt{2}$ and $A(D) = 1$, so $0 \leq \iint_D \sqrt{x^3 + y^3} \, dA \leq \sqrt{2}$.

49. Since $m \leq f(x, y) \leq M$, $\iint_D m \, dA \leq \iint_D f(x, y) \, dA \leq \iint_D M \, dA$ by (8) $\Rightarrow$

$m \iint_D 1 \, dA \leq \iint_D f(x, y) \, dA \leq M \iint_D 1 \, dA$ by (7) $\Rightarrow$ $mA(D) \leq \iint_D f(x, y) \, dA \leq MA(D)$ by (10).

51. $\iint_D \left(x^2 \tan x + y^3 + 4\right) dA = \iint_D x^2 \tan x \, dA + \iint_D y^3 \, dA + \iint_D 4 \, dA$. But $x^2 \tan x$ is an odd function of

x and D is symmetric with respect to the y-axis, so $\iint_D x^2 \tan x \, dA = 0$. Similarly, y^3 is an odd function of y and

D is symmetric with respect to the x-axis, so $\iint_D y^3 \, dA = 0$. Thus

$$\iint_D \left(x^2 \tan x + y^3 + 4\right) dA = 4 \iint_D dA = 4 \, (\text{area of } D) = 4 \cdot \pi \left(\sqrt{2}\right)^2 = 8\pi$$

53. Since $\sqrt{1 - x^2 - y^2} \geq 0$, we can interpret $\iint_D \sqrt{1 - x^2 - y^2} \, dA$ as the volume of the solid that lies below the

graph of $z = \sqrt{1 - x^2 - y^2}$ and above the region D in the xy-plane. $z = \sqrt{1 - x^2 - y^2}$ is equivalent to

$x^2 + y^2 + z^2 = 1$, $z \geq 0$ which meets the xy-plane in the circle $x^2 + y^2 = 1$, the boundary of D. Thus, the solid is

an upper hemisphere of radius 1 which has volume $\frac{1}{2} \left[\frac{4}{3} \pi (1)^3\right] = \frac{2}{3}\pi$.

16.4 Double Integrals in Polar Coordinates ET 15.4

1. The region R is more easily described by polar coordinates: $R = \{(r, \theta) \mid 0 \leq r \leq 2, 0 \leq \theta \leq 2\pi\}$. Thus

$\iint_R f(x, y) \, dA = \int_0^{2\pi} \int_0^2 f(r \cos \theta, r \sin \theta) \, r \, dr \, d\theta$.

3. The region R is more easily described by rectangular coordinates: $R = \{(x, y) \mid -2 \leq x \leq 2, x \leq y \leq 2\}$. Thus

$\iint_R f(x, y) \, dA = \int_{-2}^2 \int_x^2 f(x, y) \, dy \, dx$.

5. The region R is more easily described by polar coordinates: $R = \{(r, \theta) \mid 2 \leq r \leq 5, 0 \leq \theta \leq 2\pi\}$. Thus

$\iint_R f(x, y) \, dA = \int_0^{2\pi} \int_2^5 f(r \cos \theta, r \sin \theta) \, r \, dr \, d\theta$.

7. The region R can be described in polar coordinates as $R = \{(r, \theta) \mid 0 \leq \theta \leq 2\pi, 0 \leq r \leq 5\}$. Thus,

$\iint_R x \, dA = \int_0^{2\pi} \int_0^5 (r \cos \theta) \, r \, dr \, d\theta = \left(\int_0^{2\pi} \cos \theta \, d\theta\right) \left(\int_0^5 r^2 \, dr\right) = [\sin \theta]_0^{2\pi} \left[\frac{1}{3} r^3\right]_0^5 = 0.$

9. $\iint_R xy \, dA = \int_0^{\pi/2} \int_2^5 (r \cos \theta)(r \sin \theta) \, r \, dr \, d\theta = \int_0^{\pi/2} \sin \theta \cos \theta \, d\theta \int_2^5 r^3 \, dr = \left[\frac{1}{2} \sin^2 \theta\right]_0^{\pi/2} \left[\frac{1}{4} r^4\right]_2^5$

$= \frac{1}{2} \cdot \frac{5^4 - 2^4}{4} = \frac{609}{8}$

11. $\iint_D e^{-x^2 - y^2} \, dA = \int_{-\pi/2}^{\pi/2} \int_0^2 e^{-r^2} \, r \, dr \, d\theta = \left(\int_{-\pi/2}^{\pi/2} d\theta\right) \left(\int_0^2 re^{-r^2} \, dr\right)$

$= [\theta]_{-\pi/2}^{\pi/2} \left[-\frac{1}{2} e^{-r^2}\right]_0^2 = \pi \left(-\frac{1}{2}\right) \left(e^{-4} - e^0\right) = \frac{\pi}{2} \left(1 - e^{-4}\right)$

13. $\int_0^{2\pi} \int_0^{2\theta} r^2 \, r \, dr \, d\theta = \int_0^{2\pi} \left[\frac{1}{4} r^4\right]_{r=0}^{r=2\theta} d\theta = \frac{1}{4} \int_0^{2\pi} 16\theta^4 \, d\theta = \frac{3}{4} \left[\theta^5\right]_0^{2\pi} = \frac{3}{4} \left(32\pi^5\right) = 24\pi^5$

15. One loop is given by the region

$D = \{(r, \theta) \mid -\pi/6 \leq \theta \leq \pi/6, 0 \leq r \leq \cos 3\theta\}$, so the area is

$\iint_D dA = \int_{-\pi/6}^{\pi/6} \int_0^{\cos 3\theta} r \, dr \, d\theta = \int_{-\pi/6}^{\pi/6} \left[\frac{1}{2} r^2\right]_{r=0}^{r=\cos 3\theta} d\theta$

$= \int_{-\pi/6}^{\pi/6} \frac{1}{2} \cos^2 3\theta \, d\theta = 2 \int_0^{\pi/6} \frac{1}{2} \left(\frac{1 + \cos 6\theta}{2}\right) d\theta$

$= \frac{1}{2} \left[\theta + \frac{1}{6} \sin 6\theta\right]_0^{\pi/6} = \frac{\pi}{12}$

17. By symmetry, the two loops of the lemniscate are equal in area, so

$A = 2 \int_{-\pi/4}^{\pi/4} \int_0^{2\sqrt{\cos 2\theta}} r \, dr \, d\theta = \int_{-\pi/4}^{\pi/4} \left[r^2\right]_{r=0}^{r=2\sqrt{\cos 2\theta}} d\theta = \int_{-\pi/4}^{\pi/4} 4 \cos 2\theta \, d\theta$

$= 8 \int_0^{\pi/4} \cos 2\theta \, d\theta = 4 \sin 2\theta]_0^{\pi/4} = 4.$

19. $V = \iint_{x^2+y^2 \le 9} \left(x^2+y^2\right) dA = \int_0^{2\pi}\int_0^3 \left(r^2\right) r\, dr\, d\theta = \int_0^{2\pi} d\theta \int_0^3 r^3\, dr = [\theta]_0^{2\pi}\left[\frac{1}{4}r^4\right]_0^3 = 2\pi\left(\frac{81}{4}\right) = \frac{81\pi}{2}$

21. By symmetry,

$$V = 2\iint_{x^2+y^2 \le a^2} \sqrt{a^2-x^2-y^2}\, dA = 2\int_0^{2\pi}\int_0^a \sqrt{a^2-r^2}\, r\, dr\, d\theta = 2\int_0^{2\pi} d\theta \int_0^a r\sqrt{a^2-r^2}\, dr$$

$$= 2\,[\theta]_0^{2\pi}\left[-\frac{1}{3}\left(a^2-r^2\right)^{3/2}\right]_0^a = 2\,(2\pi)\left(0+\frac{1}{3}a^3\right) = \frac{4\pi}{3}a^3$$

23. The cone $z = \sqrt{x^2+y^2}$ intersects the sphere $x^2+y^2+z^2 = 1$ when $x^2+y^2+\left(\sqrt{x^2+y^2}\right)^2 = 1$ or $x^2+y^2 = \frac{1}{2}$. So

$$V = \iint_{x^2+y^2 \le 1/2}\left(\sqrt{1-x^2-y^2}-\sqrt{x^2+y^2}\right)dA = \int_0^{2\pi}\int_0^{1/\sqrt{2}}\left(\sqrt{1-r^2}-r\right)r\, dr\, d\theta$$

$$= \int_0^{2\pi} d\theta \int_0^{1/\sqrt{2}}\left(r\sqrt{1-r^2}-r^2\right)dr = [\theta]_0^{2\pi}\left[-\frac{1}{3}\left(1-r^2\right)^{3/2}-\frac{1}{3}r^3\right]_0^{1/\sqrt{2}}$$

$$= 2\pi\left(-\frac{1}{3}\right)\left(\frac{1}{\sqrt{2}}-1\right) = \frac{\pi}{3}\left(2-\sqrt{2}\right)$$

25. The given solid is the region inside the cylinder $x^2+y^2 = 4$ between the surfaces $z = \sqrt{64-4x^2-4y^2}$ and $z = -\sqrt{64-4x^2-4y^2}$. So

$$V = \iint_{x^2+y^2 \le 4}\left[\sqrt{64-4x^2-4y^2}-\left(-\sqrt{64-4x^2-4y^2}\right)\right]dA$$

$$= \iint_{x^2+y^2 \le 4} 2\sqrt{64-4x^2-4y^2}\, dA = 4\int_0^{2\pi}\int_0^2 \sqrt{16-r^2}\, r\, dr\, d\theta$$

$$= 4\int_0^{2\pi} d\theta \int_0^2 r\sqrt{16-r^2}\, dr = 4\,[\theta]_0^{2\pi}\left[-\frac{1}{3}\left(16-r^2\right)^{3/2}\right]_0^2$$

$$= 8\pi\left(-\frac{1}{3}\right)\left(12^{3/2}-16^{2/3}\right) = \frac{8\pi}{3}\left(64-24\sqrt{3}\right)$$

27.

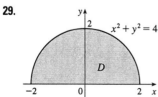

$$\int_0^1\int_0^{\sqrt{1-x^2}} e^{x^2+y^2}\, dy\, dx = \int_0^{\pi/2}\int_0^1 e^{r^2} r\, dr\, d\theta = \int_0^{\pi/2} d\theta \int_0^1 re^{r^2}\, dr$$

$$= [\theta]_0^{\pi/2}\left[\frac{1}{2}e^{r^2}\right]_0^1 = \frac{1}{4}\pi\,(e-1)$$

29.

$$\int_0^\pi\int_0^2 \left(r\cos\theta\right)^2\left(r\sin\theta\right)^2 r\, dr\, d\theta = \int_0^\pi \left(\sin\theta\cos\theta\right)^2 d\theta \int_0^2 r^5\, dr$$

$$= \int_0^\pi \left(\frac{1}{2}\sin 2\theta\right)^2 d\theta \int_0^2 r^5\, dr$$

$$= \frac{1}{4}\left[\frac{1}{2}\theta-\frac{1}{8}\sin 4\theta\right]_0^\pi \left[\frac{1}{6}r^6\right]_0^2$$

$$= \frac{1}{4}\left(\frac{\pi}{2}\right)\left(\frac{64}{6}\right) = \frac{4\pi}{3}$$

31. The surface of the water in the pool is a circular disk D with radius 20 ft. If we place D on coordinate axes with the origin at the center of D and define $f(x, y)$ to be the depth of the water at (x, y), then the volume of water in the pool is the volume of the solid that lies above $D = \{(x, y) \mid x^2 + y^2 \le 400\}$ and below the graph of $f(x, y)$. We can associate north with the positive y-direction, so we are given that the depth is constant in the x-direction and the depth increases linearly in the y-direction from $f(0, -20) = 2$ to $f(0, 20) = 7$. The trace in the yz-plane is a line segment from $(0, -20, 2)$ to $(0, 20, 7)$. The slope of this line is $\frac{7-2}{20-(-20)} = \frac{1}{8}$, so an equation of the line is $z - 7 = \frac{1}{8}(y - 20) \implies z = \frac{1}{8}y + \frac{9}{2}$. Since $f(x, y)$ is independent of x, $f(x, y) = \frac{1}{8}y + \frac{9}{2}$. Thus the volume is given by $\iint_D f(x, y)\, dA$, which is most conveniently evaluated using polar coordinates. Then $D = \{(r, \theta) \mid 0 \le r \le 20, 0 \le \theta \le 2\pi\}$ and substituting $x = r\cos\theta$, $y = r\sin\theta$ the integral becomes

$$\int_0^{2\pi} \int_0^{20} \left(\tfrac{1}{8}r\sin\theta + \tfrac{9}{2}\right) r\, dr\, d\theta = \int_0^{2\pi} \left[\tfrac{1}{24}r^3 \sin\theta + \tfrac{9}{4}r^2\right]_{r=0}^{r=20} d\theta$$

$$= \int_0^{2\pi} \left(\tfrac{1000}{3}\sin\theta + 900\right) d\theta = \left[-\tfrac{1000}{3}\cos\theta + 900\theta\right]_0^{2\pi}$$

$$= 1800\pi$$

Thus the pool contains $1800\pi \approx 5655$ ft^3 of water.

33. $\int_{1/\sqrt{2}}^1 \int_{\sqrt{1-x^2}}^x xy\, dy\, dx + \int_1^{\sqrt{2}} \int_0^x xy\, dy\, dx + \int_{\sqrt{2}}^2 \int_0^{\sqrt{4-x^2}} xy\, dy\, dx = \int_0^{\pi/4} \int_1^2 r^3 \cos\theta \sin\theta\, dr\, d\theta$

$$= \int_0^{\pi/4} \left[\frac{r^4}{4} \cos\theta \sin\theta\right]_{r=1}^{r=2} d\theta$$

$$= \frac{15}{4} \int_0^{\pi/4} \sin\theta \cos\theta\, d\theta$$

$$= \frac{15}{4} \left[\frac{\sin^2\theta}{2}\right]_0^{\pi/4} = \frac{15}{16}$$

35. (a) We integrate by parts with $u = x$ and $dv = xe^{-x^2} dx$. Then $du = dx$ and $v = -\frac{1}{2}e^{-x^2}$, so

$$\int_0^\infty x^2 e^{-x^2}\, dx = \lim_{t\to\infty} \int_0^t x^2 e^{-x^2}\, dx = \lim_{t\to\infty} \left(\left[-\tfrac{1}{2}xe^{-x^2}\right]_0^t + \int_0^t \tfrac{1}{2}e^{-x^2}\, dx\right)$$

$$= \lim_{t\to\infty} \left(-\tfrac{1}{2}te^{-t^2}\right) + \tfrac{1}{2}\int_0^\infty e^{-x^2}\, dx = 0 + \tfrac{1}{2}\int_0^\infty e^{-x^2}\, dx \text{ (by l'Hospital's Rule)}$$

$$= \tfrac{1}{4}\int_{-\infty}^\infty e^{-x^2}\, dx \text{ (since } e^{-x^2} \text{ is an even function)} = \tfrac{1}{4}\sqrt{\pi} \text{ [by Exercise 34(c)]}$$

(b) Let $u = \sqrt{x}$. Then $u^2 = x \implies dx = 2u\, du \implies$

$$\int_0^\infty \sqrt{x}e^{-x}\, dx = \lim_{t\to\infty} \int_0^t \sqrt{x}e^{-x}\, dx = \lim_{t\to\infty} \int_0^{\sqrt{t}} ue^{-u^2} 2u\, du = 2\int_0^\infty u^2 e^{-u^2}\, du$$

$$= 2\left(\tfrac{1}{4}\sqrt{\pi}\right) \text{ [by part(a)]} = \tfrac{1}{2}\sqrt{\pi}$$

16.5 Applications of Double Integrals   ET 15.5

1. $Q = \iint_D \left(x^2 + 3y^2\right) dA = \int_0^2 \int_1^2 \left(x^2 + 3y^2\right) dy \, dx = \int_0^2 \left[x^2 y + y^3\right]_{y=1}^{y=2} dx = \int_0^2 \left(x^2 + 7\right) dx$

$= \left[\frac{1}{3}x^3 + 7x\right]_0^2 = \frac{8}{3} + 14 = \frac{50}{3}$ C

3. $m = \iint_D \rho(x,y) \, dA = \int_{-1}^1 \int_0^1 x^2 \, dy \, dx = \int_{-1}^1 x^2 \, dx \int_0^1 dy = \left[\frac{1}{3}x^3\right]_{-1}^1 [y]_0^1 = \frac{2}{3}$,

$\bar{x} = \frac{1}{m} \iint_D x\rho(x,y) \, dA = \frac{3}{2} \int_{-1}^1 \int_0^1 x^3 \, dy \, dx = \frac{3}{2} \int_{-1}^1 x^3 \, dx \int_0^1 dy = \frac{3}{2} \left[\frac{1}{4}x^4\right]_{-1}^1 [y]_0^1 = 0$,

$\bar{y} = \frac{1}{m} \iint_D y\rho(x,y) \, dA = \frac{3}{2} \int_{-1}^1 \int_0^1 x^2 y \, dy \, dx = \frac{3}{2} \int_{-1}^1 x^2 \, dx \int_0^1 y \, dy = \frac{3}{2} \left[\frac{1}{3}x^3\right]_{-1}^1 \left[\frac{1}{2}y^2\right]_0^1 = \frac{1}{2}$. Hence

$(\bar{x}, \bar{y}) = \left(0, \frac{1}{2}\right)$.

5. $m = \int_0^2 \int_{x/2}^{3-x} (x+y) \, dy \, dx = \int_0^2 \left[xy + \frac{1}{2}y^2\right]_{y=x/2}^{y=3-x} dx = \int_0^2 \left[x\left(3 - \frac{3}{2}x\right) + \frac{1}{2}(3-x)^2 - \frac{1}{8}x^2\right] dx$

$= \int_0^2 \left(-\frac{9}{8}x^2 + \frac{9}{2}\right) dx = \left[-\frac{9}{8}\left(\frac{1}{3}x^3\right) + \frac{9}{2}x\right]_0^2 = 6$,

$M_y = \int_0^2 \int_{x/2}^{3-x} \left(x^2 + xy\right) dy \, dx = \int_0^2 \left[x^2 y + \frac{1}{2}xy^2\right]_{y=x/2}^{y=3-x} dx = \int_0^2 \left(\frac{9}{2}x - \frac{9}{8}x^3\right) dx = \frac{9}{2}$, and

$M_x = \int_0^2 \int_{x/2}^{3-y} \left(xy + y^2\right) dy \, dx = \int_0^2 \left[\frac{1}{2}xy^2 + \frac{1}{3}y^3\right]_{y=x/2}^{y=3-x} dx = \int_0^2 \left(9 - \frac{9}{2}x\right) dx = 9$. Hence $m = 6$,

$(\bar{x}, \bar{y}) = \left(\dfrac{M_y}{m}, \dfrac{M_x}{m}\right) = \left(\dfrac{3}{4}, \dfrac{3}{2}\right)$.

7. $m = \int_0^1 \int_{x^2}^1 xy \, dy \, dx = \int_0^1 \left(\frac{1}{2}x - \frac{1}{2}x^5\right) dx = \frac{1}{4} - \frac{1}{12} = \frac{1}{6}$,

$M_y = \int_0^1 \int_{x^2}^1 x^2 y \, dy \, dx = \int_0^1 \left(\frac{1}{2}x^2 - \frac{1}{2}x^6\right) dx = \frac{1}{6} - \frac{1}{14} = \frac{2}{21}$ and

$M_x = \int_0^1 \int_{x^2}^1 xy^2 \, dy \, dx = \int_0^1 \left(\frac{1}{3}x - \frac{1}{3}x^7\right) dx = \frac{1}{6} - \frac{1}{24} = \frac{1}{8}$. Hence $m = \frac{1}{6}$, $(\bar{x}, \bar{y}) = \left(\frac{4}{7}, \frac{3}{4}\right)$.

9.

$m = \int_{-1}^2 \int_{y^2}^{y+2} 3 \, dx \, dy = \int_{-1}^2 \left(3y + 6 - 3y^2\right) dy = \frac{27}{2}$,

$M_y = \int_{-1}^2 \int_{y^2}^{y+2} 3x \, dx \, dy = \int_{-1}^2 \frac{3}{2} \left[(y+2)^2 - y^4\right] dy$

$= \left[\frac{1}{2}(y+2)^3 - \frac{3}{10}y^5\right]_{-1}^2 = \frac{108}{5}$

and

$M_x = \int_{-1}^2 \int_{y^2}^{y+2} 3y \, dx \, dy = \int_{-1}^2 \left(3y^2 + 6y - 3y^3\right) dy$

$= \left[y^3 + 3y^2 - \frac{3}{4}y^4\right]_{-1}^2 = \frac{27}{4}$

Hence $m = \frac{27}{2}$, $(\bar{x}, \bar{y}) = \left(\frac{8}{5}, \frac{1}{2}\right)$.

11. $\rho(x,y) = ky = kr \sin\theta$, $m = \int_0^{\pi/2} \int_0^1 kr^2 \sin\theta \, dr \, d\theta = \frac{1}{3}k \int_0^{\pi/2} \sin\theta \, d\theta = \frac{1}{3}k \left[-\cos\theta\right]_0^{\pi/2} = \frac{1}{3}k$,

$M_y = \int_0^{\pi/2} \int_0^1 kr^3 \sin\theta \cos\theta \, dr \, d\theta = \frac{1}{4}k \int_0^{\pi/2} \sin\theta \cos\theta \, d\theta = \frac{1}{8}k \left[-\cos 2\theta\right]_0^{\pi/2} = \frac{1}{8}k$,

$M_x = \int_0^{\pi/2} \int_0^1 kr^3 \sin^2\theta \, dr \, d\theta = \frac{1}{4}k \int_0^{\pi/2} \sin^2\theta \, d\theta = \frac{1}{8}k \left[\theta + \sin 2\theta\right]_0^{\pi/2} = \frac{\pi}{16}k$. Hence $(\bar{x}, \bar{y}) = \left(\frac{3}{8}, \frac{3\pi}{16}\right)$.

13. Placing the vertex opposite the hypotenuse at $(0,0)$, $\rho(x,y) = k\left(x^2 + y^2\right)$. Then

$$m = \int_0^a \int_0^{a-x} k\left(x^2 + y^2\right) dy \, dx = k \int_0^a \left[ax^2 - x^3 + \frac{1}{3}(a-x)^3\right] dx$$

$$= k \left[\frac{1}{3}ax^3 - \frac{1}{4}x^4 - \frac{1}{12}(a-x)^4\right]_0^a = \frac{1}{6}ka^4$$

By symmetry,

$$M_y = M_x = \int_0^a \int_0^{a-x} ky\left(x^2 + y^2\right) dy\, dx = k\int_0^a \left[\tfrac{1}{2}\left(a-x\right)^2 x^2 + \tfrac{1}{4}\left(a-x\right)^4\right] dx$$

$$= k\left[\tfrac{1}{6}a^2 x^3 - \tfrac{1}{4}ax^4 + \tfrac{1}{10}x^5 - \tfrac{1}{20}\left(a-x\right)^5\right]_0^a = \tfrac{1}{15}ka^5$$

Hence $(\overline{x}, \overline{y}) = \left(\tfrac{2}{5}a, \tfrac{2}{5}a\right)$.

15. $I_x = \iint_D y^2 \rho\left(x,y\right) dA = \int_0^1 \int_{x^2}^1 y^2\left(xy\right) dy\, dx = \int_0^1 \left[\tfrac{1}{4}xy^4\right]_{y=x^2}^{y=1} dx = \int_0^1 \tfrac{1}{4}\left(x - x^9\right) dx = \tfrac{1}{8} - \tfrac{1}{40} = \tfrac{1}{10}$,

$I_y = \iint_D x^2 \rho\left(x,y\right) dA = \int_0^1 \int_{x^2}^1 x^3 y\, dy\, dx = \int_0^1 \left[\tfrac{1}{2}x^3 y^2\right]_{y=x^2}^{y=1} dx = \int_0^1 \tfrac{1}{2}\left(x^3 - x^7\right) dx = \tfrac{1}{8} - \tfrac{1}{16} = \tfrac{1}{16}$,

$I_0 = I_x + I_y = \tfrac{13}{80}$.

17. $I_x = \int_{-1}^2 \int_{y^2}^{y+2} 3y^2 \, dx\, dy = \int_{-1}^2 \left(3y^3 + 6y^2 - 3y^4\right) dy = \left[\tfrac{3}{4}y^4 + 2y^3 - \tfrac{3}{5}y^5\right]_{-1}^2 = \tfrac{189}{20}$,

$I_y = \int_{-1}^2 \int_{y^2}^{y+2} 3x^2 \, dx\, dy = \int_{-1}^2 \left[\left(y+2\right)^3 - y^6\right] dy = \left[\tfrac{1}{4}\left(y+2\right)^4 - \tfrac{1}{7}y^7\right]_{-1}^2 = \tfrac{1269}{28}$, and

$I_0 = I_x + I_y = \tfrac{1917}{35}$.

19. Using a CAS, we find $m = \iint_D \rho\left(x,y\right) dA = \int_0^\pi \int_0^{\sin x} xy\, dy\, dx = \dfrac{\pi^2}{8}$. Then

$$\overline{x} = \frac{1}{m}\iint_D x\rho\left(x,y\right) dA = \frac{8}{\pi^2}\int_0^\pi \int_0^{\sin x} x^2 y\, dy\, dx = \frac{2\pi}{3} - \frac{1}{\pi} \text{ and}$$

$$\overline{y} = \frac{1}{m}\iint_D y\rho\left(x,y\right) dA = \frac{8}{\pi^2}\int_0^\pi \int_0^{\sin x} xy^2 \, dy\, dx = \frac{16}{9\pi}, \text{ so } (\overline{x}, \overline{y}) = \left(\frac{2\pi}{3} - \frac{1}{\pi}, \frac{16}{9\pi}\right).$$

The moments of inertia are $I_x = \iint_D y^2 \rho\left(x,y\right) dA = \int_0^\pi \int_0^{\sin x} xy^3 \, dy\, dx = \dfrac{3\pi^2}{64}$,

$I_y = \iint_D x^2 \rho\left(x,y\right) dA = \int_0^\pi \int_0^{\sin x} x^3 y\, dy\, dx = \dfrac{\pi^2}{16}\left(\pi^2 - 3\right)$, and $I_0 = I_x + I_y = \dfrac{\pi^2}{64}\left(4\pi^2 - 9\right)$.

21. $I_x = \int_0^a \int_0^a \rho y^2 \, dx\, dy = \rho\int_0^a dx \int_0^a y^2 \, dy = \rho\left[x\right]_0^a \left[\tfrac{1}{3}y^3\right]_0^a = \rho a\left(\tfrac{1}{3}a^3\right) = \tfrac{1}{3}\rho a^4 = I_y$ by symmetry, and

$m = \rho a^2$ since the lamina is homogeneous. Hence $\overline{\overline{x}}^2 = \dfrac{I_y}{m} \ \Rightarrow \ \overline{\overline{x}} = \left[\left(\tfrac{1}{3}\rho a^4\right) / \left(\rho a^2\right)\right]^{1/2} = \tfrac{1}{\sqrt{3}}a$ and

$\overline{\overline{y}}^2 = \dfrac{I_x}{m} \ \Rightarrow \ \overline{\overline{y}} = \tfrac{1}{\sqrt{3}}a$.

23. (a) $f\left(x,y\right)$ is a joint density function, so we know $\iint_{\mathbb{R}^2} f\left(x,y\right) dA = 1$. Since $f\left(x,y\right) = 0$ outside the rectangle $[0,1] \times [0,2]$, we can say

$$\iint_{\mathbb{R}^2} f\left(x,y\right) dA = \int_{-\infty}^\infty \int_{-\infty}^\infty f\left(x,y\right) dy\, dx = \int_0^1 \int_0^2 Cx\left(1 + y\right) dy\, dx$$

$$= C\int_0^1 x\left[y + \tfrac{1}{2}y^2\right]_{y=0}^{y=2} dx = C\int_0^1 4x\, dx = C\left[2x^2\right]_0^1 = 2C$$

Then $2C = 1 \ \Rightarrow \ C = \tfrac{1}{2}$.

(b) $P\left(X \le 1, Y \le 1\right) = \int_{-\infty}^1 \int_{-\infty}^1 f\left(x,y\right) dy\, dx = \int_0^1 \int_0^1 \tfrac{1}{2}x\left(1 + y\right) dy\, dx$

$$= \int_0^1 \tfrac{1}{2}x\left[y + \tfrac{1}{2}y^2\right]_{y=0}^{y=1} dx = \int_0^1 \tfrac{1}{2}x\left(\tfrac{3}{2}\right) dx = \tfrac{3}{4}\left[\tfrac{1}{2}x^2\right]_0^1 = \tfrac{3}{8} \text{ or } 0.375$$

(c) $P(X+Y \leq 1) = P((X,Y) \in D)$ where D is the triangular region shown in the figure. Thus

$$P(X+Y \leq 1) = \iint_D f(x,y)\, dA = \int_0^1 \int_0^{1-x} \tfrac{1}{2}x(1+y)\, dy\, dx$$

$$= \int_0^1 \tfrac{1}{2}x\left[y + \tfrac{1}{2}y^2\right]_{y=0}^{y=1-x} dx = \int_0^1 \tfrac{1}{2}x\left(\tfrac{1}{2}x^2 - 2x + \tfrac{3}{2}\right) dx$$

$$= \tfrac{1}{4}\int_0^1 \left(x^3 - 4x^2 + 3x\right) dx = \tfrac{1}{4}\left[\tfrac{x^4}{4} - 4\tfrac{x^3}{3} + 3\tfrac{x^2}{2}\right]_0^1$$

$$= \tfrac{5}{48} \approx 0.1042$$

25. (a) $f(x,y) \geq 0$, so f is a joint density function if $\iint_{\mathbb{R}^2} f(x,y)\, dA = 1$. Here, $f(x,y) = 0$ outside the first quadrant, so

$$\iint_{\mathbb{R}^2} f(x,y)\, dA = \int_0^\infty \int_0^\infty 0.1 e^{-(0.5x+0.2y)}\, dy\, dx = 0.1 \int_0^\infty \int_0^\infty e^{-0.5x} e^{-0.2y}\, dy\, dx$$

$$= 0.1 \int_0^\infty e^{-0.5x}\, dx \int_0^\infty e^{-0.2y}\, dy = 0.1 \lim_{t\to\infty} \int_0^t e^{-0.5x}\, dx \lim_{t\to\infty} \int_0^t e^{-0.2y}\, dy$$

$$= 0.1 \lim_{t\to\infty}\left[-2e^{-0.5x}\right]_0^t \lim_{t\to\infty}\left[-5e^{-0.2y}\right]_0^t$$

$$= 0.1 \lim_{t\to\infty}\left[-2\left(e^{-0.5t}-1\right)\right] \lim_{t\to\infty}\left[-5\left(e^{-0.2t}-1\right)\right]$$

$$= (0.1)\cdot(-2)(0-1)\cdot(-5)(0-1) = 1$$

Thus $f(x,y)$ is a joint density function.

(b) (i) No restriction is placed on X, so

$$P(Y \geq 1) = \int_{-\infty}^\infty \int_1^\infty f(x,y)\, dy\, dx = \int_0^\infty \int_1^\infty 0.1 e^{-(0.5x+0.2y)}\, dy\, dx$$

$$= 0.1 \int_0^\infty e^{-0.5x}\, dx \int_1^\infty e^{-0.2y}\, dy = 0.1 \lim_{t\to\infty} \int_0^t e^{-0.5x}\, dx \lim_{t\to\infty} \int_1^t e^{-0.2y}\, dy$$

$$= 0.1 \lim_{t\to\infty}\left[-2e^{-0.5x}\right]_0^t \lim_{t\to\infty}\left[-5e^{-0.2y}\right]_1^t$$

$$= 0.1 \lim_{t\to\infty}\left[-2\left(e^{-0.5t}-1\right)\right] \lim_{t\to\infty}\left[-5\left(e^{-0.2t}-e^{-0.2}\right)\right]$$

$$= (0.1)\cdot(-2)(0-1)\cdot(-5)\left(0-e^{-0.2}\right) = e^{-0.2} \approx 0.8187$$

(ii) $P(X \leq 2, Y \leq 4) = \int_{-\infty}^2 \int_{-\infty}^4 f(x,y)\, dy\, dx = \int_0^2 \int_0^4 0.1 e^{-(0.5x+0.2y)}\, dy\, dx$

$$= 0.1 \int_0^2 e^{-0.5x}\, dx \int_0^4 e^{-0.2y}\, dy$$

$$= 0.1\left[-2e^{-0.5x}\right]_0^2 \left[-5e^{-0.2y}\right]_0^4$$

$$= (0.1)\cdot(-2)\left(e^{-1}-1\right)\cdot(-5)\left(e^{-0.8}-1\right)$$

$$= \left(e^{-1}-1\right)\left(e^{-0.8}-1\right) = 1 + e^{-1.8} - e^{-0.8} - e^{-1}$$

$$\approx 0.3481$$

(c) The expected value of X is given by

$$\mu_1 = \iint_{\mathbb{R}^2} x f(x,y)\, dA = \int_0^\infty \int_0^\infty x\left[0.1 e^{-(0.5x+0.2y)}\right] dy\, dx$$

$$= 0.1 \int_0^\infty x e^{-0.5x}\, dx \int_0^\infty e^{-0.2y}\, dy$$

$$= 0.1 \lim_{t\to\infty} \int_0^t x e^{-0.5x}\, dx \lim_{t\to\infty} \int_0^t e^{-0.2y}\, dy$$

To evaluate the first integral, we integrate by parts with $u = x$ and $dv = e^{-0.5x} \, dx$ (or we can use Formula 96 in the Table of Integrals):

$\int xe^{-0.5x} \, dx = -2xe^{-0.5x} - \int -2e^{-0.5x} \, dx = -2xe^{-0.5x} - 4e^{-0.5x} = -2(x+2)e^{-0.5x}$. Thus

$$\mu_1 = 0.1 \lim_{t \to \infty} \left[-2(x+2)e^{-0.5x} \right]_0^t \lim_{t \to \infty} \left[-5e^{-0.2y} \right]_0^t$$

$$= 0.1 \lim_{t \to \infty} (-2)\left[(t+2)e^{-0.5t} - 2 \right] \lim_{t \to \infty} (-5)\left[e^{-0.2t} - 1 \right]$$

$$= 0.1(-2)\left(\lim_{t \to \infty} \frac{t+2}{e^{0.5t}} - 2 \right)(-5)(-1) = 2 \text{ (by l'Hospital's Rule)}$$

The expected value of Y is given by

$$\mu_2 = \iint_{\mathbb{R}^2} yf(x,y) \, dA = \int_0^\infty \int_0^\infty y\left[0.1e^{-(0.5x+0.2y)} \right] dy \, dx$$

$$= 0.1 \int_0^\infty e^{-0.5x} \, dx \int_0^\infty ye^{-0.2y} \, dy$$

$$= 0.1 \lim_{t \to \infty} \int_0^t e^{-0.5x} \, dx \lim_{t \to \infty} \int_0^t ye^{-0.2y} \, dy$$

To evaluate the second integral, we integrate by parts with $u = y$ and $dv = e^{-0.2y} \, dy$ (or again we can use Formula 96 in the Table of Integrals) which gives

$\int ye^{-0.2y} \, dy = -5ye^{-0.2y} + \int 5e^{-0.2y} \, dy = -5(y+5)e^{-0.2y}$. Then

$$\mu_2 = 0.1 \lim_{t \to \infty} \left[-2e^{-0.5x} \right]_0^t \lim_{t \to \infty} \left[-5(y+5)e^{-0.2y} \right]_0^t$$

$$= 0.1 \lim_{t \to \infty} \left[-2(e^{-0.5t} - 1) \right] \lim_{t \to \infty} \left(-5\left[(t+5)e^{-0.2t} - 5 \right] \right)$$

$$= 0.1(-2)(-1) \cdot (-5)\left(\lim_{t \to \infty} \frac{t+5}{e^{0.2t}} - 5 \right) = 5 \text{ (by l'Hospital's Rule)}$$

27. The random variables X and Y are normally distributed with $\mu_1 = 45$, $\mu_2 = 20$, $\sigma_1 = 0.5$, and $\sigma_2 = 0.1$. The individual density functions for X and Y, then, are $f_1(x) = \dfrac{1}{0.5\sqrt{2\pi}} e^{-(x-45)^2/0.5}$ and

$f_2(y) = \dfrac{1}{0.1\sqrt{2\pi}} e^{-(y-20)^2/0.02}$. Since X and Y are independent, the joint density function is the product

$$f(x,y) = f_1(x) f_2(y) = \frac{1}{0.5\sqrt{2\pi}} e^{-(x-45)^2/0.5} \frac{1}{0.1\sqrt{2\pi}} e^{-(y-20)^2/0.02}$$

$$= \tfrac{10}{\pi} e^{-2(x-45)^2 - 50(y-20)^2}$$

Then

$$P(40 \le X \le 50, 20 \le Y \le 25) = \int_{40}^{50} \int_{20}^{25} f(x,y) \, dy \, dx$$

$$= \tfrac{10}{\pi} \int_{40}^{50} \int_{20}^{25} e^{-2(x-45)^2 - 50(y-20)^2} \, dy \, dx$$

Using a CAS or calculator to evaluate the integral, we get $P(40 \le X \le 50, 20 \le Y \le 25) \approx 0.500$.

29. (a) If $f(P, A)$ is the probability that an individual at A will be infected by an individual at P, and $k\,dA$ is the number of infected individuals in an element of area dA, then $f(P, A)\,k\,dA$ is the number of infections that should result from exposure of the individual at A to infected people in the element of area dA. Integration over D gives the number of infections of the person at A due to all the infected people in D. In rectangular coordinates (with the origin at the city's center), the exposure of a person at A is

$$E = \iint_D kf(P, A)\,dA = k\iint_D \frac{20 - d(P, A)}{20}\,dA$$

$$= k\iint_D \left[1 - \frac{\sqrt{(x - x_0)^2 + (y - y_0)^2}}{20}\right]dx\,dy$$

(b) If $A = (0, 0)$, then

$$E = k\iint_D \left[1 - \frac{1}{20}\sqrt{x^2 + y^2}\right]dx\,dy = k\int_0^{2\pi}\int_0^{10}\left(1 - \frac{r}{20}\right)r\,dr\,d\theta$$

$$= 2\pi k\left[\frac{r^2}{2} - \frac{r^3}{60}\right]_0^{10} = 2\pi k\left(50 - \frac{50}{3}\right) = \frac{200}{3}\pi k \approx 209k$$

For A at the edge of the city, it is convenient to use a polar coordinate system centered at A. Then the polar equation for the circular boundary of the city becomes $r = 20\cos\theta$ instead of $r = 10$, and the distance from A to a point P in the city is again r (see the figure). So

$$E = k\int_{-\pi/2}^{\pi/2}\int_0^{20\cos\theta}\left(1 - \frac{r}{20}\right)r\,dr\,d\theta$$

$$= k\int_{-\pi/2}^{\pi/2}\left[\frac{r^2}{2} - \frac{r^3}{60}\right]_{r=0}^{r=20\cos\theta}d\theta$$

$$= k\int_{-\pi/2}^{\pi/2}\left(200\cos^2\theta - \frac{400}{3}\cos^3\theta\right)d\theta$$

$$= 200k\int_{-\pi/2}^{\pi/2}\left[\frac{1}{2} + \frac{1}{2}\cos 2\theta - \frac{2}{3}\left(1 - \sin^2\theta\right)\cos\theta\right]d\theta$$

$$= 200k\left[\frac{1}{2}\theta + \frac{1}{4}\sin 2\theta - \frac{2}{3}\sin\theta + \frac{2}{3}\cdot\frac{1}{3}\sin^3\theta\right]_{-\pi/2}^{\pi/2}$$

$$= 200k\left[\frac{\pi}{4} + 0 - \frac{2}{3} + \frac{2}{9} + \frac{\pi}{4} + 0 - \frac{2}{3} + \frac{2}{9}\right]$$

$$= 200k\left(\frac{\pi}{2} - \frac{8}{9}\right) \approx 136k$$

$r = 20\cos\theta$

Therefore the risk of infection is much lower at the edge of the city than in the middle, so it is better to live at the edge.

≡16.6 Surface Area

1. Here $z = f(x, y) = 2 + 3x + 4y$ and D is the rectangle $[0, 5] \times [1, 4]$, so by Formula 2 the area of the surface is

$$A(S) = \iint_D \sqrt{3^2 + 4^2 + 1}\, dA = \sqrt{26} \iint_D dA = \sqrt{26}\, A(D)$$

$$= \sqrt{26}\, (5)\, (3) = 15\sqrt{26}$$

3. $z = f(x, y) = 6 - 3x - 2y$ which intersects the xy-plane in the line $3x + 2y = 6$, so D is the triangular region given by $\{(x, y) \mid 0 \le x \le 2, 0 \le y \le 3 - \frac{3}{2}x\}$. Thus

$$A(S) = \iint_D \sqrt{(-3)^2 + (-2)^2 + 1}\, dA = \sqrt{14} \iint_D dA = \sqrt{14}\, A(D)$$

$$= \sqrt{14}\left(\frac{1}{2} \cdot 2 \cdot 3\right) = 3\sqrt{14}$$

5. $y^2 + z^2 = 9 \;\Rightarrow\; z = \sqrt{9 - y^2}.\; f_x = 0,\; f_y = -y(9 - y^2)^{-1/2} \;\Rightarrow$

$$A(S) = \int_0^4 \int_0^2 \sqrt{0^2 + \left[-y(9 - y^2)^{-1/2}\right]^2 + 1}\, dy\, dx = \int_0^4 \int_0^2 \sqrt{\frac{y^2}{9 - y^2} + 1}\, dy\, dx$$

$$= \int_0^4 \int_0^2 \frac{3}{\sqrt{9 - y^2}}\, dy\, dx = 3 \int_0^4 \left[\sin^{-1}\frac{y}{3}\right]_{y=0}^{y=2} dx = 3\left[\left(\sin^{-1}\frac{2}{3}\right)x\right]_0^4 = 12\sin^{-1}\frac{2}{3}$$

7. $z = f(x, y) = y^2 - x^2$ with $1 \le x^2 + y^2 \le 4$. Then

$$A(S) = \iint_D \sqrt{1 + 4x^2 + 4y^2}\, dA = \int_0^{2\pi} \int_1^2 \sqrt{1 + 4r^2}\, r\, dr\, d\theta = \int_0^{2\pi} d\theta \int_1^2 \sqrt{1 + 4r^2}\, r\, dr$$

$$= [\theta]_0^{2\pi} \left[\frac{1}{12}(1 + 4r^2)^{3/2}\right]_1^2 = \frac{\pi}{6}\left(17\sqrt{17} - 5\sqrt{5}\right)$$

9. $z = f(x, y) = xy$ with $0 \le x^2 + y^2 \le 1$, so $f_x = y,\; f_y = x \;\Rightarrow$

$$A(S) = \iint_D \sqrt{y^2 + x^2 + 1}\, dA = \int_0^{2\pi} \int_0^1 \sqrt{r^2 + 1}\, r\, dr\, d\theta = \int_0^{2\pi} \left[\frac{1}{3}(r^2 + 1)^{3/2}\right]_{r=0}^{r=1} d\theta$$

$$= \int_0^{2\pi} \frac{1}{3}(2\sqrt{2} - 1)\, d\theta = \frac{2\pi}{3}(2\sqrt{2} - 1).$$

11. $z = \sqrt{a^2 - x^2 - y^2},\; z_x = -x(a^2 - x^2 - y^2)^{-1/2},\; z_y = -y(a^2 - x^2 - y^2)^{-1/2},$

$$A(S) = \iint_D \sqrt{\frac{x^2 + y^2}{a^2 - x^2 - y^2} + 1}\, dA$$

$$= \int_{-\pi/2}^{\pi/2} \int_0^{a\cos\theta} \sqrt{\frac{r^2}{a^2 - r^2} + 1}\, r\, dr\, d\theta$$

$$= \int_{-\pi/2}^{\pi/2} \int_0^{a\cos\theta} \frac{ar}{\sqrt{a^2 - r^2}}\, dr\, d\theta$$

$$= \int_{-\pi/2}^{\pi/2} \left[-a\sqrt{a^2 - r^2}\right]_{r=0}^{r=a\cos\theta} d\theta$$

$$= \int_{-\pi/2}^{\pi/2} -a\left(\sqrt{a^2 - a^2\cos^2\theta} - a\right) d\theta = 2a^2 \int_0^{\pi/2}\left(1 - \sqrt{1 - \cos^2\theta}\right) d\theta$$

$$= 2a^2 \int_0^{\pi/2} d\theta - 2a^2 \int_0^{\pi/2} \sqrt{\sin^2\theta}\, d\theta = a^2\pi - 2a^2 \int_0^{\pi/2} \sin\theta\, d\theta = a^2(\pi - 2)$$

13. (a) The midpoints of the four squares are $\left(\frac{1}{4}, \frac{1}{4}\right)$, $\left(\frac{1}{4}, \frac{3}{4}\right)$, $\left(\frac{3}{4}, \frac{1}{4}\right)$, and $\left(\frac{3}{4}, \frac{3}{4}\right)$; the derivatives of the function
$f(x, y) = x^2 + y^2$ are $f_x(x, y) = 2x$ and $f_y(x, y) = 2y$, so the Midpoint Rule gives

$$A(S) = \int_0^1 \int_0^1 \sqrt{[f_x(x, y)]^2 + [f_y(x, y)]^2 + 1} \, dy \, dx$$

$$\approx \frac{1}{4}\left(\sqrt{\left[2\left(\frac{1}{4}\right)\right]^2 + \left[2\left(\frac{1}{4}\right)\right]^2 + 1} + \sqrt{\left[2\left(\frac{1}{4}\right)\right]^2 + \left[2\left(\frac{3}{4}\right)\right]^2 + 1}\right.$$

$$\left. + \sqrt{\left[2\left(\frac{3}{4}\right)\right]^2 + \left[2\left(\frac{1}{4}\right)\right]^2 + 1} + \sqrt{\left[2\left(\frac{3}{4}\right)\right]^2 + \left[2\left(\frac{3}{4}\right)\right]^2 + 1}\right)$$

$$= \frac{1}{4}\left(\sqrt{\frac{3}{2}} + 2\sqrt{\frac{7}{2}} + \sqrt{\frac{11}{2}}\right) \approx 1.8279$$

(b) A CAS estimates the integral to be

$A(S) = \int_0^1 \int_0^1 \sqrt{f_x^2 + f_y^2 + 1} \, dy \, dx = \int_0^1 \int_0^1 \sqrt{4x^2 + 4y^2 + 1} \, dy \, dx \approx 1.8616$. This agrees with the
Midpoint estimate only in the first decimal place.

15. Since $f(x, y) = x^2 + 2y$, we have $f_x = 2x$, $f_y = 2$. We use a CAS to calculate the integral
$A(S) = \int_0^1 \int_0^1 \sqrt{f_x^2 + f_y^2 + 1} \, dy \, dx = \int_0^1 \int_0^1 \sqrt{4x^2 + 5} \, dy \, dx = \int_0^1 \sqrt{4x^2 + 5} \, dx$, and find that
$A(S) = \frac{3}{2} + \frac{5}{8}\ln 5$.

17. $f(x, y) = 1 + x^2 y^2 \quad \Rightarrow \quad f_x = 2xy^2$, $f_y = 2x^2 y$. We use a CAS (with precision reduced to five significant
digits, to speed up the calculation) to estimate the integral

$A(S) = \int_{-1}^1 \int_{-\sqrt{1-x^2}}^{\sqrt{1-x^2}} \sqrt{f_x^2 + f_y^2 + 1} \, dy \, dx = \int_{-1}^1 \int_{-\sqrt{1-x^2}}^{\sqrt{1-x^2}} \sqrt{4x^2 y^4 + 4x^4 y^2 + 1} \, dy \, dx$, and find that
$A(S) \approx 3.3213$.

19. Here $z = f(x, y) = ax + by + c$, $f_x(x, y) = a$, $f_y(x, y) = b$, so
$A(S) = \iint_D \sqrt{a^2 + b^2 + 1} \, dA = \sqrt{a^2 + b^2 + 1} \iint_D dA = \sqrt{a^2 + b^2 + 1}\, A(D)$.

21. If we project the surface onto the xz-plane, then the surface lies "above" the disk $x^2 + z^2 \leq 25$ in the xz-plane. We
have $y = f(x, z) = x^2 + z^2$ and, adapting Formula 2, the area of the surface is

$$A(S) = \iint_{x^2+z^2\leq 25} \sqrt{[f_x(x, z)]^2 + [f_z(x, z)]^2 + 1} \, dA = \iint_{x^2+z^2\leq 25} \sqrt{4x^2 + 4z^2 + 1} \, dA$$

Converting to polar coordinates $x = r\cos\theta$, $z = r\sin\theta$ we have

$$A(S) = \int_0^{2\pi} \int_0^5 \sqrt{4r^2 + 1}\, r \, dr \, d\theta = \int_0^{2\pi} d\theta \int_0^5 r\left(4r^2 + 1\right)^{1/2} dr$$

$$= [\theta]_0^{2\pi} \left[\frac{1}{12}\left(4r^2 + 1\right)^{3/2}\right]_0^5 = \frac{\pi}{6}\left(101\sqrt{101} - 1\right)$$

16.7 Triple Integrals

1. $\int_0^1 \int_{-1}^2 \int_0^3 xyz^2 \, dz \, dx \, dy = \int_0^1 \int_{-1}^2 xy \left[\frac{1}{3}z^3\right]_{z=0}^{z=3} dx \, dy = \int_0^1 \int_{-1}^2 9xy \, dx \, dy$

$= \int_0^1 \left[\frac{9}{2}x^2 y\right]_{x=-1}^{x=2} dy = \int_0^1 \frac{27}{2}y \, dy = \frac{27}{4}y^2\big]_0^1 = \frac{27}{4}$

3. $\int_0^1 \int_0^z \int_0^{x+z} 6xz \, dy \, dx \, dz = \int_0^1 \int_0^z [6xyz]_{y=0}^{y=x+z} dx \, dz = \int_0^1 \int_0^z 6xz \, (x+z) \, dx \, dz$

$= \int_0^1 \left[2x^3 z + 3x^2 z^2\right]_{x=0}^{x=z} dz = \int_0^1 \left(2z^4 + 3z^4\right) dz = \int_0^1 5z^4 \, dz = z^5\big]_0^1 = 1$

5. $\int_0^3 \int_0^1 \int_0^{\sqrt{1-z^2}} ze^y \, dx \, dz \, dy = \int_0^3 \int_0^1 [xze^y]_{x=0}^{x=\sqrt{1-z^2}} dz \, dy = \int_0^3 \int_0^1 ze^y \sqrt{1-z^2} \, dz \, dy$

$= \int_0^3 \left[-\frac{1}{3}\left(1-z^2\right)^{3/2} e^y\right]_{z=0}^{z=1} dy = \int_0^3 \frac{1}{3}e^y \, dy = \frac{1}{3}e^y\big]_0^3 = \frac{1}{3}\left(e^3 - 1\right)$

7. $\iiint_E 2x \, dV = \int_0^2 \int_0^{\sqrt{4-y^2}} \int_0^y 2x \, dz \, dx \, dy = \int_0^2 \int_0^{\sqrt{4-y^2}} [2xz]_{z=0}^{z=y} dx \, dy = \int_0^2 \int_0^{\sqrt{4-y^2}} 2xy \, dx \, dy$

$= \int_0^2 \left[x^2 y\right]_{x=0}^{x=\sqrt{4-y^2}} dy = \int_0^2 \left(4 - y^2\right) y \, dy = \left[2y^2 - \frac{1}{4}y^4\right]_0^2 = 4$

9. Here $E = \{(x, y, z) \mid 0 \le x \le 1, 0 \le y \le \sqrt{x}, 0 \le z \le 1 + x + y\}$, so

$\iiint_E 6xy \, dV = \int_0^1 \int_0^{\sqrt{x}} \int_0^{1+x+y} 6xy \, dz \, dy \, dx = \int_0^1 \int_0^{\sqrt{x}} [6xyz]_{z=0}^{z=1+x+y} dy \, dx$

$= \int_0^1 \int_0^{\sqrt{x}} 6xy \, (1 + x + y) \, dy \, dx = \int_0^1 \left[3xy^2 + 3x^2 y^2 + 2xy^3\right]_{y=0}^{y=\sqrt{x}} dx$

$= \int_0^1 \left(3x^2 + 3x^3 + 2x^{5/2}\right) dx = \left[x^3 + \frac{3}{4}x^4 + \frac{4}{7}x^{7/2}\right]_0^1 = \frac{65}{28}$

11. Here E is the region that lies below the plane with x-, y-, and z-intercepts 1, 2, and 3 respectively, that is, below the plane $2z + 6x + 3y = 6$ and above the region in the xy-plane bounded by the lines $x = 0$, $y = 0$ and $6x + 3y = 6$. So

$\iiint_E xy \, dV = \int_0^1 \int_0^{2-2x} \int_0^{3-3x-3y/2} xy \, dz \, dy \, dx = \int_0^1 \int_0^{2-2x} \left(3xy - 3x^2 y - \frac{3}{2}xy^2\right) dy \, dx$

$= \int_0^1 \left[\frac{3}{2}xy^2 - \frac{3}{2}x^2 y^2 - \frac{1}{2}xy^3\right]_{y=0}^{y=2-2x} dx = \int_0^1 \left(2x - 6x^2 + 6x^3 - 2x^4\right) dx$

$= \left[x^2 - 2x^3 + \frac{3}{2}x^4 - \frac{2}{5}x^5\right]_0^1 = \frac{1}{10}.$

13.

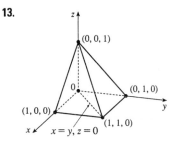

(0, 0, 1)
(0, 1, 0)
(1, 0, 0)
(1, 1, 0)
$x = y, z = 0$

By symmetry $\iiint_E z \, dV = 2 \iiint_{E'} z \, dV$ where E' is the part of E
to the left [as viewed from $(10, 10, 0)$] of the plane $x = y$. So

$\iiint_E z \, dV = \int_0^1 \int_y^1 \int_0^{1-x} 2z \, dz \, dx \, dy = \int_0^1 \int_y^1 (1 - x)^2 \, dx \, dy$

$= \int_0^1 \left[-\frac{1}{3}(1 - x)^3\right]_{x=y}^{x=1} dy = \int_0^1 \frac{1}{3}(1 - y)^3 \, dy$

$= \frac{1}{12}(1 - y)^4\big]_0^1 = \frac{1}{12}$

15.

The projection E on the yz-plane is the disk $y^2 + z^2 \le 1$. Using polar coordinates $y = r\cos\theta$ and $z = r\sin\theta$, we get

$$\iiint_E x\, dV = \iint_D \left[\int_{4y^2 + 4z^2}^4 x\, dx \right] dA$$

$$= \tfrac{1}{2} \iint_D \left[4^2 - \left(4y^2 + 4z^2 \right)^2 \right] dA = 8 \int_0^{2\pi} \int_0^1 \left(1 - r^4 \right) r\, dr\, d\theta$$

$$= 8 \int_0^{2\pi} d\theta \int_0^1 \left(r - r^5 \right) dr = 8\,(2\pi) \left[\tfrac{1}{2} r^2 - \tfrac{1}{6} r^6 \right]_0^1 = \tfrac{16\pi}{3}$$

17. The plane $2x + 3y + 6z = 12$ intersects the xy-plane when $2x + 3y + 6\,(0) = 12 \;\Rightarrow\; y = 4 - \tfrac{2}{3}x$. So $E = \left\{ (x,y,z) \mid 0 \le x \le 6,\, 0 \le y \le 4 - \tfrac{2}{3}x,\, 0 \le z \le \tfrac{1}{6}\,(12 - 2x - 3y) \right\}$ and

$$V = \int_0^6 \int_0^{4 - 2x/3} \int_0^{(12 - 2x - 3y)/6} dz\, dy\, dx = \tfrac{1}{6} \int_0^6 \int_0^{4 - 2x/3} (12 - 2x - 3y)\, dy\, dx$$

$$= \frac{1}{6} \int_0^6 \left[12y - 2xy - \frac{3}{2} y^2 \right]_{y=0}^{y=4-2x/3} dx = \frac{1}{6} \int_0^6 \left[\frac{(12 - 2x)^2}{3} - \frac{3}{2}\frac{12 - 2x}{9} \right] dx$$

$$= \tfrac{1}{36} \int_0^6 (12 - 2x)^2\, dx = \left[\tfrac{1}{36} \left(-\tfrac{1}{6} \right) (12 - 2x)^3 \right]_0^6 = 8$$

19.

$$V = \int_0^1 \int_{-\sqrt{x}}^{\sqrt{x}} \int_0^{1-x} dz\, dy\, dx$$

$$= \int_0^1 \int_{-\sqrt{x}}^{\sqrt{x}} (1 - x)\, dy\, dx$$

$$= \int_0^1 2\sqrt{x}\,(1 - x)\, dx$$

$$= \int_0^1 2 \left(\sqrt{x} - x^{3/2} \right) dx$$

$$= 2 \left[\tfrac{2}{3} x^{3/2} - \tfrac{2}{5} x^{5/2} \right]_0^1 = 2 \left(\tfrac{2}{3} - \tfrac{2}{5} \right) = \tfrac{8}{15}$$

21. (a) The wedge can be described as the region

$$D = \left\{ (x,y,z) \mid y^2 + z^2 \le 1,\, 0 \le x \le 1,\, 0 \le y \le x \right\}$$

$$= \left\{ (x,y,z) \mid 0 \le x \le 1,\, 0 \le y \le x,\, 0 \le z \le \sqrt{1 - y^2} \right\}$$

So the integral expressing the volume of the wedge is

$$\iiint_D dV = \int_0^1 \int_0^x \int_0^{\sqrt{1 - y^2}} dz\, dy\, dx.$$

(b) A CAS gives $\int_0^1 \int_0^x \int_0^{\sqrt{1 - y^2}} dz\, dy\, dx = \tfrac{\pi}{4} - \tfrac{1}{3}$. (Or use Formulas 30 and 87 from the Table of Integrals.)

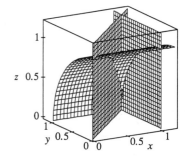

23. $E = \left\{ (x,y,z) \mid 0 \le x \le 1,\, 0 \le z \le 1 - x,\, 0 \le y \le 2 - 2z \right\}$, the solid bounded by the three coordinate planes and the planes $z = 1 - x,\, y = 2 - 2z$.

25.

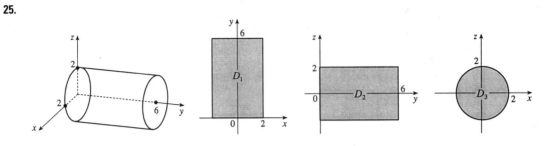

If D_1, D_2, D_3 are the projections of E on the xy-, yz-, and xz-planes, then

$$D_1 = \{(x, y) \mid -2 \le x \le 2, 0 \le y \le 6\}$$

$$D_2 = \{(y, z) \mid -2 \le z \le 2, 0 \le y \le 6\}$$

$$D_3 = \{(x, z) \mid x^2 + z^2 \le 4\}$$

Therefore

$$E = \left\{(x, y, z) \mid -\sqrt{4 - x^2} \le z \le \sqrt{4 - x^2},\ -2 \le x \le 2, 0 \le y \le 6\right\}$$

$$= \left\{(x, y, z) \mid -\sqrt{4 - z^2} \le x \le \sqrt{4 - z^2},\ -2 \le z \le 2, 0 \le y \le 6\right\}$$

$$\iiint_E f(x, y, z)\, dV = \int_{-2}^{2} \int_{0}^{6} \int_{-\sqrt{4 - x^2}}^{\sqrt{4 - x^2}} f(x, y, z)\, dz\, dy\, dx = \int_{0}^{6} \int_{-2}^{2} \int_{-\sqrt{4 - x^2}}^{\sqrt{4 - x^2}} f(x, y, z)\, dz\, dx\, dy$$

$$= \int_{0}^{6} \int_{-2}^{2} \int_{-\sqrt{4 - z^2}}^{\sqrt{4 - z^2}} f(x, y, z)\, dx\, dz\, dy = \int_{-2}^{2} \int_{0}^{6} \int_{-\sqrt{4 - z^2}}^{\sqrt{4 - z^2}} f(x, y, z)\, dx\, dy\, dz$$

$$= \int_{-2}^{2} \int_{-\sqrt{4 - x^2}}^{\sqrt{4 - x^2}} \int_{0}^{6} f(x, y, z)\, dy\, dz\, dx = \int_{-2}^{2} \int_{-\sqrt{4 - z^2}}^{\sqrt{4 - z^2}} \int_{0}^{6} f(x, y, z)\, dy\, dx\, dz$$

27.

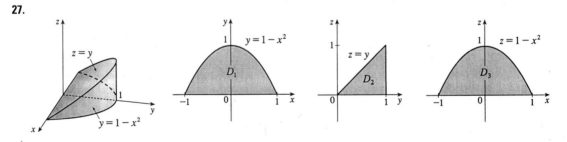

If D_1, D_2, and D_3 are the projections of E on the xy-, yz-, and xz-planes, then

$$D_1 = \{(x, y) \mid -1 \le x \le 1, 0 \le y \le 1 - x^2\} = \left\{(x, y) \mid 0 \le y \le 1, -\sqrt{1 - y} \le x \le \sqrt{1 - y}\right\},$$

$$D_2 = \{(y, z) \mid 0 \le y \le 1, 0 \le z \le y\} = \{(y, z) \mid 0 \le z \le 1, z \le y \le 1\},\ \text{and}$$

$$D_3 = \{(x, z) \mid -1 \le x \le 1, 0 \le z \le 1 - x^2\} = \left\{(x, z) \mid 0 \le z \le 1, -\sqrt{1 - z} \le x \le \sqrt{1 - z}\right\}.$$

Therefore

$$E = \{(x, y, z) \mid -1 \le x \le 1, 0 \le y \le 1 - x^2, 0 \le z \le y\}$$
$$= \left\{(x, y, z) \mid 0 \le y \le 1, -\sqrt{1-y} \le x \le \sqrt{1-y}, 0 \le z \le y\right\}$$
$$= \left\{(x, y, z) \mid 0 \le y \le 1, 0 \le z \le y, -\sqrt{1-y} \le x \le \sqrt{1-y}\right\}$$
$$= \left\{(x, y, z) \mid 0 \le z \le 1, z \le y \le 1, -\sqrt{1-y} \le x \le \sqrt{1-y}\right\}$$
$$= \{(x, y, z) \mid -1 \le x \le 1, 0 \le z \le 1 - x^2, z \le y \le 1 - x^2\}$$
$$= \left\{(x, y, z) \mid 0 \le z \le 1, -\sqrt{1-z} \le x \le \sqrt{1-z}, z \le y \le 1 - x^2\right\}$$

Then

$$\iiint_E f(x, y, z)\, dV = \int_{-1}^{1} \int_0^{1-x^2} \int_0^y f(x, y, z)\, dz\, dy\, dx = \int_0^1 \int_{-\sqrt{1-y}}^{\sqrt{1-y}} \int_0^y f(x, y, z)\, dz\, dx\, dy$$

$$= \int_0^1 \int_0^y \int_{-\sqrt{1-y}}^{\sqrt{1-y}} f(x, y, z)\, dx\, dz\, dy = \int_0^1 \int_z^1 \int_{-\sqrt{1-y}}^{\sqrt{1-y}} f(x, y, z)\, dx\, dy\, dz$$

$$= \int_{-1}^1 \int_0^{1-x^2} \int_z^{1-x^2} f(x, y, z)\, dy\, dz\, dx = \int_0^1 \int_{-\sqrt{1-z}}^{\sqrt{1-z}} \int_z^{1-x^2} f(x, y, z)\, dy\, dx\, dz$$

29.

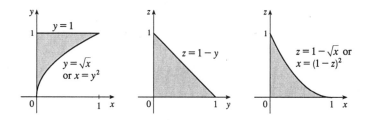

The diagrams show the projections of E on the xy-, yz-, and xz-planes. Therefore

$$\int_0^1 \int_{\sqrt{x}}^1 \int_0^{1-y} f(x, y, z)\, dz\, dy\, dx = \int_0^1 \int_0^{y^2} \int_0^{1-y} f(x, y, z)\, dz\, dx\, dy$$

$$= \int_0^1 \int_0^{1-z} \int_0^{y^2} f(x, y, z)\, dx\, dy\, dz$$

$$= \int_0^1 \int_0^{1-y} \int_0^{y^2} f(x, y, z)\, dx\, dz\, dy$$

$$= \int_0^1 \int_0^{1-\sqrt{x}} \int_{\sqrt{x}}^{1-z} f(x, y, z)\, dy\, dz\, dx$$

$$= \int_0^1 \int_0^{(1-z)^2} \int_{\sqrt{x}}^{1-z} f(x, y, z)\, dy\, dx\, dz$$

31.

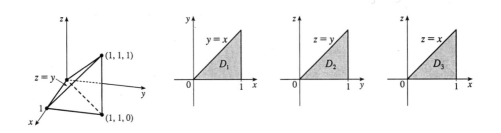

$\int_0^1 \int_y^1 \int_0^y f(x, y, z) \, dz \, dx \, dy = \iiint_E f(x, y, z) \, dV$ where $E = \{(x, y, z) \mid 0 \leq z \leq y, y \leq x \leq 1, 0 \leq y \leq 1\}$.

If D_1, D_2, and D_3 are the projections of E on the xy-, yz- and xz-planes then

$D_1 = \{(x, y) \mid 0 \leq y \leq 1, y \leq x \leq 1\} = \{(x, y) \mid 0 \leq x \leq 1, 0 \leq y \leq x\}$,

$D_2 = \{(y, z) \mid 0 \leq y \leq 1, 0 \leq z \leq y\} = \{(y, z) \mid 0 \leq z \leq 1, z \leq y \leq 1\}$, and

$D_3 = \{(x, z) \mid 0 \leq x \leq 1, 0 \leq z \leq x\} = \{(x, z) \mid 0 \leq z \leq 1, z \leq x \leq 1\}$.

Thus we also have

$E = \{(x, y, z) \mid 0 \leq x \leq 1, 0 \leq y \leq x, 0 \leq z \leq y\} = \{(x, y, z) \mid 0 \leq y \leq 1, 0 \leq z \leq y, y \leq x \leq 1\}$

$ = \{(x, y, z) \mid 0 \leq z \leq 1, z \leq y \leq 1, y \leq x \leq 1\} = \{(x, y, z) \mid 0 \leq x \leq 1, 0 \leq z \leq x, z \leq y \leq x\}$

$ = \{(x, y, z) \mid 0 \leq z \leq 1, z \leq x \leq 1, z \leq y \leq x\}$.

Then

$\int_0^1 \int_y^1 \int_0^y f(x, y, z) \, dz \, dx \, dy = \int_0^1 \int_0^x \int_0^y f(x, y, z) \, dz \, dy \, dx = \int_0^1 \int_0^y \int_y^1 f(x, y, z) \, dx \, dz \, dy$

$ = \int_0^1 \int_z^1 \int_y^1 f(x, y, z) \, dx \, dy \, dz = \int_0^1 \int_0^x \int_z^x f(x, y, z) \, dy \, dz \, dx$

$ = \int_0^1 \int_z^1 \int_z^x f(x, y, z) \, dy \, dx \, dz$

33. $m = \iiint_E \rho(x, y, z) \, dV = \int_0^1 \int_0^{\sqrt{x}} \int_0^{1+x+y} 2 \, dz \, dy \, dx$

$ = \int_0^1 \int_0^{\sqrt{x}} 2(1 + x + y) \, dy \, dx = \int_0^1 \left[2y + 2xy + y^2\right]_{y=0}^{y=\sqrt{x}} \, dx$

$ = \int_0^1 \left(2\sqrt{x} + 2x^{3/2} + x\right) \, dx = \left[\frac{4}{3}x^{3/2} + \frac{4}{5}x^{5/2} + \frac{1}{2}x^2\right]_0^1 = \frac{79}{30}$

$M_{yz} = \iiint_E x\rho(x, y, z) \, dV = \int_0^1 \int_0^{\sqrt{x}} \int_0^{1+x+y} 2x \, dz \, dy \, dx$

$\phantom{M_{yz}} = \int_0^1 \int_0^{\sqrt{x}} 2x(1 + x + y) \, dy \, dx = \int_0^1 \left[2xy + 2x^2y + xy^2\right]_{y=0}^{y=\sqrt{x}} \, dx$

$\phantom{M_{yz}} = \int_0^1 \left(2x^{3/2} + 2x^{5/2} + x^2\right) \, dx = \left[\frac{4}{5}x^{5/2} + \frac{4}{7}x^{7/2} + \frac{1}{3}x^3\right]_0^1 = \frac{179}{105}$

$M_{xz} = \iiint_E y\rho(x, y, z) \, dV = \int_0^1 \int_0^{\sqrt{x}} \int_0^{1+x+y} 2y \, dz \, dy \, dx$

$\phantom{M_{xz}} = \int_0^1 \int_0^{\sqrt{x}} 2y(1 + x + y) \, dy \, dx = \int_0^1 \left[y^2 + xy^2 + \frac{2}{3}y^3\right]_{y=0}^{y=\sqrt{x}} \, dx$

$\phantom{M_{xz}} = \int_0^1 \left(x + x^2 + \frac{2}{3}x^{3/2}\right) \, dx = \left[\frac{1}{2}x^2 + \frac{1}{3}x^3 + \frac{4}{15}x^{5/2}\right]_0^1 = \frac{11}{10}$

$$M_{xy} = \iiint_E z\rho\,(x, y, z)\,dV = \int_0^1 \int_0^{\sqrt{x}} \int_0^{1+x+y} 2z\,dz\,dy\,dx$$

$$= \int_0^1 \int_0^{\sqrt{x}} [z^2]_{z=0}^{z=1+x+y}\,dy\,dx = \int_0^1 \int_0^{\sqrt{x}} (1+x+y)^2\,dy\,dx$$

$$= \int_0^1 \int_0^{\sqrt{x}} (1 + 2x + 2y + 2xy + x^2 + y^2)\,dy\,dx$$

$$= \int_0^1 \left[y + 2xy + y^2 + xy^2 + x^2 y + \tfrac{1}{3}y^3\right]_{y=0}^{y=\sqrt{x}}\,dx = \int_0^1 \left(\sqrt{x} + \tfrac{7}{3}x^{3/2} + x + x^2 + x^{5/2}\right)dx$$

$$= \left[\tfrac{2}{3}x^{3/2} + \tfrac{14}{15}x^{5/2} + \tfrac{1}{2}x^2 + \tfrac{1}{3}x^3 + \tfrac{2}{7}x^{7/2}\right]_0^1 = \tfrac{571}{210}$$

Thus the mass is $\frac{79}{30}$ and the center of mass is $(\overline{x}, \overline{y}, \overline{z}) = \left(\dfrac{M_{yz}}{m}, \dfrac{M_{xz}}{m}, \dfrac{M_{xy}}{m}\right) = \left(\dfrac{358}{553}, \dfrac{33}{79}, \dfrac{571}{553}\right).$

35. $m = \int_0^a \int_0^a \int_0^a (x^2 + y^2 + z^2)\,dx\,dy\,dz = \int_0^a \int_0^a \left[\tfrac{1}{3}x^3 + xy^2 + xz^2\right]_{x=0}^{x=a}\,dy\,dz$

$$= \int_0^a \int_0^a \left(\tfrac{1}{3}a^3 + ay^2 + az^2\right)\,dy\,dz = \int_0^a \left[\tfrac{1}{3}a^3 y + \tfrac{1}{3}ay^3 + ayz^2\right]_{y=0}^{y=a}\,dz$$

$$= \int_0^a \left(\tfrac{2}{3}a^4 + a^2 z^2\right)\,dz = \left[\tfrac{2}{3}a^4 z + \tfrac{1}{3}a^2 z^3\right]_0^a = \tfrac{2}{3}a^5 + \tfrac{1}{3}a^5 = a^5$$

$$M_{yz} = \int_0^a \int_0^a \int_0^a \left[x^3 + x\left(y^2 + z^2\right)\right]\,dx\,dy\,dz = \int_0^a \int_0^a \left[\tfrac{1}{4}a^4 + \tfrac{1}{2}a^2\left(y^2 + z^2\right)\right]\,dy\,dz$$

$$= \int_0^a \left(\tfrac{1}{4}a^5 + \tfrac{1}{6}a^5 + \tfrac{1}{2}a^3 z^2\right)\,dz = \tfrac{1}{4}a^6 + \tfrac{1}{3}a^6 = \tfrac{7}{12}a^6$$

$$= M_{xz} = M_{xy} \text{ by symmetry of } E \text{ and } \rho\,(x, y, z)$$

Hence $(\overline{x}, \overline{y}, \overline{z}) = \left(\tfrac{7}{12}a, \tfrac{7}{12}a, \tfrac{7}{12}a\right).$

37. (a) $m = \int_{-1}^1 \int_{-\sqrt{1-y^2}}^{\sqrt{1-y^2}} \int_{4y^2+4z^2}^4 (x^2 + y^2 + z^2)\,dx\,dz\,dy$

(b) $(\overline{x}, \overline{y}, \overline{z})$ where $\overline{x} = m^{-1} \int_{-1}^1 \int_{-\sqrt{1-y^2}}^{\sqrt{1-y^2}} \int_{4y^2+4z^2}^4 x\,(x^2 + y^2 + z^2)\,dx\,dz\,dy,$

$$\overline{y} = m^{-1} \int_{-1}^1 \int_{-\sqrt{1-y^2}}^{\sqrt{1-y^2}} \int_{4y^2+4z^2}^4 y\,(x^2 + y^2 + z^2)\,dx\,dz\,dy, \text{ and}$$

$$\overline{z} = m^{-1} \int_{-1}^1 \int_{-\sqrt{1-y^2}}^{\sqrt{1-y^2}} \int_{4y^2+4z^2}^4 z\,(x^2 + y^2 + z^2)\,dx\,dz\,dy$$

(c) $I_z = \int_{-1}^1 \int_{-\sqrt{1-y^2}}^{\sqrt{1-y^2}} \int_{4y^2+4z^2}^4 (x^2 + y^2)\,(x^2 + y^2 + z^2)\,dx\,dz\,dy$

39. (a) $m = \int_0^1 \int_0^{\sqrt{1-x^2}} \int_0^y (1 + x + y + z)\,dz\,dy\,dx = \tfrac{3\pi}{32} + \tfrac{11}{24}$

(b) $(\overline{x}, \overline{y}, \overline{z}) = \left(m^{-1} \int_0^1 \int_0^{\sqrt{1-x^2}} \int_0^y x\,(1 + x + y + z)\,dz\,dy\,dx,\right.$

$$m^{-1} \int_0^1 \int_0^{\sqrt{1-x^2}} \int_0^y y\,(1 + x + y + z)\,dz\,dy\,dx,$$

$$\left.m^{-1} \int_0^1 \int_0^{\sqrt{1-x^2}} \int_0^y z\,(1 + x + y + z)\,dz\,dy\,dx\right)$$

$$= \left(\dfrac{28}{9\pi + 44}, \dfrac{30\pi + 128}{45\pi + 220}, \dfrac{45\pi + 208}{135\pi + 660}\right)$$

(c) $I_z = \int_0^1 \int_0^{\sqrt{1-x^2}} \int_0^y (x^2 + y^2)\,(1 + x + y + z)\,dz\,dy\,dx = \dfrac{68 + 15\pi}{240}$

41. $I_x = \int_0^L \int_0^L \int_0^L k\left(y^2 + z^2\right) dz\, dy\, dx = k \int_0^L \int_0^L \left(Ly^2 + \frac{1}{3}L^3\right) dy\, dx = k \int_0^L \frac{2}{3}L^4\, dx = \frac{2}{3}kL^5.$

By symmetry, $I_x = I_y = I_z = \frac{2}{3}kL^5.$

43. (a) $f(x, y, z)$ is a joint density function, so we know $\iiint_{\mathbb{R}^3} f(x, y, z)\, dV = 1$. Here we have

$$\iiint_{\mathbb{R}^3} f(x, y, z)\, dV = \int_{-\infty}^{\infty} \int_{-\infty}^{\infty} \int_{-\infty}^{\infty} f(x, y, z)\, dz\, dy\, dx = \int_0^2 \int_0^2 \int_0^2 Cxyz\, dz\, dy\, dx$$

$$= C \int_0^2 x\, dx \int_0^2 y\, dy \int_0^2 z\, dz = C \left[\frac{x^2}{2}\right]_0^2 \left[\frac{y^2}{2}\right]_0^2 \left[\frac{z^2}{2}\right]_0^2$$

$$= 8C$$

Then we must have $8C = 1 \Rightarrow C = \frac{1}{8}.$

(b) $P(X \le 1, Y \le 1, Z \le 1) = \int_{-\infty}^{1} \int_{-\infty}^{1} \int_{-\infty}^{1} f(x, y, z)\, dz\, dy\, dx$

$$= \int_0^1 \int_0^1 \int_0^1 \frac{1}{8}xyz\, dz\, dy\, dx = \frac{1}{8} \int_0^1 x\, dx \int_0^1 y\, dy \int_0^1 z\, dz$$

$$= \frac{1}{8} \left[\frac{x^2}{2}\right]_0^1 \left[\frac{y^2}{2}\right]_0^1 \left[\frac{z^2}{2}\right]_0^1 = \frac{1}{8} \left(\frac{1}{2}\right)^3 = \frac{1}{64}$$

(c) $P(X + Y + Z \le 1) = P((X, Y, Z) \in E)$ where E is the solid region in the first octant bounded by the coordinate planes and the plane $x + y + z = 1$. The plane $x + y + z = 1$ meets the xy-plane in the line $x + y = 1$, so we have

$$P(X + Y + Z \le 1) = \iiint_E f(x, y, z)\, dV = \int_0^1 \int_0^{1-x} \int_0^{1-x-y} \frac{1}{8}xyz\, dz\, dy\, dx$$

$$= \frac{1}{8} \int_0^1 \int_0^{1-x} xy \left[\frac{1}{2}z^2\right]_{z=0}^{z=1-x-y} dy\, dx$$

$$= \frac{1}{16} \int_0^1 \int_0^{1-x} xy(1 - x - y)^2\, dy\, dx$$

$$= \frac{1}{16} \int_0^1 \int_0^{1-x} \left[\left(x^3 - 2x^2 + x\right)y + \left(2x^2 - 2x\right)y^2 + xy^3\right] dy\, dx$$

$$= \frac{1}{16} \int_0^1 \left[\left(x^3 - 2x^2 + x\right)\frac{1}{2}y^2 + \left(2x^2 - 2x\right)\frac{1}{3}y^3 + x\left(\frac{1}{4}y^4\right)\right]_{y=0}^{y=1-x} dx$$

$$= \frac{1}{192} \int_0^1 \left(x - 4x^2 + 6x^3 - 4x^4 + x^5\right) dx = \frac{1}{192} \left(\frac{1}{30}\right) = \frac{1}{5760}$$

45. $V(E) = L^3,$

$$f_{\text{ave}} = \frac{1}{L^3} \int_0^L \int_0^L \int_0^L xyz\, dx\, dy\, dz = \frac{1}{L^3} \int_0^L x\, dx \int_0^L y\, dy \int_0^L z\, dz$$

$$= \frac{1}{L^3} \left[\frac{x^2}{2}\right]_0^L \left[\frac{y^2}{2}\right]_0^L \left[\frac{z^2}{2}\right]_0^L = \frac{1}{L^3} \frac{L^2}{2} \frac{L^2}{2} \frac{L^2}{2} = \frac{L^3}{8}$$

47. The triple integral will attain its maximum when the integrand $1 - x^2 - 2y^2 - 3z^2$ is positive in the region E and negative everywhere else. For if E contains some region F where the integrand is negative, the integral could be increased by excluding F from E, and if E fails to contain some part G of the region where the integrand is positive, the integral could be increased by including G in E. So we require that $x^2 + 2y^2 + 3z^2 \le 1$. This describes the region bounded by the ellipsoid $x^2 + 2y^2 + 3z^2 = 1$.

16.8 Triple Integrals in Cylindrical and Spherical Coordinates ET 15.8

1. The region of integration is given in cylindrical coordinates by

$E = \{(r, \theta, z) \mid 0 \le \theta \le 2\pi, 0 \le r \le 2, 0 \le z \le 4 - r^2\}$. This represents the solid region bounded above by

$z = 4 - r^2 = 4 - x^2 - y^2$, a paraboloid, and below by the xy-plane.

$$\int_0^{2\pi} \int_0^2 \int_0^{4-r^2} r \, dz \, dr \, d\theta = \int_0^{2\pi} \int_0^2 \left(4r - r^3\right) dr \, d\theta$$

$$= \int_0^{2\pi} \left[2r^2 - \tfrac{1}{4}r^4\right]_{r=0}^{r=2} d\theta$$

$$= \int_0^{2\pi} (8 - 4) \, d\theta = 4\theta]_0^{2\pi} = 8\pi$$

3. The region of integration is given in spherical coordinates by

$E = \{(\rho, \theta, \phi) \mid 0 \le \rho \le 1, 0 \le \theta \le \frac{\pi}{2}, 0 \le \phi \le \frac{\pi}{2}\}$. Thus E is the solid in the first octant bounded by the sphere

$\rho = x^2 + y^2 + z^2 = 1$ and the three coordinate planes.

$$\int_0^{\pi/2} \int_0^{\pi/2} \int_0^1 \rho^2 \sin \phi \, d\rho \, d\theta \, d\phi = \int_0^{\pi/2} \int_0^{\pi/2} \left[\tfrac{1}{3}\rho^3 \sin \phi\right]_{\rho=0}^{\rho=1} d\theta \, d\phi$$

$$= \int_0^{\pi/2} \int_0^{\pi/2} \tfrac{1}{3} \sin \phi \, d\theta \, d\phi = \int_0^{\pi/2} \tfrac{1}{3} \sin \phi \, [\theta]_{\theta=0}^{\theta=\pi/2} \, d\phi$$

$$= \tfrac{1}{3} \int_0^{\pi/2} \tfrac{\pi}{2} \sin \phi \, d\phi = \tfrac{\pi}{6} \left[-\cos \phi\right]_0^{\pi/2} = \tfrac{\pi}{6}$$

5. The solid E is most conveniently described if we use cylindrical coordinates:

$E = \{(r, \theta, z) \mid 0 \le \theta \le \frac{\pi}{2}, 0 \le r \le 3, 0 \le z \le 2\}$. Then

$\iiint_E f(x, y, z) \, dV = \int_0^{\pi/2} \int_0^3 \int_0^2 f(r \cos \theta, r \sin \theta, z) \, r \, dz \, dr \, d\theta$.

7. In cylindrical coordinates, E is given by $\{(r, \theta, z) \mid 0 \le \theta \le 2\pi, 0 \le r \le 4, -5 \le z \le 4\}$. So

$$\iiint_E \sqrt{x^2 + y^2} \, dV = \int_0^{2\pi} \int_0^4 \int_{-5}^4 \sqrt{r^2} \, r \, dz \, dr \, d\theta = \int_0^{2\pi} d\theta \int_0^4 r^2 \, dr \int_{-5}^4 dz$$

$$= [\theta]_0^{2\pi} \left[\tfrac{1}{3}r^3\right]_0^4 [z]_{-5}^4 = (2\pi) \left(\tfrac{64}{3}\right)(9) = 384\pi$$

9. In cylindrical coordinates E is bounded by the cylinders $r = 1$ and $r = 2$, the plane $z = x + 2 = r \cos \theta + 2$, and

the xy-plane, so E is given by $\{(r, \theta, z) \mid 0 \le \theta \le 2\pi, 1 \le r \le 2, 0 \le z \le r \cos \theta + 2\}$. Thus

$$\iiint_E y \, dV = \int_0^{2\pi} \int_1^2 \int_0^{2+r \cos \theta} (r \sin \theta) \, r \, dz \, dr \, d\theta = \int_0^{2\pi} \int_1^2 r^2 \sin \theta \, [z]_{z=0}^{z=2+r \cos \theta} \, dr \, d\theta$$

$$= \int_0^{2\pi} \int_1^2 \left(2r^2 + r^3 \cos \theta\right) \sin \theta \, dr \, d\theta = \int_0^{2\pi} \left[\tfrac{2}{3}r^3 + \tfrac{1}{4}r^4 \cos \theta\right]_{r=1}^{r=2} \sin \theta \, d\theta$$

$$= \int_0^{2\pi} \left(\tfrac{14}{3} + \tfrac{15}{4} \cos \theta\right) \sin \theta \, d\theta = \left[-\tfrac{14}{3} \cos \theta - \tfrac{15}{8} \cos^2 \theta\right]_0^{2\pi} = 0$$

11. In cylindrical coordinates, E is bounded by the cylinder $r = 1$, the plane $z = 0$, and the cone $z = 2r$. So
$E = \{(r, \theta, z) \mid 0 \le \theta \le 2\pi, 0 \le r \le 1, 0 \le z \le 2r\}$ and

$$\iiint_E x^2 \, dV = \int_0^{2\pi} \int_0^1 \int_0^{2r} r^2 \cos^2\theta \, r \, dz \, dr \, d\theta = \int_0^{2\pi} \int_0^1 \left[r^3 \cos^2\theta \, z\right]_{z=0}^{z=2r} dr \, d\theta$$

$$= \int_0^{2\pi} \int_0^1 2r^4 \cos^2\theta \, dr \, d\theta = \int_0^{2\pi} \left[\tfrac{2}{5}r^5 \cos^2\theta\right]_{r=0}^{r=1} d\theta = \tfrac{2}{5} \int_0^{2\pi} \cos^2\theta \, d\theta$$

$$= \frac{2}{5} \int_0^{2\pi} \frac{1 + \cos 2\theta}{2} \, d\theta = \frac{1}{5} \left[\theta + \frac{1}{2} \sin 2\theta\right]_0^{2\pi} = \frac{2\pi}{5}$$

13. The paraboloids intersect when $x^2 + y^2 = 36 - 3x^2 - 3y^2 \Rightarrow D = \{(x, y) \mid x^2 + y^2 \le 9\}$. So, in cylindrical
coordinates, $E = \{(r, \theta, z) \mid r^2 \le z \le 36 - r^2, 0 \le r \le 3, 0 \le \theta \le 2\pi\}$ and

$$V = \int_0^{2\pi} \int_0^3 \int_{r^2}^{36 - 3r^2} r \, dz \, dr \, d\theta = \int_0^{2\pi} \int_0^3 \left(36r - 4r^3\right) dr \, d\theta$$

$$= \int_0^{2\pi} \left[18r^2 - r^4\right]_{r=0}^{r=3} d\theta = \int_0^{2\pi} 81 \, d\theta = 162\pi$$

15. The paraboloid $z = 4x^2 + 4y^2$ intersects the plane $z = a$ when $a = 4x^2 + 4y^2$ or $x^2 + y^2 = \tfrac{1}{4}a$. So, in cylindrical
coordinates, $E = \{(r, \theta, z) \mid 0 \le r \le \tfrac{1}{2}\sqrt{a}, 0 \le \theta \le 2\pi, 4r^2 \le z \le a\}$. Thus

$$m = \int_0^{2\pi} \int_0^{\sqrt{a}/2} \int_{4r^2}^a Kr \, dz \, dr \, d\theta = K \int_0^{2\pi} \int_0^{\sqrt{a}/2} \left(ar - 4r^3\right) dr \, d\theta$$

$$= K \int_0^{2\pi} \left[\tfrac{1}{2}ar^2 - r^4\right]_{r=0}^{r=\sqrt{a}/2} d\theta = K \int_0^{2\pi} \tfrac{1}{16}a^2 \, d\theta = \tfrac{1}{8}a^2\pi K$$

Since the region is homogeneous and symmetric, $M_{yz} = M_{xz} = 0$ and

$$M_{xy} = \int_0^{2\pi} \int_0^{\sqrt{a}/2} \int_{4r^2}^a Krz \, dz \, dr \, d\theta = K \int_0^{2\pi} \int_0^{\sqrt{a}/2} \left(\tfrac{1}{2}a^2 r - 8r^5\right) dr \, d\theta$$

$$= K \int_0^{2\pi} \left[\tfrac{1}{4}a^2 r^2 - \tfrac{4}{3}r^6\right]_{r=0}^{r=\sqrt{a}/2} d\theta = K \int_0^{2\pi} \tfrac{1}{24}a^3 \, d\theta = \tfrac{1}{12}a^3\pi K$$

Hence $(\bar{x}, \bar{y}, \bar{z}) = \left(0, 0, \tfrac{2}{3}a\right)$.

17. In spherical coordinates, B is represented by $\{(\rho, \theta, \phi) \mid 0 \le \rho \le 1, 0 \le \theta \le 2\pi, 0 \le \phi \le \pi\}$. Thus

$$\iiint_B \left(x^2 + y^2 + z^2\right) dV = \int_0^\pi \int_0^{2\pi} \int_0^1 \left(\rho^2\right) \rho^2 \sin\phi \, d\rho \, d\theta \, d\phi = \int_0^\pi \sin\phi \, d\phi \int_0^{2\pi} d\theta \int_0^1 \rho^4 \, d\rho$$

$$= \left[-\cos\phi\right]_0^\pi \left[\theta\right]_0^{2\pi} \left[\tfrac{1}{5}\rho^5\right]_0^1 = (2)(2\pi)\left(\tfrac{1}{5}\right) = \tfrac{4\pi}{5}$$

19. In spherical coordinates, E is represented by $\{(\rho, \theta, \phi) \mid 1 \le \rho \le 2, 0 \le \theta \le \tfrac{\pi}{2}, 0 \le \phi \le \tfrac{\pi}{2}\}$. Thus

$$\iiint_E z \, dV = \int_0^{\pi/2} \int_0^{\pi/2} \int_1^2 (\rho\cos\phi) \rho^2 \sin\phi \, d\rho \, d\theta \, d\phi$$

$$= \int_0^{\pi/2} \cos\phi \sin\phi \, d\phi \int_0^{\pi/2} d\theta \int_1^2 \rho^3 \, d\rho = \left[\tfrac{1}{2}\sin^2\phi\right]_0^{\pi/2} \left[\theta\right]_0^{\pi/2} \left[\tfrac{1}{4}\rho^4\right]_1^2$$

$$= \left(\tfrac{1}{2}\right)\left(\tfrac{\pi}{2}\right)\left(\tfrac{15}{4}\right) = \tfrac{15\pi}{16}$$

21. $\iiint_E \sqrt{x^2 + y^2 + z^2} \, dV = \int_0^{2\pi} \int_0^{\pi/6} \int_0^2 (\rho) \rho^2 \sin\phi \, d\rho \, d\phi \, d\theta$

$$= \int_0^{2\pi} d\theta \int_0^{\pi/6} \sin\phi \, d\phi \int_0^2 \rho^3 \, d\rho = [\theta]_0^{2\pi} \left[-\cos\phi\right]_0^{\pi/6} \left[\tfrac{1}{4}\rho^4\right]_0^2$$

$$= (2\pi)\left(1 - \tfrac{\sqrt{3}}{2}\right)(4) = 8\pi\left(1 - \tfrac{\sqrt{3}}{2}\right) = 4\pi\left(2 - \sqrt{3}\right)$$

23. Since $\rho = 4\cos\phi$ implies $\rho^2 = 4\rho\cos\phi$, the equation is that of a sphere of radius 2 with center at $(0, 0, 2)$. Thus

$$V = \int_0^{2\pi}\int_0^{\pi/3}\int_0^{4\cos\phi}\rho^2\sin\phi\,d\rho\,d\phi\,d\theta = \int_0^{2\pi}\int_0^{\pi/3}\left[\tfrac{1}{3}\rho^3\right]_{\rho=0}^{\rho=4\cos\phi}\sin\phi\,d\phi\,d\theta$$

$$= \int_0^{2\pi}\int_0^{\pi/3}\left(\tfrac{64}{3}\cos^3\phi\right)\sin\phi\,d\phi\,d\theta = \int_0^{2\pi}\left[-\tfrac{16}{3}\cos^4\phi\right]_{\phi=0}^{\phi=\pi/3}d\theta$$

$$= \int_0^{2\pi}-\tfrac{16}{3}\left(\tfrac{1}{16}-1\right)d\theta = 5\theta\Big]_0^{2\pi} = 10\pi$$

25. By the symmetry of the problem $M_{yz} = M_{xz} = 0$. Then

$$M_{xy} = \int_0^{2\pi}\int_0^{\pi/3}\int_0^{4\cos\phi}\rho^3\cos\phi\sin\phi\,d\rho\,d\phi\,d\theta = \int_0^{2\pi}\int_0^{\pi/3}\cos\phi\sin\phi\left(64\cos^4\phi\right)d\phi\,d\theta$$

$$= \int_0^{2\pi}64\left[-\tfrac{1}{6}\cos^6\phi\right]_{\phi=0}^{\phi=\pi/3}d\theta = \int_0^{2\pi}\tfrac{21}{2}\,d\theta = 21\pi$$

Hence $(\bar{x}, \bar{y}, \bar{z}) = (0, 0, 2.1)$.

27. (a) The density function is $\rho(x, y, z) = K$, a constant, and by the symmetry of the problem $M_{xz} = M_{yz} = 0$.
Then $M_{xy} = \int_0^{2\pi}\int_0^{\pi/2}\int_0^a K\rho^3\sin\phi\cos\phi\,d\rho\,d\phi\,d\theta = \tfrac{1}{2}\pi K a^4\int_0^{\pi/2}\sin\phi\cos\phi\,d\phi = \tfrac{1}{8}\pi K a^4$. But the mass is
K (volume of the hemisphere) $= \tfrac{2}{3}\pi K a^3$, so the centroid is $\left(0, 0, \tfrac{3}{8}a\right)$.

(b) Place the center of the base at $(0, 0, 0)$; the density function is $\rho(x, y, z) = K$. By symmetry, the moments of
inertia about any two such diameters will be equal, so we just need to find I_x:

$$I_x = \int_0^{2\pi}\int_0^{\pi/2}\int_0^a \left(K\rho^2\sin\phi\right)\rho^2\left(\sin^2\phi\sin^2\theta + \cos^2\phi\right)d\rho\,d\phi\,d\theta$$

$$= K\int_0^{2\pi}\int_0^{\pi/2}\left(\sin^3\phi\sin^2\theta + \sin\phi\cos^2\phi\right)\left(\tfrac{1}{5}a^5\right)d\phi\,d\theta$$

$$= \tfrac{1}{5}K a^5\int_0^{2\pi}\left[\sin^2\theta\left(-\cos\phi + \tfrac{1}{3}\cos^3\phi\right) + \left(-\tfrac{1}{3}\cos^3\phi\right)\right]_{\phi=0}^{\phi=\pi/2}d\theta$$

$$= \tfrac{1}{5}K a^5\int_0^{2\pi}\left[\tfrac{2}{3}\sin^2\theta + \tfrac{1}{3}\right]d\theta = \tfrac{1}{5}K a^5\left[\tfrac{2}{3}\left(\tfrac{1}{2}\theta - \tfrac{1}{4}\sin 2\theta\right) + \tfrac{1}{3}\theta\right]_0^{2\pi}$$

$$= \tfrac{1}{5}K a^5\left[\tfrac{2}{3}(\pi - 0) + \tfrac{1}{3}(2\pi - 0)\right] = \tfrac{4}{15}K a^5\pi$$

29. In spherical coordinates $z = \sqrt{x^2 + y^2}$ becomes $\cos\phi = \sin\phi$ or $\phi = \tfrac{\pi}{4}$. Then
$V = \int_0^{2\pi}\int_0^{\pi/4}\int_0^1\rho^2\sin\phi\,d\rho\,d\phi\,d\theta = \int_0^{2\pi}d\theta\int_0^{\pi/4}\sin\phi\,d\phi\int_0^1\rho^2\,d\rho = \tfrac{1}{3}\pi\left(2 - \sqrt{2}\right)$,
$M_{xy} = \int_0^{2\pi}\int_0^{\pi/4}\int_0^1\rho^3\sin\phi\cos\phi\,d\rho\,d\phi\,d\theta = 2\pi\left[-\tfrac{1}{4}\cos 2\phi\right]_0^{\pi/4}\left(\tfrac{1}{4}\right) = \tfrac{\pi}{8}$ and by symmetry $M_{yz} = M_{xz} = 0$.
Hence $(\bar{x}, \bar{y}, \bar{z}) = \left(0, 0, \dfrac{3}{8(2 - \sqrt{2})}\right)$.

31. In cylindrical coordinates the paraboloid is given by $z = r^2$ and the plane by $z = 2r\sin\theta$ and they intersect in the
circle $r = 2\sin\theta$. Then $\iiint_E z\,dV = \int_0^\pi\int_0^{2\sin\theta}\int_{r^2}^{2r\sin\theta} rz\,dz\,dr\,d\theta = \tfrac{5\pi}{6}$ (using a CAS).

33. The region E of integration is the region above the paraboloid $z = x^2 + y^2$, or $z = r^2$, and below the paraboloid
$z = 2 - x^2 - y^2$, or $z = 2 - r^2$. Also, we have $-1 \le x \le 1$ with $-\sqrt{1 - x^2} \le y \le \sqrt{1 - x^2}$ which describes the
unit circle in the xy-plane. Thus,

$$\int_{-1}^1\int_{-\sqrt{1-x^2}}^{\sqrt{1-x^2}}\int_{x^2+y^2}^{2-x^2-y^2}\left(x^2 + y^2\right)^{3/2}dz\,dy\,dx = \int_0^{2\pi}\int_0^1\int_{r^2}^{2-r^2}\left(r^2\right)^{3/2}r\,dz\,dr\,d\theta$$

$$= \int_0^{2\pi}\int_0^1\left[r^4 z\right]_{z=r^2}^{z=2-r^2}dr\,d\theta = \int_0^{2\pi}\int_0^1\left(2r^4 - r^6 - r^6\right)dr\,d\theta = \int_0^{2\pi}\left(\tfrac{2}{5} - \tfrac{2}{7}\right)d\theta = \tfrac{8\pi}{35}$$

35. The region of integration E is the top half of the sphere $x^2 + y^2 + z^2 = 9$. So

$$\int_{-3}^{3} \int_{-\sqrt{9-x^2}}^{\sqrt{9-x^2}} \int_{0}^{\sqrt{9-x^2-y^2}} z\sqrt{x^2+y^2+z^2}\, dz\, dy\, dx = \iiint_E z\sqrt{x^2+y^2+z^2}\, dV$$

$$= \int_0^{2\pi} \int_0^{\pi/2} \int_0^3 (\rho^2 \cos\phi)(\rho^2 \sin\phi)\, d\rho\, d\phi\, d\theta = \int_0^{2\pi} d\theta \int_0^{\pi/2} \cos\phi \sin\phi\, d\phi \int_0^3 \rho^4\, d\rho$$

$$= [\theta]_0^{2\pi} \left[\tfrac{1}{2}\sin^2\phi\right]_0^{\pi/2} \left[\tfrac{1}{5}\rho^5\right]_0^3 = (2\pi)\left(\tfrac{1}{2}\right)\left(\tfrac{243}{5}\right) = \tfrac{243}{5}\pi$$

37. If E is the solid enclosed by the surface $\rho = 1 + \tfrac{1}{5}\sin 6\theta \sin 5\phi$, it can be described in spherical coordinates as $E = \{(\rho, \theta, \phi) \mid 0 \leq \rho \leq 1 + \tfrac{1}{5}\sin 6\theta \sin 5\phi, 0 \leq \theta \leq 2\pi, 0 \leq \phi \leq \pi\}$. Its volume is given by $V(E) = \iiint_E dV = \int_0^\pi \int_0^{2\pi} \int_0^{1 + (\sin 6\theta \sin 5\phi)/5} \rho^2 \sin\phi\, d\rho\, d\theta\, d\phi = \tfrac{136\pi}{99}$ (using a CAS).

39. (a) From the diagram, $z = r \cot\phi_0$ to $z = \sqrt{a^2 - r^2}$, $r = 0$ to $r = a\sin\phi_0$ (or use $a^2 - r^2 = r^2 \cot^2\phi_0$). Thus

$$V = \int_0^{2\pi} \int_0^{a\sin\phi_0} \int_{r\cot\phi_0}^{\sqrt{a^2-r^2}} r\, dz\, dr\, d\theta$$

$$= 2\pi \int_0^{a\sin\phi_0} \left(r\sqrt{a^2-r^2} - r^2 \cot\phi_0\right) dr$$

$$= \tfrac{2\pi}{3} \left[-(a^2-r^2)^{3/2} - r^3 \cot\phi_0\right]_0^{a\sin\phi_0}$$

$$= \tfrac{2\pi}{3} \left[-(a^2 - a^2 \sin^2\phi_0)^{3/2} - a^3 \sin^3\phi_0 \cot\phi_0 + a^3\right]$$

$$= \tfrac{2}{3}\pi a^3 \left[1 - (\cos^3\phi_0 + \sin^2\phi_0 \cos\phi_0)\right] = \tfrac{2}{3}\pi a^3 (1 - \cos\phi_0)$$

(b) The wedge in question is the shaded area rotated from $\theta = \theta_1$ to $\theta = \theta_2$. Letting

V_{ij} = volume of the region bounded by the sphere of radius ρ_i
    and the cone with angle ϕ_j ($\theta = \theta_1$ to θ_2)

and letting V be the volume of the wedge, we have

$$V = (V_{22} - V_{21}) - (V_{12} - V_{11})$$

$$= \tfrac{1}{3}(\theta_2 - \theta_1)\left[\rho_2^3(1 - \cos\phi_2) - \rho_2^3(1 - \cos\phi_1) - \rho_1^3(1 - \cos\phi_2) + \rho_1^3(1 - \cos\phi_1)\right]$$

$$= \tfrac{1}{3}(\theta_2 - \theta_1)\left[(\rho_2^3 - \rho_1^3)(1 - \cos\phi_2) - (\rho_2^3 - \rho_1^3)(1 - \cos\phi_1)\right]$$

$$= \tfrac{1}{3}(\theta_2 - \theta_1)\left[(\rho_2^3 - \rho_1^3)(\cos\phi_1 - \cos\phi_2)\right]$$

Or: Show that $V = \displaystyle\int_{\theta_1}^{\theta_2} \int_{\rho_1 \sin\phi_1}^{\rho_2 \sin\phi_2} \int_{r\cot\phi_2}^{r\cot\phi_1} r\, dz\, dr\, d\theta$.

(c) By the Mean Value Theorem with $f(\rho) = \rho^3$ there exists some $\tilde{\rho}$ with $\rho_1 \leq \tilde{\rho} \leq \rho_2$ such that $f(\rho_2) - f(\rho_1) = f'(\tilde{\rho})(\rho_2 - \rho_1)$ or $\rho_1^3 - \rho_2^3 = 3\tilde{\rho}^2 \Delta\rho$. Similarly there exists ϕ with $\phi_1 \leq \tilde{\phi} \leq \phi_2$ such that $\cos\phi_2 - \cos\phi_1 = \left(-\sin\tilde{\phi}\right)\Delta\phi$. Substituting into the result from (b) gives

$$\Delta V = (\tilde{\rho}^2 \Delta\rho)(\theta_2 - \theta_1)\left(\sin\tilde{\phi}\right)\Delta\phi = \tilde{\rho}^2 \sin\tilde{\phi}\, \Delta\rho\, \Delta\phi\, \Delta\theta.$$

1. $x = u + 4v$, $y = 3u - 2v$.

The Jacobian is $\dfrac{\partial (x, y)}{\partial (u, v)} = \begin{vmatrix} \partial x/\partial u & \partial x/\partial v \\ \partial y/\partial u & \partial y/\partial v \end{vmatrix} = \begin{vmatrix} 1 & 4 \\ 3 & -2 \end{vmatrix} = 1\,(-2) - 4\,(3) = -14.$

3. $\dfrac{\partial (x, y)}{\partial (u, v)} = \begin{vmatrix} \dfrac{\partial x}{\partial u} & \dfrac{\partial x}{\partial v} \\ \dfrac{\partial y}{\partial u} & \dfrac{\partial y}{\partial v} \end{vmatrix} = \begin{vmatrix} \dfrac{v}{(u+v)^2} & -\dfrac{u}{(u+v)^2} \\ -\dfrac{v}{(u-v)^2} & \dfrac{u}{(u-v)^2} \end{vmatrix} = \dfrac{uv}{(u+v)^2\,(u-v)^2} - \dfrac{uv}{(u+v)^2\,(u-v)^2} = 0$

5. $\dfrac{\partial (x, y, z)}{\partial (u, v, w)} = \begin{vmatrix} \partial x/\partial u & \partial x/\partial v & \partial x/\partial w \\ \partial y/\partial u & \partial y/\partial v & \partial y/\partial w \\ \partial z/\partial u & \partial z/\partial v & \partial z/\partial w \end{vmatrix} = \begin{vmatrix} v & u & 0 \\ 0 & w & v \\ w & 0 & u \end{vmatrix}$

$= v \begin{vmatrix} w & v \\ 0 & u \end{vmatrix} - u \begin{vmatrix} 0 & v \\ w & u \end{vmatrix} + 0 \begin{vmatrix} 0 & w \\ w & 0 \end{vmatrix} = v\,(uw - 0) - u\,(0 - vw) = 2uvw$

7. The transformation maps the boundary of S to the boundary of the image R, so we first look at side S_1 in the uv-plane. S_1 is described by $v = 0$ $(0 \le u \le 3)$, so $x = 2u + 3v = 2u$ and $y = u - v = u$. Eliminating u, we have $x = 2y$, $0 \le x \le 6$. S_2 is the line segment $u = 3$, $0 \le v \le 2$, so $x = 6 + 3v$ and $y = 3 - v$. Then $v = 3 - y$ $\Rightarrow$ $x = 6 + 3\,(3 - y) = 15 - 3y$, $6 \le x \le 12$. S_3 is the line segment $v = 2$, $0 \le u \le 3$, so $x = 2u + 6$ and $y = u - 2$, giving $u = y + 2$ $\Rightarrow$ $x = 2y + 10$, $6 \le x \le 12$. Finally, S_4 is the segment $u = 0$, $0 \le v \le 2$, so $x = 3v$ and $y = -v$ $\Rightarrow$ $x = -3y$, $0 \le x \le 6$. The image of set S is the region R shown in the xy-plane, a parallelogram bounded by these four segments.

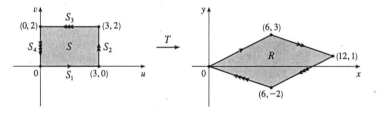

9. S_1 is the line segment $u = v$, $0 \le u \le 1$, so $y = v = u$ and $x = u^2 = y^2$. Since $0 \le u \le 1$, the image is the portion of the parabola $x = y^2$, $0 \le y \le 1$. S_2 is the segment $v = 1$, $0 \le u \le 1$, thus $y = v = 1$ and $x = u^2$, so $0 \le x \le 1$. The image is the line segment $y = 1$, $0 \le x \le 1$. S_3 is the segment $u = 0$, $0 \le v \le 1$, so $x = u^2 = 0$ and $y = v$ $\Rightarrow$ $0 \le y \le 1$. The image is the segment $x = 0$, $0 \le y \le 1$. Thus, the image of S is the region R in the first quadrant bounded by the parabola $x = y^2$, the y-axis, and the line $y = 1$.

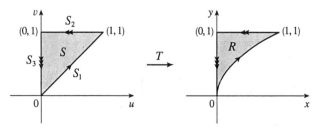

11. $\dfrac{\partial (x, y)}{\partial (u, v)} = \begin{vmatrix} 1/3 & 1/3 \\ -2/3 & 1/3 \end{vmatrix} = \frac{1}{3}$ and $3x + 4y = (u + v) + \frac{4}{3}(v - 2u) = \frac{1}{3}(7v - 5u)$. To find the region S in the

uv-plane that corresponds to R we first find the corresponding boundary under the given tranformation: The line
$y = x$ is the image of $\frac{1}{3}(v - 2u) = \frac{1}{3}(u + v)$ or $u = 0$, $y = x - 2$ ⇒

$\frac{1}{3}(v - 2u) = \frac{1}{3}(u + v) - 2$ or $u = 2$, $y = -2x$ ⇒ $\frac{1}{3}(v - 2u) = -\frac{2}{3}(u + v)$ or $v = 0$, and

$y = 3 - 2x$ ⇒ $\frac{1}{3}(v - 2u) = 3 - \frac{2}{3}(u + v)$ or $v = 3$. Thus S is the rectangle $[0, 2] \times [0, 3]$ in the uv-plane
and

$$\iint_R (3x + 4y)\, dA = \iint_S \frac{1}{3}(7v - 5u)\left|\frac{\partial (x, y)}{\partial (u, v)}\right| du\, dv = \int_0^3 \int_0^2 \frac{1}{3}(7v - 5u)\left(\frac{1}{3}\right) du\, dv$$

$$= \frac{1}{9}\int_0^3 (14v - 10)\, dv = \frac{1}{9}(33) = \frac{11}{3}$$

13. $\dfrac{\partial (x, y)}{\partial (u, v)} = \begin{vmatrix} 2 & 0 \\ 0 & 3 \end{vmatrix} = 6$, $x^2 = 4u^2$ and the planar ellipse $9x^2 + 4y^2 \le 36$ is the image of the disk $u^2 + v^2 \le 1$.
Thus

$$\iint_R x^2\, dA = \iint_{u^2 + v^2 \le 1} (4u^2)\,(6)\, du\, dv = \int_0^{2\pi} \int_0^1 (24r^2 \cos^2 \theta)\, r\, dr\, d\theta$$

$$= 24 \int_0^{2\pi} \cos^2 \theta\, d\theta \int_0^1 r^3\, dr = 24\left[\frac{1}{2}x + \frac{1}{4}\sin 2x\right]_0^{2\pi} \left[\frac{1}{4}r^4\right]_0^1$$

$$= 24\,(\pi)\left(\frac{1}{4}\right) = 6\pi$$

15. $\dfrac{\partial (x, y)}{\partial (u, v)} = \begin{vmatrix} 1/v & -u/v^2 \\ 0 & 1 \end{vmatrix} = \frac{1}{v}$, $xy = u$, $y = x$ is the image of the parabola $v^2 = u$, $y = 3x$ is the image of the

parabola $v^2 = 3u$, and the hyperbolas $xy = 1$, $xy = 3$ are the images of the lines $u = 1$ and $u = 3$ respectively.
Thus

$$\iint_R xy\, dA = \int_1^3 \int_{\sqrt{u}}^{\sqrt{3u}} u\left(\frac{1}{v}\right) dv\, du = \int_1^3 u\left(\ln \sqrt{3u} - \ln \sqrt{u}\right) du$$

$$= \int_1^3 u \ln \sqrt{3}\, du = 4 \ln \sqrt{3} = 2 \ln 3$$

17. (a) $\dfrac{\partial (x, y, z)}{\partial (u, v, w)} = \begin{vmatrix} a & 0 & 0 \\ 0 & b & 0 \\ 0 & 0 & c \end{vmatrix} = abc$ and since $u = \dfrac{x}{a}$, $v = \dfrac{y}{b}$, $w = \dfrac{z}{c}$ the solid enclosed by the

ellipsoid is the image of the ball $u^2 + v^2 + w^2 \le 1$. So
$$\iiint_E dV = \iiint_{u^2 + v^2 + w^2 \le 1} abc\, du\, dv\, dw = (abc)\,(\text{volume of the ball}) = \frac{4}{3}\pi abc.$$

(b) If we approximate the surface of Earth by the ellipsoid $\dfrac{x^2}{6378^2} + \dfrac{y^2}{6378^2} + \dfrac{z^2}{6356^2} = 1$, then we can estimate the
volume of Earth by finding the volume of the solid E enclosed by the ellipsoid. From part (a), this is
$$\iiint_E dV = \frac{4}{3}\pi\,(6378)\,(6378)\,(6356) \approx 1.083 \times 10^{12}\text{ km}^3.$$

19. Letting $u = 2x - y$ and $v = 3x + y$, we have $x = \frac{1}{5}(u + v)$, $y = \frac{1}{5}(2v - 3u)$. Then

$\dfrac{\partial (x, y)}{\partial (u, v)} = \begin{vmatrix} 1/5 & 1/5 \\ -3/5 & 2/5 \end{vmatrix} = \frac{1}{5}$ and

$$\iint_R xy\, dA = \int_{-2}^1 \int_{-3}^1 \frac{(u + v)(2v - 3u)}{25}\left(\frac{1}{5}\right) du\, dv = \frac{1}{125}\int_{-2}^1 \int_{-3}^1 (2v^2 - uv - 3u^2)\, du\, dv$$

$$= \frac{1}{125}\int_{-2}^1 (8v^2 + 4v - 28)\, dv = -\frac{66}{125}$$

21. Letting $u = y - x$, $v = y + x$, we have $y = \frac{1}{2}(u + v)$, $x = \frac{1}{2}(v - u)$. Then $\dfrac{\partial(x,y)}{\partial(u,v)} = \begin{vmatrix} -1/2 & 1/2 \\ 1/2 & 1/2 \end{vmatrix} = -\dfrac{1}{2}$

and R is the image of the trapezoidal region with vertices $(-1,1)$, $(-2,2)$, $(2,2)$, and $(1,1)$. Thus

$$\iint_R \cos\frac{y-x}{y+x}\, dA = \int_1^2 \int_{-v}^v \cos\frac{u}{v}\left|-\frac{1}{2}\right| du\, dv = \frac{1}{2}\int_1^2 \left[v\sin\frac{u}{v}\right]_{u=-v}^{u=v} dv$$

$$= \frac{1}{2}\int_1^2 2v\sin(1)\, dv = \frac{3}{2}\sin 1$$

23. Let $u = x + y$ and $v = -x + y$. Then $u + v = 2y \;\Rightarrow\; y = \frac{1}{2}(u + v)$ and

$u - v = 2x \;\Rightarrow\; x = \frac{1}{2}(u - v)$. $\dfrac{\partial(x,y)}{\partial(u,v)} = \begin{vmatrix} 1/2 & -1/2 \\ 1/2 & 1/2 \end{vmatrix} = \dfrac{1}{2}$. Now

$|u| = |x + y| \le |x| + |y| \le 1 \;\Rightarrow\; -1 \le u \le 1$, and

$|v| = |-x + y| \le |x| + |y| \le 1 \Rightarrow\; -1 \le v \le 1$.

R is the image of the square region with vertices $(1,1)$, $(1,-1)$, $(-1,-1)$, and

$(-1,1)$. So $\iint_R e^{x+y}\, dA = \frac{1}{2}\int_{-1}^1 \int_{-1}^1 e^u\, du\, dv = \frac{1}{2}\left[e^u\right]_{-1}^1 \left[v\right]_{-1}^1 = e - e^{-1}$.

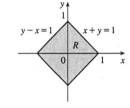

▆16 Review

CONCEPT CHECK

1. (a) A double Riemann sum of f is $\sum_{i=1}^m \sum_{j=1}^n f\left(x_{ij}^*, y_{ij}^*\right) \Delta A$, where ΔA is the area of each subrectangle and

$f\left(x_{ij}^*, y_{ij}^*\right)$ is a sample point in each subrectangle. If $f(x,y) \ge 0$, this sum represents an approximation to the volume of the solid that lies above the rectangle R and below the graph of f.

(b) $\iint_R f(x,y)\, dA = \lim_{m,n\to\infty} \sum_{i=1}^m \sum_{j=1}^n f\left(x_{ij}^*, y_{ij}^*\right) \Delta A$

(c) If $f(x,y) \ge 0$, $\iint_R f(x,y)\, dA$ represents the volume of the solid that lies above the rectangle R and below the surface $z = f(x,y)$. If f takes on both positive and negative values, $\iint_R f(x,y)\, dA$ is the difference of the volume above R but below the surface $z = f(x,y)$ and the volume below R but above the surface $z = f(x,y)$.

(d) We usually evaluate $\iint_R f(x,y)\, dA$ as an iterated integral according to Fubini's Theorem (see Theorem 16.2.4 [ET 15.2.4]).

(e) The Midpoint Rule for Double Integrals says that we approximate the double integral $\iint_R f(x,y)\, dA$ by the double Riemann sum $\sum_{i=1}^m \sum_{j=1}^n f\left(\bar{x}_i, \bar{y}_j\right) \Delta A$ where the sample points $\left(\bar{x}_i, \bar{y}_j\right)$ are the centers of the subrectangles.

(f) $f_{\text{ave}} = \dfrac{1}{A(R)} \iint_R f(x,y)\, dA$ where $A(R)$ is the area of R.

2. (a) See (1) and (2) and the accompanying discussion in Section 16.3 [ET 15.3].

(b) See (1) and (2) and the accompanying discussion in Section 16.3 [ET 15.3].

(c) See (5) and the preceding discussion in Section 16.3 [ET 15.3].

(d) See (6)–(11) in Section 16.3 [ET 15.3].

3. We may want to change from rectangular to polar coordinates in a double integral if the region R of integration is more easily described in polar coordinates. To accomplish this, we use

$\iint_R f(x, y) \, dA = \int_\alpha^\beta \int_a^b f(r \cos \theta, r \sin \theta) \, r \, dr \, d\theta$ where R is given by $0 \leq a \leq r \leq b, \alpha \leq \theta \leq \beta$.

4. (a) $m = \iint_D \rho(x, y) \, dA$

(b) $M_x = \iint_D y\rho(x, y) \, dA$, $M_y = \iint_D x\rho(x, y) \, dA$

(c) The center of mass is $(\overline{x}, \overline{y})$ where $\overline{x} = \dfrac{M_y}{m}$ and $\overline{y} = \dfrac{M_x}{m}$.

(d) $I_x = \iint_D y^2 \rho(x, y) \, dA$, $I_y = \iint_D x^2 \rho(x, y) \, dA$, $I_0 = \iint_D (x^2 + y^2) \rho(x, y) \, dA$

5. (a) $P(a \leq X \leq b, c \leq Y \leq d) = \int_a^b \int_c^d f(x, y) \, dy \, dx$

(b) $f(x, y) \geq 0$ and $\iint_{\mathbb{R}^2} f(x, y) \, dA = 1$.

(c) The expected value of X is $\mu_1 = \iint_{\mathbb{R}^2} xf(x, y) \, dA$; the expected value of Y is $\mu_2 = \iint_{\mathbb{R}^2} yf(x, y) \, dA$.

6. $A(S) = \iint_D \sqrt{[f_x(x, y)]^2 + [f_y(x, y)]^2 + 1} \, dA$

7. (a) $\iiint_B f(x, y, z) \, dV = \lim\limits_{l,m,n \to \infty} \sum\limits_{i=1}^l \sum\limits_{j=1}^m \sum\limits_{k=1}^n f(x_{ijk}^*, y_{ijk}^*, z_{ijk}^*) \, \Delta V$

(b) We usually evaluate $\iiint_B f(x, y, z) \, dV$ as an iterated integral according to Fubini's Theorem for Triple Integrals (see Theorem 16.7.4 [ET 15.7.4]).

(c) See the paragraph following Example 16.7.1 [ET 15.7.1].

(d) See (5) and (6) and the accompanying discussion in Section 16.7 [ET 15.7].

(e) See (10) and the accompanying discussion in Section 16.7 [ET 15.7].

(f) See (11) and the preceding discussion in Section 16.7 [ET 15.7].

8. (a) $m = \iiint_E \rho(x, y, z) \, dV$

(b) $M_{yz} = \iiint_E x\rho(x, y, z) \, dV$, $M_{xz} = \iiint_E y\rho(x, y, z) \, dV$, $M_{xy} = \iiint_E z\rho(x, y, z) \, dV$.

(c) The center of mass is $(\overline{x}, \overline{y}, \overline{z})$ where $\overline{x} = \dfrac{M_{yz}}{m}$, $\overline{y} = \dfrac{M_{xz}}{m}$, and $\overline{z} = \dfrac{M_{xy}}{m}$.

(d) $I_x = \iiint_E (y^2 + z^2) \rho(x, y, z) \, dV$, $I_y = \iiint_E (x^2 + z^2) \rho(x, y, z) \, dV$,

$I_z = \iiint_E (x^2 + y^2) \rho(x, y, z) \, dV$.

9. (a) See Formula 16.8.2 [ET 15.8.2] and the accompanying discussion.

(b) See Formula 16.8.4 [ET 15.8.4] and the accompanying discussion.

(c) We may want to change from rectangular to cylindrical or spherical coordinates in a triple integral if the region E of integration is more easily described in cylindrical or spherical coordinates or if the triple integral is easier to evaluate using cylindrical or spherical coordinates.

10. (a) $\dfrac{\partial(x, y)}{\partial(u, v)} = \begin{vmatrix} \partial x/\partial u & \partial x/\partial v \\ \partial y/\partial u & \partial y/\partial v \end{vmatrix} = \dfrac{\partial x}{\partial u}\dfrac{\partial y}{\partial v} - \dfrac{\partial x}{\partial v}\dfrac{\partial y}{\partial u}$

(b) See (9) and the accompanying discussion in Section 16.9 [ET 15.9].

(c) See (13) and the accompanying discussion in Section 16.9 [ET 15.9].

TRUE-FALSE QUIZ

1. This is true by Fubini's Theorem.

3. True:

$$\iint_D \sqrt{4 - x^2 - y^2}\, dA = \text{ the volume under the surface } x^2 + y^2 + z^2 = 4 \text{ and above the } xy\text{-plane}$$

$$= \tfrac{1}{2}\ \big(\text{the volume of the sphere } x^2 + y^2 + z^2 = 4\big) = \tfrac{1}{2} \cdot \tfrac{4}{3}\pi\,(2)^3 = \tfrac{16}{3}\pi$$

5. The volume enclosed by the cone $z = \sqrt{x^2 + y^2}$ and the plane $z = 2$ is, in cylindrical coordinates,

$V = \int_0^{2\pi}\int_0^2\int_r^2 r\,dz\,dr\,d\theta \neq \int_0^{2\pi}\int_0^2\int_r^2 dz\,dr\,d\theta$, so the assertion is false.

EXERCISES

1. As shown in the contour map, we divide R into 9 equally sized subsquares, each with area $\Delta A = 1$. Then we approximate $\iint_R f(x, y)\, dA$ by a Riemann sum with $m = n = 3$ and the sample points the upper right corners of each square, so

$$\iint_R f(x, y)\, dA \approx \sum_{i=1}^{3}\sum_{j=1}^{3} f(x_i, y_j)\,\Delta A$$

$$= \Delta A\,[f(1,1) + f(1,2) + f(1,3) + f(2,1) + f(2,2)$$

$$+ f(2,3) + f(3,1) + f(3,2) + f(3,3)]$$

Using the contour lines to estimate the function values, we have

$$\iint_R f(x, y)\, dA \approx 1\,[2.7 + 4.7 + 8.0 + 4.7 + 6.7 + 10.0 + 6.7 + 8.6 + 11.9] \approx 64.0$$

3. $\int_1^2\int_0^2 (y + 2xe^y)\, dx\,dy = \int_1^2 \big[xy + x^2 e^y\big]_{x=0}^{x=2}\, dy = \int_1^2 (2y + 4e^y)\, dy = \big[y^2 + 4e^y\big]_1^2$

$$= 4 + 4e^2 - 1 - 4e = 4e^2 - 4e + 3$$

5. $\int_0^1\int_0^x \cos(x^2)\, dy\,dx = \int_0^1 \big[\cos(x^2)\,y\big]_{y=0}^{y=x}\, dx = \int_0^1 x\cos(x^2)\, dx = \tfrac{1}{2}\sin(x^2)\big]_0^1 = \tfrac{1}{2}\sin 1$

7. $\int_0^\pi\int_0^1\int_0^{\sqrt{1-y^2}} y\sin x\, dz\,dy\,dx = \int_0^\pi\int_0^1 \big[(y\sin x)\,z\big]_{z=0}^{z=\sqrt{1-y^2}}\, dy\,dx = \int_0^\pi\int_0^1 y\sqrt{1-y^2}\,\sin x\, dy\,dx$

$$= \int_0^\pi \Big[-\tfrac{1}{3}\,(1 - y^2)^{3/2}\sin x\Big]_{y=0}^{y=1}\, dx = \int_0^\pi \tfrac{1}{3}\sin x\, dx = -\tfrac{1}{3}\cos x\big]_0^\pi = \tfrac{2}{3}$

9. The region R is more easily described by polar coordinates: $R = \{(r, \theta) \mid 2 \le r \le 4, 0 \le \theta \le \pi\}$. Thus

$\iint_R f(x, y)\, dA = \int_0^\pi\int_2^4 f(r\cos\theta, r\sin\theta)\, r\,dr\,d\theta$.

11.

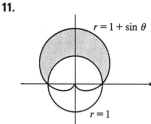

$r = 1 + \sin\theta$

$r = 1$

The region whose area is given by $\int_0^\pi\int_1^{1+\sin\theta} r\,dr\,d\theta$ is

$\{(r, \theta) \mid 0 \le \theta \le \pi, 1 \le r \le 1 + \sin\theta\}$, which is the region outside the

circle $r = 1$ and inside the cardioid $r = 1 + \sin\theta$.

13.

$$\int_0^1 \int_x^1 e^{x/y}\, dy\, dx = \int_0^1 \int_0^y e^{x/y}\, dx\, dy$$

$$= \int_0^1 \left[y e^{x/y} \right]_{x=0}^{x=y} dy$$

$$= \int_0^1 (ey - y)\, dy = \left[\tfrac{e}{2} y^2 - \tfrac{1}{2} y^2 \right]_0^1$$

$$= \tfrac{1}{2}(e - 1)$$

15. $\int_2^4 \int_0^1 \dfrac{1}{(x-y)^2}\, dx\, dy = \int_2^4 \left[-(x-y)^{-1} \right]_{x=0}^{x=1} dy = \int_2^4 \left(-\dfrac{1}{y} - \dfrac{1}{1-y} \right) dy$

$$= \left[-\ln y + \ln|1-y| \right]_2^4 = -\ln 4 + \ln 3 + \ln 2 = \ln \tfrac{3}{2}$$

17. The curves $y^2 = x^3$ and $y = x$ intersect when $x^3 = x$, that is when $x = 0$ and $x = 1$ (note that $x \neq -1$ since $x^3 = y^2 \;\Rightarrow\; x \geq 0$.) So $\int_0^1 \int_{x^{3/2}}^x xy\, dy\, dx = \int_0^1 \left[\tfrac{1}{2} x^3 - \tfrac{1}{2} x^4 \right] dx = \left[\tfrac{1}{8} x^4 - \tfrac{1}{10} x^5 \right]_0^1 = \tfrac{1}{40}$.

19. $\int_0^1 \int_0^{1-y^2} (xy + 2x + 3y)\, dx\, dy = \int_0^1 \left[\tfrac{1}{2} x^2 y + x^2 + 3xy \right]_{x=0}^{x=1-y^2} dy$

$$= \int_0^1 \left[\tfrac{1}{2} y (1-y^2)^2 + (1-y^2)^2 + 3y(1-y^2) \right] dy$$

$$= \tfrac{1}{2} y^5 + y^4 - 4y^3 - 2y^2 + \tfrac{7}{2} y + 1 = \tfrac{1}{5} + \tfrac{1}{12} - \tfrac{2}{3} + \tfrac{7}{4} = \tfrac{41}{30}$$

21.

$$\iint_D (x^2 + y^2)^{3/2}\, dA = \int_0^{\pi/3} \int_0^3 (r^2)^{3/2}\, r\, dr\, d\theta$$

$$= \int_0^{\pi/3} d\theta \int_0^3 r^4\, dr = [\theta]_0^{\pi/3} \left[\tfrac{1}{5} r^5 \right]_0^3$$

$$= \tfrac{\pi}{3} \tfrac{3^5}{5} = \tfrac{81\pi}{5}$$

23. $\iiint_E x^2 z\, dV = \int_0^2 \int_0^{2x} \int_0^x x^2 z\, dz\, dy\, dx = \int_0^2 \int_0^{2x} \tfrac{1}{2} x^4\, dy\, dx = \int_0^2 x^5\, dx = \tfrac{1}{6} \cdot 2^6 = \tfrac{32}{3}$

25. $\iiint_E y^2 z^2\, dV = \int_{-1}^1 \int_{-\sqrt{1-y^2}}^{\sqrt{1-y^2}} \int_0^{1-y^2-z^2} y^2 z^2\, dx\, dz\, dy = \int_{-1}^1 \int_{-\sqrt{1-y^2}}^{\sqrt{1-y^2}} y^2 z^2 (1-y^2-z^2)\, dz\, dy$

$$= \int_0^{2\pi} \int_0^1 (r^2 \cos^2 \theta)(r^2 \sin^2 \theta)(1-r^2)\, r\, dr\, d\theta = \int_0^{2\pi} \int_0^1 \tfrac{1}{4} \sin^2 2\theta\, (r^5 - r^7)\, dr\, d\theta$$

$$= \int_0^{2\pi} \tfrac{1}{8} (1 - \cos 4\theta) \left[\tfrac{1}{6} r^6 - \tfrac{1}{8} r^8 \right]_{r=0}^{r=1} d\theta = \tfrac{1}{192} \left[\theta - \tfrac{1}{4} \sin 4\theta \right]_0^{2\pi} = \tfrac{2\pi}{192} = \tfrac{\pi}{96}$$

27. $\iiint_E yz\, dV = \int_{-2}^2 \int_0^{\sqrt{4-x^2}} \int_0^y yz\, dz\, dy\, dx = \int_{-2}^2 \int_0^{\sqrt{4-x^2}} \tfrac{1}{2} y^3\, dy\, dx = \int_0^\pi \int_0^2 \tfrac{1}{2} r^3 \sin^3 \theta\, r\, dr\, d\theta$

$$= \tfrac{16}{5} \int_0^\pi \sin^3 \theta\, d\theta = \tfrac{16}{5} \left[-\cos \theta + \tfrac{1}{3} \cos^3 \theta \right]_0^\pi = \tfrac{64}{15}$$

29. $V = \int_0^2 \int_1^4 (x^2 + 4y^2)\, dy\, dx = \int_0^2 \left[x^2 y + \tfrac{4}{3} y^3 \right]_{y=1}^{y=4} dx = \int_0^2 (3x^2 + 84)\, dx = 176$

31.

$$V = \int_0^2 \int_0^y \int_0^{(2-y)/2} dz\, dx\, dy = \int_0^2 \int_0^y \left(1 - \tfrac{1}{2}y\right) dx\, dy$$

$$= \int_0^2 \left(y - \tfrac{1}{2}y^2\right) dy = \tfrac{2}{3}$$

33. Using the wedge above the plane $z = 0$ and below the plane $z = mx$ and noting that we have the same volume for $m < 0$ as for $m > 0$ (so use $m > 0$), we have

$$V = 2 \int_0^{a/3} \int_0^{\sqrt{a^2 - 9y^2}} mx\, dx\, dy = 2 \int_0^{a/3} \tfrac{1}{2}m\left(a^2 - 9y^2\right) dy = m\left[a^2 y - 3y^3\right]_0^{a/3}$$

$$= m\left(\tfrac{1}{3}a^3 - \tfrac{1}{9}a^3\right) = \tfrac{2}{9}ma^3$$

35. (a) $m = \int_0^1 \int_0^{1-y^2} y\, dx\, dy = \int_0^1 \left(y - y^3\right) dy = \tfrac{1}{2} - \tfrac{1}{4} = \tfrac{1}{4}$

(b) $M_y = \int_0^1 \int_0^{1-y^2} xy\, dx\, dy = \int_0^1 \tfrac{1}{2}y\left(1 - y^2\right)^2 dy = -\tfrac{1}{12}\left(1 - y^2\right)^3\Big]_0^1 = \tfrac{1}{12}$,

$M_x = \int_0^1 \int_0^{1-y^2} y^2\, dx\, dy = \int_0^1 \left(y^2 - y^4\right) dy = \tfrac{2}{15}$. Hence $(\overline{x}, \overline{y}) = \left(\tfrac{1}{3}, \tfrac{8}{15}\right)$.

(c) $I_x = \int_0^1 \int_0^{1-y^2} y^3\, dx\, dy = \int_0^1 \left(y^3 - y^5\right) dy = \tfrac{1}{12}$,

$I_y = \int_0^1 \int_0^{1-y^2} yx^2\, dx\, dy = \int_0^1 \tfrac{1}{3}y\left(1 - y^2\right)^3 dy = -\tfrac{1}{24}\left(1 - y^2\right)^4\Big]_0^1 = \tfrac{1}{24}$, $I_0 = I_x + I_y = \tfrac{1}{8}$,

$\overline{\overline{y}}^2 = \tfrac{1/12}{1/4} = \tfrac{1}{3} \quad \Rightarrow \quad \overline{\overline{y}} = \tfrac{1}{\sqrt{3}}$, and $\overline{\overline{x}}^2 = \tfrac{1/24}{1/4} = \tfrac{1}{6} \quad \Rightarrow \quad \overline{\overline{x}} = \tfrac{1}{\sqrt{6}}$.

37. (a) The equation of the cone with the suggested orientation is $(h - z) = \tfrac{h}{a}\sqrt{x^2 + y^2}, 0 \le z \le h$. Then $V = \tfrac{1}{3}\pi a^2 h$ is the volume of one frustum of a cone; by symmetry $M_{yz} = M_{xz} = 0$; and

$$M_{xy} = \iint_{x^2 + y^2 \le a^2} \int_0^{h - (h/a)\sqrt{x^2 + y^2}} z\, dz\, dA = \int_0^{2\pi} \int_0^a \int_0^{(h/a)(a - r)} rz\, dz\, dr\, d\theta$$

$$= \pi \int_0^a r\frac{h^2}{a^2}(a - r)^2\, dr = \frac{\pi h^2}{a^2} \int_0^a \left(a^2 r - 2ar^2 + r^3\right) dr = \frac{\pi h^2}{a^2}\left(\frac{a^4}{2} - \frac{2a^4}{3} + \frac{a^4}{4}\right) = \frac{\pi h^2 a^2}{12}$$

Hence the centroid is $(\overline{x}, \overline{y}, \overline{z}) = \left(0, 0, \tfrac{1}{4}h\right)$.

(b) $I_z = \int_0^{2\pi} \int_0^a \int_0^{(h/a)(a - r)} r^3\, dz\, dr\, d\theta = 2\pi \int_0^a \frac{h}{a}\left(ar^3 - r^4\right) dr = \frac{2\pi h}{a}\left(\frac{a^5}{4} - \frac{a^5}{5}\right) = \frac{\pi a^4 h}{10}$

39. Let D represent the given triangle; then D can be described as the area enclosed by the x- and y-axes and the line $y = 2 - 2x$, or equivalently $D = \{(x, y) \mid 0 \le x \le 1, 0 \le y \le 2 - 2x\}$. We want to find the surface area of the part of the graph of $z = x^2 + y$ that lies over D, so using Equation 16.6.3 [ET 15.6.3] we have

$$A(S) = \iint_D \sqrt{1 + \left(\frac{\partial z}{\partial x}\right)^2 + \left(\frac{\partial z}{\partial y}\right)^2}\, dA = \iint_D \sqrt{1 + (2x)^2 + (1)^2}\, dA$$

$$= \int_0^1 \int_0^{2 - 2x} \sqrt{2 + 4x^2}\, dy\, dx = \int_0^1 \sqrt{2 + 4x^2}\, [y]_0^{2 - 2x}\, dx = \int_0^1 (2 - 2x)\sqrt{2 + 4x^2}\, dx$$

$$= \int_0^1 2\sqrt{2 + 4x^2}\, dx - \int_0^1 2x\sqrt{2 + 4x^2}\, dx$$

Using Formula 21 in the Table of Integrals with $a = \sqrt{2}$, $u = 2x$, and $du = 2\,dx$, we have

$\int 2\sqrt{2 + 4x^2}\,dx = x\sqrt{2 + 4x^2} + \ln\left(2x + \sqrt{2 + 4x^2}\right)$. If we substitute $u = 2 + 4x^2$ in the second integral, then

$du = 8x\,dx$ and $\int 2x\sqrt{2 + 4x^2}\,dx = \frac{1}{4}\int \sqrt{u}\,du = \frac{1}{4} \cdot \frac{2}{3}u^{3/2} = \frac{1}{6}\left(2 + 4x^2\right)^{3/2}$. Thus

$$A(S) = \left[x\sqrt{2 + 4x^2} + \ln\left(2x + \sqrt{2 + 4x^2}\right) - \frac{1}{6}\left(2 + 4x^2\right)^{3/2}\right]_0^1$$

$$= \sqrt{6} + \ln\left(2 + \sqrt{6}\right) - \frac{1}{6}(6)^{3/2} - \ln\sqrt{2} + \frac{\sqrt{2}}{3}$$

$$= \ln\frac{2 + \sqrt{6}}{\sqrt{2}} + \frac{\sqrt{2}}{3} = \ln\left(\sqrt{2} + \sqrt{3}\right) + \frac{\sqrt{2}}{3} \approx 1.6176$$

41.

$$\int_0^{\sqrt{2}} \int_y^{\sqrt{4 - y^2}} \frac{1}{1 + x^2 + y^2}\,dx\,dy = \int_0^{\pi/4} \int_0^2 \frac{1}{1 + r^2}\,r\,dr\,d\theta$$

$$= \int_0^{\pi/4} d\theta \int_0^2 \frac{r}{1 + r^2}\,dr$$

$$= [\theta]_0^{\pi/4}\,\left[\frac{1}{2}\ln|1 + r^2|\right]_0^2$$

$$= \frac{\pi}{4}\left(\frac{1}{2}\ln 5\right) = \frac{\pi}{8}\ln 5$$

43. From the graph, it appears that $1 - x^2 = e^x$ at $x \approx -0.71$ and at

$x = 0$, with $1 - x^2 > e^x$ on $(-0.71, 0)$. So the desired integral is

$$\iint_D y^2\,dA \approx \int_{-0.71}^0 \int_{e^x}^{1 - x^2} y^2\,dy\,dx$$

$$= \frac{1}{3}\int_{-0.71}^0 \left[(1 - x^2)^3 - e^{3x}\right]\,dx$$

$$= \frac{1}{3}\left[x - x^3 + \frac{3}{5}x^5 - \frac{1}{7}x^7 - \frac{1}{3}e^{3x}\right]_{-0.71}^0 \approx 0.0512$$

45. (a) $f(x, y)$ is a joint density function, so we know that $\iint_{\mathbb{R}^2} f(x, y)\,dA = 1$. Since $f(x, y) = 0$ outside the

rectangle $[0, 3] \times [0, 2]$, we can say

$$\iint_{\mathbb{R}^2} f(x, y)\,dA = \int_{-\infty}^\infty \int_{-\infty}^\infty f(x, y)\,dy\,dx = \int_0^3 \int_0^2 C(x + y)\,dy\,dx$$

$$= C\int_0^3 \left[xy + \frac{1}{2}y^2\right]_{y=0}^{y=2}\,dx = C\int_0^3 (2x + 2)\,dx = C\left[x^2 + 2x\right]_0^3 = 15C$$

Then $15C = 1 \Rightarrow C = \frac{1}{15}$.

(b) $P(X \le 2, Y \ge 1) = \int_{-\infty}^2 \int_1^\infty f(x, y)\,dy\,dx = \int_0^2 \int_1^2 \frac{1}{15}(x + y)\,dy\,dx = \frac{1}{15}\int_0^2 \left[xy + \frac{1}{2}y^2\right]_{y=1}^{y=2}\,dx$

$= \frac{1}{15}\int_0^2 \left(x + \frac{3}{2}\right)\,dx = \frac{1}{15}\left[\frac{1}{2}x^2 + \frac{3}{2}x\right]_0^2 = \frac{1}{3}$

(c) $P(X + Y \le 1) = P((X, Y) \in D)$ where D is the triangular region

shown in the figure. Thus

$$P(X + Y \le 1) = \iint_D f(x, y)\,dA = \int_0^1 \int_0^{1 - x} \frac{1}{15}(x + y)\,dy\,dx$$

$$= \frac{1}{15}\int_0^1 \left[xy + \frac{1}{2}y^2\right]_{y=0}^{y=1-x}\,dx$$

$$= \frac{1}{15}\int_0^1 \left[x(1 - x) + \frac{1}{2}(1 - x)^2\right]\,dx$$

$$= \frac{1}{30}\int_0^1 (1 - x^2)\,dx = \frac{1}{30}\left[x - \frac{1}{3}x^3\right]_0^1 = \frac{1}{45}$$

47.

$$\int_{-1}^{1} \int_{x^2}^{1} \int_{0}^{1-y} f(x,y,z)\, dz\, dy\, dx = \int_{0}^{1} \int_{0}^{1-z} \int_{-\sqrt{y}}^{\sqrt{y}} f(x,y,z)\, dx\, dy\, dz$$

49. Since $u = x - y$ and $v = x + y$, $x = \frac{1}{2}(u+v)$ and $y = \frac{1}{2}(v-u)$. Thus $\dfrac{\partial(x,y)}{\partial(u,v)} = \begin{vmatrix} 1/2 & 1/2 \\ -1/2 & 1/2 \end{vmatrix} = \dfrac{1}{2}$ and

$$\iint_{R} \frac{x-y}{x+y}\, dA = \int_{2}^{4} \int_{-2}^{0} \frac{u}{v}\left(\frac{1}{2}\right) du\, dv = -\int_{2}^{4} \frac{dv}{v} = -\ln 2.$$

51. Let $u = y - x$ and $v = y + x$ so $x = y - u = (v - x) - u \;\Rightarrow\; x = \frac{1}{2}(v - u)$ and

$y = v - \frac{1}{2}(v-u) = \frac{1}{2}(v+u)$. $\left| \dfrac{\partial(x,y)}{\partial(u,v)} \right| = \left| \dfrac{\partial x}{\partial u}\dfrac{\partial y}{\partial v} - \dfrac{\partial x}{\partial v}\dfrac{\partial y}{\partial u} \right| = \left| -\frac{1}{2}\left(\frac{1}{2}\right) - \frac{1}{2}\left(\frac{1}{2}\right) \right| = \left| -\frac{1}{2} \right| = \frac{1}{2}$. R is the

image under this transformation of the square with vertices $(u, v) = (0, 0)$, $(-2, 0)$, $(0, 2)$, and $(-2, 2)$. So

$$\iint_{R} xy\, dA = \int_{0}^{2} \int_{-2}^{0} \frac{v^2 - u^2}{4}\left(\frac{1}{2}\right) du\, dv = \frac{1}{8}\int_{0}^{2}\left[v^2 u - \frac{1}{3}u^3\right]_{u=-2}^{u=0} dv = \frac{1}{8}\int_{0}^{2}\left(2v^2 - \frac{8}{3}\right) dv$$

$$= \frac{1}{8}\left[\frac{2}{3}v^3 - \frac{8}{3}v\right]_{0}^{2} = 0$$

This result could have been anticipated by symmetry, since the integrand is an odd function of y and R is symmetric
about the x-axis.

53. For each r such that D_r lies within the domain, $A(D_r) = \pi r^2$, and by the Mean Value Theorem for Double

Integrals there exists (x_r, y_r) in D_r such that $f(x_r, y_r) = \dfrac{1}{\pi r^2} \iint_{D_r} f(x,y)\, dA$. But $\lim\limits_{r \to 0^+}(x_r, y_r) = (a, b)$, so

$$\lim_{r \to 0^+} \frac{1}{\pi r^2} \iint_{D_r} f(x,y)\, dA = \lim_{r \to 0^+} f(x_r, y_r) = f(a, b) \text{ by the continuity of } f.$$

Problems Plus

1.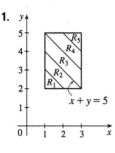

Let $R = \bigcup_{i=1}^{5} R_i$, where

$R_i = \{(x,y) \mid x + y \geq i + 2, x + y < i + 3, 1 \leq x \leq 3, 2 \leq y \leq 5\}.$

$\iint_R [\![x+y]\!]\, dA = \sum_{i=1}^{5} \iint_{R_i} [\![x+y]\!]\, dA = \sum_{i=1}^{5} [\![x+y]\!] \iint_{R_i} dA$, since

$[\![x+y]\!] = \text{constant} = i + 2$ for $(x,y) \in R_i$. Therefore

$$\iint_R [\![x+y]\!]\, dA = \sum_{i=1}^{5} (i+2)\,[A\,(R_i)]$$

$$= 3A\,(R_1) + 4A\,(R_2) + 5A\,(R_3) + 6A\,(R_4) + 7A\,(R_5)$$

$$= 3\left(\tfrac{1}{2}\right) + 4\left(\tfrac{3}{2}\right) + 5\,(2) + 6\left(\tfrac{3}{2}\right) + 7\left(\tfrac{1}{2}\right) = 30$$

3. $f_{\text{ave}} = \dfrac{1}{b-a} \int_a^b f(x)\,dx = \dfrac{1}{1-0} \int_0^1 \left[\int_x^1 \cos\left(t^2\right)\, dt \right]\, dx$

$= \int_0^1 \int_x^1 \cos\left(t^2\right)\, dt\, dx$

$= \int_0^1 \int_0^t \cos\left(t^2\right)\, dx\, dt$ (changing the order of integration)

$= \int_0^1 t \cos\left(t^2\right)\, dt = \tfrac{1}{2} \sin\left(t^2\right)\Big]_0^1 = \tfrac{1}{2}\sin 1$

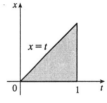

5. Since $|xy| < 1$, except at $(1,1)$, the formula for the sum of a geometric series gives $\dfrac{1}{1-xy} = \displaystyle\sum_{n=0}^{\infty} (xy)^n$, so

$$\int_0^1 \int_0^1 \frac{1}{1-xy}\, dx\, dy = \int_0^1 \int_0^1 \sum_{n=0}^{\infty} (xy)^n\, dx\, dy = \sum_{n=0}^{\infty} \int_0^1 \int_0^1 (xy)^n\, dx\, dy$$

$$= \sum_{n=0}^{\infty} \left[\int_0^1 x^n\, dx \right] \left[\int_0^1 y^n\, dy \right] = \sum_{n=0}^{\infty} \frac{1}{n+1} \cdot \frac{1}{n+1}$$

$$= \sum_{n=0}^{\infty} \frac{1}{(n+1)^2} = \frac{1}{1^2} + \frac{1}{2^2} + \frac{1}{3^2} + \cdots = \sum_{n=1}^{\infty} \frac{1}{n^2}$$

7. (a) Since $|xyz| < 1$ except at $(1,1,1)$, the formula for the sum of a geometric series gives $\dfrac{1}{1-xyz} = \displaystyle\sum_{n=0}^{\infty} (xyz)^n$,

so

$$\int_0^1 \int_0^1 \int_0^1 \frac{1}{1-xyz}\, dx\, dy\, dz = \int_0^1 \int_0^1 \int_0^1 \sum_{n=0}^{\infty} (xyz)^n\, dx\, dy\, dz = \sum_{n=0}^{\infty} \int_0^1 \int_0^1 \int_0^1 (xyz)^n\, dx\, dy\, dz$$

$$= \sum_{n=0}^{\infty} \left[\int_0^1 x^n\, dx \right] \left[\int_0^1 y^n\, dy \right] \left[\int_0^1 z^n\, dz \right] = \sum_{n=0}^{\infty} \frac{1}{n+1} \cdot \frac{1}{n+1} \cdot \frac{1}{n+1}$$

$$= \sum_{n=0}^{\infty} \frac{1}{(n+1)^3} = \frac{1}{1^3} + \frac{1}{2^3} + \frac{1}{3^3} + \cdots = \sum_{n=1}^{\infty} \frac{1}{n^3}.$$

(b) Since $|-xyz| < 1$, except at $(1, 1, 1)$, the formula for the sum of a geometric series gives

$$\frac{1}{1+xyz} = \sum_{n=0}^{\infty} (-xyz)^n, \text{ so}$$

$$\int_0^1 \int_0^1 \int_0^1 \frac{1}{1+xyz} \, dx \, dy \, dz = \int_0^1 \int_0^1 \int_0^1 \sum_{n=0}^{\infty} (-xyz)^n \, dx \, dy \, dz = \sum_{n=0}^{\infty} \int_0^1 \int_0^1 \int_0^1 (-xyz)^n \, dx \, dy \, dz$$

$$= \sum_{n=0}^{\infty} (-1)^n \left[\int_0^1 x^n \, dx \right] \left[\int_0^1 y^n \, dy \right] \left[\int_0^1 z^n \, dz \right]$$

$$= \sum_{n=0}^{\infty} (-1)^n \frac{1}{n+1} \cdot \frac{1}{n+1} \cdot \frac{1}{n+1}$$

$$= \sum_{n=0}^{\infty} \frac{(-1)^n}{(n+1)^3} = \frac{1}{1^3} - \frac{1}{2^3} + \frac{1}{3^3} - \cdots = \sum_{n=1}^{\infty} \frac{(-1)^{n-1}}{n^3}$$

To evaluate this sum, we first write out a few terms: $s = 1 - \frac{1}{2^3} + \frac{1}{3^3} - \frac{1}{4^3} + \frac{1}{5^3} - \frac{1}{6^3} \approx 0.8998$. Notice that

$a_7 = \frac{1}{7^3} < 0.003$. By the Alternating Series Estimation Theorem from Section 12.5 [ET 11.5], we have

$|s - s_6| \le a_7 < 0.003$. This error of 0.003 will not affect the second decimal place, so we have $s \approx 0.90$.

9. $\int_0^x \int_0^y \int_0^z f(t) \, dt \, dz \, dy = \iiint_E f(t) \, dV$, where

$E = \{(t, z, y) \mid 0 \le t \le z, 0 \le z \le y, 0 \le y \le x\}$.

If we let D be the projection of E on the yt-plane then

$D = \{(y, t) \mid 0 \le t \le x, t \le y \le x\}$. And we see from the diagram

that $E = \{(t, z, y) \mid t \le z \le y, t \le y \le x, 0 \le t \le x\}$. So

$$\int_0^x \int_0^y \int_0^z f(t) \, dt \, dz \, dy = \int_0^x \int_t^x \int_t^y f(t) \, dz \, dy \, dt = \int_0^x \left[\int_t^x (y - t) f(t) \, dy \right] dt$$

$$= \int_0^x \left[\left(\tfrac{1}{2} y^2 - ty \right) f(t) \right]_{y=t}^{y=x} dt = \int_0^x \left[\tfrac{1}{2} x^2 - tx - \tfrac{1}{2} t^2 + t^2 \right] f(t) \, dt$$

$$= \int_0^x \left[\tfrac{1}{2} x^2 - tx + \tfrac{1}{2} t^2 \right] f(t) \, dt = \int_0^x \tfrac{1}{2} \left(x^2 - 2tx + t^2 \right) f(t) \, dt$$

$$= \tfrac{1}{2} \int_0^x (x - t)^2 f(t) \, dt$$

≡17.1 Vector Fields

1. $\mathbf{F}(x, y) = x\,\mathbf{i} + y\,\mathbf{j}$

The length of the vector $x\,\mathbf{i} + y\,\mathbf{j}$ is the distance from $(0, 0)$ to (x, y). Each vector points away from the origin.

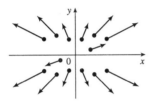

3. $\mathbf{F}(x, y) = y\,\mathbf{i} + \mathbf{j}$

The length of the vector $y\,\mathbf{i} + \mathbf{j}$ is $\sqrt{y^2 + 1}$. Vectors are tangent to parabolas opening about the x-axis.

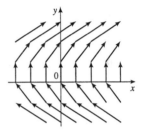

5. $\mathbf{F}(x, y) = \dfrac{y\,\mathbf{i} + x\,\mathbf{j}}{\sqrt{x^2 + y^2}}$

The length of the vector $\dfrac{y\,\mathbf{i} + x\,\mathbf{j}}{\sqrt{x^2 + y^2}}$ is 1.

7. $\mathbf{F}(x, y, z) = \mathbf{j}$

All vectors in this field are parallel to the y-axis and have length 1.

9. $\mathbf{F}(x, y, z) = y\,\mathbf{j}$

The length of $\mathbf{F}(x, y, z)$ is $|y|$. No vectors emanate from the xz-plane since $y = 0$ there. In each plane $y = b$, all the vectors are identical.

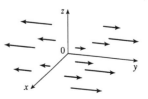

11. $\mathbf{F}(x, y) = \langle y, x \rangle$ corresponds to graph III, since in the first quadrant all the vectors have positive x- and y-components, in the second quadrant all vectors have positive x-components and negative y-components, in the third quadrant all vectors have negative x- and y-components, and in the fourth quadrant all vectors have negative x-components and positive y-components.

13. $\mathbf{F}(x, y) = \langle \sin x, \sin y \rangle$ corresponds to graph II, since the vector field is the same on each square of the form $[2n\pi, 2(n+1)\pi] \times [2m\pi, 2(m+1)\pi]$, m, n any integers.

15. $\mathbf{F}(x, y, z) = \mathbf{i} + 2\mathbf{j} + 3\mathbf{k}$ corresponds to graph IV, since all vectors have identical length and direction.

17. $\mathbf{F}(x, y, z) = x\mathbf{i} + y\mathbf{j} + 3\mathbf{k}$ corresponds to graph III; the projection of each vector onto the xy-plane is $x\mathbf{i} + y\mathbf{j}$, which points away from the origin, and the vectors point generally upward because their z-components are all 3.

19.

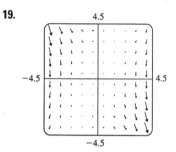

The vector field seems to have very short vectors near the line $y = 2x$. For $\mathbf{F}(x, y) = \langle 0, 0 \rangle$ we must have $y^2 - 2xy = 0$ and $3xy - 6x^2 = 0$. The first equation holds if $y = 0$ or $y = 2x$, and the second holds if $x = 0$ or $y = 2x$. So both equations hold [and thus $\mathbf{F}(x, y) = 0$] along the line $y = 2x$.

21. $\nabla f(x, y) = f_x(x, y)\mathbf{i} + f_y(x, y)\mathbf{j} = \dfrac{1}{x + 2y}\mathbf{i} + \dfrac{2}{x + 2y}\mathbf{j}$

23. $\nabla f(x, y, z) = f_x(x, y, z)\mathbf{i} + f_y(x, y, z)\mathbf{j} + f_z(x, y, z)\mathbf{k}$

$$= \frac{x}{\sqrt{x^2 + y^2 + z^2}}\mathbf{i} + \frac{y}{\sqrt{x^2 + y^2 + z^2}}\mathbf{j} + \frac{z}{\sqrt{x^2 + y^2 + z^2}}\mathbf{k}$$

25. $f(x, y) = x^2 - \frac{1}{2}y^2$, $\nabla f(x, y) = 2x\mathbf{i} - y\mathbf{j}$

The length of $\nabla f(x, y)$ is $\sqrt{4x^2 + y^2}$, and $\nabla f(x, y)$ terminates on the x-axis at the point $(3x, 0)$.

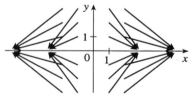

27. We graph ∇f along with a contour map of f.

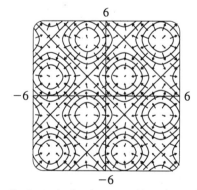

The graph shows that the gradient vectors are perpendicular to the level curves. Also, the gradient vectors point in the direction in which f is increasing and are longer where the level curves are closer together.

29. $f(x, y) = xy \implies \nabla f(x, y) = y\mathbf{i} + x\mathbf{j}$. In the first quadrant, both components of each vector are positive, while in the third quadrant both components are negative. However, in the second quadrant each vector's x-component is positive while its y-component is negative (and vice versa in the fourth quadrant). Thus, ∇f is graph IV.

31. $f(x, y) = x^2 + y^2 \implies \nabla f(x, y) = 2x\mathbf{i} + 2y\mathbf{j}$. Thus, each vector $\nabla f(x, y)$ has the same direction and twice the length of the position vector of the point (x, y), so the vectors all point directly away from the origin and their lengths increase as we move away from the origin. Hence, ∇f is graph II.

33. (a) We sketch the vector field $\mathbf{F}(x, y) = x\mathbf{i} - y\mathbf{j}$ along with several approximate flow lines. The flow lines appear to be hyperbolas with shape similar to the graph of $y = \pm 1/x$, so we might guess that the flow lines have equations

$$y = C/x.$$

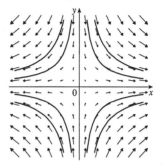

(b) If $x = x(t)$ and $y = y(t)$ are parametric equations of a flow line, then the velocity vector of the flow line at the point (x, y) is $x'(t)\,\mathbf{i} + y'(t)\,\mathbf{j}$. Since the velocity vectors coincide with the vectors in the vector field, we have $x'(t)\,\mathbf{i} + y'(t)\,\mathbf{j} = x\mathbf{i} - y\mathbf{j} \implies dx/dt = x, dy/dt = -y$. To solve these differential equations, we know $dx/dt = x \implies dx/x = dt \implies \ln|x| = t + C \implies x = \pm e^{t+C} = Ae^t$ for some constant A, and $dy/dt = -y \implies dy/y = -dt \implies \ln|y| = -t + K \implies y = \pm e^{-t+K} = Be^{-t}$ for some constant B. Therefore $xy = Ae^t Be^{-t} = AB = $ constant. If the flow line passes through $(1, 1)$ then $(1)(1) = $ constant $= 1 \implies xy = 1 \implies y = 1/x, x > 0$.

17.2 Line Integrals   ET 16.2

1. $x = t^2$ and $y = t$, $0 \le t \le 2$, so by Formula 3

$$\int_C y\,ds = \int_0^2 t\sqrt{\left(\frac{dx}{dt}\right)^2 + \left(\frac{dy}{dt}\right)^2}\,dt = \int_0^2 t\sqrt{(2t)^2 + (1)^2}\,dt$$

$$= \int_0^2 t\sqrt{4t^2 + 1}\,dt = \tfrac{1}{12}\left(4t^2 + 1\right)^{3/2}\Big]_0^2 = \tfrac{1}{12}\left(17\sqrt{17} - 1\right)$$

3. Parametric equations for C are $x = 4\cos t$, $y = 4\sin t$, $-\frac{\pi}{2} \le t \le \frac{\pi}{2}$. Then

$$\int_C xy^4\,ds = \int_{-\pi/2}^{\pi/2} (4\cos t)(4\sin t)^4 \sqrt{(-4\sin t)^2 + (4\cos t)^2}\,dt$$

$$= \int_{-\pi/2}^{\pi/2} 4^5 \cos t \sin^4 t \sqrt{16\left(\sin^2 t + \cos^2 t\right)}\,dt$$

$$= 4^5 \int_{-\pi/2}^{\pi/2} \left(\sin^4 t \cos t\right)(4)\,dt = (4)^6 \left[\tfrac{1}{5}\sin^5 t\right]_{-\pi/2}^{\pi/2} = \frac{2 \cdot 4^6}{5} = 1638.4$$

5. If we choose x as the parameter, parametric equations for C are $x = x$, $y = x^2$ for $1 \le x \le 3$ and

$$\int_C (xy + \ln x)\,dy = \int_1^3 \left(x \cdot x^2 + \ln x\right) 2x\,dx = \int_1^3 2\left(x^4 + x\ln x\right)dx$$

$$= 2\left[\tfrac{1}{5}x^5 + \tfrac{1}{2}x^2\ln x - \tfrac{1}{4}x^2\right]_1^3 \quad \text{(by integrating by parts in the second term)}$$

$$= 2\left(\tfrac{243}{5} + \tfrac{9}{2}\ln 3 - \tfrac{9}{4} - \tfrac{1}{5} + \tfrac{1}{4}\right) = \tfrac{464}{5} + 9\ln 3$$

7.

$C = C_1 + C_2$

On C_1: $x = x, y = 0 \implies dy = 0\, dx, 0 \le x \le 2$.

On C_2: $x = x, y = 2x - 4 \implies dy = 2\, dx, 2 \le x \le 3$.

Then $\int_C xy\, dx + (x - y)\, dy = \int_{C_1} xy\, dx + (x - y)\, dy + \int_{C_2} xy\, dx + (x - y)\, dy$

$$= \int_0^2 (0 + 0)\, dx + \int_2^3 \left[(2x^2 - 4x) + (-x + 4)(2) \right] dx$$

$$= \int_2^3 (2x^2 - 6x + 8)\, dx = \tfrac{17}{3}$$

9. $x = 4\sin t, y = 4\cos t, z = 3t, 0 \le t \le \tfrac{\pi}{2}$. Then by Formula 9,

$$\int_C xy^3\, ds = \int_0^{\pi/2} (4\sin t)(4\cos t)^3 \sqrt{\left(\frac{dx}{dt}\right)^2 + \left(\frac{dy}{dt}\right)^2 + \left(\frac{dz}{dt}\right)^2}\, dt$$

$$= \int_0^{\pi/2} 4^4 \cos^3 t \sin t \sqrt{(4\cos t)^2 + (-4\sin t)^2 + (3)^2}\, dt$$

$$= \int_0^{\pi/2} 256 \cos^3 t \sin t \sqrt{16 \left(\cos^2 t + \sin^2 t\right) + 9}\, dt$$

$$= 1280 \int_0^{\pi/2} \cos^3 t \sin t\, dt = -320 \cos^4 t \Big]_0^{\pi/2} = 320$$

11. Parametric equations for C are $x = t, y = 2t, z = 3t, 0 \le t \le 1$. Then

$$\int_C xe^{yz}\, ds = \int_0^1 te^{(2t)(3t)} \sqrt{1^2 + 2^2 + 3^2}\, dt = \sqrt{14} \int_0^1 te^{6t^2}\, dt$$

$$= \sqrt{14} \left[\tfrac{1}{12} e^{6t^2} \right]_0^1 = \tfrac{\sqrt{14}}{12} \left(e^6 - 1 \right)$$

13. $\int_C x^3 y^2 z\, dz = \int_0^1 (2t)^3 \left(t^2\right)^2 \left(t^2\right) 2t\, dt = \int_0^1 \left(8t^3\right) \left(t^4\right) \left(t^2\right) (2t)\, dt = \int_0^1 16t^{10}\, dt = \tfrac{16}{11} t^{11} \Big]_0^1 = \tfrac{16}{11}$

15. On C_1: $x = 0 \implies dx = 0\, dt, y = t \implies dy = dt, z = t \implies$
$dz = dt, 0 \le t \le 1$.
On C_2: $x = t \implies dx = dt, y = t + 1 \implies dy = dt, z = 2t + 1 \implies$
$dz = 2\, dt, 0 \le t \le 1$.
On C_3: $x = 1 \implies dx = 0\, dt, y = 2 \implies dy = 0\, dt, z = t + 3 \implies$
$dz = dt, 0 \le t \le 1$.
Then

$\int_C z^2\, dx - z\, dy + 2y\, dz = \int_0^1 (0 - t + 2t)\, dt + \int_0^1 \left[(2t + 1)^2 - (2t + 1) + 2(t + 1)(2) \right] dt + \int_0^1 (0 + 0 + 4)\, dt$

$$= \tfrac{1}{2} + \left[\tfrac{4}{3} t^3 + 3t^2 + 4t \right]_0^1 + 4 = \tfrac{77}{6}$$

17. (a) Along the line $x = -3$, the vectors of $\mathbf{F}$ have positive y-components, so since the path goes upward, the integrand $\mathbf{F} \cdot \mathbf{T}$ is always positive. Therefore $\int_{C_1} \mathbf{F} \cdot d\mathbf{r} = \int_{C_1} \mathbf{F} \cdot \mathbf{T}\, ds$ is positive.

(b) All of the (nonzero) field vectors along the circle with radius 3 are pointed in the clockwise direction, that is, opposite the direction to the path. So $\mathbf{F} \cdot \mathbf{T}$ is negative, and therefore $\int_{C_2} \mathbf{F} \cdot d\mathbf{r} = \int_{C_2} \mathbf{F} \cdot \mathbf{T}\, ds$ is negative.

19. $\mathbf{r}(t) = t^2\, \mathbf{i} - t^3\, \mathbf{j}$, so $\mathbf{F}(\mathbf{r}(t)) = \left(t^2\right)^2 \left(-t^3\right)^3 \mathbf{i} - \left(-t^3\right) \sqrt{t^2}\, \mathbf{j} = -t^{13}\, \mathbf{i} + t^4\, \mathbf{j}$ and $\mathbf{r}'(t) = 2t\, \mathbf{i} - 3t^2\, \mathbf{j}$. Thus

$$\int_C \mathbf{F} \cdot d\mathbf{r} = \int_0^1 \mathbf{F}(\mathbf{r}(t)) \cdot \mathbf{r}'(t)\, dt = \int_0^1 \left(-2t^{14} - 3t^6\right) dt = \left[-\tfrac{2}{15} t^{15} - \tfrac{3}{7} t^7 \right]_0^1 = -\tfrac{59}{105}$$

21. $\int_C \mathbf{F} \cdot d\mathbf{r} = \int_0^1 \langle \sin t^3, \cos(-t^2), t^4 \rangle \cdot \langle 3t^2, -2t, 1 \rangle \, dt$

$= \int_0^1 \left(3t^2 \sin t^3 - 2t \cos t^2 + t^4 \right) dt = \left[-\cos t^3 - \sin t^2 + \tfrac{1}{5} t^5 \right]_0^1 = \tfrac{6}{5} - \cos 1 - \sin 1$

23. We graph $\mathbf{F}(x, y) = (x - y)\mathbf{i} + xy\mathbf{j}$ and the curve C. We see that most of the vectors starting on C point in roughly the same direction as C, so for these portions of C the tangential component $\mathbf{F} \cdot \mathbf{T}$ is positive. Although some vectors in the third quadrant which start on C point in roughly the opposite direction, and hence give negative tangential components, it seems reasonable that the effect of these portions of C is outweighed by the positive tangential components. Thus, we would expect $\int_C \mathbf{F} \cdot d\mathbf{r} = \int_C \mathbf{F} \cdot \mathbf{T} \, ds$ to be positive.

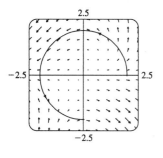

To verify, we evaluate $\int_C \mathbf{F} \cdot d\mathbf{r}$. The curve C can be represented by $\mathbf{r}(t) = 2\cos t \, \mathbf{i} + 2\sin t \, \mathbf{j}$, $0 \le t \le \frac{3\pi}{2}$, so $\mathbf{F}(\mathbf{r}(t)) = (2\cos t - 2\sin t)\mathbf{i} + 4\cos t \sin t \, \mathbf{j}$ and $\mathbf{r}'(t) = -2\sin t \, \mathbf{i} + 2\cos t \, \mathbf{j}$. Then

$\int_C \mathbf{F} \cdot d\mathbf{r} = \int_0^{3\pi/2} \mathbf{F}(\mathbf{r}(t)) \cdot \mathbf{r}'(t) \, dt$

$= \int_0^{3\pi/2} \left[-2\sin t (2\cos t - 2\sin t) + 2\cos t (4\cos t \sin t) \right] dt$

$= 4 \int_0^{3\pi/2} \left(\sin^2 t - \sin t \cos t + 2\sin t \cos^2 t \right) dt$

$= 3\pi + \tfrac{2}{3}$ (using a CAS)

25. (a) $\int_C \mathbf{F} \cdot d\mathbf{r} = \int_0^1 \left\langle e^{t^2 - 1}, t^5 \right\rangle \cdot \langle 2t, 3t^2 \rangle \, dt = \int_0^1 \left(2te^{t^2-1} + 3t^7 \right) dt = \left[e^{t^2 - 1} + \tfrac{3}{8} t^8 \right]_0^1 = \tfrac{11}{8} - 1/e$

(b) $\mathbf{r}(0) = \mathbf{0}$, $\mathbf{F}(\mathbf{r}(0)) = \langle e^{-1}, 0 \rangle$; $\mathbf{r}\left(\tfrac{1}{\sqrt{2}} \right) = \left\langle \tfrac{1}{2}, \tfrac{1}{2\sqrt{2}} \right\rangle$, $\mathbf{F}\left(\mathbf{r}\left(\tfrac{1}{\sqrt{2}} \right) \right) = \left\langle e^{-1/2}, \tfrac{1}{4\sqrt{2}} \right\rangle$; $\mathbf{r}(1) = \langle 1, 1 \rangle$, $\mathbf{F}(\mathbf{r}(1)) = \langle 1, 1 \rangle$.

In order to generate the graph with Maple, we use the PLOT command (not to be confused with the plot command) to define each of the vectors. For example,

```
v1:=PLOT(CURVES([[0,0],[evalf(1/exp(1)),0]]));
```

generates the vector from the vector field at the point $(0, 0)$ (but without an arrowhead) and gives it the name v1. To show everything on the same screen, we use the display command.

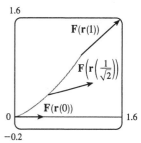

In Mathematica, we use ListPlot (with the PlotJoined -> True option) to generate the vectors, and then Show to show everything on the same screen.

27. The part of the astroid that lies in the quadrant is parametrized by $x = \cos^3 t$, $y = \sin^3 t$, $0 \le t \le \frac{\pi}{2}$. Now $\dfrac{dx}{dt} = 3\cos^2 t \, (-\sin t)$ and $\dfrac{dy}{dt} = 3\sin^2 t \cos t$, so

$\sqrt{\left(\dfrac{dx}{dt} \right)^2 + \left(\dfrac{dy}{dt} \right)^2} = \sqrt{9\cos^4 t \sin^2 t + 9\sin^4 t \cos^2 t} = 3\cos t \sin t \sqrt{\cos^2 t + \sin^2 t} = 3\cos t \sin t$.

Therefore $\int_C x^3 y^5 \, ds = \int_0^{\pi/2} \cos^9 t \sin^{15} t \, (3\cos t \sin t) \, dt = \dfrac{945}{16,777,216} \pi$.

29. A calculator or CAS gives $\int_C x \sin y \, ds = \int_1^2 \ln t \sin \left(e^{-t}\right) \sqrt{(1/t)^2 + (-e^{-t})^2} \, dt \approx 0.052$.

31. We use the parametrization $x = 2\cos t$, $y = 2\sin t$, $-\frac{\pi}{2} \le t \le \frac{\pi}{2}$.

Then $ds = \sqrt{\left(\dfrac{dx}{dt}\right)^2 + \left(\dfrac{dy}{dt}\right)^2} \, dt = \sqrt{(-2\sin t)^2 + (2\cos t)^2} \, dt = 2 \, dt$, so

$m = \int_C k \, ds = 2k \int_{-\pi/2}^{\pi/2} dt = 2k\,(\pi)$, $\bar{x} = \frac{1}{2\pi k} \int_C xk \, ds = \frac{1}{2\pi} \int_{-\pi/2}^{\pi/2} (2\cos t)\,2 \, dt = \frac{1}{2\pi} \left[4\sin t\right]_{-\pi/2}^{\pi/2} = \frac{4}{\pi}$,

$\bar{y} = \frac{1}{2\pi k} \int_C yk \, ds = \frac{1}{2\pi} \int_{-\pi/2}^{\pi/2} (2\sin t)\,2 \, dt = 0$. Hence $(\bar{x}, \bar{y}) = \left(\frac{4}{\pi}, 0\right)$.

33. (a) $\bar{x} = \dfrac{1}{m} \displaystyle\int_C x\rho\,(x,y,z)\,ds$, $\bar{y} = \dfrac{1}{m} \displaystyle\int_C y\rho\,(x,y,z)\,ds$, $\bar{z} = \dfrac{1}{m} \displaystyle\int_C z\rho\,(x,y,z)\,ds$ where

$m = \int_C \rho\,(x,y,z)\,ds$.

(b) $m = \int_C k \, ds = k \int_0^{2\pi} \sqrt{4\sin^2 t + 4\cos^2 t + 9} \, dt = k\sqrt{13} \int_0^{2\pi} dt = 2\pi k\sqrt{13}$,

$\bar{x} = \dfrac{1}{2\pi k\sqrt{13}} \displaystyle\int_0^{2\pi} k2\sqrt{13}\sin t \, dt = 0$, $\bar{y} = \dfrac{1}{2\pi k\sqrt{13}} \displaystyle\int_0^{2\pi} k2\sqrt{13}\cos t \, dt = 0$,

$\bar{z} = \dfrac{1}{2\pi k\sqrt{13}} \displaystyle\int_0^{2\pi} \left(k\sqrt{13}\right)(3t)\,dt = \dfrac{3}{2\pi}\left(2\pi^2\right) = 3\pi$. Hence $(\bar{x}, \bar{y}, \bar{z}) = (0, 0, 3\pi)$.

35. From Example 3, $\rho\,(x,y) = k\,(1 - y)$, $x = \cos t$, $y = \sin t$, and $ds = dt$, $0 \le t \le \pi$ $\Rightarrow$

$$I_x = \int_C y^2 \rho\,(x,y)\,ds = \int_0^\pi \sin^2 t \left[k\,(1 - \sin t)\right] dt = k \int_0^\pi \left(\sin^2 t - \sin^3 t\right) dt$$

$$= \tfrac{1}{2}k \int_0^\pi (1 - \cos 2t)\,dt - k \int_0^\pi \left(1 - \cos^2 t\right) \sin t \, dt$$

(Let $u = \cos t$, $du = -\sin t \, dt$
in the second integral)

$$= k\left[\tfrac{\pi}{2} + \int_1^{-1} \left(1 - u^2\right) du\right] = k\left(\tfrac{\pi}{2} - \tfrac{4}{3}\right)$$

$$I_y = \int_C x^2 \rho\,(x,y)\,ds = k \int_0^\pi \cos^2 t\,(1 - \sin t)\,dt = \tfrac{k}{2} \int_0^\pi (1 + \cos 2t)\,dt - k \int_0^\pi \cos^2 t \sin t \, dt$$

$$= k\left(\tfrac{\pi}{2} - \tfrac{2}{3}\right), \text{ using the same substitution as above.}$$

37. $W = \int_C \mathbf{F} \cdot d\mathbf{r} = \int_0^{2\pi} \langle t - \sin t, 3 - \cos t\rangle \cdot \langle 1 - \cos t, \sin t\rangle \, dt$

$= \int_0^{2\pi} (t - t\cos t - \sin t + \sin t \cos t + 3\sin t - \sin t \cos t)\,dt$

$= \int_0^{2\pi} (t - t\cos t + 2\sin t)\,dt = \left[\tfrac{1}{2}t^2 - (t\sin t + \cos t) - 2\cos t\right]_0^{2\pi}$

(by integrating by parts in the second term)

$= 2\pi^2$

39. $W = \int_C \mathbf{F} \cdot d\mathbf{r} = \int_0^1 \langle t^6, -t^5, -t^7\rangle \cdot \langle 2t, -3t^2, 4t^3\rangle \, dt = \int_0^1 \left(5t^7 - 4t^{10}\right) dt = \tfrac{5}{8} - \tfrac{4}{11} = \tfrac{23}{88}$

41. Let $\mathbf{F} = 185 \, \mathbf{k}$. To parametrize the staircase, let

$$x = 20 \cos t, \; y = 20 \sin t, \; z = \tfrac{90}{6\pi} t = \tfrac{15}{\pi} t, \; 0 \le t \le 6\pi \;\; \Rightarrow$$

$$W = \int_C \mathbf{F} \cdot d\mathbf{r} = \int_0^{6\pi} \langle 0, 0, 185 \rangle \cdot \langle -20 \sin t, 20 \cos t, \tfrac{15}{\pi} \rangle \, dt = (185) \, \tfrac{15}{\pi} \int_0^{6\pi} \, dt = (185)(90)$$

$$\approx 1.67 \times 10^4 \text{ ft-lb}$$

43. The work done in moving the object is $\int_C \mathbf{F} \cdot d\mathbf{r} = \int_C \mathbf{F} \cdot \mathbf{T} \, ds$. We can approximate this integral by dividing C into 7 segments of equal length $\Delta s = 2$ and approximating $\mathbf{F} \cdot \mathbf{T}$, that is, the tangential component of force, at a point (x_i^*, y_i^*) on each segment. Since C is composed of straight line segments, $\mathbf{F} \cdot \mathbf{T}$ is the scalar projection of each force vector onto C. If we choose (x_i^*, y_i^*) to be the point on the segment closest to the origin, then the work done is

$$\int_C \mathbf{F} \cdot \mathbf{T} \, ds \approx \sum_{i=1}^{7} [\mathbf{F}(x_i^*, y_i^*) \cdot \mathbf{T}(x_i^*, y_i^*)] \, \Delta s$$

$$= [2 + 2 + 2 + 2 + 1 + 1 + 1](2) = 22$$

Thus, we estimate the work done to be approximately 22 J.

17.3 The Fundamental Theorem for Line Integrals   ET 16.3

1. C appears to be a smooth curve, and since ∇f is continuous, we know f is differentiable. Then Theorem 2 says that the value of $\int_C \nabla f \cdot d\mathbf{r}$ is simply the difference of the values of f at the terminal and initial points of C. From the graph, this is $50 - 10 = 40$.

3. $\partial (6x + 5y) / \partial y = 5 = \partial (5x + 4y) / \partial x$ and the domain of $\mathbf{F}$ is $\mathbb{R}^2$ which is open and simply-connected, so by Theorem 6 $\mathbf{F}$ is conservative. Thus, there exists a function f such that $\nabla f = \mathbf{F}$, that is, $f_x(x, y) = 6x + 5y$ and $f_y(x, y) = 5x + 4y$. But $f_x(x, y) = 6x + 5y$ implies $f(x, y) = 3x^2 + 5xy + g(y)$ and differentiating both sides of this equation with respect to y gives $f_y(x, y) = 5x + g'(y)$. Thus $5x + 4y = 5x + g'(y)$ so $g'(y) = 4y$ and $g(y) = 2y^2 + K$ where K is a constant. Hence $f(x, y) = 3x^2 + 5xy + 2y^2 + K$ is a potential function for $\mathbf{F}$.

5. $\partial (xe^y) / \partial y = xe^y, \; \partial (ye^x) / \partial x = ye^x$. Since these are not equal, $\mathbf{F}$ is not conservative.

7. $\partial (2x \cos y - y \cos x) / \partial y = -2x \sin y - \cos x = \partial (-x^2 \sin y - \sin x) / \partial x$ and the domain of $\mathbf{F}$ is $\mathbb{R}^2$. Hence $\mathbf{F}$ is conservative so there exists a function f such that $\nabla f = \mathbf{F}$. Then $f_x(x, y) = 2x \cos y - y \cos x$ implies $f(x, y) = x^2 \cos y - y \sin x + g(y)$ and $f_y(x, y) = -x^2 \sin y - \sin x + g'(y)$. But $f_y(x, y) = -x^2 \sin y - \sin x$ so $g'(y) = 0 \;\; \Rightarrow \;\; g(y) = K$. Then $f(x, y) = x^2 \cos y - y \sin x + K$ is a potential function for $\mathbf{F}$.

9. $\partial (ye^x + \sin y) / \partial y = e^x + \cos y = \partial (e^x + x \cos y) / \partial x$ and the domain of $\mathbf{F}$ is $\mathbb{R}^2$. Hence $\mathbf{F}$ is conservative so there exists a function f such that $\nabla f = \mathbf{F}$. Then $f_x(x, y) = ye^x + \sin y$ implies $f(x, y) = ye^x + x \sin y + g(y)$ and $f_y(x, y) = e^x + x \cos y + g'(y)$. But $f_y(x, y) = e^x + x \cos y$ so $g(y) = K$ and $f(x, y) = ye^x + x \sin y + K$ is a potential function for $\mathbf{F}$.

11. (a) $\mathbf{F}$ has continuous first-order partial derivatives and $\dfrac{\partial}{\partial y}(2xy) = 2x = \dfrac{\partial}{\partial x}(x^2)$ on $\mathbb{R}^2$, which is open and

simply-connected. Thus, $\mathbf{F}$ is conservative by Theorem 6. Then we know that the line integral of $\mathbf{F}$ is independent of path; in particular, the value of $\int_C \mathbf{F} \cdot d\mathbf{r}$ depends only on the endpoints of C. Since all three curves have the same initial and terminal points, $\int_C \mathbf{F} \cdot d\mathbf{r}$ will have the same value for each curve.

(b) We first find a potential function f, so that $\nabla f = \mathbf{F}$. We know $f_x(x, y) = 2xy$ and $f_y(x, y) = x^2$. Integrating $f_x(x, y)$ with respect to x, we have $f(x, y) = x^2 y + g(y)$. Differentiating both sides with respect to y gives $f_y(x, y) = x^2 + g'(y)$, so we must have $x^2 + g'(y) = x^2 \Rightarrow g'(y) = 0 \Rightarrow g(y) = K$, a constant. Thus $f(x, y) = x^2 y + K$. All three curves start at $(1, 2)$ and end at $(3, 2)$, so by Theorem 2, $\int_C \mathbf{F} \cdot d\mathbf{r} = f(3, 2) - f(1, 2) = 18 - 2 = 16$ for each curve.

13. (a) $f_x(x, y) = x^3 y^4$ implies $f(x, y) = \frac{1}{4}x^4 y^4 + g(y)$ and $f_y(x, y) = x^4 y^3 + g'(y)$. But $f_y(x, y) = x^4 y^3$ so $g'(y) = 0 \Rightarrow g(y) = K$, a constant. We can take $K = 0$, so $f(x, y) = \frac{1}{4}x^4 y^4$.

(b) The initial point of C is $\mathbf{r}(0) = (0, 1)$ and the terminal point is $\mathbf{r}(1) = (1, 2)$, so $\int_C \mathbf{F} \cdot d\mathbf{r} = f(1, 2) - f(0, 1) = 4 - 0 = 4$.

15. (a) $f_x(x, y, z) = y$ implies $f(x, y, z) = xy + g(y, z)$ and $f_y(x, y, z) = x + \partial g/\partial y$. But $f_y(x, y, z) = x + z$ so $\partial g/\partial y = z$ and $g(y, z) = yz + h(z)$. Thus $f(x, y, z) = xy + yz + h(z)$ and $f_z(x, y, z) = y + h'(z)$. But $f_z(x, y, z) = y$ so $h'(z) = 0$ or $h(z) = K$. Hence $f(x, y, z) = xy + yz$ (setting $K = 0$).

(b) $\int_C \mathbf{F} \cdot d\mathbf{r} = f(8, 3, -1) - f(2, 1, 4) = 21 - 6 = 15$

17. (a) $f_x(x, y, z) = 2xz + \sin y$ implies $f(x, y, z) = x^2 z + x \sin y + g(y, z)$ and $f_y(x, y, z) = x \cos y + g_y(y, z)$. But $f_y(x, y, z) = x \cos y$ so $g_y(y, z) = 0$ and $f(x, y, z) = x^2 z + x \sin y + h(z)$. Thus $f_z(x, y, z) = x^2 + h'(z)$. But $f_z(x, y, z) = x^2$ so $h'(z) = 0$ and $f(x, y, z) = x^2 z + x \sin y$ (setting $K = 0$).

(b) $\mathbf{r}(0) = \langle 1, 0, 0 \rangle$, $\mathbf{r}(2\pi) = \langle 1, 0, 2\pi \rangle$. Thus $\int_C \mathbf{F} \cdot d\mathbf{r} = f(1, 0, 2\pi) - f(1, 0, 0) = 2\pi$.

19. Here $\mathbf{F}(x, y) = (2x \sin y)\,\mathbf{i} + (x^2 \cos y - 3y^2)\,\mathbf{j}$. Then $f(x, y) = x^2 \sin y - y^3$ is a potential function for $\mathbf{F}$, that is, $\nabla f = \mathbf{F}$ so $\mathbf{F}$ is conservative and thus its line integral is independent of path. Hence $\int_C 2x \sin y \, dx + (x^2 \cos y - 3y^2) \, dy = \int_C \mathbf{F} \cdot d\mathbf{r} = f(5, 1) - f(-1, 0) = 25 \sin 1 - 1$.

21. $\mathbf{F}(x, y) = x^2 y^3\,\mathbf{i} + x^3 y^2\,\mathbf{j}$, $W = \int_C \mathbf{F} \cdot d\mathbf{r}$. Since $\partial(x^2 y^3)/\partial y = 3x^2 y^2 = \partial(x^3 y^2)/\partial x$, there exists a function f such that $\nabla f = \mathbf{F}$. In fact, $f_x = x^2 y^3 \Rightarrow f(x, y) = \frac{1}{3}x^3 y^3 + g(y) \Rightarrow f_y = x^3 y^2 + g'(y) \Rightarrow g'(y) = 0$, so we can take $f(x, y) = \frac{1}{3}x^3 y^3$. Thus $W = \int_C \mathbf{F} \cdot d\mathbf{r} = f(2, 1) - f(0, 0) = \frac{1}{3}(2^3)(1^3) - 0 = \frac{8}{3}$.

23. We know that if the vector field (call it $\mathbf{F}$) is conservative, then around any closed path C, $\int_C \mathbf{F} \cdot d\mathbf{r} = 0$. But take C to be some circle centered at the origin, oriented counterclockwise. All of the field vectors along C oppose motion along C, so the integral around C will be negative. Therefore the field is not conservative.

25.

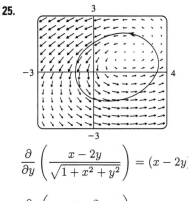

From the graph, it appears that $\mathbf{F}$ is not conservative. For example, any closed curve containing the point $(2, 1)$ seems to have many field vectors pointing counterclockwise along it, and none pointing clockwise. So along this path the integral $\int \mathbf{F} \cdot d\mathbf{r} \neq 0$. To confirm our guess, we calculate

$$\frac{\partial}{\partial y}\left(\frac{x - 2y}{\sqrt{1 + x^2 + y^2}}\right) = (x - 2y)\left[\frac{-y}{(1 + x^2 + y^2)^{3/2}}\right] - \frac{2}{\sqrt{1 + x^2 + y^2}} = \frac{-2 - 2x^2 - xy}{(1 + x^2 + y^2)^{3/2}},$$

$$\frac{\partial}{\partial x}\left(\frac{x - 2}{\sqrt{1 + x^2 + y^2}}\right) = (x - 2)\left[\frac{-x}{(1 + x^2 + y^2)^{3/2}}\right] + \frac{1}{\sqrt{1 + x^2 + y^2}} = \frac{1 + y^2 + 2x}{(1 + x^2 + y^2)^{3/2}}. \text{ These are not}$$

equal, so the field is not conservative, by Theorem 5.

27. Since $\mathbf{F}$ is conservative, there exists a function f such that $\mathbf{F} = \nabla f$, that is, $P = f_x, Q = f_y$, and $R = f_z$. Since P, Q and R have continuous first order partial derivatives, Clairaut's Theorem says that $\partial P/\partial y = f_{xy} = f_{yx} = \partial Q/\partial x, \partial P/\partial z = f_{xz} = f_{zx} = \partial R/\partial x$, and $\partial Q/\partial z = f_{yz} = f_{zy} = \partial R/\partial y$.

29. $D = \{(x, y) \mid x > 0, y > 0\} = $ the first quadrant (excluding the axes).

(a) D is open because around every point in D we can put a disk that lies in D.

(b) D is connected because the straight line segment joining any two points in D lies in D.

(c) D is simply-connected because it's connected and has no holes.

31. $D = \{(x, y) \mid 1 < x^2 + y^2 < 4\} = $ the annular region between the circles with center $(0, 0)$ and radii 1 and 2.

(a) D is open.

(b) D is connected.

(c) D is not simply-connected. For example, $x^2 + y^2 = (1.5)^2$ is simple and closed and lies within D but encloses points that are not in D. (Or we can say, D has a hole, so is not simply-connected.)

33. (a) $P = -\dfrac{y}{x^2 + y^2}, \dfrac{\partial P}{\partial y} = \dfrac{y^2 - x^2}{(x^2 + y^2)^2}$ and $Q = \dfrac{x}{x^2 + y^2}, \dfrac{\partial Q}{\partial x} = \dfrac{y^2 - x^2}{(x^2 + y^2)^2}$. Thus $\dfrac{\partial P}{\partial y} = \dfrac{\partial Q}{\partial x}$.

(b) C_1: $x = \cos t, y = \sin t, 0 \leq t \leq \pi, C_2$: $x = \cos t, y = \sin t, t = 2\pi$ to $t = \pi$. Then

$$\int_{C_1} \mathbf{F} \cdot d\mathbf{r} = \int_0^\pi \frac{(-\sin t)(-\sin t) + (\cos t)(\cos t)}{\cos^2 t + \sin^2 t} dt = \int_0^\pi dt = \pi \text{ and } \int_{C_2} \mathbf{F} \cdot d\mathbf{r} = \int_{2\pi}^\pi dt = -\pi.$$

Since these aren't equal, the line integral of $\mathbf{F}$ isn't independent of path. (Or notice that $\int_{C_3} \mathbf{F} \cdot d\mathbf{r} = \int_0^{2\pi} dt = 2\pi$ where C_3 is the circle $x^2 + y^2 = 1$, and apply the contrapositive of Theorem 3.) This doesn't contradict Theorem 6, since the domain of $\mathbf{F}$, which is $\mathbb{R}^2$ except the origin, isn't simply-connected.

17.4 Green's Theorem

1. (a)

$C_1: x = t \quad \Rightarrow \quad dx = dt, y = 0 \quad \Rightarrow \quad dy = 0\,dt, 0 \le t \le 2.$

$C_2: x = 2 \quad \Rightarrow \quad dx = 0\,dt, y = t \quad \Rightarrow \quad dy = dt, 0 \le t \le 3.$

$C_3: x = 2 - t \quad \Rightarrow \quad dx = -dt, y = 3 \quad \Rightarrow \quad dy = 0\,dt, 0 \le t \le 2$

$C_4: x = 0 \quad \Rightarrow \quad dx = 0\,dt, y = 3 - t \quad \Rightarrow \quad dy = -dt, 0 \le t \le 3.$

Thus $\oint_C xy^2\,dx + x^3\,dy = \oint_{C_1 + C_2 + C_3 + C_4} xy^2\,dx + x^3\,dy$

$$= \int_0^2 0\,dt + \int_0^3 8\,dt + \int_0^2 -9\,(2-t)\,dt + \int_0^3 0\,dt$$

$$= 0 + 24 - 18 + 0 = 6$$

(b) $\oint_C xy^2\,dx + x^3\,dy = \iint_D \left[\frac{\partial}{\partial x}(x^3) - \frac{\partial}{\partial y}(xy^2) \right] dA = \int_0^2 \int_0^3 (3x^2 - 2xy)\,dy\,dx$

$$= \int_0^2 (9x^2 - 9x)\,dx = 24 - 18 = 6$$

3. (a)

$C_1: x = t \quad \Rightarrow \quad dx = dt, y = 0 \quad \Rightarrow \quad dy = 0\,dt, 0 \le t \le 1.$

$C_2: x = 1 \quad \Rightarrow \quad dx = 0\,dt, y = t \quad \Rightarrow \quad dy = dt, 0 \le t \le 2.$

$C_3: x = 1 - t \quad \Rightarrow \quad dx = -dt, y = 2 - 2t \quad \Rightarrow \quad dy = -2\,dt, 0 \le t \le 1.$

Thus $\oint_C xy\,dx + x^2 y^3\,dy = \oint_{C_1 + C_2 + C_3} xy\,dx + x^2 y^3\,dy$

$$= \int_0^1 0\,dt + \int_0^2 t^3\,dt + \int_0^1 \left[-(1-t)(2-2t) - 2(1-t)^2(2-2t)^3 \right] dt$$

$$= 0 + \left[\tfrac{1}{4}t^4 \right]_0^2 + \left[\tfrac{2}{3}(1-t)^3 + \tfrac{8}{3}(1-t)^6 \right]_0^1 = 4 - \tfrac{10}{3} = \tfrac{2}{3}$$

(b) $\oint_C xy\,dx + x^2 y^3\,dy = \iint_D \left[\frac{\partial}{\partial x}(x^2 y^3) - \frac{\partial}{\partial y}(xy) \right] dA = \int_0^1 \int_0^{2x} (2xy^3 - x)\,dy\,dx$

$$= \int_0^1 \left[\tfrac{1}{2}xy^4 - xy \right]_{y=0}^{y=2x} dx = \int_0^1 (8x^5 - 2x^2)\,dx = \tfrac{4}{3} - \tfrac{2}{3} = \tfrac{2}{3}$$

5. We can parametrize C as $x = \cos\theta, y = \sin\theta, 0 \le \theta \le 2\pi$. Then the line integral is

$\oint_C P\,dx + Q\,dy = \int_0^{2\pi} \cos^4\theta \sin^5\theta\,(-\sin\theta)\,d\theta + \int_0^{2\pi} (-\cos^7\theta \sin^6\theta)\cos\theta\,d\theta = -\frac{29\pi}{1024}$, according to a CAS.

The double integral is $\iint_D \left(\frac{\partial Q}{\partial x} - \frac{\partial P}{\partial y} \right) dA = \int_{-1}^{1} \int_{-\sqrt{1-x^2}}^{\sqrt{1-x^2}} (-7x^6 y^6 - 5x^4 y^4)\,dy\,dx = -\frac{29\pi}{1024}$, verifying

Green's Theorem in this case.

7. The region D enclosed by C is $[0,1] \times [0,1]$, so

$$\int_C e^y\,dx + 2xe^y\,dy = \iint_D \left[\frac{\partial}{\partial x}(2xe^y) - \frac{\partial}{\partial y}(e^y) \right] dA = \int_0^1 \int_0^1 (2e^y - e^y)\,dy\,dx$$

$$= \int_0^1 dx \int_0^1 e^y\,dy = (1)(e^1 - e^0) = e - 1$$

9. $\int_C \left(y + e^{\sqrt{x}} \right) dx + (2x + \cos y^2)\,dy = \iint_D \left[\frac{\partial}{\partial x}(2x + \cos y^2) - \frac{\partial}{\partial y}\left(y + e^{\sqrt{x}} \right) \right] dA$

$$= \int_0^1 \int_{y^2}^{\sqrt{y}} (2 - 1)\,dx\,dy = \int_0^1 \left(y^{1/2} - y^2 \right) dy = \tfrac{1}{3}$$

11. $\int_C y^3 \, dx - x^3 \, dy = \iint_D \left[\frac{\partial}{\partial x} \left(-x^3 \right) - \frac{\partial}{\partial y} \left(y^3 \right) \right] dA = \iint_D \left(-3x^2 - 3y^2 \right) dA = \int_0^{2\pi} \int_0^2 \left(-3r^2 \right) r \, dr \, d\theta$

$$= -3 \int_0^{2\pi} d\theta \int_0^2 r^3 \, dr = -3 \, (2\pi) \, (4) = -24\pi$$

13. The region D enclosed by C is given by $\left\{ (x, y) \mid -2 \le x \le 2, -\sqrt{4 - x^2} \le y \le \sqrt{4 - x^2} \right\}$ or, in polar coordinates, $\{ (r, \theta) \mid 0 \le \theta \le \pi, 0 \le r \le 2 \}$. Thus,

$\int_C xy \, dx + 2x^2 \, dy = \iint_D \left[\frac{\partial}{\partial x} \left(2x^2 \right) - \frac{\partial}{\partial y} \left(xy \right) \right] dA = \iint_D \left(4x - x \right) dA = \int_0^\pi \int_0^2 \left(3r \cos \theta \right) r \, dr \, d\theta$

$$= 3 \int_0^\pi \cos \theta \, d\theta \int_0^2 r^2 \, dr = 3 \left[\sin \theta \right]_0^\pi \left[\tfrac{1}{3} r^3 \right]_0^2 = 3 \, (0) \left(\tfrac{8}{3} \right) = 0$$

15. The region D enclosed by C is given, in polar coordinates, by $\left\{ (r, \theta) \mid 0 \le \theta \le \frac{\pi}{4}, 0 \le r \le 2 \right\}$. Thus

$\int_C \mathbf{F} \cdot d\mathbf{r} = \int_C \left(y^2 - x^2 y \right) dx + xy^2 \, dy = \iint_D \left(y^2 - 2y + x^2 \right) dA$

$$= \int_0^{\pi/4} \int_0^2 \left(r^2 - 2r \sin \theta \right) r \, dr \, d\theta = \int_0^{\pi/4} \left[4 - \tfrac{16}{3} \sin \theta \right] d\theta$$

$$= \left[4\theta + \tfrac{16}{3} \cos \theta \right]_0^{\pi/4} = \pi + \tfrac{8}{3} \left(\sqrt{2} - 2 \right)$$

17. By Green's Theorem, $W = \int_C \mathbf{F} \cdot d\mathbf{r} = \int_C x \, (x + y) \, dx + xy^2 \, dy = \iint_D \left(y^2 - x \right) dy \, dx$ where C is the path described in the question and D is the triangle bounded by C. So

$$W = \int_0^1 \int_0^{1-x} \left(y^2 - x \right) dy \, dx = \int_0^1 \left[\tfrac{1}{3} y^3 - xy \right]_{y=0}^{y=1-x} dx = \int_0^1 \left(\tfrac{1}{3} \left(1 - x \right)^3 - x \left(1 - x \right) \right) dx$$

$$= \left[-\tfrac{1}{12} \left(1 - x \right)^4 - \tfrac{1}{2} x^2 + \tfrac{1}{3} x^3 \right]_0^1 = \left(-\tfrac{1}{2} + \tfrac{1}{3} \right) - \left(-\tfrac{1}{12} \right) = -\tfrac{1}{12}$$

19. $A = \oint_C x \, dy = \int_0^{2\pi} \left(\cos^3 t \right) \left(3 \sin^2 t \cos t \right) dt = 3 \int_0^{2\pi} \left(\cos^4 t \sin^2 t \right) dt$

$$= 3 \left[-\tfrac{1}{6} \left(\sin t \cos^5 t \right) + \tfrac{1}{6} \left[\tfrac{1}{4} \left(\sin t \cos^3 t \right) + \tfrac{3}{8} \left(\cos t \sin t \right) + \tfrac{3}{8} t \right] \right]_0^{2\pi} = 3 \left(\tfrac{1}{6} \right) \left(\tfrac{6}{8} \pi \right) = \tfrac{3}{8} \pi$$

Or: $3 \int_0^{2\pi} \left(\cos^4 t \sin^2 t \right) dt = 3 \int_0^{2\pi} \tfrac{1}{8} \left[\tfrac{1}{2} \left(1 - \cos 4t \right) + \sin^2 2t \cos 2t \right] dt = \tfrac{3}{8} \pi$

21. (a) Using Equation 17.2.8 [ET 16.2.8], we write parametric equations of the line segment as $x = (1 - t) x_1 + t x_2$, $y = (1 - t) y_1 + t y_2$, $0 \le t \le 1$. Then $dx = (x_2 - x_1) \, dt$ and $dy = (y_2 - y_1) \, dt$, so

$\int_C x \, dy - y \, dx = \int_0^1 \left[(1 - t) x_1 + t x_2 \right] (y_2 - y_1) \, dt + \left[(1 - t) y_1 + t y_2 \right] (x_2 - x_1) \, dt$

$$= \int_0^1 \left(x_1 \left(y_2 - y_1 \right) - y_1 \left(x_2 - x_1 \right) + t \left[\left(y_2 - y_1 \right) \left(x_2 - x_1 \right) - \left(x_2 - x_1 \right) \left(y_2 - y_1 \right) \right] \right) dt$$

$$= \int_0^1 \left(x_1 y_2 - x_2 y_1 \right) dt = x_1 y_2 - x_2 y_1$$

(b) We apply Green's Theorem to the path $C = C_1 \cup C_2 \cup \cdots \cup C_n$, where C_i is the line segment that joins (x_i, y_i) to (x_{i+1}, y_{i+1}) for $i = 1, 2, \ldots, n - 1$, and C_n is the line segment that joins (x_n, y_n) to (x_1, y_1). From (5), $\tfrac{1}{2} \int_C x \, dy - y \, dx = \iint_D dA$, where D is the polygon bounded by C. Therefore

area of polygon $= A(D) = \iint_D dA = \tfrac{1}{2} \int_C x \, dy - y \, dx$

$$= \tfrac{1}{2} \left(\int_{C_1} x \, dy - y \, dx + \int_{C_2} x \, dy - y \, dx + \cdots + \int_{C_{n-1}} x \, dy - y \, dx + \int_{C_n} x \, dy - y \, dx \right)$$

To evaluate these integrals we use the formula from (a) to get

$A(D) = \tfrac{1}{2} \left[(x_1 y_2 - x_2 y_1) + (x_2 y_3 - x_3 y_2) + \cdots + (x_{n-1} y_n - x_n y_{n-1}) + (x_n y_1 - x_1 y_n) \right]$.

(c) $A = \tfrac{1}{2} \left[(0 \cdot 1 - 2 \cdot 0) + (2 \cdot 3 - 1 \cdot 1) + (1 \cdot 2 - 0 \cdot 3) + (0 \cdot 1 - (-1) \cdot 2) + (-1 \cdot 0 - 0 \cdot 1) \right]$

$$= \tfrac{1}{2} \left(0 + 5 + 2 + 2 \right) = \tfrac{9}{2}$$

23. Here $A = \frac{1}{2}(1)(1) = \frac{1}{2}$ and $C = C_1 + C_2 + C_3$, where C_1: $x = x$, $y = 0$, $0 \le x \le 1$;
C_2: $x = x$, $y = 1 - x$, $x = 1$ to $x = 0$; and C_3: $x = 0$, $y = 1$ to $y = 0$. Then
$\bar{x} = \frac{1}{2A} \oint_C x^2 \, dy = \int_{C_1} x^2 \, dy + \int_{C_2} x^2 \, dy + \int_{C_3} x^2 \, dy = 0 + \int_1^0 (x^2)(-dx) + 0 = \frac{1}{3}$. Similarly,
$\bar{y} = -\frac{1}{2A} \oint_C y^2 \, dx = \int_{C_1} y^2 \, dx + \int_{C_2} y^2 \, dx + \int_{C_3} y^2 \, dx = 0 + \int_1^0 (1-x)^2 (-dx) + 0 = \frac{1}{3}$.
Therefore $(\bar{x}, \bar{y}) = \left(\frac{1}{3}, \frac{1}{3}\right)$.

25. By Green's Theorem, $-\frac{1}{3}\rho \oint_C y^3 \, dx = -\frac{1}{3}\rho \iint_D (-3y^2) \, dA = \iint_D y^2 \rho \, dA = I_x$ and
$\frac{1}{3}\rho \oint_C x^3 \, dy = \frac{1}{3}\rho \iint_D (3x^2) \, dA = \iint_D x^2 \rho \, dA = I_y$.

27. Since C is a simple closed path which doesn't pass through or enclose the origin, there exists an open region that
doesn't contain the origin but does contain D. Thus $P = -y/(x^2 + y^2)$ and $Q = x/(x^2 + y^2)$ have continuous
partial derivatives on this open region containing D and we can apply Green's Theorem. But by Exercise 17.3.33(a)
[ET 16.3.33(a)], $\partial P/\partial y = \partial Q/\partial x$, so $\oint_C \mathbf{F} \cdot d\mathbf{r} = \iint_D 0 \, dA = 0$.

29. Using the first part of (5), we have that $\iint_R dx \, dy = A(R) = \int_{\partial R} x \, dy$. But $x = g(u, v)$, and
$dy = \dfrac{\partial h}{\partial u} \, du + \dfrac{\partial h}{\partial v} \, dv$, and we orient ∂S by taking the positive direction to be that which corresponds, under the
mapping, to the positive direction along ∂R, so

$$\int_{\partial R} x \, dy = \int_{\partial S} g(u, v) \left(\frac{\partial h}{\partial u} \, du + \frac{\partial h}{\partial v} \, dv \right) = \int_{\partial S} g(u, v) \frac{\partial h}{\partial u} \, du + g(u, v) \frac{\partial h}{\partial v} \, dv$$

$$= \pm \iint_S \left[\frac{\partial}{\partial u} \left(g(u, v) \frac{\partial h}{\partial v} \right) - \frac{\partial}{\partial v} \left(g(u, v) \frac{\partial h}{\partial u} \right) \right] dA \quad \text{(using Green's Theorem in the } uv\text{-plane)}$$

$$= \pm \iint_S \left(\frac{\partial g}{\partial u} \frac{\partial h}{\partial v} + g(u, v) \frac{\partial^2 h}{\partial u \partial v} - \frac{\partial g}{\partial v} \frac{\partial h}{\partial u} - g(u, v) \frac{\partial^2 h}{\partial v \partial u} \right) dA \quad \text{(using the Chain Rule)}$$

$$= \pm \iint_S \left(\frac{\partial x}{\partial u} \frac{\partial y}{\partial v} - \frac{\partial x}{\partial v} \frac{\partial y}{\partial u} \right) dA \text{ (by the equality of mixed partials)} = \pm \iint_S \frac{\partial(x, y)}{\partial(u, v)} \, du \, dv$$

The sign is chosen to be positive if the orientation that we gave to ∂S corresponds to the usual positive orientation,
and it is negative otherwise. In either case, since $A(R)$ is positive, the sign chosen must be the same as the sign of
$\dfrac{\partial(x, y)}{\partial(u, v)}$. Therefore $A(R) = \iint_R dx \, dy = \iint_S \left| \dfrac{\partial(x, y)}{\partial(u, v)} \right| \, du \, dv$.

17.5 Curl and Divergence   ET 16.5

1. (a) curl $\mathbf{F} = \nabla \times \mathbf{F} = \begin{vmatrix} \mathbf{i} & \mathbf{j} & \mathbf{k} \\ \partial/\partial x & \partial/\partial y & \partial/\partial z \\ xy & yz & zx \end{vmatrix}$

$$= \left[\frac{\partial}{\partial y}(zx) - \frac{\partial}{\partial z}(yz) \right] \mathbf{i} - \left[\frac{\partial}{\partial x}(zx) - \frac{\partial}{\partial z}(xy) \right] \mathbf{j} + \left[\frac{\partial}{\partial x}(yz) - \frac{\partial}{\partial y}(xy) \right] \mathbf{k}$$

$$= (0 - y)\mathbf{i} - (z - 0)\mathbf{j} + (0 - x)\mathbf{k} = -y\mathbf{i} - z\mathbf{j} - x\mathbf{k}$$

(b) div $\mathbf{F} = \nabla \cdot \mathbf{F} = \dfrac{\partial}{\partial x}(xy) + \dfrac{\partial}{\partial y}(yz) + \dfrac{\partial}{\partial z}(zx) = y + z + x = x + y + z$

3. (a) curl $\mathbf{F} = \nabla \times \mathbf{F} = \begin{vmatrix} \mathbf{i} & \mathbf{j} & \mathbf{k} \\ \partial/\partial x & \partial/\partial y & \partial/\partial z \\ xyz & 0 & -x^2y \end{vmatrix} = \left(-x^2 - 0\right)\mathbf{i} - \left(-2xy - xy\right)\mathbf{j} + \left(0 - xz\right)\mathbf{k}$

$= -x^2\,\mathbf{i} + 3xy\,\mathbf{j} - xz\,\mathbf{k}$

(b) div $\mathbf{F} = \nabla \cdot \mathbf{F} = \dfrac{\partial}{\partial x}\left(xyz\right) + \dfrac{\partial}{\partial y}\left(0\right) + \dfrac{\partial}{\partial z}\left(-x^2y\right) = yz + 0 + 0 = yz$

5. (a) curl $\mathbf{F} = \nabla \times \mathbf{F} = \begin{vmatrix} \mathbf{i} & \mathbf{j} & \mathbf{k} \\ \partial/\partial x & \partial/\partial y & \partial/\partial z \\ e^x \sin y & e^x \cos y & z \end{vmatrix} = \left(0 - 0\right)\mathbf{i} - \left(0 - 0\right)\mathbf{j} + \left(e^x \cos y - e^x \cos y\right)\mathbf{k} = \mathbf{0}$

(b) div $\mathbf{F} = \nabla \cdot \mathbf{F} = \dfrac{\partial}{\partial x}\left(e^x \sin y\right) + \dfrac{\partial}{\partial y}\left(e^x \cos y\right) + \dfrac{\partial}{\partial z}\left(z\right) = e^x \sin y - e^x \sin y + 1 = 1$

7. (a) curl $\mathbf{F} = \nabla \times \mathbf{F} = \begin{vmatrix} \mathbf{i} & \mathbf{j} & \mathbf{k} \\ \partial/\partial x & \partial/\partial y & \partial/\partial z \\ x/z & y/z & -1/z \end{vmatrix} = \left(0 + \dfrac{y}{z^2}\right)\mathbf{i} - \left(0 + \dfrac{x}{z^2}\right)\mathbf{j} + \left(0 - 0\right)\mathbf{k} = \dfrac{y}{z^2}\,\mathbf{i} - \dfrac{x}{z^2}\,\mathbf{j}$

(b) div $\mathbf{F} = \nabla \cdot \mathbf{F} = \dfrac{\partial}{\partial x}\left(\dfrac{x}{z}\right) + \dfrac{\partial}{\partial y}\left(\dfrac{y}{z}\right) + \dfrac{\partial}{\partial z}\left(-\dfrac{1}{z}\right) = \dfrac{1}{z} + \dfrac{1}{z} + \dfrac{1}{z^2} = \dfrac{2z + 1}{z^2}$

9. If the vector field is $\mathbf{F} = P\,\mathbf{i} + Q\,\mathbf{j} + R\,\mathbf{k}$, then we know $R = 0$. In addition, the x-component of each vector of $\mathbf{F}$ is 0, so $P = 0$, hence $\dfrac{\partial P}{\partial x} = \dfrac{\partial P}{\partial y} = \dfrac{\partial P}{\partial z} = \dfrac{\partial R}{\partial x} = \dfrac{\partial R}{\partial y} = \dfrac{\partial R}{\partial z} = 0$. Q decreases as y increases, so $\dfrac{\partial Q}{\partial y} < 0$, but Q doesn't change in the x- or z-directions, so $\dfrac{\partial Q}{\partial x} = \dfrac{\partial Q}{\partial z} = 0$.

(a) div $\mathbf{F} = \dfrac{\partial P}{\partial x} + \dfrac{\partial Q}{\partial y} + \dfrac{\partial R}{\partial z} = 0 + \dfrac{\partial Q}{\partial y} + 0 < 0$

(b) curl $\mathbf{F} = \left(\dfrac{\partial R}{\partial y} - \dfrac{\partial Q}{\partial z}\right)\mathbf{i} + \left(\dfrac{\partial P}{\partial z} - \dfrac{\partial R}{\partial x}\right)\mathbf{j} + \left(\dfrac{\partial Q}{\partial x} - \dfrac{\partial P}{\partial y}\right)\mathbf{k} = \left(0 - 0\right)\mathbf{i} + \left(0 - 0\right)\mathbf{j} + \left(0 - 0\right)\mathbf{k} = \mathbf{0}$

11. If the vector field is $\mathbf{F} = P\,\mathbf{i} + Q\,\mathbf{j} + R\,\mathbf{k}$, then we know $R = 0$. In addition, the y-component of each vector of $\mathbf{F}$ is 0, so $Q = 0$, hence $\dfrac{\partial Q}{\partial x} = \dfrac{\partial Q}{\partial y} = \dfrac{\partial Q}{\partial z} = \dfrac{\partial R}{\partial x} = \dfrac{\partial R}{\partial y} = \dfrac{\partial R}{\partial z} = 0$. P increases as y increases, so $\dfrac{\partial P}{\partial y} > 0$, but P doesn't change in the x- or z-directions, so $\dfrac{\partial P}{\partial x} = \dfrac{\partial P}{\partial z} = 0$.

(a) div $\mathbf{F} = \dfrac{\partial P}{\partial x} + \dfrac{\partial Q}{\partial y} + \dfrac{\partial R}{\partial z} = 0 + 0 + 0 = 0$

(b) curl $\mathbf{F} = \left(\dfrac{\partial R}{\partial y} - \dfrac{\partial Q}{\partial z}\right)\mathbf{i} + \left(\dfrac{\partial P}{\partial z} - \dfrac{\partial R}{\partial x}\right)\mathbf{j} + \left(\dfrac{\partial Q}{\partial x} - \dfrac{\partial P}{\partial y}\right)\mathbf{k}$

$= \left(0 - 0\right)\mathbf{i} + \left(0 - 0\right)\mathbf{j} + \left(0 - \dfrac{\partial P}{\partial y}\right)\mathbf{k} = -\dfrac{\partial P}{\partial y}\,\mathbf{k}$

Since $\dfrac{\partial P}{\partial y} > 0$, $-\dfrac{\partial P}{\partial y}\,\mathbf{k}$ is a vector pointing in the negative z-direction.

13. $\text{curl } \mathbf{F} = \nabla \times \mathbf{F} = \begin{vmatrix} \mathbf{i} & \mathbf{j} & \mathbf{k} \\ \partial/\partial x & \partial/\partial y & \partial/\partial z \\ yz & xz & xy \end{vmatrix} = (x - x)\mathbf{i} - (y - y)\mathbf{j} + (z - z)\mathbf{k} = \mathbf{0}$ and $\mathbf{F}$ is defined on all of $\mathbb{R}^3$

with component functions which have continuous partial derivatives, so by Theorem 4, $\mathbf{F}$ is conservative. Thus, there exists a function f such that $\mathbf{F} = \nabla f$. Then $f_x(x, y, z) = yz$ implies $f(x, y, z) = xyz + g(y, z)$ and $f_y(x, y, z) = xz + g_y(y, z)$. But $f_y(x, y, z) = xz$, so $g(y, z) = h(z)$ and $f(x, y, z) = xyz + h(z)$. Thus $f_z(x, y, z) = xy + h'(z)$ but $f_z(x, y, z) = xy$ so $h(z) = K$, a constant. Hence a potential function for $\mathbf{F}$ is $f(x, y, z) = xyz + K$.

15. $\text{curl } \mathbf{F} = \nabla \times \mathbf{F} = \begin{vmatrix} \mathbf{i} & \mathbf{j} & \mathbf{k} \\ \partial/\partial x & \partial/\partial y & \partial/\partial z \\ 2xy & x^2 + 2yz & y^2 \end{vmatrix} = (2y - 2y)\mathbf{i} - (0 - 0)\mathbf{j} + (2x - 2x)\mathbf{k} = \mathbf{0}$, $\mathbf{F}$ is defined on all

of $\mathbb{R}^3$, and the partial derivatives of the component functions are continuous, so $\mathbf{F}$ is conservative. Thus there exists a function f such that $\nabla f = \mathbf{F}$. Then $f_x(x, y, z) = 2xy$ implies $f(x, y, z) = x^2 y + g(y, z)$ and $f_y(x, y, z) = x^2 + g_y(y, z)$. But $f_y(x, y, z) = x^2 + 2yz$, so $g(y, z) = y^2 z + h(z)$ and $f(x, y, z) = x^2 y + y^2 z + h(z)$. Thus $f_z(x, y, z) = y^2 + h'(z)$ but $f_z(x, y, z) = y^2$ so $h(z) = K$ and $f(x, y, z) = x^2 y + y^2 z + K$.

17. $\text{curl } \mathbf{F} = \nabla \times \mathbf{F} = \begin{vmatrix} \mathbf{i} & \mathbf{j} & \mathbf{k} \\ \partial/\partial x & \partial/\partial y & \partial/\partial z \\ e^x & e^z & e^y \end{vmatrix} = (e^y - e^z)\mathbf{i} - (0 - 0)\mathbf{j} + (0 - 0)\mathbf{k} \neq \mathbf{0}$, so $\mathbf{F}$ isn't conservative.

19. No. Assume there is such a $\mathbf{G}$. Then $\text{div}(\text{curl } \mathbf{G}) = y^2 + z^2 + x^2 \neq 0$, which contradicts Theorem 11.

21. $\text{curl } \mathbf{F} = \begin{vmatrix} \mathbf{i} & \mathbf{j} & \mathbf{k} \\ \partial/\partial x & \partial/\partial y & \partial/\partial z \\ f(x) & g(y) & h(z) \end{vmatrix} = (0 - 0)\mathbf{i} + (0 - 0)\mathbf{j} + (0 - 0)\mathbf{k} = \mathbf{0}$. Hence $\mathbf{F} = f(x)\mathbf{i} + g(y)\mathbf{j} + h(z)\mathbf{k}$

is irrotational.

For Exercises 23–29, let $\mathbf{F}(x, y, z) = P_1\mathbf{i} + Q_1\mathbf{j} + R_1\mathbf{k}$ and $\mathbf{G}(x, y, z) = P_2\mathbf{i} + Q_2\mathbf{j} + R_2\mathbf{k}$.

23. $\text{div}(\mathbf{F} + \mathbf{G}) = \dfrac{\partial(P_1 + P_2)}{\partial x} + \dfrac{\partial(Q_1 + Q_2)}{\partial y} + \dfrac{\partial(R_1 + R_2)}{\partial z}$

$= \left(\dfrac{\partial P_1}{\partial x} + \dfrac{\partial Q_1}{\partial y} + \dfrac{\partial R_1}{\partial z} \right) + \left(\dfrac{\partial P_2}{\partial x} + \dfrac{\partial Q_2}{\partial y} + \dfrac{\partial R_3}{\partial z} \right) = \text{div } \mathbf{F} + \text{div } \mathbf{G}$

25. $\text{div}(f\mathbf{F}) = \dfrac{\partial(fP_1)}{\partial x} + \dfrac{\partial(fQ_1)}{\partial y} + \dfrac{\partial(fR_1)}{\partial z}$

$= \left(f\dfrac{\partial P_1}{\partial x} + P_1\dfrac{\partial f}{\partial x} \right) + \left(f\dfrac{\partial Q_1}{\partial y} + Q_1\dfrac{\partial f}{\partial y} \right) + \left(f\dfrac{\partial R_1}{\partial z} + R_1\dfrac{\partial f}{\partial z} \right)$

$= f\left(\dfrac{\partial P_1}{\partial x} + \dfrac{\partial Q_1}{\partial y} + \dfrac{\partial R_1}{\partial z} \right) + \langle P_1, Q_1, R_1 \rangle \cdot \left\langle \dfrac{\partial f}{\partial x}, \dfrac{\partial f}{\partial y}, \dfrac{\partial f}{\partial z} \right\rangle = f \, \text{div } \mathbf{F} + \mathbf{F} \cdot \nabla f$

27. $\operatorname{div}(\mathbf{F} \times \mathbf{G}) = \nabla \cdot (\mathbf{F} \times \mathbf{G}) = \begin{vmatrix} \partial/\partial x & \partial/\partial y & \partial/\partial z \\ P_1 & Q_1 & R_1 \\ P_2 & Q_2 & R_2 \end{vmatrix} = \frac{\partial}{\partial x} \begin{vmatrix} Q_1 & R_1 \\ Q_2 & R_2 \end{vmatrix} - \frac{\partial}{\partial y} \begin{vmatrix} P_1 & R_1 \\ P_2 & R_2 \end{vmatrix} + \frac{\partial}{\partial z} \begin{vmatrix} P_1 & Q_1 \\ P_2 & Q_2 \end{vmatrix}$

$$= \left[Q_1 \frac{\partial R_2}{\partial x} + R_2 \frac{\partial Q_1}{\partial x} - Q_2 \frac{\partial R_1}{\partial x} - R_1 \frac{\partial Q_2}{\partial x} \right]$$

$$- \left[P_1 \frac{\partial R_2}{\partial y} + R_2 \frac{\partial P_1}{\partial y} - P_2 \frac{\partial R_1}{\partial y} - R_1 \frac{\partial P_2}{\partial y} \right]$$

$$+ \left[P_1 \frac{\partial Q_2}{\partial z} + Q_2 \frac{\partial P_1}{\partial z} - P_2 \frac{\partial Q_1}{\partial z} - Q_1 \frac{\partial P_2}{\partial z} \right]$$

$$= \left[P_2 \left(\frac{\partial R_1}{\partial y} - \frac{\partial Q_1}{\partial z} \right) + Q_2 \left(\frac{\partial P_1}{\partial z} - \frac{\partial R_1}{\partial x} \right) + R_2 \left(\frac{\partial Q_1}{\partial x} - \frac{\partial P_1}{\partial y} \right) \right]$$

$$- \left[P_1 \left(\frac{\partial R_2}{\partial y} - \frac{\partial Q_2}{\partial z} \right) + Q_1 \left(\frac{\partial P_2}{\partial z} - \frac{\partial R_2}{\partial x} \right) + R_1 \left(\frac{\partial Q_2}{\partial x} - \frac{\partial P_2}{\partial y} \right) \right]$$

$$= \mathbf{G} \cdot \operatorname{curl} \mathbf{F} - \mathbf{F} \cdot \operatorname{curl} \mathbf{G}$$

29. $\operatorname{curl} \operatorname{curl} \mathbf{F} = \nabla \times (\nabla \times \mathbf{F}) = \begin{vmatrix} \mathbf{i} & \mathbf{j} & \mathbf{k} \\ \partial/\partial x & \partial/\partial y & \partial/\partial z \\ \partial R_1/\partial y - \partial Q_1/\partial z & \partial P_1/\partial z - \partial R_1/\partial x & \partial Q_1/\partial x - \partial P_1/\partial y \end{vmatrix}$

$$= \left(\frac{\partial^2 Q_1}{\partial y \partial x} - \frac{\partial^2 P_1}{\partial y^2} - \frac{\partial^2 P_1}{\partial z^2} + \frac{\partial^2 R_1}{\partial z \partial x} \right) \mathbf{i} + \left(\frac{\partial^2 R_1}{\partial z \partial y} - \frac{\partial^2 Q_1}{\partial z^2} - \frac{\partial^2 Q_1}{\partial x^2} + \frac{\partial^2 P_1}{\partial x \partial y} \right) \mathbf{j}$$

$$+ \left(\frac{\partial^2 P_1}{\partial x \partial z} - \frac{\partial^2 R_1}{\partial x^2} - \frac{\partial^2 R_1}{\partial y^2} + \frac{\partial^2 Q_1}{\partial y \partial z} \right) \mathbf{k}$$

Now let's consider grad div $\mathbf{F} - \nabla^2 \mathbf{F}$ and compare with the above.
(Note that $\nabla^2 \mathbf{F}$ is defined on page 1113 [ET 1079].)

$$\text{grad div } \mathbf{F} - \nabla^2 \mathbf{F} = \left[\left(\frac{\partial^2 P_1}{\partial x^2} + \frac{\partial^2 Q_1}{\partial x \partial y} + \frac{\partial^2 R_1}{\partial x \partial z} \right) \mathbf{i} + \left(\frac{\partial^2 P_1}{\partial y \partial x} + \frac{\partial^2 Q_1}{\partial y^2} + \frac{\partial^2 R_1}{\partial y \partial z} \right) \mathbf{j} \right.$$

$$\left. + \left(\frac{\partial^2 P_1}{\partial z \partial x} + \frac{\partial^2 Q_1}{\partial z \partial y} + \frac{\partial^2 R_1}{\partial z^2} \right) \mathbf{k} \right]$$

$$- \left[\left(\frac{\partial^2 P_1}{\partial x^2} + \frac{\partial^2 P_1}{\partial y^2} + \frac{\partial^2 P_1}{\partial z^2} \right) \mathbf{i} + \left(\frac{\partial^2 Q_1}{\partial x^2} + \frac{\partial^2 Q_1}{\partial y^2} + \frac{\partial^2 Q_1}{\partial z^2} \right) \mathbf{j} \right.$$

$$\left. + \left(\frac{\partial^2 R_1}{\partial x^2} + \frac{\partial^2 R_1}{\partial y^2} + \frac{\partial^2 R_1}{\partial z^2} \right) \mathbf{k} \right]$$

$$= \left(\frac{\partial^2 Q_1}{\partial x \partial y} + \frac{\partial^2 R_1}{\partial x \partial z} - \frac{\partial^2 P_1}{\partial y^2} - \frac{\partial^2 P_1}{\partial z^2} \right) \mathbf{i} + \left(\frac{\partial^2 P_1}{\partial y \partial x} + \frac{\partial^2 R_1}{\partial y \partial z} - \frac{\partial^2 Q_1}{\partial x^2} - \frac{\partial^2 Q_1}{\partial z^2} \right) \mathbf{j}$$

$$+ \left(\frac{\partial^2 P_1}{\partial z \partial x} + \frac{\partial^2 Q_1}{\partial z \partial y} - \frac{\partial^2 R_1}{\partial x^2} - \frac{\partial^2 R_2}{\partial y^2} \right) \mathbf{k}$$

Then applying Clairaut's Theorem to reverse the order of differentiation in the second partial derivatives as needed and comparing, we have $\operatorname{curl} \operatorname{curl} \mathbf{F} = \operatorname{grad} \operatorname{div} \mathbf{F} - \nabla^2 \mathbf{F}$ as desired.

31. (a) $\nabla r = \nabla \sqrt{x^2 + y^2 + z^2} = \dfrac{x}{\sqrt{x^2 + y^2 + z^2}}\, \mathbf{i} + \dfrac{y}{\sqrt{x^2 + y^2 + z^2}}\, \mathbf{j} + \dfrac{z}{\sqrt{x^2 + y^2 + z^2}}\, \mathbf{k}$

$$= \dfrac{x\mathbf{i} + y\mathbf{j} + z\mathbf{k}}{\sqrt{x^2 + y^2 + z^2}} = \dfrac{\mathbf{r}}{r}$$

(b) $\nabla \times \mathbf{r} = \begin{vmatrix} \mathbf{i} & \mathbf{j} & \mathbf{k} \\ \dfrac{\partial}{\partial x} & \dfrac{\partial}{\partial y} & \dfrac{\partial}{\partial z} \\ x & y & z \end{vmatrix}$

$$= \left[\frac{\partial}{\partial y}(z) - \frac{\partial}{\partial z}(y) \right]\mathbf{i} + \left[\frac{\partial}{\partial z}(x) - \frac{\partial}{\partial x}(z) \right]\mathbf{j} + \left[\frac{\partial}{\partial x}(y) - \frac{\partial}{\partial y}(x) \right]\mathbf{k} = \mathbf{0}$$

(c) $\nabla \left(\dfrac{1}{r} \right) = \nabla \left(\dfrac{1}{\sqrt{x^2 + y^2 + z^2}} \right)$

$$= \dfrac{-\dfrac{1}{2\sqrt{x^2+y^2+z^2}}(2x)}{x^2+y^2+z^2}\,\mathbf{i} - \dfrac{\dfrac{1}{2\sqrt{x^2+y^2+z^2}}(2y)}{x^2+y^2+z^2}\,\mathbf{j} - \dfrac{\dfrac{1}{2\sqrt{x^2+y^2+z^2}}(2z)}{x^2+y^2+z^2}\,\mathbf{k}$$

$$= -\dfrac{x\mathbf{i} + y\mathbf{j} + z\mathbf{k}}{(x^2 + y^2 + z^2)^{3/2}} = -\dfrac{\mathbf{r}}{r^3}$$

(d) $\nabla \ln r = \nabla \ln \left(x^2 + y^2 + z^2 \right)^{1/2} = \tfrac{1}{2} \nabla \ln \left(x^2 + y^2 + z^2 \right)$

$$= \dfrac{x}{x^2 + y^2 + z^2}\,\mathbf{i} + \dfrac{y}{x^2 + y^2 + z^2}\,\mathbf{j} + \dfrac{z}{x^2 + y^2 + z^2}\,\mathbf{k} = \dfrac{x\mathbf{i} + y\mathbf{j} + z\mathbf{k}}{x^2 + y^2 + z^2} = \dfrac{\mathbf{r}}{r^2}$$

33. By (13), $\oint_C f(\nabla g) \cdot \mathbf{n}\, ds = \iint_D \operatorname{div}(f \nabla g)\, dA = \iint_D [f \operatorname{div}(\nabla g) + \nabla g \cdot \nabla f]\, dA$ by Exercise 25. But $\operatorname{div}(\nabla g) = \nabla^2 g$. Hence $\iint_D f\nabla^2 g\, dA = \oint_C f(\nabla g) \cdot \mathbf{n}\, ds - \iint_D \nabla g \cdot \nabla f\, dA$.

35. (a) We know that $\omega = v/d$, and from the diagram $\sin\theta = d/r \;\Rightarrow\; v = d\omega = (\sin\theta)\, r\omega = |\mathbf{w} \times \mathbf{r}|$. But $\mathbf{v}$ is perpendicular to both $\mathbf{w}$ and $\mathbf{r}$, so that $\mathbf{v} = \mathbf{w} \times \mathbf{r}$.

(b) From (a), $\mathbf{v} = \mathbf{w} \times \mathbf{r} = \begin{vmatrix} \mathbf{i} & \mathbf{j} & \mathbf{k} \\ 0 & 0 & \omega \\ x & y & z \end{vmatrix} = (0 \cdot z - \omega y)\mathbf{i} + (\omega x - 0 \cdot z)\mathbf{j} + (0 \cdot y - x \cdot 0)\mathbf{k} = -\omega y\, \mathbf{i} + \omega x\, \mathbf{j}$

(c) $\operatorname{curl} \mathbf{v} = \nabla \times \mathbf{v} = \begin{vmatrix} \mathbf{i} & \mathbf{j} & \mathbf{k} \\ \partial/\partial x & \partial/\partial y & \partial/\partial z \\ -\omega y & \omega x & 0 \end{vmatrix}$

$$= \left[\frac{\partial}{\partial y}(0) - \frac{\partial}{\partial z}(\omega x) \right]\mathbf{i} + \left[\frac{\partial}{\partial z}(-\omega y) - \frac{\partial}{\partial x}(0) \right]\mathbf{j} + \left[\frac{\partial}{\partial x}(\omega x) - \frac{\partial}{\partial y}(-\omega y) \right]\mathbf{k}$$

$$= [\omega - (-\omega)]\mathbf{k} = 2\omega\, \mathbf{k} = 2\mathbf{w}$$

1. $\mathbf{r}(u, v) = u \cos v \, \mathbf{i} + u \sin v \, \mathbf{j} + u^2 \, \mathbf{k}$, so the corresponding parametric equations for the surface are
$x = u \cos v, y = u \sin v, z = u^2$. For any point (x, y, z) on the surface, we have
$x^2 + y^2 = u^2 \cos^2 v + u^2 \sin^2 v = u^2 = z$. Since no restrictions are placed on the parameters, the surface is
$z = x^2 + y^2$, which we recognize as a circular paraboloid opening upward whose axis is the z-axis.

3. $\mathbf{r}(x, \theta) = \langle x, \cos \theta, \sin \theta \rangle$, so the corresponding parametric equations for the surface are
$x = x, y = \cos \theta, z = \sin \theta$. For any point (x, y, z) on the surface, we have $y^2 + z^2 = \cos^2 \theta + \sin^2 \theta = 1$, so any
vertical trace in $x = k$ is the circle $y^2 + z^2 = 1, x = k$. Since $x = x$ with no restriction, the surface is a circular
cylinder with radius 1 whose axis is the x-axis.

5. $\mathbf{r}(u, v) = \langle u^2 + 1, v^3 + 1, u + v \rangle, -1 \le u \le 1, -1 \le v \le 1$.

The surface has parametric equations $x = u^2 + 1, y = v^3 + 1,$
$z = u + v, -1 \le u \le 1, -1 \le v \le 1$. If we keep u constant at u_0,
$x = u_0^2 + 1$, a constant, so the corresponding grid curves must be the
curves parallel to the yz-plane. If v is constant, we have $y = v_0^3 + 1$, a
constant, so these grid curves are the curves parallel to the xz-plane.

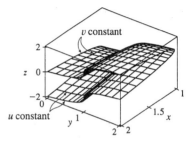

7. $\mathbf{r}(u, v) = \langle \cos^3 u \cos^3 v, \sin^3 u \cos^3 v, \sin^3 v \rangle$.

The surface has parametric equations $x = \cos^3 u \cos^3 v, y = \sin^3 u \cos^3 v, z = \sin^3 v, 0 \le u \le \pi, 0 \le v \le 2\pi$.
Note that if $v = v_0$ is constant then $z = \sin^3 v_0$ is constant, so the corresponding grid curves must be the curves
parallel to the xy-plane. The vertically oriented grid curves, then, correspond to $u = u_0$ being held constant, giving
$x = \cos^3 u_0 \cos^3 v, y = \sin^3 u_0 \cos^3 v, z = \sin^3 v$. These curves lie in vertical planes that contain the z-axis.

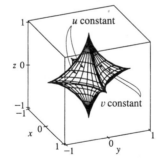

9. $x = \cos u \sin 2v, y = \sin u \sin 2v, z = \sin v$.

The complete graph of the surface is given by the parametric domain
$0 \le u \le \pi, 0 \le v \le 2\pi$. Note that if $v = v_0$ is constant, the
parametric equations become $x = \cos u \sin 2v_0, y = \sin u \sin 2v_0,$
$z = \sin v_0$ which represent a circle of radius $\sin 2v_0$ in the plane
$z = \sin v_0$. So the circular grid curves we see lying horizontally are
the grid curves which have v constant. The vertical grid curves, then,
correspond to $u = u_0$ being held constant, giving $x = \cos u_0 \sin 2v$
and $y = \sin u_0 \sin 2v$ with $z = \sin v$ which has a "figure-eight"
shape.

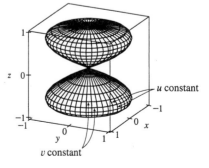

11. $\mathbf{r}(u, v) = \cos v \, \mathbf{i} + \sin v \, \mathbf{j} + u \, \mathbf{k}$. The parametric equations for the surface are $x = \cos v$, $y = \sin v$, $z = u$. Then $x^2 + y^2 = \cos^2 v + \sin^2 v = 1$ and $z = u$ with no restriction on u, so we have a circular cylinder, graph IV. The grid curves with u constant are the horizontal circles we see in the plane $z = u$. If v is constant, both x and y are constant with z free to vary, so the corresponding grid curves are the lines on the cylinder parallel to the z-axis.

13. $\mathbf{r}(u, v) = u \cos v \, \mathbf{i} + u \sin v \, \mathbf{j} + v \, \mathbf{k}$. The parametric equations for the surface are $x = u \cos v$, $y = u \sin v$, $z = v$. We look at the grid curves first; if we fix v, then x and y parametrize a straight line in the plane $z = v$ which intersects the z-axis. If u is held constant, the projection onto the xy-plane is circular; with $z = v$, each grid curve is a helix. The surface is a spiraling ramp, graph I.

15. $x = (u - \sin u) \cos v$, $y = (1 - \cos u) \sin v$, $z = u$. If u is held constant, x and y give an equation of an ellipse in the plane $z = u$, thus the grid curves are horizontally oriented ellipses. Note that when $u = 0$, the "ellipse" is the single point $(0, 0, 0)$, and when $u = \pi$, we have $y = 0$ while x ranges from $-\pi$ to π, a line segment parallel to the x-axis in the plane $z = \pi$. This is the upper "seam" we see in graph II. When v is held constant, $z = u$ is free to vary, so the corresponding grid curves are the curves we see running up and down along the surface.

17. From Example 3, parametric equations for the plane through the point $(1, 2, -3)$ that contains the vectors $\mathbf{a} = \langle 1, 1, -1 \rangle$ and $\mathbf{b} = \langle 1, -1, 1 \rangle$ are $x = 1 + u(1) + v(1) = 1 + u + v$, $y = 2 + u(1) + v(-1) = 2 + u - v$, $z = -3 + u(-1) + v(1) = -3 - u + v$.

19. $x = x$, $y = 6 - 3x^2 - 2z^2$, $z = z$ where $3x^2 + 2z^2 \leq 6$ since $y \geq 0$. Then the associated vector equation is $\mathbf{r}(x, y) = x \, \mathbf{i} + (6 - 3x^2 - 2z^2) \, \mathbf{j} + z \, \mathbf{k}$.

21. Since the cone intersects the sphere in the circle $x^2 + y^2 = 2$, $z = 2$ and we want the portion of the sphere above this, we can parametrize the surface as $x = x$, $y = y$, $z = \sqrt{4 - x^2 - y^2}$ where $2 \leq x^2 + y^2 \leq 4$.
Alternate Solution: Using spherical coordinates, $x = 2 \sin \phi \cos \theta$, $y = 2 \sin \phi \sin \theta$, $z = 2 \cos \phi$ where $0 \leq \phi \leq \frac{\pi}{4}$ and $0 \leq \theta \leq 2\pi$.

23. The surface is a disc with radius 4 and center $(0, 0, 5)$. Thus, $x = r \cos \theta$, $y = r \sin \theta$, $z = 5$ where $0 \leq r \leq 4$, $0 \leq \theta \leq 2\pi$ is a parametric representation of the surface.
Alternate Solution: In rectangular coordinates we could represent the surface as $x = x$, $y = y$, $z = 5$ where $0 \leq x^2 + y^2 \leq 16$.

25. Using Equations 3, we have the parametrization $x = x$, $y = e^{-x} \cos \theta$, $z = e^{-x} \sin \theta$, $0 \leq x \leq 3$, $0 \leq \theta \leq 2\pi$.

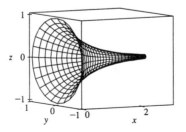

27. (a) Replacing $\cos u$ by $\sin u$ and $\sin u$ by $\cos u$ gives parametric equations

$x = (2 + \sin v) \sin u$, $y = (2 + \sin v) \cos u$, $z = u + \cos v$. From the graph, it appears that the direction of the spiral is reversed. We can verify this observation by noting that the projection of the spiral grid curves onto the xy-plane, given by $x = (2 + \sin v) \sin u$, $y = (2 + \sin v) \cos u$, $z = 0$, draws a circle in the clockwise direction for each value of v. The original equations, on the other hand, give circular projections drawn in the counterclockwise direction. The equation for z is identical in both surfaces, so as z increases, these grid curves spiral up in opposite directions for the two surfaces.

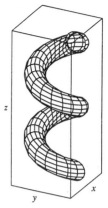

(b) Replacing $\cos u$ by $\cos 2u$ and $\sin u$ by $\sin 2u$ gives parametric equations

$x = (2 + \sin v) \cos 2u$, $y = (2 + \sin v) \sin 2u$, $z = u + \cos v$. From the graph, it appears that the number of coils in the surface doubles within the same parametric domain. We can verify this observation by noting that the projection of the spiral grid curves onto the xy-plane, given by $x = (2 + \sin v) \cos 2u$, $y = (2 + \sin v) \sin 2u$, $z = 0$ (where v is constant), complete circular revolutions for $0 \le u \le \pi$ while the original surface requires $0 \le u \le 2\pi$ for a complete revolution. Thus, the new surface winds around twice as fast as the original surface, and since the equation for z is identical in both surfaces, we observe twice as many circular coils in the same z-interval.

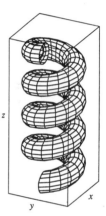

29. $\mathbf{r}(u, v) = (u + v)\mathbf{i} + 3u^2\mathbf{j} + (u - v)\mathbf{k}$.

$\mathbf{r}_u = \mathbf{i} + 6u\mathbf{j} + \mathbf{k}$ and $\mathbf{r}_v = \mathbf{i} - \mathbf{k}$, so

$\mathbf{r}_u \times \mathbf{r}_v = -6u\mathbf{i} + 2\mathbf{j} - 6u\mathbf{k}$. Since the point $(2, 3, 0)$ corresponds to $u = 1$, $v = 1$, a normal vector to the surface at $(2, 3, 0)$ is $-6\mathbf{i} + 2\mathbf{j} - 6\mathbf{k}$, and an equation of the tangent plane is $-6x + 2y - 6z = -6$ or $3x - y + 3z = 3$.

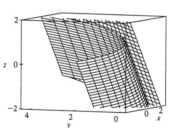

31. $\mathbf{r}(u, v) = uv\mathbf{i} + ue^v\mathbf{j} + ve^u\mathbf{k}$.

$\mathbf{r}_u = \langle v, e^v, ve^u \rangle$, $\mathbf{r}_v = \langle u, ue^v, e^u \rangle$, and

$\mathbf{r}_u \times \mathbf{r}_v = e^{u+v}(1 - uv)\mathbf{i} + e^u(uv - v)\mathbf{j} + e^v(uv - u)\mathbf{k}$. The point $(0, 0, 0)$ corresponds to $u = 0$, $v = 0$. Thus a normal vector to the surface at $(0, 0, 0)$ is $\mathbf{i}$, and an equation of the tangent plane is $x = 0$.

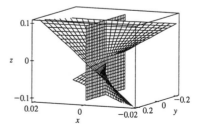

33. Here $z = f(x, y) = 4 - x - 2y$ and D is the disk $x^2 + y^2 \le 4$. Thus, by Formula 9,

$A(S) = \iint_D \sqrt{1 + (-1)^2 + (-2)^2}\, dA = \sqrt{6} \iint_D dA = \sqrt{6}A(D) = 4\sqrt{6}\pi$.

35. $z = f(x, y) = y^2 - x^2$ with $1 \le x^2 + y^2 \le 4$. Then

$$A(S) = \iint_D \sqrt{1 + 4x^2 + 4y^2}\, dA = \int_0^{2\pi} \int_1^2 \sqrt{1 + 4r^2}\, r\, dr\, d\theta = \int_0^{2\pi} d\theta \int_1^2 \sqrt{1 + 4r^2}\, r\, dr$$

$$= [\theta]_0^{2\pi} \left[\tfrac{1}{12}(1 + 4r^2)^{3/2} \right]_1^2 = \tfrac{\pi}{6} \left(17\sqrt{17} - 5\sqrt{5} \right)$$

37. A parametric representation of the surface is $x = x$, $y = 4x + z^2$, $z = z$ with $0 \le x \le 1$, $0 \le z \le 1$. Hence
$\mathbf{r}_x \times \mathbf{r}_z = (\mathbf{i} + 4\mathbf{j}) \times (2z\,\mathbf{j} + \mathbf{k}) = 4\mathbf{i} - \mathbf{j} + 2z\,\mathbf{k}$.

Note: In general, if $y = f(x, z)$ then $\mathbf{r}_z \times \mathbf{r}_x = -\dfrac{\partial f}{\partial x}\mathbf{i} + \mathbf{j} - \dfrac{\partial f}{\partial x}\mathbf{k}$ and

$$A(S) = \iint_D \sqrt{1 + \left(\frac{\partial f}{\partial x}\right)^2 + \left(\frac{\partial f}{\partial z}\right)^2}\, dA. \text{ Then}$$

$$A(S) = \int_0^1 \int_0^1 \sqrt{17 + 4z^2}\, dx\, dz = \int_0^1 \sqrt{17 + 4z^2}\, dz$$

$$= \tfrac{1}{2} \left(z\sqrt{17 + 4z^2} + \tfrac{17}{2} \ln\left| 2z + \sqrt{4z^2 + 17} \right| \right) \Big|_0^1 = \tfrac{\sqrt{21}}{2} + \tfrac{17}{4} \left[\ln\left(2 + \sqrt{21}\right) - \ln\sqrt{17} \right]$$

39. Let $A(S_1)$ be the surface area of that portion of the surface which lies above the plane $z = 0$. Then
$A(S) = 2A(S_1)$. Following Example 10, a parametric representation of S_1 is $x = a\sin\phi\cos\theta$, $y = a\sin\phi\sin\theta$,
$z = a\cos\phi$ and $|\mathbf{r}_\phi \times \mathbf{r}_\theta| = a^2 \sin\phi$. For D, $0 \le \phi \le \tfrac{\pi}{2}$ and for each fixed ϕ, $\left(x - \tfrac{1}{2}a\right)^2 + y^2 \le \left(\tfrac{1}{2}a\right)^2$ or
$\left[a\sin\phi\cos\theta - \tfrac{1}{2}a\right]^2 + a^2 \sin^2\phi \sin^2\theta \le (a/2)^2$ implies $a^2 \sin^2\phi - a^2 \sin\phi\cos\theta \le 0$ or
$\sin\phi\,(\sin\phi - \cos\theta) \le 0$. But $0 \le \phi \le \tfrac{\pi}{2}$, so $\cos\theta \ge \sin\phi$ or $\sin\left(\tfrac{\pi}{2} + \theta\right) \ge \sin\phi$ or $\phi - \tfrac{\pi}{2} \le \theta \le \tfrac{\pi}{2} - \phi$.
Hence $D = \left\{(\phi, \theta) \mid 0 \le \phi \le \tfrac{\pi}{2}, \phi - \tfrac{\pi}{2} \le \theta \le \tfrac{\pi}{2} - \phi\right\}$. Then

$$A(S_1) = \int_0^{\pi/2} \int_{\phi - (\pi/2)}^{(\pi/2) - \phi} a^2 \sin\phi\, d\theta\, d\phi = a^2 \int_0^{\pi/2} (\pi - 2\phi)\sin\phi\, d\phi$$

$$= a^2 \left[(-\pi\cos\phi) - 2(-\phi\cos\phi + \sin\phi) \right]_0^{\pi/2} = a^2(\pi - 2)$$

Thus $A(S) = 2a^2(\pi - 2)$.

Alternate Solution: Working on S_1 we could parametrize the portion of the sphere by $x = x$, $y = y$,

$z = \sqrt{a^2 - x^2 - y^2}$. Then $|\mathbf{r}_x \times \mathbf{r}_y| = \sqrt{1 + \dfrac{x^2}{a^2 - x^2 - y^2} + \dfrac{y^2}{a^2 - x^2 - y^2}} = \dfrac{a}{\sqrt{a^2 - x^2 - y^2}}$ and

$$A(S_1) = \iint_{0 \le (x - (a/2))^2 + y^2 \le (a/2)^2} \frac{a}{\sqrt{a^2 - x^2 - y^2}}\, dA = \int_{-\pi/2}^{\pi/2} \int_0^{a\cos\theta} \frac{a}{\sqrt{a^2 - r^2}}\, r\, dr\, d\theta$$

$$= \int_{-\pi/2}^{\pi/2} -a\left(a^2 - r^2\right)^{1/2} \Big|_{r=0}^{r = a\cos\theta} d\theta = \int_{-\pi/2}^{\pi/2} a^2 \left[1 - \left(1 - \cos^2\theta\right)^{1/2} \right] d\theta$$

$$= \int_{-\pi/2}^{\pi/2} a^2 (1 - |\sin\theta|)\, d\theta = 2a^2 \int_0^{\pi/2} (1 - \sin\theta)\, d\theta = 2a^2\left(\tfrac{\pi}{2} - 1\right)$$

Thus $A(S) = 4a^2\left(\tfrac{\pi}{2} - 1\right) = 2a^2(\pi - 2)$.

Notes:

(1) Perhaps working in spherical coordinates is the most obvious approach here. However, you must be careful in setting up D.

(2) In the alternate solution, you can avoid having to use $|\sin\theta|$ by working in the first octant and then multiplying by 4. However, if you set up S_1 as above and arrived at $A(S_1) = a^2\pi$, you now see your error.

41. $r_u = \langle v, 1, 1 \rangle$, $r_v = \langle u, 1, -1 \rangle$ and $r_u \times r_v = \langle -2, u + v, v - u \rangle$. Then

$$A(S) = \iint_{u^2 + v^2 \leq 1} \sqrt{4 + 2u^2 + 2v^2}\, dA = \int_0^{2\pi} \int_0^1 r\sqrt{4 + 2r^2}\, dr\, d\theta = \int_0^{2\pi} d\theta \int_0^1 r\sqrt{4 + 2r^2}\, dr$$

$$= 2\pi \left[\tfrac{1}{6}(4 + 2r^2)^{3/2} \right]_0^1 = \tfrac{\pi}{3}\left(6\sqrt{6} - 8 \right) = \pi\left(2\sqrt{6} - \tfrac{8}{3} \right)$$

43. (a) The midpoints of the four squares are $\left(\tfrac{1}{4}, \tfrac{1}{4}\right)$, $\left(\tfrac{1}{4}, \tfrac{3}{4}\right)$, $\left(\tfrac{3}{4}, \tfrac{1}{4}\right)$, and $\left(\tfrac{3}{4}, \tfrac{3}{4}\right)$; the derivatives of the function $f(x, y) = x^2 + y^2$ are $f_x(x, y) = 2x$ and $f_y(x, y) = 2y$, so the Midpoint Rule gives

$$A(S) = \int_0^1 \int_0^1 \sqrt{[f_x(x, y)]^2 + [f_y(x, y)]^2 + 1}\, dy\, dx$$

$$\approx \tfrac{1}{4}\left(\sqrt{\left[2\left(\tfrac{1}{4}\right)\right]^2 + \left[2\left(\tfrac{1}{4}\right)\right]^2 + 1} + \sqrt{\left[2\left(\tfrac{1}{4}\right)\right]^2 + \left[2\left(\tfrac{3}{4}\right)\right]^2 + 1} \right.$$

$$\left. + \sqrt{\left[2\left(\tfrac{3}{4}\right)\right]^2 + \left[2\left(\tfrac{1}{4}\right)\right]^2 + 1} + \sqrt{\left[2\left(\tfrac{3}{4}\right)\right]^2 + \left[2\left(\tfrac{3}{4}\right)\right]^2 + 1} \right)$$

$$= \tfrac{1}{4}\left(\sqrt{\tfrac{3}{2}} + 2\sqrt{\tfrac{7}{2}} + \sqrt{\tfrac{11}{2}} \right) \approx 1.8279$$

(b) A CAS estimates the integral to be

$$A(S) = \int_0^1 \int_0^1 \sqrt{f_x^2 + f_y^2 + 1}\, dy\, dx = \int_0^1 \int_0^1 \sqrt{4x^2 + 4y^2 + 1}\, dy\, dx \approx 1.8616.$$ This agrees with the Midpoint estimate only in the first decimal place.

45. (a) $x = a \sin u \cos v$, $y = b \sin u \sin v$, $z = c \cos u$ $\Rightarrow$

(b)

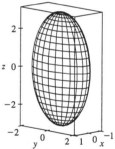

$$\frac{x^2}{a^2} + \frac{y^2}{b^2} + \frac{z^2}{c^2} = (\sin u \cos v)^2 + (\sin u \sin v)^2 + (\cos u)^2$$

$$= \sin^2 u + \cos^2 u = 1$$

and since the ranges of u and v are sufficient to generate the entire graph, the parametric equations represent an ellipsoid.

(c) From the parametric equations (with $a = 1$, $b = 2$, and $c = 3$), we calculate $r_u = \cos u \cos v\, \mathbf{i} + 2 \cos u \sin v\, \mathbf{j} - 3 \sin u\, \mathbf{k}$ and $r_v = -\sin u \sin v\, \mathbf{i} + 2 \sin u \cos v\, \mathbf{j}$. So $r_u \times r_v = 6 \sin^2 u \cos v\, \mathbf{i} + 3 \sin^2 u \sin v\, \mathbf{j} + 2 \sin u \cos u\, \mathbf{k}$, and the surface area is given by

$$A(S) = \int_0^{2\pi} \int_0^\pi |r_u \times r_v|\, du\, dv$$

$$= \int_0^{2\pi} \int_0^\pi \sqrt{36 \sin^4 u \cos^2 v + 9 \sin^4 u \sin^2 v + 4 \cos^2 u \sin^2 u}\, du\, dv$$

47. $r(u, v) = \langle \cos^3 u \cos^3 v, \sin^3 u \cos^3 v, \sin^3 v \rangle$, so $r_u = \langle -3 \cos^2 u \sin u \cos^3 v, 3 \sin^2 u \cos u \cos^3 v, 0 \rangle$, $r_v = \langle -3 \cos^3 u \cos^2 v \sin v, -3 \sin^3 u \cos^2 v \sin v, 3 \sin^2 v \cos v \rangle$, and $r_u \times r_v = \langle 9 \cos u \sin^2 u \cos^4 v \sin^2 v, 9 \cos^2 u \sin u \cos^4 v \sin^2 v, 9 \cos^2 u \sin^2 u \cos^5 v \sin v \rangle$. Then

$$|r_u \times r_v| = 9\sqrt{\cos^2 u \sin^4 u \cos^8 v \sin^4 v + \cos^4 u \sin^2 u \cos^8 v \sin^4 v + \cos^4 u \sin^4 u \cos^{10} v \sin^2 v}$$

$$= 9\sqrt{\cos^2 u \sin^2 u \cos^8 v \sin^2 v \left(\sin^2 v + \cos^2 u \sin^2 u \cos^2 v \right)}$$

$$= 9 \cos^4 v\, |\cos u \sin u \sin v|\, \sqrt{\sin^2 v + \cos^2 u \sin^2 u \cos^2 v}$$

Using a CAS, we have

$$A(S) = \int_0^\pi \int_0^{2\pi} 9 \cos^4 v\, |\cos u \sin u \sin v|\, \sqrt{\sin^2 v + \cos^2 u \sin^2 u \cos^2 v}\, dv\, du \approx 4.4506.$$

17.7 Surface Integrals

1. Each face of the cube has surface area $2^2 = 4$, and the points P_{ij}^* are the points where the cube intersects the coordinate axes. Here, $f(x, y, z) = \sqrt{x^2 + 2y^2 + 3z^2}$, so by Definition 1,

$$\iint_S f(x, y, z)\, dS \approx [f(1, 0, 0)]\,(4) + [f(-1, 0, 0)]\,(4) + [f(0, 1, 0)]\,(4) + [f(0, -1, 0)]\,(4)$$
$$+ [f(0, 0, 1)]\,(4) + [f(0, 0, -1)]\,(4)$$
$$= 4\left(1 + 1 + 2\sqrt{2} + 2\sqrt{3}\right) = 8\left(1 + \sqrt{2} + \sqrt{3}\right) \approx 33.170$$

3. We can use the xz- and yz-planes to divide H into four patches of equal size, each with surface area equal to $\frac{1}{8}$ the surface area of a sphere with radius $\sqrt{50}$, so $\Delta S = \frac{1}{8}\,(4)\,\pi\left(\sqrt{50}\right)^2 = 25\pi$. Then $(\pm 3, \pm 4, 5)$ are sample points in the four patches, and using a Riemann sum as in Definition 1, we have

$$\iint_H f(x, y, z)\, dS \approx f(3, 4, 5)\,\Delta S + f(3, -4, 5)\,\Delta S + f(-3, 4, 5)\,\Delta S + f(-3, -4, 5)\,\Delta S$$
$$= (7 + 8 + 9 + 12)\,(25\pi) = 900\pi \approx 2827$$

5. $z = 1 + 2x + 3y$ so $\dfrac{\partial z}{\partial x} = 2$ and $\dfrac{\partial z}{\partial y} = 3$. Then by Formula 2,

$$\iint_S x^2 yz\, dS = \iint_D x^2 yz \sqrt{\left(\frac{\partial z}{\partial x}\right)^2 + \left(\frac{\partial z}{\partial y}\right)^2 + 1}\, dA$$
$$= \int_0^3 \int_0^2 x^2 y\,(1 + 2x + 3y)\,\sqrt{4 + 9 + 1}\, dy\, dx$$
$$= \sqrt{14} \int_0^3 \int_0^2 \left(x^2 y + 2x^3 y + 3x^2 y^2\right)\, dy\, dx$$
$$= \sqrt{14} \int_0^3 \left[\tfrac{1}{2}x^2 y^2 + x^3 y^2 + x^2 y^3\right]_{y=0}^{y=2}\, dx$$
$$= \sqrt{14} \int_0^3 \left(10x^2 + 4x^3\right)\, dx = \sqrt{14}\left[\tfrac{10}{3}x^3 + x^4\right]_0^3 = 171\sqrt{14}$$

7. S is the part of the plane $z = 1 - x - y$ over the region $D = \{(x, y) \mid 0 \le x \le 1, 0 \le y \le 1 - x\}$. Thus

$$\iint_S yz\, dS = \iint_D y\,(1 - x - y)\,\sqrt{(-1)^2 + (-1)^2 + 1}\, dA$$
$$= \sqrt{3} \int_0^1 \int_0^{1-x} \left(y - xy - y^2\right)\, dy\, dx = \sqrt{3} \int_0^1 \left[\tfrac{1}{2}y^2 - \tfrac{1}{2}xy^2 - \tfrac{1}{3}y^3\right]_{y=0}^{y=1-x}\, dx$$
$$= \sqrt{3} \int_0^1 \tfrac{1}{6}\,(1 - x)^3\, dx = -\tfrac{\sqrt{3}}{24}\,(1 - x)^4\Big|_0^1 = \tfrac{\sqrt{3}}{24}$$

9. Using x and z as parameters, we have $\mathbf{r}(x, z) = x\,\mathbf{i} + \left(x^2 + 4z\right)\mathbf{j} + z\,\mathbf{k}$, $0 \le x \le 2$, $0 \le z \le 2$. Then $\mathbf{r}_x \times \mathbf{r}_z = (\mathbf{i} + 2x\,\mathbf{j}) \times (4\mathbf{j} + \mathbf{k}) = 2x\,\mathbf{i} - \mathbf{j} + 4\mathbf{k}$ and $|\mathbf{r}_x \times \mathbf{r}_z| = \sqrt{4x^2 + 17}$. Thus

$$\iint_S x\, dS = \int_0^2 \int_0^2 x\sqrt{4x^2 + 17}\, dx\, dz = \int_0^2 dz \int_0^2 x\sqrt{4x^2 + 17}\, dx$$
$$= 2\left[\frac{1}{12}\,(4x^2 + 17)^{3/2}\right]_0^2 = \frac{33\sqrt{33} - 17\sqrt{17}}{6}$$

11. S is the part of the plane $z = y + 3$ over the disk $D = \{(x, y) \mid x^2 + y^2 \le 1\}$. Thus

$$\iint_S yz\, dS = \iint_D y\,(y + 3)\,\sqrt{(0)^2 + (1)^2 + 1}\, dA = \sqrt{2} \int_0^{2\pi} \int_0^1 r\sin\theta\,(r\sin\theta + 3)\,r\, dr\, d\theta$$
$$= \sqrt{2} \int_0^{2\pi} \left[\tfrac{1}{4}r^4 \sin^2\theta + r^3 \sin\theta\right]_{r=0}^{r=1}\, d\theta = \sqrt{2} \int_0^{2\pi} \left(\tfrac{1}{4}\sin^2\theta + \sin\theta\right)\, d\theta$$
$$= \sqrt{2}\left[\tfrac{1}{4}\left(\tfrac{1}{2}\theta - \tfrac{1}{4}\sin 2\theta\right) - \cos\theta\right]_0^{2\pi} = \tfrac{\pi}{2\sqrt{2}}$$

13. Using spherical coordinates and Example 17.6.10 [ET 16.6.10] we have

$\mathbf{r}(\phi, \theta) = 2 \sin \phi \cos \theta \, \mathbf{i} + 2 \sin \phi \sin \theta \, \mathbf{j} + 2 \cos \phi \, \mathbf{k}$ and $|\mathbf{r}_\phi \times \mathbf{r}_\theta| = 4 \sin \phi$. Then

$\iint_S (x^2 z + y^2 z) \, dS = \int_0^{2\pi} \int_0^{\pi/2} (4 \sin^2 \phi)(2 \cos \phi)(4 \sin \phi) \, d\phi \, d\theta = 16\pi \sin^4 \phi \big]_0^{\pi/2} = 16\pi$.

15. Using cylindrical coordinates, we have $\mathbf{r}(\theta, z) = 3 \cos \theta \, \mathbf{i} + 3 \sin \theta \, \mathbf{j} + z \, \mathbf{k}$, $0 \le \theta \le 2\pi$, $0 \le z \le 2$, and $|\mathbf{r}_\theta \times \mathbf{r}_z| = 3$.

$\iint_S (x^2 y + z^2) \, dS = \int_0^{2\pi} \int_0^2 (27 \cos^2 \theta \sin \theta + z^2) \, 3 \, dz \, d\theta = \int_0^{2\pi} (162 \cos^2 \theta \sin \theta + 8) \, d\theta = 16\pi$

17. $\mathbf{r}(u, v) = uv \, \mathbf{i} + (u + v) \, \mathbf{j} + (u - v) \, \mathbf{k}$, $u^2 + v^2 \le 1$ and $|\mathbf{r}_u \times \mathbf{r}_v| = \sqrt{4 + 2u^2 + 2v^2}$ (see Exercise 17.6.41 [ET 16.6.41]). Then

$\iint_S yx \, dS = \iint_{u^2 + v^2 \le 1} (u^2 - v^2) \sqrt{4 + 2u^2 + 2v^2} \, dA = \int_0^{2\pi} \int_0^1 r^2 (\cos^2 \theta - \sin^2 \theta) \sqrt{4 + 2r^2} \, r \, dr \, d\theta$

$= \left[\int_0^{2\pi} (\cos^2 \theta - \sin^2 \theta) \, d\theta \right] \left[\int_0^1 r^3 \sqrt{4 + 2r^2} \, dr \right] = 0$

since the first integral is 0.

19. $\mathbf{F}(x, y, z) = xy \, \mathbf{i} + yz \, \mathbf{j} + zx \, \mathbf{k}$, $z = g(x, y) = 4 - x^2 - y^2$, and D is the square $[0, 1] \times [0, 1]$, so by Equation 8

$\iint_S \mathbf{F} \cdot d\mathbf{S} = \iint_D [-xy(-2x) - yz(-2y) + zx] \, dA$

$= \int_0^1 \int_0^1 [2x^2 y + 2y^2 (4 - x^2 - y^2) + x(4 - x^2 - y^2)] \, dy \, dx$

$= \int_0^1 \left(\tfrac{1}{3} x^2 + \tfrac{11}{3} x - x^3 + \tfrac{34}{15} \right) dx = \tfrac{713}{180}$

21. $\mathbf{F}(x, y, z) = xze^y \, \mathbf{i} - xze^y \, \mathbf{j} + z \, \mathbf{k}$, $z = g(x, y) = 1 - x - y$, and $D = \{(x, y) \mid 0 \le x \le 1, 0 \le y \le 1 - x\}$. Since S has downward orientation, we have

$\iint_S \mathbf{F} \cdot d\mathbf{S} = - \iint_D [-xze^y(-1) - (-xze^y)(-1) + z] \, dA = - \int_0^1 \int_0^{1-x} (1 - x - y) \, dy \, dx$

$= - \int_0^1 \left(\tfrac{1}{2} x^2 - x + \tfrac{1}{2} \right) dx = -\tfrac{1}{6}$

23. $\mathbf{F}(\mathbf{r}(\phi, \theta)) = 3 \sin \phi \cos \theta \, \mathbf{i} + 3 \sin \phi \sin \theta \, \mathbf{j} + 3 \cos \phi \, \mathbf{k}$ and

$\mathbf{r}_\phi \times \mathbf{r}_\theta = 9 \sin^2 \phi \cos \theta \, \mathbf{i} + 9 \sin^2 \phi \sin \theta \, \mathbf{j} + 9 \sin \phi \cos \phi \, \mathbf{k}$. Then

$\mathbf{F}(\mathbf{r}(\phi, \theta)) \cdot (\mathbf{r}_\phi \times \mathbf{r}_\theta) = 27 \sin^3 \phi \cos^2 \theta + 27 \sin^3 \phi \sin^2 \theta + 27 \sin \phi \cos^2 \phi = 27 \sin \phi$ and

$\iint_S \mathbf{F} \cdot d\mathbf{S} = \int_0^{2\pi} \int_0^\pi 27 \sin \phi \, d\phi \, d\theta = (2\pi)(54) = 108\pi$.

25. Let S_1 be the paraboloid $y = x^2 + z^2$, $0 \le y \le 1$ and S_2 the disk $x^2 + z^2 \le 1$, $y = 1$. Since S is a closed surface, we use the outward orientation. On S_1: $\mathbf{F}(\mathbf{r}(x, z)) = (x^2 + z^2) \, \mathbf{j} - z \, \mathbf{k}$ and $\mathbf{r}_x \times \mathbf{r}_z = 2x \, \mathbf{i} - \mathbf{j} + 2z \, \mathbf{k}$ (since the $\mathbf{j}$-component must be negative on S_1). Then

$\iint_{S_1} \mathbf{F} \cdot d\mathbf{S} = \iint_{x^2 + z^2 \le 1} [-(x^2 + z^2) - 2z^2] \, dA = - \int_0^{2\pi} \int_0^1 (r^2 + 2r^2 \cos^2 \theta) \, r \, dr \, d\theta$

$= - \int_0^{2\pi} \tfrac{1}{4} (1 + 2 \cos^2 \theta) \, d\theta = - \left(\tfrac{\pi}{2} + \tfrac{\pi}{2} \right) = -\pi$

On S_2: $\mathbf{F}(\mathbf{r}(x, z)) = \mathbf{j} - z \, \mathbf{k}$ and $\mathbf{r}_z \times \mathbf{r}_x = \mathbf{j}$. Then $\iint_{S_2} \mathbf{F} \cdot d\mathbf{S} = \iint_{x^2 + z^2 \le 1} (1) \, dA = \pi$. Hence

$\iint_S \mathbf{F} \cdot d\mathbf{S} = -\pi + \pi = 0$.

27. Here S consists of the six faces of the cube as labeled in the figure. On S_1:

$\mathbf{F} = \mathbf{i} + 2y\mathbf{j} + 3z\mathbf{k}$, $\mathbf{r}_y \times \mathbf{r}_z = \mathbf{i}$ and $\iint_{S_1} \mathbf{F} \cdot d\mathbf{S} = \int_{-1}^{1} \int_{-1}^{1} dy\, dz = 4$;

S_2: $\mathbf{F} = x\mathbf{i} + 2\mathbf{j} + 3z\mathbf{k}$, $\mathbf{r}_z \times \mathbf{r}_x = \mathbf{j}$ and $\iint_{S_2} \mathbf{F} \cdot d\mathbf{S} = \int_{-1}^{1} \int_{-1}^{1} 2\, dx\, dz = 8$;

S_3: $\mathbf{F} = x\mathbf{i} + 2y\mathbf{j} + 3\mathbf{k}$, $\mathbf{r}_x \times \mathbf{r}_y = \mathbf{k}$ and $\iint_{S_3} \mathbf{F} \cdot d\mathbf{S} = \int_{-1}^{1} \int_{-1}^{1} 3\, dx\, dy = 12$;

S_4: $\mathbf{F} = -\mathbf{i} + 2y\mathbf{j} + 3z\mathbf{k}$, $\mathbf{r}_z \times \mathbf{r}_y = -\mathbf{i}$ and $\iint_{S_4} \mathbf{F} \cdot d\mathbf{S} = 4$;

S_5: $\mathbf{F} = x\mathbf{i} - 2\mathbf{j} + 3z\mathbf{k}$, $\mathbf{r}_x \times \mathbf{r}_z = -\mathbf{j}$ and $\iint_{S_5} \mathbf{F} \cdot d\mathbf{S} = 8$;

S_6: $\mathbf{F} = x\mathbf{i} + 2y\mathbf{j} - 3\mathbf{k}$, $\mathbf{r}_y \times \mathbf{r}_x = -\mathbf{k}$ and $\iint_{S_6} \mathbf{F} \cdot d\mathbf{S} = \int_{-1}^{1} \int_{-1}^{1} 3\, dx\, dy = 12$.

Hence $\iint_S \mathbf{F} \cdot d\mathbf{S} = \sum_{i=1}^{6} \iint_{S_i} \mathbf{F} \cdot d\mathbf{S} = 48$.

29. $z = xy \quad \Rightarrow \quad \partial z / \partial x = y$, $\partial z / \partial y = x$, so by Formula 2, a CAS gives

$\iint_S xyz\, dS = \int_0^1 \int_0^1 xy\,(xy)\, \sqrt{y^2 + x^2 + 1}\, dx\, dy \approx 0.1642$.

31. We use Formula 2 with $z = 3 - 2x^2 - y^2 \quad \Rightarrow \quad \partial z / \partial x = -4x$, $\partial z / \partial y = -2y$. The boundaries of the region

$3 - 2x^2 - y^2 \geq 0$ are $-\sqrt{\frac{3}{2}} \leq x \leq \sqrt{\frac{3}{2}}$ and $-\sqrt{3 - 2x^2} \leq y \leq \sqrt{3 - 2x^2}$, so we use a CAS (with precision

reduced to seven or fewer digits; otherwise the calculation takes a very long time) to calculate

$\iint_S x^2 y^2 z^2\, dS = \int_{-\sqrt{3/2}}^{\sqrt{3/2}} \int_{-\sqrt{3 - 2x^2}}^{\sqrt{3 - 2x^2}} x^2 y^2 \left(3 - 2x^2 - y^2\right)^2 \sqrt{16x^2 + 4y^2 + 1}\, dy\, dx \approx 3.4895$.

33. If S is given by $y = h\,(x, z)$, then S is also the level surface $f\,(x, y, z) = y - h\,(x, z) = 0$.

$\mathbf{n} = \dfrac{\nabla f\,(x, y, z)}{|\nabla f\,(x, y, z)|} = \dfrac{-h_x\,\mathbf{i} + \mathbf{j} - h_z\,\mathbf{k}}{\sqrt{h_x^2 + 1 + h_z^2}}$, and $-\mathbf{n}$ is the unit normal that points to the left. Now we proceed as in

the derivation of (8), using Formula 2 to evaluate

$$\iint_S \mathbf{F} \cdot d\mathbf{S} = \iint_S \mathbf{F} \cdot \mathbf{n}\, dS = \iint_D (P\mathbf{i} + Q\mathbf{j} + R\mathbf{k}) \, \dfrac{\dfrac{\partial h}{\partial x}\,\mathbf{i} - \mathbf{j} + \dfrac{\partial h}{\partial z}\,\mathbf{k}}{\sqrt{\left(\dfrac{\partial h}{\partial x}\right)^2 + 1 + \left(\dfrac{\partial h}{\partial z}\right)^2}} \, \sqrt{\left(\dfrac{\partial h}{\partial x}\right)^2 + 1 + \left(\dfrac{\partial h}{\partial z}\right)^2}\, dA$$

where D is the projection of $f\,(x, y, z)$ onto the xz-plane. Therefore

$$\iint_S \mathbf{F} \cdot d\mathbf{S} = \iint_D \left(P \dfrac{\partial h}{\partial x} - Q + R \dfrac{\partial h}{\partial z} \right) dA.$$

35. $m = \iint_S K\, dS = K \cdot 4\pi \left(\frac{1}{2} a^2\right) = 2\pi a^2 K$; by symmetry $M_{xz} = M_{yz} = 0$, and

$M_{xy} = \iint_S zK\, dS = K \int_0^{2\pi} \int_0^{\pi/2} (a \cos \phi)\,(a^2 \sin \phi)\, d\phi\, d\theta = 2\pi K a^3 \left[-\frac{1}{4} \cos 2\phi\right]_0^{\pi/2} = \pi K a^3$. Hence

$(\bar{x}, \bar{y}, \bar{z}) = \left(0, 0, \frac{1}{2} a\right)$.

37. (a) $I_z = \iint_S \left(x^2 + y^2\right) \rho\,(x, y, z)\, dS$

(b) $I_z = \iint_S \left(x^2 + y^2\right) \left(10 - \sqrt{x^2 + y^2}\right) dS = \iint_{1 \leq x^2 + y^2 \leq 16} \left(x^2 + y^2\right) \left(10 - \sqrt{x^2 + y^2}\right) \sqrt{2}\, dA$

$= \int_0^{2\pi} \int_1^4 \sqrt{2}\,(10r^3 - r^4)\, dr\, d\theta = 2\sqrt{2}\pi \left(\frac{4329}{10}\right) = \frac{4329}{5} \sqrt{2}\pi$

39. $\rho(x,y,z) = 1200$, $\mathbf{V} = y\,\mathbf{i} + \mathbf{j} + z\,\mathbf{k}$, $\mathbf{F} = \rho\mathbf{V} = (1200)(y\,\mathbf{i} + \mathbf{j} + z\,\mathbf{k})$. S is given by
$\mathbf{r}(x,y) = x\,\mathbf{i} + y\,\mathbf{j} + \left[9 - \frac{1}{4}(x^2 + y^2)\right]\mathbf{k}$, $0 \le x^2 + y^2 \le 36$ and $\mathbf{r}_x \times \mathbf{r}_y = \frac{1}{2}x\,\mathbf{i} + \frac{1}{2}y\,\mathbf{j} + \mathbf{k}$. Thus the rate of
flow is given by

$$\iint_S \mathbf{F} \cdot d\mathbf{S} = \iint_{0 \le x^2 + y^2 \le 36} (1200)\left(\tfrac{1}{2}xy + \tfrac{1}{2}y + \left[9 - \tfrac{1}{4}(x^2 + y^2)\right]\right) dA$$

$$= 1200 \int_0^6 \int_0^{2\pi} \left[\tfrac{1}{2}r^2 \sin\theta\cos\theta + \tfrac{1}{2}r\sin\theta + 9 - \tfrac{1}{4}r^2\right] r\,d\theta\,dr$$

$$= 1200 \int_0^6 2\pi\left(9r - \tfrac{1}{4}r^3\right) dr = (1200)(2\pi)(81) = 194{,}400\pi$$

41. S consists of the hemisphere S_1 given by $z = \sqrt{a^2 - x^2 - y^2}$ and the disk S_2 given by
$0 \le x^2 + y^2 \le a^2$, $z = 0$. On S_1: $\mathbf{E} = a\sin\phi\cos\theta\,\mathbf{i} + a\sin\phi\sin\theta\,\mathbf{j} + 2a\cos\phi\,\mathbf{k}$,
$\mathbf{T}_\phi \times \mathbf{T}_\theta = a^2\sin^2\phi\cos\theta\,\mathbf{i} + a^2\sin^2\phi\sin\theta\,\mathbf{j} + a^2\sin\phi\cos\phi\,\mathbf{k}$. Thus

$$\iint_{S_1} \mathbf{E} \cdot d\mathbf{S} = \int_0^{2\pi}\int_0^{\pi/2} \left(a^3\sin^3\phi + 2a^3\sin\phi\cos^2\phi\right) d\phi\,d\theta$$

$$= \int_0^{2\pi}\int_0^{\pi/2} \left(a^3\sin\phi + a^3\sin\phi\cos^2\phi\right) d\phi\,d\theta = (2\pi)\,a^3\left(1 + \tfrac{1}{3}\right) = \tfrac{8}{3}\pi a^3$$

On S_2: $\mathbf{E} = x\,\mathbf{i} + y\,\mathbf{j}$, and $\mathbf{r}_y \times \mathbf{r}_x = -\mathbf{k}$ so $\iint_{S_2} \mathbf{E} \cdot d\mathbf{S} = 0$. Hence the total charge is
$q = \epsilon_0 \iint_S \mathbf{E} \cdot d\mathbf{S} = \tfrac{8}{3}\pi a^3 \epsilon_0$.

43. $K\nabla u = 6.5\,(4y\,\mathbf{j} + 4z\,\mathbf{k})$. S is given by $\mathbf{r}(x,\theta) = x\,\mathbf{i} + \sqrt{6}\cos\theta\,\mathbf{j} + \sqrt{6}\sin\theta\,\mathbf{k}$ and since we want the inward
heat flow, we use $\mathbf{r}_x \times \mathbf{r}_\theta = -\sqrt{6}\cos\theta\,\mathbf{j} - \sqrt{6}\sin\theta\,\mathbf{k}$. Then the rate of heat flow inward is given by
$\iint_S (-K\nabla u) \cdot d\mathbf{S} = \int_0^{2\pi}\int_0^4 -(6.5)(-24)\,dx\,d\theta = (2\pi)(156)(4) = 1248\pi$.

17.8 Stokes' Theorem   ET 16.8

1. Both H and P are oriented piecewise-smooth surfaces that are bounded by the simple, closed, smooth curve
$x^2 + y^2 = 4$, $z = 0$ (which we can take to be oriented positively for both surfaces). Then H and P satisfy the
hypotheses of Stokes' Theorem, so by (3) we know $\iint_H \operatorname{curl}\mathbf{F} \cdot d\mathbf{S} = \int_C \mathbf{F} \cdot d\mathbf{r} = \iint_P \operatorname{curl}\mathbf{F} \cdot d\mathbf{S}$ (where C is
the boundary curve).

3. The boundary curve C is the circle $x^2 + y^2 = 4$, $z = 0$ oriented in the counterclockwise direction. The vector
equation is $\mathbf{r}(t) = 2\cos t\,\mathbf{i} + 2\sin t\,\mathbf{j}$, $0 \le t \le 2\pi$, so $\mathbf{r}'(t) = -2\sin t\,\mathbf{i} + 2\cos t\,\mathbf{j}$ and
$\mathbf{F}(\mathbf{r}(t)) = (2\cos t)^2\, e^{(2\sin t)(0)}\,\mathbf{i} + (2\sin t)^2\, e^{(2\cos t)(0)}\,\mathbf{j} + (0)^2\, e^{(2\cos t)(2\sin t)}\,\mathbf{k} = 4\cos^2 t\,\mathbf{i} + 4\sin^2 t\,\mathbf{j}$. Then,
by Stokes' Theorem,

$$\iint_S \operatorname{curl}\mathbf{F} \cdot d\mathbf{S} = \int_C \mathbf{F} \cdot d\mathbf{r} = \int_0^{2\pi} \mathbf{F}(\mathbf{r}(t)) \cdot \mathbf{r}'(t)\,dt = \int_0^{2\pi} \left(-8\cos^2 t\sin t + 8\sin^2 t\cos t\right) dt$$

$$= 8\left[\tfrac{1}{3}\cos^3 t + \tfrac{1}{3}\sin^3 t\right]_0^{2\pi} = 0$$

5. C is the square in the plane $z = -1$. By (3), $\iint_{S_1} \operatorname{curl}\mathbf{F} \cdot d\mathbf{S} = \oint_C \mathbf{F} \cdot d\mathbf{r} = \iint_{S_2} \operatorname{curl}\mathbf{F} \cdot d\mathbf{S}$ where
S_1 is the original cube without the bottom and S_2 is the bottom face of the cube.
$\operatorname{curl}\mathbf{F} = x^2 z\,\mathbf{i} + (xy - 2xyz)\,\mathbf{j} + (y - xz)\,\mathbf{k}$. For S_2, we choose $\mathbf{n} = \mathbf{k}$ so that C has the same orientation for
both surfaces. Then $\operatorname{curl}\mathbf{F} \cdot \mathbf{n} = y - xz = x + y$ on S_2, where $z = -1$. Thus
$\iint_{S_2} \operatorname{curl}\mathbf{F} \cdot d\mathbf{S} = \int_{-1}^1 \int_{-1}^1 (x + y)\,dx\,dy = 0$ so $\iint_{S_1} \operatorname{curl}\mathbf{F} \cdot d\mathbf{S} = 0$.

7. curl $\mathbf{F} = -2z\,\mathbf{i} - 2x\,\mathbf{j} - 2y\,\mathbf{k}$ and we take the surface S to be the planar region enclosed by C, so S is the portion of the plane $x + y + z = 1$ over $D = \{(x, y) \mid 0 \le x \le 1, 0 \le y \le 1 - x\}$. Since C is oriented counterclockwise, we orient S upward. Using Equation 17.7.8 [ET 16.7.8], we have $z = g(x, y) = 1 - x - y$, $P = -2z$, $Q = -2x$, $R = -2y$, and

$$\int_C \mathbf{F} \cdot d\mathbf{r} = \iint_S \text{curl } \mathbf{F} \cdot d\mathbf{S} = \iint_D [-(-2z)(-1) - (-2x)(-1) + (-2y)]\, dA$$
$$= \int_0^1 \int_0^{1-x} (-2)\, dy\, dx = -2 \int_0^1 (1 - x)\, dx = -1$$

9. The curve of intersection is an ellipse in the plane $z = x + 4$ with unit normal $\mathbf{n} = \frac{1}{\sqrt{2}}(-\mathbf{i} + \mathbf{k})$ and

curl $\mathbf{F} = 5\mathbf{i} + 2\mathbf{j} + 4\mathbf{k}$ so curl $\mathbf{F} \cdot \mathbf{n} = -\frac{1}{\sqrt{2}}$. Then

$\oint_C \mathbf{F} \cdot d\mathbf{r} = -\iint_S \frac{1}{\sqrt{2}}\, dS = -\frac{1}{\sqrt{2}}$ (surface area of planar ellipse) $= -\frac{1}{\sqrt{2}}\pi\,(2)\,(2\sqrt{2}) = -4\pi$. (Recall that the area of an ellipse with semiaxes a and b is $\pi a b$.)

11. (a) The curve of intersection is an ellipse in the plane $x + y + z = 1$ with unit normal $\mathbf{n} = \frac{1}{\sqrt{3}}(\mathbf{i} + \mathbf{j} + \mathbf{k})$,

curl $\mathbf{F} = x^2\,\mathbf{j} + y^2\,\mathbf{k}$ and curl $\mathbf{F} \cdot \mathbf{n} = \frac{1}{\sqrt{3}}(x^2 + y^2)$. Then

$$\oint_C \mathbf{F} \cdot d\mathbf{r} = \iint_S \frac{1}{\sqrt{3}}(x^2 + y^2)\, dS = \iint_{x^2 + y^2 \le 9}(x^2 + y^2)\, dx\, dy$$
$$= \int_0^{2\pi} \int_0^3 r^3\, dr\, d\theta = 2\pi\left(\frac{81}{4}\right) = \frac{81\pi}{2}$$

(b)

(c) One possible parametrization is $x = 3\cos t$, $y = 3\sin t$, $z = 1 - 3\cos t - 3\sin t$, $0 \le t \le 2\pi$.

13. The boundary curve C is the circle $x^2 + y^2 = 9$, $z = 0$ oriented in the counterclockwise direction as viewed from $(0, 0, 1)$. Then $\mathbf{r}(t) = 3\cos t\,\mathbf{i} + 3\sin t\,\mathbf{j}$, $0 \le t \le 2\pi$, so $\mathbf{F}(\mathbf{r}(t)) = 9\sin t\,\mathbf{i} - 18\cos t\,\mathbf{k}$ and $\mathbf{F} \cdot \mathbf{r}'(t) = -27\sin^2 t$. Thus $\oint_C \mathbf{F} \cdot d\mathbf{r} = \int_0^{2\pi}(-27\sin^2 t)\, dt = -27\pi$. Now curl $\mathbf{F} = -4\mathbf{i} + 6\mathbf{j} - 3\mathbf{k}$, $\mathbf{r}_x \times \mathbf{r}_y = 2x\,\mathbf{i} + 2y\,\mathbf{j} + \mathbf{k}$, so

$$\iint_S \text{curl } \mathbf{F} \cdot d\mathbf{S} = \iint_{x^2 + y^2 \le 9}(-8x + 12y - 3)\, dA = \int_0^{2\pi}\int_0^3 (-8r\cos\theta + 12r\sin\theta - 3)\,r\, dr\, d\theta$$
$$= \int_0^3 (-3r)(2\pi)\, dr = -27\pi$$

15. The x-, y-, and z-intercepts of the plane are all 1, so C consists of the three line segments
C_1: $\mathbf{r}_1(t) = (1 - t)\mathbf{i} + t\mathbf{j}$, $0 \le t \le 1$, C_2: $\mathbf{r}_2(t) = (1 - t)\mathbf{j} + t\mathbf{k}$, $0 \le t \le 1$, and C_3: $\mathbf{r}_3(t) = t\mathbf{i} + (1 - t)\mathbf{k}$, $0 \le t \le 1$. Then

$$\oint_C \mathbf{F} \cdot d\mathbf{r} = \int_0^1 [t\mathbf{i} + (1 - t)\mathbf{k}] \cdot (-\mathbf{i} + \mathbf{j})\, dt + \int_0^1 [(1 - t)\mathbf{i} + t\mathbf{j}] \cdot (-\mathbf{j} + \mathbf{k})\, dt + \int_0^1 [(1 - t)\mathbf{j} + t\mathbf{k}] \cdot (\mathbf{i} - \mathbf{k})\, dt$$
$$= \int_0^1 (-3t)\, dt = -\frac{3}{2}$$

Now curl $\mathbf{F} = -\mathbf{i} - \mathbf{j} - \mathbf{k}$ and $\mathbf{r}_x \times \mathbf{r}_y = \mathbf{i} + \mathbf{j} + \mathbf{k}$. Hence $\iint_S \text{curl } \mathbf{F} \cdot d\mathbf{S} = \int_0^1\int_0^{1-x}(-3)\, dy\, dx = -\frac{3}{2}$.

17. $\operatorname{curl} \mathbf{F} = \begin{vmatrix} \mathbf{i} & \mathbf{j} & \mathbf{k} \\ \partial/\partial x & \partial/\partial y & \partial/\partial z \\ x^x + z^2 & y^y + x^2 & z^z + y^2 \end{vmatrix} = 2y\,\mathbf{i} + 2z\,\mathbf{j} + 2x\,\mathbf{k}$ and $W = \int_C \mathbf{F} \cdot d\mathbf{r} = \iint_S \operatorname{curl} \mathbf{F} \cdot d\mathbf{S}$.

To parametrize the surface, let $x = 2\cos\theta\sin\phi$, $y = 2\sin\theta\sin\phi$, $z = 2\cos\phi$, so that

$\mathbf{r}(\phi, \theta) = 2\sin\phi\cos\theta\,\mathbf{i} + 2\sin\phi\sin\theta\,\mathbf{j} + 2\cos\phi\,\mathbf{k}$, $0 \le \phi \le \frac{\pi}{2}$, $0 \le \theta \le \frac{\pi}{2}$, and

$\mathbf{r}_\phi \times \mathbf{r}_\theta = 4\sin^2\phi\cos\theta\,\mathbf{i} + 4\sin^2\phi\sin\theta\,\mathbf{j} + 4\sin\phi\cos\phi\,\mathbf{k}$. Then

$\operatorname{curl}\mathbf{F}(\mathbf{r}(\phi,\theta)) = 4\sin\phi\sin\theta\,\mathbf{i} + 4\cos\phi\,\mathbf{j} + 4\sin\phi\cos\theta\,\mathbf{k}$, and

$\operatorname{curl}\mathbf{F} \cdot (\mathbf{r}_\phi \times \mathbf{r}_\theta) = 16\sin^3\phi\sin\theta\cos\theta + 16\cos\phi\sin^2\phi\sin\theta + 16\sin^2\phi\cos\phi\cos\theta$. Therefore

$$\iint_S \operatorname{curl}\mathbf{F} \cdot d\mathbf{S} = \iint_D \operatorname{curl}\mathbf{F} \cdot (\mathbf{r}_\phi \times \mathbf{r}_\theta)\,dA$$

$$= 16\left[\int_0^{\pi/2}\sin\theta\cos\theta\,d\theta\right]\left[\int_0^{\pi/2}\sin^3\phi\,d\phi\right] + 16\left[\int_0^{\pi/2}\sin\theta\,d\theta\right]\left[\int_0^{\pi/2}\sin^2\phi\cos\phi\,d\phi\right]$$

$$+16\left[\int_0^{\pi/2}\cos\theta\,d\theta\right]\left[\int_0^{\pi/2}\sin^2\phi\cos\phi\,d\phi\right]$$

$$= 8\left[-\cos\phi + \tfrac{1}{3}\cos^3\phi\right]_0^{\pi/2} + 16\,(1)\left[\tfrac{1}{3}\sin^3\phi\right]_0^{\pi/2} + 16\,(1)\left[\tfrac{1}{3}\sin^3\phi\right]_0^{\pi/2}$$

$$= 8\left[0 + 1 + 0 - \tfrac{1}{3}\right] + 16\left(\tfrac{1}{3}\right) + 16\left(\tfrac{1}{3}\right) = \tfrac{16}{3} + \tfrac{16}{3} + \tfrac{16}{3} = 16$$

19. Assume S is centered at the origin with radius a and let H_1 and H_2 be the upper and lower hemispheres, respectively, of S. Then $\iint_S \operatorname{curl}\mathbf{F} \cdot d\mathbf{S} = \iint_{H_1} \operatorname{curl}\mathbf{F} \cdot d\mathbf{S} + \iint_{H_2} \operatorname{curl}\mathbf{F} \cdot d\mathbf{S} = \oint_{C_1}\mathbf{F}\cdot d\mathbf{r} + \oint_{C_2}\mathbf{F}\cdot d\mathbf{r}$ by Stokes' Theorem. But C_1 is the circle $x^2 + y^2 = a^2$ oriented in the counterclockwise direction while C_2 is the same circle oriented in the clockwise direction. Hence $\oint_{C_2}\mathbf{F}\cdot d\mathbf{r} = -\oint_{C_1}\mathbf{F}\cdot d\mathbf{r}$ so $\iint_S \operatorname{curl}\mathbf{F}\cdot d\mathbf{S} = 0$ as desired.

17.9 The Divergence Theorem   ET 16.9

1. The vectors that end near P_1 are longer than the vectors that start near P_1, so the net flow is inward near P_1 and $\operatorname{div}\mathbf{F}(P_1)$ is negative. The vectors that end near P_2 are shorter than the vectors that start near P_2, so the net flow is outward near P_2 and $\operatorname{div}\mathbf{F}(P_2)$ is positive.

3. $\operatorname{div}\mathbf{F} = 3 + x + 2x = 3 + 3x$, so

$\iiint_E \operatorname{div}\mathbf{F}\,dV = \int_0^1\int_0^1\int_0^1 (3x + 3)\,dx\,dy\,dz = \frac{9}{2}$ (notice the triple integral is three times the volume of the cube plus three times $\overline{x}$).

To compute $\iint_S \mathbf{F}\cdot d\mathbf{S}$, on S_1: $\mathbf{n} = \mathbf{i}$, $\mathbf{F} = 3\,\mathbf{i} + y\,\mathbf{j} + 2z\,\mathbf{k}$, and

$\iint_{S_1}\mathbf{F}\cdot d\mathbf{S} = \iint_{S_1} 3\,dS = 3$;

S_2: $\mathbf{F} = 3x\,\mathbf{i} + x\,\mathbf{j} + 2xz\,\mathbf{k}$, $\mathbf{n} = \mathbf{j}$ and $\iint_{S_2}\mathbf{F}\cdot d\mathbf{S} = \iint_{S_2} x\,dS = \frac{1}{2}$;

S_3: $\mathbf{F} = 3x\,\mathbf{i} + xy\,\mathbf{j} + 2x\,\mathbf{k}$, $\mathbf{n} = \mathbf{k}$ and $\iint_{S_3}\mathbf{F}\cdot d\mathbf{S} = \iint_{S_3} 2x\,dS = 1$;

S_4: $\mathbf{F} = 0$, $\iint_{S_4}\mathbf{F}\cdot d\mathbf{S} = 0$; S_5: $\mathbf{F} = 3x\,\mathbf{i} + 2x\,\mathbf{k}$, $\mathbf{n} = -\mathbf{j}$ and $\iint_{S_5}\mathbf{F}\cdot d\mathbf{S} = \iint_{S_5} 0\,dS = 0$;

S_6: $\mathbf{F} = 3x\,\mathbf{i} + xy\,\mathbf{j}$, $\mathbf{n} = -\mathbf{k}$ and $\iint_{S_6}\mathbf{F}\cdot d\mathbf{S} = \iint_{S_6} 0\,dS = 0$. Thus $\iint_S\mathbf{F}\cdot d\mathbf{S} = \frac{9}{2}$.

5. $\operatorname{div} \mathbf{F} = x + y + z$, so

$$\iiint_E \operatorname{div} \mathbf{F}\, dV = \int_0^{2\pi} \int_0^1 \int_0^1 (r\cos\theta + r\sin\theta + z)\, r\, dz\, dr\, d\theta = \int_0^{2\pi} \int_0^1 (r^2 \cos\theta + r^2 \sin\theta + \tfrac{1}{2} r)\, dr\, d\theta$$

$$= \int_0^{2\pi} \left(\tfrac{1}{3} \cos\theta + \tfrac{1}{3} \sin\theta + \tfrac{1}{4} \right) d\theta = \tfrac{1}{4}(2\pi) = \tfrac{\pi}{2}$$

Let S_1 be the top of the cylinder, S_2 the bottom, and S_3 the vertical edge.

On S_1, $z = 1$, $\mathbf{n} = \mathbf{k}$, and $\mathbf{F} = xy\,\mathbf{i} + y\,\mathbf{j} + x\,\mathbf{k}$, so

$$\iint_{S_1} \mathbf{F} \cdot d\mathbf{S} = \iint_{S_1} \mathbf{F} \cdot \mathbf{n}\, dS = \iint_{S_1} x\, dS = \int_0^{2\pi} \int_0^1 (r\cos\theta)\, r\, dr\, d\theta = [\sin\theta]_0^{2\pi} \left[\tfrac{1}{3} r^3\right]_0^1 = 0. \text{ On } S_2, z = 0,$$

$\mathbf{n} = -\mathbf{k}$, and $\mathbf{F} = xy\,\mathbf{i}$ so $\iint_{S_2} \mathbf{F} \cdot d\mathbf{S} = \iint_{S_2} 0\, dS = 0$. S_3 is given by $\mathbf{r}(\theta, z) = \cos\theta\,\mathbf{i} + \sin\theta\,\mathbf{j} + z\,\mathbf{k}$,

$0 \le \theta \le 2\pi$, $0 \le z \le 1$. Then $\mathbf{r}_\theta \times \mathbf{r}_z = \cos\theta\,\mathbf{i} + \sin\theta\,\mathbf{j}$ and

$$\iint_{S_3} \mathbf{F} \cdot d\mathbf{S} = \iint_D \mathbf{F} \cdot (\mathbf{r}_\theta \times \mathbf{r}_z)\, dA = \int_0^{2\pi} \int_0^1 (\cos^2\theta \sin\theta + z \sin^2\theta)\, dz\, d\theta$$

$$= \int_0^{2\pi} (\cos^2\theta \sin\theta + \tfrac{1}{2} \sin^2\theta)\, d\theta = \left[-\tfrac{1}{3} \cos^3\theta + \tfrac{1}{4}(\theta - \tfrac{1}{2} \sin 2\theta) \right]_0^{2\pi} = \tfrac{\pi}{2}$$

Thus $\iint_S \mathbf{F} \cdot d\mathbf{S} = 0 + 0 + \tfrac{\pi}{2} = \tfrac{\pi}{2}$.

7. $\operatorname{div} \mathbf{F} = \dfrac{\partial}{\partial x}(3y^2 z^3) + \dfrac{\partial}{\partial y}(9x^2 yz^2) + \dfrac{\partial}{\partial z}(4xy^2) = 9x^2 z^2$, so by the Divergence Theorem,

$$\iint_S \mathbf{F} \cdot d\mathbf{S} = \iiint_E 9x^2 z^2\, dV = \int_{-1}^1 \int_{-1}^1 \int_{-1}^1 9x^2 z^2\, dx\, dy\, dz = 8.$$

9. $\operatorname{div} \mathbf{F} = \dfrac{\partial}{\partial x}(-xz) + \dfrac{\partial}{\partial y}(-yz) + \dfrac{\partial}{\partial z}(z^2) = -z - z + 2z = 0$,

so $\iint_S \mathbf{F} \cdot d\mathbf{S} = \iiint_E \operatorname{div} \mathbf{F}\, dV = \iiint_E 0\, dV = 0.$

11. $\operatorname{div} \mathbf{F} = 3y^2 + 0 + 3z^2$, so using cylindrical coordinates with $y = r\cos\theta$, $z = r\sin\theta$, $x = x$ we have

$$\iint_S \mathbf{F} \cdot d\mathbf{S} = \iiint_E (3y^2 + 3z^2)\, dV = \int_0^{2\pi} \int_0^1 \int_{-1}^2 (3r^2 \cos^2\theta + 3r^2 \sin^2\theta)\, r\, dx\, dr\, d\theta$$

$$= 3 \int_0^{2\pi} d\theta \int_0^1 r^3\, dr \int_{-1}^2 dx = 3(2\pi)\left(\tfrac{1}{4}\right)(3) = \tfrac{9\pi}{2}$$

13. $\iint_S \mathbf{F} \cdot d\mathbf{S} = \iiint_E 3(x^2 + y^2 + z^2)\, dV = \int_0^{2\pi} \int_0^\pi \int_0^1 3\rho^4 \sin\phi\, d\rho\, d\phi\, d\theta = 2\pi \int_0^\pi \tfrac{3}{5} \sin\phi\, d\phi = \tfrac{12}{5}\pi$

15. $\iint_S \mathbf{F} \cdot d\mathbf{S} = \iiint_E 2y\, dV = \iint_{x^2 + y^2 \le 9} \int_{y-3}^0 2y\, dz\, dA = \int_0^{2\pi} \int_0^3 \int_{-3 + r\sin\theta}^0 (2r^2 \sin\theta)\, dz\, dr\, d\theta$

$$= \int_0^{2\pi} \int_0^3 (6r^2 \sin\theta - 2r^3 \sin^2\theta)\, dr\, d\theta = \int_0^{2\pi} \left[54 \sin\theta - \tfrac{81}{2} \sin^2\theta \right] d\theta = -\tfrac{81}{2}\pi$$

17. $\iint_S \mathbf{F} \cdot d\mathbf{S} = \iiint_E \sqrt{3 - x^2}\, dV = \int_{-1}^1 \int_{-1}^1 \int_0^{2 - x^4 - y^4} \sqrt{3 - x^2}\, dz\, dy\, dx = \tfrac{341}{60}\sqrt{2} + \tfrac{81}{20} \sin^{-1}\left(\tfrac{\sqrt{3}}{3}\right)$

19. For S_1 we have $\mathbf{n} = -\mathbf{k}$, so $\mathbf{F} \cdot \mathbf{n} = \mathbf{F} \cdot (-\mathbf{k}) = -x^2 z - y^2 = -y^2$ (since $z = 0$ on S_1). So if D is the unit disk,

we get $\iint_{S_1} \mathbf{F} \cdot d\mathbf{S} = \iint_{S_1} \mathbf{F} \cdot \mathbf{n}\, dS = \iint_D (-y^2)\, dA = -\int_0^{2\pi} \int_0^1 r^2 \sin^2\theta\, r\, dr\, d\theta = -\tfrac{1}{4}\pi$. Now

since S_2 is closed, we can use the Divergence Theorem. Since

$\operatorname{div} \mathbf{F} = \dfrac{\partial}{\partial x}(z^2 x) + \dfrac{\partial}{\partial y}(\tfrac{1}{3} y^3 + \tan z) + \dfrac{\partial}{\partial z}(x^2 z + y^2) = z^2 + y^2 + x^2$, we use spherical coordinates to get

$\iint_{S_2} \mathbf{F} \cdot d\mathbf{S} = \iiint_E \operatorname{div} \mathbf{F}\, dV = \int_0^{2\pi} \int_0^{\pi/2} \int_0^1 \rho^2 \cdot \rho^2 \sin\phi\, d\rho\, d\phi\, d\theta = \tfrac{2}{5}\pi$. Finally

$\iint_S \mathbf{F} \cdot d\mathbf{S} = \iint_{S_2} \mathbf{F} \cdot d\mathbf{S} - \iint_{S_1} \mathbf{F} \cdot d\mathbf{S} = \tfrac{2}{5}\pi - (-\tfrac{1}{4}\pi) = \tfrac{13}{20}\pi$.

21. Since $\dfrac{\mathbf{x}}{|\mathbf{x}|^3} = \dfrac{x\,\mathbf{i} + y\,\mathbf{j} + z\,\mathbf{k}}{(x^2 + y^2 + z^2)^{3/2}}$ and $\dfrac{\partial}{\partial x}\left(\dfrac{x}{(x^2 + y^2 + z^2)^{3/2}}\right) = \dfrac{(x^2 + y^2 + z^2) - 3x^2}{(x^2 + y^2 + z^2)^{5/2}}$ with similar

expressions for $\dfrac{\partial}{\partial y}\left(\dfrac{y}{(x^2 + y^2 + z^2)^{3/2}}\right)$ and $\dfrac{\partial}{\partial z}\left(\dfrac{z}{(x^2 + y^2 + z^2)^{3/2}}\right)$, we have

$$\operatorname{div}\left(\dfrac{\mathbf{x}}{|\mathbf{x}|^3}\right) = \dfrac{3\left(x^2 + y^2 + z^2\right) - 3\left(x^2 + y^2 + z^2\right)}{(x^2 + y^2 + z^2)^{5/2}} = 0, \text{ except at } (0,0,0) \text{ where it is undefined.}$$

23. $\iint_S \mathbf{a} \cdot \mathbf{n}\,dS = \iiint_E \operatorname{div} \mathbf{a}\,dV = 0$ since $\operatorname{div} \mathbf{a} = 0$.

25. $\iint_S \operatorname{curl} \mathbf{F} \cdot d\mathbf{S} = \iiint_E \operatorname{div}(\operatorname{curl} \mathbf{F})\,dV = 0$ by Theorem 17.5.11 [ET 16.5.11].

27. $\iint_S (f\nabla g) \cdot \mathbf{n}\,dS = \iiint_E \operatorname{div}(f\nabla g)\,dV = \iiint_E (f\nabla^2 g + \nabla g \cdot \nabla f)\,dV$ by Exercise 17.5.25 [ET 16.5.25].

29. If $\mathbf{c} = c_1\,\mathbf{i} + c_2\,\mathbf{j} + c_3\,\mathbf{k}$ is an arbitrary constant vector, we define $\mathbf{F} = f\mathbf{c} = fc_1\,\mathbf{i} + fc_2\,\mathbf{j} + fc_3\,\mathbf{k}$. Then

$$\operatorname{div} \mathbf{F} = \operatorname{div} f\mathbf{c} = \frac{\partial f}{\partial x}c_1 + \frac{\partial f}{\partial y}c_2 + \frac{\partial f}{\partial z}c_3 = \nabla f \cdot \mathbf{c} \text{ and the divergence theorem says}$$

$$\iint_S \mathbf{F} \cdot d\mathbf{S} = \iiint_E \operatorname{div} \mathbf{F}\,dV \quad \Rightarrow \quad \iint_S \mathbf{F} \cdot \mathbf{n}\,dS = \iiint_E \nabla f \cdot \mathbf{c}\,dV. \text{ In particular, if } \mathbf{c} = \mathbf{i} \text{ then}$$

$$\iint_S f\mathbf{i} \cdot \mathbf{n}\,dS = \iiint_E \nabla f \cdot \mathbf{i}\,dV \quad \Rightarrow \quad \iint_S fn_1\,dS = \iiint_E \frac{\partial f}{\partial x}\,dV \text{ (where } \mathbf{n} = n_1\,\mathbf{i} + n_2\,\mathbf{j} + n_3\,\mathbf{k}).$$

Similarly, if $\mathbf{c} = \mathbf{j}$ we have $\displaystyle\iint_S fn_2\,dS = \iiint_E \frac{\partial f}{\partial y}\,dV$, and $\mathbf{c} = \mathbf{k}$ gives $\displaystyle\iint_S fn_3\,dS = \iiint_E \frac{\partial f}{\partial z}\,dV$. Then

$$\iint_S f\mathbf{n}\,dS = \left(\iint_S fn_1\,dS\right)\mathbf{i} + \left(\iint_S fn_2\,dS\right)\mathbf{j} + \left(\iint_S fn_3\,dS\right)\mathbf{k}$$

$$= \left(\iiint_E \frac{\partial f}{\partial x}\,dV\right)\mathbf{i} + \left(\iiint_E \frac{\partial f}{\partial y}\,dV\right)\mathbf{j} + \left(\iiint_E \frac{\partial f}{\partial z}\,dV\right)\mathbf{k}$$

$$= \iiint_E \left(\frac{\partial f}{\partial x}\,\mathbf{i} + \frac{\partial f}{\partial y}\,\mathbf{j} + \frac{\partial f}{\partial z}\,\mathbf{k}\right)dV = \iiint_E \nabla f\,dV$$

as desired.

≡17 Review   ET 16

CONCEPT CHECK

1. See Definitions 1 and 2 in Section 17.1 [ET 16.1]. A vector field can represent, for example, the wind velocity at any location in space, the speed and direction of the ocean current at any location, or the force vectors of Earth's gravitational field at a location in space.

2. (a) A conservative vector field $\mathbf{F}$ is a vector field which is the gradient of some scalar function f.

(b) The function f in part (a) is called a potential function for $\mathbf{F}$, that is, $\mathbf{F} = \nabla f$.

3. (a) See Definition 17.2.2 [ET 16.2.2].

(b) We normally evaluate the line integral using Formula 17.2.3 [ET 16.2.3].

(c) The mass is $m = \int_C \rho(x, y)\,ds$, and the center of mass is $(\bar{x}, \bar{y})$ where $\bar{x} = \frac{1}{m}\int_C x\rho(x, y)\,ds$, $\bar{y} = \frac{1}{m}\int_C y\rho(x, y)\,ds$.

(d) See (5) and (6) in Section 17.2 [ET 16.2] for plane curves; we have similar definitions when C is a space curve (see the equation preceding (10) in Section 17.2 [ET 16.2]).

(e) For plane curves, see Equations 17.2.7 [ET 16.2.7]. We have similar results for space curves (see the equation preceding (10) in Section 17.2 [ET 16.2]).

4. (a) See Definition 17.2.13 [ET 16.2.13].

(b) If $\mathbf{F}$ is a force field, $\int_C \mathbf{F} \cdot d\mathbf{r}$ represents the work done by $\mathbf{F}$ in moving a particle along the curve C.

(c) $\int_C \mathbf{F} \cdot d\mathbf{r} = \int_C P\,dx + Q\,dy + R\,dz$

5. See Theorem 17.3.2 [ET 16.3.2].

6. (a) $\int_C \mathbf{F} \cdot d\mathbf{r}$ is independent of path if the line integral has the same value for any two curves that have the same initial and terminal points.

(b) See Theorem 17.3.4 [ET 16.3.4].

7. See the statement of Green's Theorem on page 1102 [ET 1068].

8. See Equations 17.4.5 [ET 16.4.5].

9. (a) $\operatorname{curl}\mathbf{F} = \left(\dfrac{\partial R}{\partial y} - \dfrac{\partial Q}{\partial z}\right)\mathbf{i} + \left(\dfrac{\partial P}{\partial z} - \dfrac{\partial R}{\partial x}\right)\mathbf{j} + \left(\dfrac{\partial Q}{\partial x} - \dfrac{\partial P}{\partial y}\right)\mathbf{k} = \nabla \times \mathbf{F}$

(b) $\operatorname{div}\mathbf{F} = \dfrac{\partial P}{\partial x} + \dfrac{\partial Q}{\partial y} + \dfrac{\partial R}{\partial z} = \nabla \cdot \mathbf{F}$

(c) For curl $\mathbf{F}$, see the discussion accompanying Figure 1 on page 1112 [ET 1078] as well as Figure 6 and the accompanying discussion on page 1142 [ET 1108]. For div $\mathbf{F}$, see the discussion following Example 5 on page 1113 [ET 1079] as well as the discussion preceding (8) on page 1149 [ET 1115].

10. See Theorem 17.3.6 [ET 16.3.6]; see Theorem 17.5.4 [ET 16.5.4].

11. (a) See (1) and (2) and the accompanying discussion in Section 17.6 [ET 16.6]; See Figure 4 and the accompanying discussion on page 1118 [ET 1084].

(b) See Definition 17.6.6 [ET 16.6.6].

(c) See Equation 17.6.9 [ET 16.6.9].

12. (a) See (1) in Section 17.7 [ET 16.7].

(b) We normally evaluate the surface integral using Formula 17.7.3 [ET 16.7.3].

(c) See Formula 17.7.2 [ET 16.7.2].

(d) The mass is $m = \iint_S \rho\,(x, y, z)\ dS$ and the center of mass is $(\bar{x}, \bar{y}, \bar{z})$ where $\bar{x} = \frac{1}{m} \iint_S x\rho\,(x, y, z)\ dS$, $\bar{y} = \frac{1}{m} \iint_S y\rho\,(x, y, z)\ dS$, $\bar{z} = \frac{1}{m} \iint_S z\rho\,(x, y, z)\ dS$.

13. (a) See Figures 7 and 8 and the accompanying discussion in Section 17.7 [ET 16.7]. A Möbius strip is a nonorientable surface; see Figures 5 and 6 and the accompanying discussion on page 1131 [ET 1097].

(b) See Definition 17.7.7 [ET 16.7.7].

(c) See Formula 17.7.9 [ET 16.7.9].

(d) See Formula 17.7.8 [ET 16.7.8].

14. See the statement of Stokes' Theorem on page 1139 [ET 1105].

15. See the statement of the Divergence Theorem on page 1145 [ET 1111].

16. In each theorem, we have an integral of a "derivative" over a region on the left side, while the right side involves the values of the original function only on the boundary of the region.

TRUE-FALSE QUIZ

1. False; div $\mathbf{F}$ is a scalar field.

3. True, by Theorem 17.5.3 [ET 16.5.3] and the fact that div $\mathbf{0} = 0$.

5. False. See Exercise 17.3.33 [ET 16.3.33]. (But the assertion is true if D is simply-connected; see Theorem 17.3.6 [ET 16.3.6].)

7. True. Apply the Divergence Theorem and use the fact that div $\mathbf{F} = 0$.

EXERCISES

1. (a) Vectors starting on C point in roughly the direction opposite to C, so the tangential component $\mathbf{F} \cdot \mathbf{T}$ is negative. Thus $\int_C \mathbf{F} \cdot d\mathbf{r} = \int_C \mathbf{F} \cdot \mathbf{T} \, ds$ is negative.

(b) The vectors that end near P are shorter than the vectors that start near P, so the net flow is outward near P and div $\mathbf{F}(P)$ is positive.

3. $\int_C x^3 z \, ds = \int_0^{\pi/2} (2 \sin t)^3 (2 \cos t) \sqrt{(2 \cos t)^2 + (1)^2 + (-2 \sin t)^2} \, dt = \int_0^{\pi/2} (16 \sin^3 t \cos t) \sqrt{5} \, dt$

$$= 4\sqrt{5} \sin^4 t \big]_0^{\pi/2} = 4\sqrt{5}$$

5. $x = \cos t \Rightarrow dx = -\sin t \, dt$, $y = \sin t \Rightarrow dy = \cos t \, dt$, $0 \le t \le 2\pi$ and

$\int_C x^3 y \, dx - x \, dy = \int_0^{2\pi} (-\cos^3 t \sin^2 t - \cos^2 t) \, dt = \int_0^{2\pi} (-\cos^3 t \sin^2 t - \cos^2 t) \, dt = -\pi$

Or: Since C is a simple closed curve, apply Green's Theorem giving

$\iint_{x^2 + y^2 \le 1} (-1 - x^3) \, dA = \int_0^1 \int_0^{2\pi} (-r - r^4 \cos^3 \theta) \, d\theta = -\pi.$

7.

C_1: $x = t$, $y = t$, $z = 2t$, $0 \le t \le 1$;

C_2: $x = 1 + 2t$, $y = 1$, $z = 2 + 2t$, $0 \le t \le 1$.

Then $\int_C y \, dx + z \, dy + x \, dz = \int_0^1 5t \, dt + \int_0^1 (4 + 4t) \, dt = \frac{17}{2}$.

9. $\mathbf{F}(\mathbf{r}(t)) = (2t + t^2) \mathbf{i} + t^4 \mathbf{j} + 4t^4 \mathbf{k}$, $\mathbf{F} \cdot \mathbf{r}'(t) = 4t + 2t^2 + 2t^5 + 16t^7$ and

$\int_C \mathbf{F} \cdot d\mathbf{r} = \int_0^1 (4t + 2t^2 + 2t^5 + 16t^7) \, dt = 5$.

11. $\dfrac{\partial (\sin y)}{\partial y} = \cos y = \dfrac{\partial (x \cos y + \sin y)}{\partial x}$ and the domain of $\mathbf{F}$ is $\mathbb{R}^2$, so $\mathbf{F}$ is conservative. Hence there exists a

function f such that $\nabla f = \mathbf{F}$. Then $f_x(x, y) = \sin y$ implies $f(x, y) = x \sin y + g(y)$ and $f_y(x, y) = x \cos y + g'(y)$. But $f_y(x, y) = x \cos y + \sin y$, so $g'(y) = \sin y$ and $f(x, y) = x \sin y - \cos y + K$ is a potential function for $\mathbf{F}$.

13. Since $\dfrac{\partial (2x + y^2 + 3x^2 y)}{\partial y} = 2y + 3x^2 = \dfrac{\partial (2xy + x^3 + 3y^2)}{\partial x}$ and the domain of $\mathbf{F}$ is $\mathbb{R}^2$, $\mathbf{F}$ is conservative.

Furthermore $f(x, y) = x^2 + xy^2 + x^3 y + y^3 + K$ is a potential function for $\mathbf{F}$. Then

$\int_C \mathbf{F} \cdot d\mathbf{r} = f(\pi, 0) - f(0, 0) = \pi^2$.

15.

$C_1: 0 \le x \le 1, y = 0; C_2: x = 1, 0 \le y \le 2; C_3: x = x, y = 2x, x = 1 \text{ to } x = 0.$

Then $\int_C xy\, dx + x^2\, dy = \int_0^1 0\, dx + \int_0^2 (0 + 1)\, dy + \int_1^0 (2x^2 + 2x^2)\, dx = \frac{2}{3}.$ Using

Green's Theorem, we have

$$\int_C xy\, dx + x^2\, dy = \iint_D \left[\frac{\partial}{\partial x}(x^2) - \frac{\partial}{\partial y}(xy) \right] dA = \iint_D (2x - x)\, dA$$

$$= \int_0^1 \int_0^{2x} x\, dy\, dx = \frac{2}{3}$$

17. $\int_C x^2 y\, dx - xy^2\, dy = \iint_{x^2 + y^2 \le 4} \left[\frac{\partial}{\partial x}(-xy^2) - \frac{\partial}{\partial y}(x^2 y) \right] dA$

$$= \iint_{x^2 + y^2 \le 4} (-y^2 - x^2)\, dA = - \int_0^{2\pi} \int_0^2 r^3\, dr\, d\theta = -8\pi$$

19. If we assume there is such a vector field **G**, then div (curl **G**) $= 2 + 3z - 2xz$. But div (curl **F**) $= 0$ for all vector fields **F**. Thus such a **G** cannot exist.

21. For any piecewise-smooth simple closed plane curve C bounding a region D, we can apply Green's Theorem to

$$\mathbf{F}(x, y) = f(x)\,\mathbf{i} + g(y)\,\mathbf{j} \text{ to get } \int_C f(x)\, dx + g(y)\, dy = \iint_D \left[\frac{\partial}{\partial x} g(y) - \frac{\partial}{\partial y} f(x) \right] dA = \iint_D 0\, dA = 0.$$

23. $\nabla^2 f = 0$ means that $\dfrac{\partial^2 f}{\partial x^2} + \dfrac{\partial^2 f}{\partial y^2} = 0$. Now if $\mathbf{F} = f_y\,\mathbf{i} - f_x\,\mathbf{j}$ and C is any closed path in D, then applying

Green's Theorem, we get

$$\int_C \mathbf{F} \cdot d\mathbf{r} = \int_C f_y\, dx - f_x\, dy = \iint_D \left[\frac{\partial}{\partial x}(-f_x) - \frac{\partial}{\partial y}(f_y) \right] dA = - \iint_D (f_{xx} + f_{yy})\, dA$$

$$= - \iint_D 0\, dA = 0$$

Therefore the line integral is independent of path, by Theorem 17.3.3 [ET 16.3.3].

25. $z = f(x, y) = x^2 + 2y$ with $0 \le x \le 1, 0 \le y \le 2x$. Thus

$A(S) = \iint_D \sqrt{1 + 4x^2 + 4}\, dA = \int_0^1 \int_0^{2x} \sqrt{5 + 4x^2}\, dy\, dx = \int_0^1 2x\sqrt{5 + 4x^2}\, dx$

$= \frac{1}{6}(5 + 4x^2)^{3/2} \Big]_0^1 = \frac{1}{6}(27 - 5\sqrt{5}).$

27. $z = f(x, y) = x^2 + y^2$ with $0 \le x^2 + y^2 \le 4$ so $\mathbf{r}_x \times \mathbf{r}_y = -2x\,\mathbf{i} - 2y\,\mathbf{j} + \mathbf{k}$ (using upward orientation). Then

$$\iint_S z\, dS = \iint_{x^2 + y^2 \le 4} (x^2 + y^2)\sqrt{4x^2 + 4y^2 + 1}\, dA = \int_0^{2\pi} \int_0^2 r^3 \sqrt{1 + 4r^2}\, dr\, d\theta$$

$$= \frac{1}{60}\pi\left(391\sqrt{17} + 1\right)$$

(Substitute $u = 1 + 4r^2$ and use tables.)

29. Since the sphere bounds a simple solid region, the Divergence Theorem applies and

$\iint_S \mathbf{F} \cdot d\mathbf{S} = \iiint_E (z - 2)\, dV = \iiint_E z\, dV - 2 \iiint_E dV = m\bar{z} - 2\left(\frac{4}{3}\pi 2^3\right) = -\frac{64}{3}\pi.$

Alternate Solution: $\mathbf{F}(\mathbf{r}(\phi, \theta)) = 4 \sin \phi \cos \theta \cos \phi\,\mathbf{i} - 4 \sin \phi \sin \theta\,\mathbf{j} + 6 \sin \phi \cos \theta\,\mathbf{k}$,

$\mathbf{r}_\phi \times \mathbf{r}_\theta = 4 \sin^2 \phi \cos \theta\,\mathbf{i} + 4 \sin^2 \phi \sin \theta\,\mathbf{j} + 4 \sin \phi \cos \phi\,\mathbf{k}$, and

$\mathbf{F} \cdot (\mathbf{r}_\phi \times \mathbf{r}_\theta) = 16 \sin^3 \phi \cos^2 \theta \cos \phi - 16 \sin^3 \phi \sin^2 \theta + 24 \sin^2 \phi \cos \phi \cos \theta$. Then

$$\iint_S \mathbf{F} \cdot d\mathbf{S} = \int_0^{2\pi} \int_0^\pi (16 \sin^3 \phi \cos \phi \cos^2 \theta - 16 \sin^3 \phi \sin^2 \theta + 24 \sin^2 \phi \cos \phi \cos \theta)\, d\phi\, d\theta$$

$$= \int_0^{2\pi} \frac{4}{3}(-16 \sin^2 \theta)\, d\theta = -\frac{64}{3}\pi$$

31. Since curl $\mathbf{F} = \mathbf{0}$, $\iint_S (\text{curl}\, \mathbf{F}) \cdot d\mathbf{S} = 0$. We parametrize C: $\mathbf{r}(t) = \cos t\, \mathbf{i} + \sin t\, \mathbf{j}$, $0 \le t \le 2\pi$ and
$\oint_C \mathbf{F} \cdot d\mathbf{r} = \int_0^{2\pi} \left(-\cos^2 t \sin t + \sin^2 t \cos t\right) dt = \frac{1}{3}\cos^3 t + \frac{1}{3}\sin^3 t \big]_0^{2\pi} = 0$.

33. The surface is given by $x + y + z = 1$ or $z = 1 - x - y$, $0 \le x \le 1$, $0 \le y \le 1 - x$ and $\mathbf{r}_x \times \mathbf{r}_y = \mathbf{i} + \mathbf{j} + \mathbf{k}$. Then

$$\oint_C \mathbf{F} \cdot d\mathbf{r} = \iint_S \text{curl}\, \mathbf{F} \cdot d\mathbf{S} = \iint_D (-y\,\mathbf{i} - z\,\mathbf{j} - x\,\mathbf{k}) \cdot (\mathbf{i} + \mathbf{j} + \mathbf{k})\, dA$$
$$= \iint_D (-1)\, dA = -(\text{area of } D) = -\frac{1}{2}$$

35. $\iiint_E \text{div}\, \mathbf{F}\, dV = \iiint_{x^2 + y^2 + z^2 \le 1} 3\, dV = 3\,(\text{volume of sphere}) = 4\pi$. Then
$\mathbf{F}\,(\mathbf{r}\,(\phi, \theta)) \cdot (\mathbf{r}_\phi \times \mathbf{r}_\theta) = \sin^3 \phi \cos^2 \theta + \sin^3 \phi \sin^2 \theta + \sin \phi \cos^2 \phi = \sin \phi$ and
$\iint_S \mathbf{F} \cdot d\mathbf{S} = \int_0^{2\pi} \int_0^\pi \sin \phi\, d\phi\, d\theta = (2\pi)(2) = 4\pi$.

37. Because curl $\mathbf{F} = \mathbf{0}$, $\mathbf{F}$ is conservative, and if $f(x, y, z) = x^3 yz - 3xy + z^2$, then $\nabla f = \mathbf{F}$. Hence
$\int_C \mathbf{F} \cdot d\mathbf{r} = \int_C \nabla f \cdot d\mathbf{r} = f(0, 3, 0) - f(0, 0, 2) = 0 - 4 = -4$.

39. By the Divergence Theorem, $\iint_S \mathbf{F} \cdot \mathbf{n}\, dS = \iiint_E \text{div}\, \mathbf{F}\, dV = 3\,(\text{volume of } E) = 3\,(8 - 1) = 21$.

Problems Plus

1. Let S_1 be the portion of $\Omega(S)$ between $S(a)$ and S, and let ∂S_1 be its boundary. Also let S_L be the lateral surface of S_1 [that is, the surface of S_1 except S and $S(a)$]. Applying the Divergence Theorem we have

$$\iint_{\partial S_1} \frac{\mathbf{r} \cdot \mathbf{n}}{r^3} \, dS = \iiint_{S_1} \nabla \cdot \frac{\mathbf{r}}{r^3} \, dV. \text{ But}$$

$$\nabla \cdot \frac{\mathbf{r}}{r^3} = \left\langle \frac{\partial}{\partial x}, \frac{\partial}{\partial y}, \frac{\partial}{\partial z} \right\rangle \cdot \left\langle \frac{x}{(x^2 + y^2 + z^2)^{3/2}}, \frac{y}{(x^2 + y^2 + z^2)^{3/2}}, \frac{z}{(x^2 + y^2 + z^2)^{3/2}} \right\rangle$$

$$= \frac{(x^2 + y^2 + z^2 - 3x^2) + (x^2 + y^2 + z^2 - 3y^2) + (x^2 + y^2 + z^2 - 3z^2)}{(x^2 + y^2 + z^2)^{5/2}} = 0$$

$$\Rightarrow \quad \iint_{\partial S_1} \frac{\mathbf{r} \cdot \mathbf{n}}{r^3} \, dS = \iiint_{S_1} 0 \, dV = 0. \text{ On the other hand, notice that for the surfaces of } \partial S_1 \text{ other than } S(a)$$

and S, $\mathbf{r} \cdot \mathbf{n} = 0 \quad \Rightarrow$

$$0 = \iint_{\partial S_1} \frac{\mathbf{r} \cdot \mathbf{n}}{r^3} \, dS = \iint_{S} \frac{\mathbf{r} \cdot \mathbf{n}}{r^3} \, dS + \iint_{S(a)} \frac{\mathbf{r} \cdot \mathbf{n}}{r^3} \, dS + \iint_{S_L} \frac{\mathbf{r} \cdot \mathbf{n}}{r^3} \, dS$$

$$= \iint_{S} \frac{\mathbf{r} \cdot \mathbf{n}}{r^3} \, dS + \iint_{S(a)} \frac{\mathbf{r} \cdot \mathbf{n}}{r^3} \, dS$$

$$\Rightarrow \quad \iint_{S} \frac{\mathbf{r} \cdot \mathbf{n}}{r^3} \, dS = -\iint_{S(a)} \frac{\mathbf{r} \cdot \mathbf{n}}{r^3} \, dS. \text{ Notice that on } S(a), r = a \quad \Rightarrow \quad \mathbf{n} = -\frac{\mathbf{r}}{r} = -\frac{\mathbf{r}}{a} \text{ and } \mathbf{r} \cdot \mathbf{r} = r^2 = a^2,$$

so that $-\iint_{S(a)} \frac{\mathbf{r} \cdot \mathbf{n}}{r^3} \, dS = \iint_{S(a)} \frac{\mathbf{r} \cdot \mathbf{r}}{a^4} \, dS = \iint_{S(a)} \frac{a^2}{a^4} \, dS = \frac{1}{a^2} \iint_{S(a)} dS = \frac{\text{area of } S(a)}{a^2} = |\Omega(S)|.$

Therefore $|\Omega(S)| = \iint_{S} \frac{\mathbf{r} \cdot \mathbf{n}}{r^3} \, dS.$

3. The given line integral $\frac{1}{2} \int_C (bz - cy) \, dx + (cx - az) \, dy + (ay - bx) \, dz$ can be expressed as $\int_C \mathbf{F} \cdot d\mathbf{r}$ if we define the vector field $\mathbf{F}$ by $\mathbf{F}(x, y, z) = P\mathbf{i} + Q\mathbf{j} + R\mathbf{k} = \frac{1}{2}(bz - cy)\mathbf{i} + \frac{1}{2}(cx - az)\mathbf{j} + \frac{1}{2}(ay - bx)\mathbf{k}$. Then define S to be the planar interior of C, so S is an oriented, smooth surface. Stokes' Theorem says $\int_C \mathbf{F} \cdot d\mathbf{r} = \iint_S \text{curl } \mathbf{F} \cdot d\mathbf{S} = \iint_S \text{curl } \mathbf{F} \cdot \mathbf{n} \, dS$. Now

$$\text{curl } \mathbf{F} = \left(\frac{\partial R}{\partial y} - \frac{\partial Q}{\partial z} \right) \mathbf{i} + \left(\frac{\partial P}{\partial z} - \frac{\partial R}{\partial x} \right) \mathbf{j} + \left(\frac{\partial Q}{\partial x} - \frac{\partial P}{\partial y} \right) \mathbf{k}$$

$$= \left(\tfrac{1}{2}a + \tfrac{1}{2}a \right)\mathbf{i} + \left(\tfrac{1}{2}b + \tfrac{1}{2}b \right)\mathbf{j} + \left(\tfrac{1}{2}c + \tfrac{1}{2}c \right)\mathbf{k} = a\mathbf{i} + b\mathbf{j} + c\mathbf{k} = \mathbf{n}$$

so curl $\mathbf{F} \cdot \mathbf{n} = \mathbf{n} \cdot \mathbf{n} = |\mathbf{n}|^2 = 1$, hence $\iint_S \text{curl } \mathbf{F} \cdot \mathbf{n} \, dS = \iint_S dS$ which is simply the surface area of S. Thus, $\int_C \mathbf{F} \cdot d\mathbf{r} = \frac{1}{2} \int_C (bz - cy) \, dx + (cx - az) \, dy + (ay - bx) \, dz$ is the plane area enclosed by C.

18

Second-Order
Differential Equations

18.1 Second-Order Linear Equations

1. The auxiliary equation is $r^2 - 6r + 8 = 0 \;\Rightarrow\; (r - 4)(r - 2) = 0 \;\Rightarrow\; r = 4, r = 2$. Then by (8) the general solution is $y = c_1 e^{4x} + c_2 e^{2x}$.

3. The auxiliary equation is $r^2 + 8r + 41 = 0 \;\Rightarrow\; r = -4 \pm 5i$. Then by (11) the general solution is $y = e^{-4x}(c_1 \cos 5x + c_2 \sin 5x)$.

5. The auxiliary equation is $r^2 - 2r + 1 = (r - 1)^2 = 0 \;\Rightarrow\; r = 1$. Then by (10), the general solution is $y = c_1 e^x + c_2 x e^x$.

7. The auxiliary equation is $4r^2 + 1 = 0 \;\Rightarrow\; r = \pm \frac{1}{2}i$, so $y = c_1 \cos\left(\frac{1}{2}x\right) + c_2 \sin\left(\frac{1}{2}x\right)$.

9. The auxiliary equation is $4r^2 + r = r(4r + 1) = 0 \;\Rightarrow\; r = 0, r = -\frac{1}{4}$, so $y = c_1 + c_2 e^{-x/4}$.

11. The auxiliary equation is $r^2 - 2r - 1 = 0 \;\Rightarrow\; r = 1 \pm \sqrt{2}$, so $y = c_1 e^{(1+\sqrt{2})t} + c_2 e^{(1-\sqrt{2})t}$.

13. The auxiliary equation is $r^2 + r + 1 = 0 \;\Rightarrow\; r = -\frac{1}{2} \pm \frac{\sqrt{3}}{2}i$, so $y = e^{-t/2}\left[c_1 \cos\left(\frac{\sqrt{3}}{2}t\right) + c_2 \sin\left(\frac{\sqrt{3}}{2}t\right)\right]$.

15. $r^2 - 8r + 16 = (r - 4)^2 = 0$ so $y = c_1 e^{4x} + c_2 x e^{4x}$.

The graphs are all asymptotic to the x-axis as $x \to -\infty$,

and as $x \to \infty$ the solutions tend to $\pm\infty$.

17. $2r^2 + 5r + 3 = (2r + 3)(r + 1) = 0$, so $r = -\frac{3}{2}, r = -1$ and the general solution is $y = c_1 e^{-3x/2} + c_2 e^{-x}$.
Then $y(0) = 3 \;\Rightarrow\; c_1 + c_2 = 3$ and $y'(0) = -4 \;\Rightarrow\; -\frac{3}{2}c_1 - c_2 = -4$, so $c_1 = 2$ and $c_2 = 1$. Thus the solution to the initial-value problem is $y = 2e^{-3x/2} + e^{-x}$.

19. $r^2 - 2r + 2 = 0 \;\Rightarrow\; r = 1 \pm i$ and the general solution is $y = e^x(c_1 \cos x + c_2 \sin x)$. But $1 = y(0) = c_1$ and $2 = y'(0) = c_1 + c_2$ so the solution to the initial-value problem is $y = e^x(\cos x + \sin x)$.

21. $r^2 - 2r - 3 = (r - 3)(r + 1) = 0$ so the general solution is $y = c_1 e^{-x} + c_2 e^{3x}$. However the conditions are given at $x = 1$ so rewrite the general solution as $y = k_1 e^{-(x-1)} + k_2 e^{3(x-1)}$. Then $3 = y(1) = k_1 + k_2$ and $1 = y'(1) = -k_1 + 3k_2$ so $k_1 = 2$, $k_2 = 1$ and the solution to the initial-value problem is $y = 2e^{-(x-1)} + e^{3(x-1)}$.

23. $r^2 + 9 = 0 \;\Rightarrow\; r = \pm 3i$ and the general solution is $y = c_1 \cos 3x + c_2 \sin 3x$. But $0 = y\left(\frac{\pi}{3}\right) = -c_1$ and $1 = y'\left(\frac{\pi}{3}\right) = -3c_2$, so the solution to the initial-value problem is $y = -\frac{1}{3}\sin 3x$.

25. $r^2 + 4r + 4 = (r + 2)^2 = 0$ so the general solution is $y = c_1 e^{-2x} + c_2 x e^{-2x}$. Then $0 = y(0) = c_1$, $3 = y(1) = c_2 e^{-2}$ so $c_2 = 3e^2$ and the solution of the boundary-value problem is $y = 3xe^{-2x+2}$.

27. $r^2 + 1 = 0$ $\Rightarrow$ $r = \pm i$ and the general solution is $y = c_1 \cos x + c_2 \sin x$. But $1 = y(0) = c_1$ and $0 = y(\pi) = -c_1$ so there is no solution.

29. $r^2 - r - 2 = (r - 2)(r + 1) = 0$ so the general solution is $y = c_1 e^{-x} + c_2 e^{2x}$. Then $1 = y(-1) = c_1 e + c_2 e^{-2}$ and $0 = y(1) = c_1 e^{-1} + c_2 e^2$ so $c_1 = \dfrac{e^5}{e^6 - 1}$ and $c_2 = \dfrac{e^2}{1 - e^6}$ so the solution to the boundary-value problem is

$$y = \frac{e^5}{e^6 - 1} e^{-x} + \frac{e^2}{1 - e^6} e^{2x} = \frac{1}{e^6 - 1} \left[e^{5-x} - e^{2(1+x)} \right].$$

31. $r^2 + 4r + 13 = 0$ $\Rightarrow$ $r = -2 \pm 3i$ and the general solution is $y = e^{-2x}(c_1 \cos 3x + c_2 \sin 3x)$. But $2 = y(0) = c_1$ and $1 = y\left(\frac{\pi}{2}\right) = e^{-\pi}(-c_2)$, so the solution to the boundary-value problem is $y = e^{-2x}(2 \cos 3x - e^{\pi} \sin 3x)$.

33. (a) *Case 1* ($\lambda = 0$): $y'' + \lambda y = 0$ $\Rightarrow$ $y'' = 0$ which has an auxiliary equation $r^2 = 0$ $\Rightarrow$ $r = 0$ $\Rightarrow$ $y = c_1 + c_2 x$ where $y(0) = 0$ and $y(L) = 0$. Thus, $0 = y(0) = c_1$ and $0 = y(L) = c_2 L$ $\Rightarrow$ $c_1 = c_2 = 0$. Thus, $y = 0$.

Case 2 ($\lambda < 0$): $y'' + \lambda y = 0$ has auxiliary equation $r^2 = -\lambda$ $\Rightarrow$ $r = \pm\sqrt{-\lambda}$ (distinct and real since $\lambda < 0$) $\Rightarrow$ $y = c_1 e^{\sqrt{-\lambda}x} + c_2 e^{-\sqrt{-\lambda}x}$ where $y(0) = 0$ and $y(L) = 0$. Thus, $0 = y(0) = c_1 + c_2$ ($\bigstar$) and $0 = y(L) = c_1 e^{\sqrt{-\lambda}L} + c_2 e^{-\sqrt{-\lambda}L}$ (†).

Multiplying ($\bigstar$) by $e^{\sqrt{-\lambda}L}$ and subtracting (†) gives $c_2 \left(e^{\sqrt{-\lambda}L} - e^{-\sqrt{-\lambda}L} \right) = 0$ $\Rightarrow$ $c_2 = 0$ and thus $c_1 = 0$ from ($\bigstar$). Thus, $y = 0$ for the cases $\lambda = 0$ and $\lambda < 0$.

(b) $y'' + \lambda y = 0$ has an auxiliary equation $r^2 + \lambda = 0$ $\Rightarrow$ $r = \pm i\sqrt{\lambda}$ $\Rightarrow$ $y = c_1 \cos \sqrt{\lambda}x + c_2 \sin \sqrt{\lambda}x$ where $y(0) = 0$ and $y(L) = 0$. Thus, $0 = y(0) = c_1$ and $0 = y(L) = c_2 \sin \sqrt{\lambda}L$ since $c_1 = 0$. Since we cannot have a trivial solution, $c_2 \neq 0$ and thus $\sin \sqrt{\lambda}L = 0$ $\Rightarrow$ $\sqrt{\lambda}L = n\pi$ where n is an integer $\Rightarrow$ $\lambda = n^2\pi^2/L^2$ and $y = c_2 \sin(n\pi x/L)$ where n is an integer.

18.2 Nonhomogeneous Linear Equations $\qquad$ ET 17.2

1. The auxiliary equation is $r^2 + 3r + 2 = (r + 2)(r + 1) = 0$, so the complementary solution is $y_c(x) = c_1 e^{-2x} + c_2 e^{-x}$. We try the particular solution $y_p(x) = Ax^2 + Bx + C$, so $y_p' = 2Ax + B$ and $y_p'' = 2A$. Substituting into the differential equation, we have $(2A) + 3(2Ax + B) + 2(Ax^2 + Bx + C) = x^2$ or $2Ax^2 + (6A + 2B)x + (2A + 3B + 2C) = x^2$. Comparing coefficients gives $2A = 1$, $6A + 2B = 0$, and $2A + 3B + 2C = 0$, so $A = \frac{1}{2}$, $B = -\frac{3}{2}$, and $C = \frac{7}{4}$. Thus the general solution is $y(x) = y_c(x) + y_p(x) = c_1 e^{-2x} + c_2 e^{-x} + \frac{1}{2}x^2 - \frac{3}{2}x + \frac{7}{4}$.

3. The auxiliary equation is $r^2 - 2r = r(r - 2) = 0$, so the complementary solution is $y_c(x) = c_1 + c_2 e^{2x}$. Try the particular solution $y_p(x) = A \cos 4x + B \sin 4x$, so $y_p' = -4A \sin 4x + 4B \cos 4x$ and $y_p'' = -16A \cos 4x - 16B \sin 4x$. Substitution into the differential equation gives $(-16A \cos 4x - 16B \sin 4x) - 2(-4A \sin 4x + 4B \cos 4x) = \sin 4x$ $\Rightarrow$ $(-16A - 8B) \cos 4x + (8A - 16B) \sin 4x = \sin 4x$. Then $-16A - 8B = 0$ and $8A - 16B = 1$ $\Rightarrow$ $A = \frac{1}{40}$ and $B = -\frac{1}{20}$. Thus the general solution is $y(x) = y_c(x) + y_p(x) = c_1 + c_2 e^{2x} + \frac{1}{40} \cos 4x - \frac{1}{20} \sin 4x$.

5. The auxiliary equation is $r^2 - 4r + 5 = 0$ with roots $r = 2 \pm i$, so the complementary solution is
$y_c(x) = e^{2x}(c_1 \cos x + c_2 \sin x)$. Try $y_p(x) = Ae^{-x}$, so $y_p' = -Ae^{-x}$ and $y_p'' = Ae^{-x}$. Substitution gives
$Ae^{-x} - 4(-Ae^{-x}) + 5(Ae^{-x}) = e^{-x} \quad \Rightarrow \quad 10Ae^{-x} = e^{-x} \quad \Rightarrow \quad A = \frac{1}{10}$. Thus the general solution is
$y(x) = e^{2x}(c_1 \cos x + c_2 \sin x) + \frac{1}{10}e^{-x}$.

7. The auxiliary equation is $r^2 + 1 = 0$ with roots $r = \pm i$, so the complementary solution is
$y_c(x) = c_1 \cos x + c_2 \sin x$. For $y'' + y = e^x$ try $y_{p_1}(x) = Ae^x$. Then $y_{p_1}' = y_{p_1}'' = Ae^x$ and substitution gives
$Ae^x + Ae^x = e^x \quad \Rightarrow \quad A = \frac{1}{2}$, so $y_{p_1}(x) = \frac{1}{2}e^x$. For $y'' + y = x^3$ try $y_{p_2}(x) = Ax^3 + Bx^2 + Cx + D$.
Then $y_{p_2}' = 3Ax^2 + 2Bx + C$ and $y_{p_2}'' = 6Ax + 2B$. Substituting, we have
$6Ax + 2B + Ax^3 + Bx^2 + Cx + D = x^3$, so $A = 1$, $B = 0$, $6A + C = 0 \quad \Rightarrow \quad C = -6$, and
$2B + D = 0 \quad \Rightarrow \quad D = 0$. Thus $y_{p_2}(x) = x^3 - 6x$ and the general solution is
$y(x) = y_c(x) + y_{p_1}(x) + y_{p_2}(x) = c_1 \cos x + c_2 \sin x + \frac{1}{2}e^x + x^3 - 6x$. But
$2 = y(0) = c_1 + \frac{1}{2} \quad \Rightarrow \quad c_1 = \frac{3}{2}$ and $0 = y'(0) = c_2 + \frac{1}{2} - 6 \quad \Rightarrow \quad c_2 = \frac{11}{2}$. Thus the solution to the
initial-value problem is $y(x) = \frac{3}{2} \cos x + \frac{11}{2} \sin x + \frac{1}{2}e^x + x^3 - 6x$.

9. $y_c(x) = c_1 e^x + c_2 e^{-x}$. Try $y_p(x) = (Ax + B)e^{3x}$. Then $y_p' = e^{3x}(A + 3Ax + 3B)$ and
$y_p'' = e^{3x}(3A + 3A + 9Ax + 9B)$. Substitution into the differential equation gives
$e^{3x}[9Ax + 9B + 6A - (Ax + B)] = xe^{3x}$, so $A = \frac{1}{8}$, $B = -\frac{3}{32}$ and the general solution is
$y(x) = c_1 e^x + c_2 e^{-x} + \left(\frac{1}{8}x - \frac{3}{32}\right)e^{3x}$. But $0 = y(0) = c_1 + c_2 - \frac{3}{32}$, $1 = y'(0) = c_1 - c_2 - \frac{9}{32} + \frac{1}{8}$ so the
solution to the initial-value problem is $y(x) = \frac{5}{8}e^x - \frac{17}{32}e^{-x} + e^{3x}\left(\frac{1}{8}x - \frac{3}{32}\right)$.

11. $y_c(x) = c_1 e^{-x/4} + c_2 e^{-x}$. Try $y_p(x) = Ae^x$. Then
$10Ae^x = e^x$, so $A = \frac{1}{10}$ and the general solution is
$y(x) = c_1 e^{-x/4} + c_2 e^{-x} + \frac{1}{10}e^x$. The solutions are all
composed of exponential curves and with the exception of the
particular solution (which approaches 0 as $x \to -\infty$), they all
approach either ∞ or $-\infty$ as $x \to -\infty$. As $x \to \infty$, all solutions
are asymptotic to $y_p = \frac{1}{10}e^x$.

13. Since the roots of the auxiliary equation are complex, we need just try
$y_p(x) = \left(Ax^4 + Bx^3 + Cx^2 + Dx + E\right)e^{2x}$.

15. Since $y_c(x) = e^x(c_1 \cos x + c_2 \sin x)$ we try $y_p(x) = xe^x(A \cos x + B \sin x)$.

Note: Solving Equations (7) and (9) in The Method of Variation of Parameters gives
$$u_1' = -\frac{Gy_2}{a(y_1 y_2' - y_2 y_1')} \quad \text{and} \quad u_2' = \frac{Gy_1}{a(y_1 y_2' - y_2 y_1')}$$
We will use these equations rather than resolving the system in each of the remaining exercises in this section.

17. (a) The complementary solution is $y_c(x) = c_1 \cos 2x + c_2 \sin 2x$. A particular solution is of the form
$y_p(x) = Ax + B$. Thus, $4Ax + 4B = x \quad \Rightarrow \quad A = \frac{1}{4}$ and $B = 0 \quad \Rightarrow \quad y_p(x) = \frac{1}{4}x$. Thus, the general
solution is $y = y_c + y_p = c_1 \cos 2x + c_2 \sin 2x + \frac{1}{4}x$.

(b) In (a), $y_c(x) = c_1 \cos 2x + c_2 \sin 2x$, so set $y_1 = \cos 2x$, $y_2 = \sin 2x$. Then

$y_1 y_2' - y_2 y_1' = 2\cos^2 2x + 2\sin^2 2x = 2$ so $u_1' = -\frac{1}{2}x \sin 2x$ $\Rightarrow$

$u_1(x) = -\frac{1}{2}\int x \sin 2x dx = -\frac{1}{4}\left(-x \cos 2x + \frac{1}{2}\sin 2x\right)$ (by parts) and $u_2' = \frac{1}{2}x \cos 2x$

$\Rightarrow u_2(x) = \frac{1}{2}\int x \cos 2x dx = \frac{1}{4}\left(x \sin 2x + \frac{1}{2}\cos 2x\right)$ (by parts). Hence

$y_p(x) = -\frac{1}{4}\left(-x \cos 2x + \frac{1}{2}\sin 2x\right)\cos 2x + \frac{1}{4}\left(x \sin 2x + \frac{1}{2}\cos 2x\right)\sin 2x = \frac{1}{4}x$. Thus

$y(x) = y_c(x) + y_p(x) = c_1 \cos 2x + c_2 \sin 2x + \frac{1}{4}x$.

19. (a) $r^2 - r = r(r-1) = 0$ $\Rightarrow$ $r = 0, 1$, so the complementary solution is $y_c(x) = c_1 e^x + c_2 x e^x$. A particular

solution is of the form $y_p(x) = Ae^{2x}$. Thus $4Ae^{2x} - 4Ae^{2x} + Ae^{2x} = e^{2x}$ $\Rightarrow$ $Ae^{2x} = e^{2x}$ $\Rightarrow$ $A = 1$

$\Rightarrow$ $y_p(x) = e^{2x}$. So a general solution is $y(x) = y_c(x) + y_p(x) = c_1 e^x + c_2 x e^x + e^{2x}$.

(b) From (a), $y_c(x) = c_1 e^x + c_2 x e^x$, so set $y_1 = e^x$, $y_2 = x e^x$. Then, $y_1 y_2' - y_2 y_1' = e^{2x}(1+x) - x e^{2x} = e^{2x}$

and so $u_1' = -xe^x$ $\Rightarrow$ $u_1(x) = -\int x e^x dx = -(x-1)e^x$ (by parts) and $u_2' = e^x$ $\Rightarrow$

$u_2(x) = \int e^x\, dx = e^x$. Hence $y_p(x) = (1-x)e^{2x} + xe^{2x} = e^{2x}$ and the general solution is

$y(x) = y_c(x) + y_p(x) = c_1 e^x + c_2 x e^x + e^{2x}$.

21. As in Example 6, $y_c(x) = c_1 \sin x + c_2 \cos x$, so set $y_1 = \sin x$, $y_2 = \cos x$. Then

$y_1 y_2' - y_2 y_1' = -\sin^2 x - \cos^2 x = -1$, so $u_1' = -\dfrac{\sec x \cos x}{-1} = 1 \Rightarrow u_1(x) = x$ and

$u_2' = \dfrac{\sec x \sin x}{-1} = -\tan x$ $\Rightarrow$ $u_2(x) = -\int \tan x dx = \ln|\cos x| = \ln(\cos x)$ on $0 < x < \frac{\pi}{2}$. Hence

$y_p(x) = x \sin x + \cos x \ln(\cos x)$ and the general solution is $y(x) = (c_1 + x)\sin x + [c_2 + \ln(\cos x)]\cos x$.

23. $y_1 = e^x$, $y_2 = e^{2x}$ and $y_1 y_2' - y_2 y_1' = e^{3x}$. So $u_1' = \dfrac{-e^{2x}}{(1+e^{-x})e^{3x}} = -\dfrac{e^{-x}}{1+e^{-x}}$ and

$u_1(x) = \int -\dfrac{e^{-x}}{1+e^{-x}} dx = \ln(1+e^{-x})$. $u_2' = \dfrac{e^x}{(1+e^{-x})e^{3x}} = \dfrac{e^x}{e^{3x}+e^{2x}}$ so

$u_2(x) = \int \dfrac{e^x}{e^{3x}+e^{2x}} dx = \ln\left(\dfrac{e^x+1}{e^x}\right) - e^{-x} = \ln(1+e^{-x}) - e^{-x}$. Hence

$y_p(x) = e^x \ln(1+e^{-x}) + e^{2x}\left[\ln(1+e^{-x}) - e^{-x}\right]$ and the general solution is

$y(x) = \left[c_1 + \ln(1+e^{-x})\right]e^x + \left[c_2 - e^{-x} + \ln(1+e^{-x})\right]e^{2x}$.

25. $y_1 = e^{-x}$, $y_2 = e^x$ and $y_1 y_2' - y_2 y_1' = 2$. So $u_1' = -\dfrac{e^x}{2x}$,

$u_2' = \dfrac{e^{-x}}{2x}$ and $y_p(x) = -e^{-x}\int \dfrac{e^x}{2x} dx + e^x \int \dfrac{e^{-x}}{2x} dx$. Hence the general solution is

$y(x) = \left(c_1 - \int \dfrac{e^x}{2x} dx\right)e^{-x} + \left(c_2 + \int \dfrac{e^{-x}}{2x} dx\right)e^x$.

18.3 Applications of Second-Order Differential Equations   ET 17.3

1. By Hooke's Law $k(0.6) = 20$ so $k = \frac{100}{3}$ is the spring constant and the differential equation is $3x'' + \frac{100}{3}x = 0$.

The general solution is $x(t) = c_1 \cos\left(\frac{10}{3}t\right) + c_2 \sin\left(\frac{10}{3}t\right)$. But $0 = x(0) = c_1$ and $1.2 = x'(0) = \frac{10}{3}c_2$, so the

position of the mass after t seconds is $x(t) = 0.36 \sin\left(\frac{10}{3}t\right)$.

3. $k(0.5) = 6$ or $k = 12$ is the spring constant, so the initial-value problem is $2x'' + 14x' + 12x = 0$, $x(0) = 1$,

$x'(0) = 0$. The general solution is $x(t) = c_1 e^{-6t} + c_2 e^{-t}$. But $1 = x(0) = c_1 + c_2$ and $0 = x'(0) = -6c_1 - c_2$.

Thus the position is given by $x(t) = -\frac{1}{5}e^{-6t} + \frac{6}{5}e^{-t}$.

5. For critical damping we need $c^2 - 4mk = 0$ or $m = c^2 / (4k) = 14^2 / (4 \cdot 12) = \frac{49}{12}$ kg.

7. We are given $m = 1$, $k = 100$, $x(0) = -0.1$ and $x'(0) = 0$. From (3), the differential equation is

$$\frac{d^2x}{dt^2} + c\frac{dx}{dt} + 100x = 0$$ with auxiliary equation $r^2 + cr + 100 = 0$. If $c = 10$, we have two complex roots

$r = -5 \pm 5\sqrt{3}i$, so the motion is underdamped and the solution is $x = e^{-5t}\left[c_1 \cos\left(5\sqrt{3}t\right) + c_2 \sin\left(5\sqrt{3}t\right)\right]$.

Then $-0.1 = x(0) = c_1$ and $0 = x'(0) = 5\sqrt{3}c_2 - 5c_1$ $\Rightarrow$ $c_2 = -\frac{1}{10\sqrt{3}}$, so

$x = e^{-5t}\left[-0.1 \cos\left(5\sqrt{3}t\right) - \frac{1}{10\sqrt{3}} \sin\left(5\sqrt{3}t\right)\right]$. If $c = 15$, we again have underdamping since the auxiliary

equation has roots $r = -\frac{15}{2} \pm \frac{5\sqrt{7}}{2}i$. The general solution is $x = e^{-15t/2}\left[c_1 \cos\left(\frac{5\sqrt{7}}{2}t\right) + c_2 \sin\left(\frac{5\sqrt{7}}{2}t\right)\right]$, so

$-0.1 = x(0) = c_1$ and $0 = x'(0) = \frac{5\sqrt{7}}{2}c_2 - \frac{15}{2}c_1$ $\Rightarrow$ $c_2 = -\frac{3}{10\sqrt{7}}$. Thus

$x = e^{-15t/2}\left[-0.1 \cos\left(\frac{5\sqrt{7}}{2}t\right) - \frac{3}{10\sqrt{7}} \sin\left(\frac{5\sqrt{7}}{2}t\right)\right]$. For $c = 20$, we have equal roots $r_1 = r_2 = -10$, so the

oscillation is critically damped and the solution is $x = (c_1 + c_2t)e^{-10t}$. Then $-0.1 = x(0) = c_1$ and

$0 = x'(0) = -10c_1 + c_2$ $\Rightarrow$ $c_2 = -1$, so $x = (-0.1 - t)e^{-10t}$. If $c = 25$ the auxiliary equation has roots

$r_1 = -5$, $r_2 = -20$, so we have overdamping and the solution is $x = c_1e^{-5t} + c_2e^{-20t}$. Then

$-0.1 = x(0) = c_1 + c_2$ and $0 = x'(0) = -5c_1 - 20c_2$ $\Rightarrow$ $c_1 = -\frac{2}{15}$ and $c_2 = \frac{1}{30}$, so

$x = -\frac{2}{15}e^{-5t} + \frac{1}{30}e^{-20t}$. If $c = 30$ we have roots $r = -15 \pm 5\sqrt{5}$, so the motion is overdamped and the solution

is $x = c_1e^{\left(-15+5\sqrt{5}\right)t} + c_2e^{\left(-15-5\sqrt{5}\right)t}$. Then $-0.1 = x(0) = c_1 + c_2$ and

$0 = x'(0) = \left(-15 + 5\sqrt{5}\right)c_1 + \left(-15 - 5\sqrt{5}\right)c_2$ $\Rightarrow$ $c_1 = \frac{-5-3\sqrt{5}}{100}$ and $c_2 = \frac{-5+3\sqrt{5}}{100}$, so

$x = \left(\frac{-5-3\sqrt{5}}{100}\right)e^{\left(-15+5\sqrt{5}\right)t} + \left(\frac{-5+3\sqrt{5}}{100}\right)e^{\left(-15-5\sqrt{5}\right)t}$.

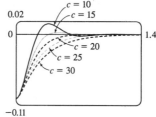

9. The differential equation is $mx'' + kx = F_0 \cos\omega_0 t$ and $\omega_0 \neq \omega = \sqrt{k/m}$.

Here the auxiliary equation is $mr^2 + k = 0$ with roots $\pm\sqrt{k/m}\,i = \pm\omega i$ so

$x_c(t) = c_1 \cos\omega t + c_2 \sin\omega t$. Since $\omega_0 \neq \omega$, try $x_p(t) = A \cos\omega_0 t + B \sin\omega_0 t$. Then we need

$(m)\left(-\omega_0^2\right)(A \cos\omega_0 t + B \sin\omega_0 t) + k(A \cos\omega_0 t + B \sin\omega_0 t) = F_0 \cos\omega_0 t$ or $A\left(k - m\omega_0^2\right) = F_0$ and

$B\left(k - m\omega_0^2\right) = 0$. Hence $B = 0$ and $A = \dfrac{F_0}{k - m\omega_0^2} = \dfrac{F_0}{m\left(\omega^2 - \omega_0^2\right)}$ since $\omega^2 = \dfrac{k}{m}$. Thus the motion of the

mass is given by $x(t) = c_1 \cos\omega t + c_2 \sin\omega t + \dfrac{F_0}{m\left(\omega^2 - \omega_0^2\right)} \cos\omega_0 t$.

11. Here the initial-value problem for the charge is $Q'' + 20Q' + 500Q = 12$, $Q(0) = Q'(0) = 0$. Then

$Q_c(t) = e^{-10t}(c_1 \cos 20t + c_2 \sin 20t)$ and try $Q_p(t) = A$ $\Rightarrow$ $500A = 12$ or $A = \frac{3}{125}$. The general solution

is $Q(t) = e^{-10t}(c_1 \cos 20t + c_2 \sin 20t) + \frac{3}{125}$. But $0 = Q(0) = c_1 + \frac{3}{125}$ and

$Q'(t) = I(t) = e^{-10t}[(-10c_1 + 20c_2) \cos 20t + (-10c_2 - 20c_1) \sin 20t]$ but $0 = Q'(0) = -10c_1 + 20c_2$.

Thus the charge is $Q(t) = -\frac{1}{250}e^{-10t}(6 \cos 20t + 3 \sin 20t) + \frac{3}{125}$ and the current is $I(t) = e^{-10t}\left(\frac{3}{5}\right) \sin 20t$.

13. As in Exercise 11, $Q_c(t) = e^{-10t}(c_1 \cos 20t + c_2 \sin 20t)$ but $E(t) = 12 \sin 10t$ so try

$Q_p(t) = A \cos 10t + B \sin 10t$. Substituting into the differential equation gives

$(-100A + 200B + 500A) \cos 10t + (-100B - 200A + 500B) \sin 10t = 12 \sin 10t,$ $\Rightarrow$ $400A + 200B = 0$

and $400B - 200A = 12$. Thus $A = -\frac{3}{250}$, $B = \frac{3}{125}$ and the general solution is

$Q(t) = e^{-10t}(c_1 \cos 20t + c_2 \sin 20t) - \frac{3}{250} \cos 10t + \frac{3}{125} \sin 10t$. But $0 = Q(0) = c_1 - \frac{3}{250}$ so $c_1 = \frac{3}{250}$. Also

$Q'(t) = \frac{3}{25} \sin 10t + \frac{6}{25} \cos 10t + e^{-10t}[(-10c_1 + 20c_2) \cos 20t + (-10c_2 - 20c_1) \sin 20t]$ and

$0 = Q'(0) = \frac{6}{25} - 10c_1 + 20c_2$ so $c_2 = -\frac{3}{500}$. Hence the charge is given by

$Q(t) = e^{-10t}\left[\frac{3}{250} \cos 20t - \frac{3}{500} \sin 20t\right] - \frac{3}{250} \cos 10t + \frac{3}{125} \sin 10t$.

15. $x(t) = A \cos(\omega t + \delta)$ $\Leftrightarrow$ $x(t) = A[\cos \omega t \cos \delta - \sin \omega t \sin \delta]$ $\Leftrightarrow$ $x(t) = A\left(\dfrac{c_1}{A} \cos \omega t + \dfrac{c_2}{A} \sin \omega t\right)$

where $\cos \delta = c_1/A$ and $\sin \delta = -c_2/A$ $\Leftrightarrow$ $x(t) = c_1 \cos \omega t + c_2 \sin \omega t$. (Note that $\cos^2 \delta + \sin^2 \delta = 1$ $\Rightarrow$

$c_1^2 + c_2^2 = A^2$.)

18.4 Series Solutions ET 17.4

1. Let $y(x) = \sum_{n=0}^{\infty} c_n x^n$. Then $y'(x) = \sum_{n=1}^{\infty} n c_n x^{n-1}$ and

$y' - y = 0$ $\Rightarrow$ $\sum_{n=1}^{\infty} n c_n x^{n-1} - \sum_{n=0}^{\infty} c_n x^n = 0$. Replacing n by $n+1$ in the first sum gives

$\sum_{n=0}^{\infty} (n+1) c_{n+1} x^n - \sum_{n=0}^{\infty} c_n x^n = 0$, so $\sum_{n=0}^{\infty} [(n+1) c_{n+1} - c_n] x^n = 0$. Equating coefficients gives

$(n+1) c_{n+1} - c_n = 0$, so the recursion relation is $c_{n+1} = \dfrac{c_n}{n+1}$, $n = 0, 1, 2, \ldots$. Then $c_1 = c_0$,

$c_2 = \dfrac{1}{2} c_1 = \dfrac{c_0}{2}$, $c_3 = \dfrac{1}{3} c_2 = \dfrac{1}{3} \cdot \dfrac{1}{2} c_0 = \dfrac{c_0}{3!}$, $c_4 = \dfrac{1}{4} c_3 = \dfrac{c_0}{4!}$, and in general, $c_n = \dfrac{c_0}{n!}$. Thus the solution is

$y(x) = \sum_{n=0}^{\infty} c_n x^n = c_0 \sum_{n=0}^{\infty} \dfrac{x^n}{n!} = c_0 e^x$.

3. Assuming $y(x) = \sum_{n=0}^{\infty} c_n x^n$, we have $y'(x) = \sum_{n=1}^{\infty} n c_n x^{n-1} = \sum_{n=0}^{\infty} (n+1) c_{n+1} x^n$ and

$-x^2 y = -\sum_{n=0}^{\infty} c_n x^{n+2} = -\sum_{n=2}^{\infty} c_{n-2} x^n$. Hence the differential equation becomes

$\sum_{n=0}^{\infty} (n+1) c_{n+1} x^n - \sum_{n=2}^{\infty} c_{n-2} x^n = 0$ or $c_1 + 2c_2 x + \sum_{n=2}^{\infty} [(n+1) c_{n+1} - c_{n-2}] x^n = 0$. Equating

coefficients gives $c_1 = c_2 = 0$ and $c_{n+1} = \dfrac{c_{n-2}}{n+1}$ for $n = 2, 3, \ldots$. But $c_1 = 0$, so $c_4 = 0$ and $c_7 = 0$

and in general $c_{3n+1} = 0$. Similarly $c_2 = 0$ so $c_{3n+2} = 0$. Finally $c_3 = \dfrac{c_0}{3}$,

$c_6 = \dfrac{c_3}{6} = \dfrac{c_0}{6 \cdot 3} = \dfrac{c_0}{3^2 \cdot 2!}$, $c_9 = \dfrac{c_6}{9} = \dfrac{c_0}{9 \cdot 6 \cdot 3} = \dfrac{c_0}{3^3 \cdot 3!}, \ldots$, and $c_{3n} = \dfrac{c_0}{3^n \cdot n!}$. Thus the solution is

$y(x) = \sum_{n=0}^{\infty} c_n x^n = \sum_{n=0}^{\infty} c_{3n} x^{3n} = c_0 \sum_{n=0}^{\infty} \dfrac{(x^3/3)^n}{n!} = c_0 e^{x^3/3}$.

5. Let $y(x) = \sum_{n=0}^{\infty} c_n x^n$ $\Rightarrow$ $y'(x) = \sum_{n=1}^{\infty} n c_n x^{n-1}$ and $y''(x) = \sum_{n=0}^{\infty} (n+2)(n+1) c_{n+2} x^n$. The

differential equation becomes $\sum_{n=0}^{\infty} (n+2)(n+1) c_{n+2} x^n + x \sum_{n=1}^{\infty} n c_n x^{n-1} + \sum_{n=0}^{\infty} c_n x^n = 0$ or

$\sum_{n=0}^{\infty} [(n+2)(n+1) c_{n+2} + n c_n + c_n] x^n$ (since $\sum_{n=1}^{\infty} n c_n x^n = \sum_{n=0}^{\infty} n c_n x^n$). Equating coefficients gives

$(n+2)(n+1) c_{n+2} + (n+1) c_n = 0$, thus the recursion relation is $c_{n+2} = \dfrac{-(n+1) c_n}{(n+2)(n+1)} = -\dfrac{c_n}{n+2}$,

$n = 0, 1, 2, \ldots$. Then the even coefficients are given by $c_2 = -\dfrac{c_0}{2}$, $c_4 = -\dfrac{c_2}{4} = \dfrac{c_0}{2 \cdot 4}$, $c_6 = -\dfrac{c_4}{6} = -\dfrac{c_0}{2 \cdot 4 \cdot 6}$,

and in general, $c_{2n} = (-1)^n \dfrac{c_0}{2 \cdot 4 \cdots \cdots 2n} = \dfrac{(-1)^n c_0}{2^n n!}$. The odd coefficients are $c_3 = -\dfrac{c_1}{3}$, $c_5 = -\dfrac{c_3}{5} = \dfrac{c_1}{3 \cdot 5}$,

$c_7 = -\dfrac{c_5}{7} = -\dfrac{c_1}{3 \cdot 5 \cdot 7}$, and in general, $c_{2n+1} = (-1)^n \dfrac{c_1}{3 \cdot 5 \cdot 7 \cdots \cdots (2n+1)} = \dfrac{(-2)^n n! c_1}{(2n+1)!}$. The solution is

$$y(x) = c_0 \sum_{n=0}^{\infty} \frac{(-1)^n}{2^n n!} x^{2n} + c_1 \sum_{n=0}^{\infty} \frac{(-2)^n n!}{(2n+1)!} x^{2n+1}.$$

7. Let $y(x) = \sum_{n=0}^{\infty} c_n x^n$. Then $y'' = \sum_{n=0}^{\infty} n(n-1) c_n x^{n-2}$, $xy' = \sum_{n=0}^{\infty} n c_n x^n$ and

$(x^2 + 1) y'' = \sum_{n=0}^{\infty} n(n-1) c_n x^n + \sum_{n=0}^{\infty} (n+2)(n+1) c_{n+2} x^n$. The differential equation becomes

$\sum_{n=0}^{\infty} [(n+2)(n+1) c_{n+2} + [n(n-1) + n - 1] c_n] x^n = 0$. The recursion relation is $c_{n+2} = -\dfrac{(n-1) c_n}{n+2}$,

$n = 0, 1, 2, \ldots$. Given c_0 and c_1, $c_2 = \dfrac{c_0}{2}$, $c_4 = -\dfrac{c_2}{4} = -\dfrac{c_0}{2^2 \cdot 2!}$, $c_6 = -\dfrac{3 c_4}{6} = (-1)^2 \dfrac{3 c_0}{2^3 \cdot 3!}$, $\ldots$,

$c_{2n} = (-1)^{n-1} \dfrac{1 \cdot 3 \cdots \cdots (2n-3) c_0}{2^n n!} = (-1)^{n-1} \dfrac{(2n-3)! c_0}{2^n 2^{n-2} n! (n-2)!} = (-1)^{n-1} \dfrac{(2n-3)! c_0}{2^{2n-2} n! (n-2)!}$ for

$n = 2, 3, \ldots$. $c_3 = \dfrac{0 \cdot c_1}{3} = 0$ $\Rightarrow$ $c_{2n+1} = 0$ for $n = 1, 2, \ldots$. Thus the solution is

$$y(x) = c_0 + c_1 x + c_0 \frac{x^2}{2} + c_0 \sum_{n=2}^{\infty} \frac{(-1)^{n-1} (2n-3)!}{2^{2n-2} n! (n-2)!} x^{2n}.$$

9. Let $y(x) = \sum_{n=0}^{\infty} c_n x^n$. Then $y''(x) = \sum_{n=0}^{\infty} (n+2)(n+1) c_{n+2} x^n$, $-xy'(x) = -\sum_{n=0}^{\infty} n c_n x^n$ and the

differential equation becomes $\sum_{n=0}^{\infty} [(n+2)(n+1) c_{n+2} - (n+1) c_n] x^n = 0$. Thus the recursion relation is

$c_{n+2} = \dfrac{c_n}{n+2}$ for $n = 0, 1, 2, \ldots$. But $c_0 = y(0) = 1$ so $c_2 = \dfrac{1}{2}$, $c_4 = \dfrac{c_2}{4} = \dfrac{1}{2 \cdot 4}$, $c_6 = \dfrac{c_4}{6} = \dfrac{1}{2 \cdot 4 \cdot 6}$, $\ldots$,

$c_{2n} = \dfrac{1}{2^n n!}$. Also $c_1 = y'(0) = 0$ and by the recursion relation $c_{2n+1} = 0$ for $n = 0, 1, 2, \ldots$. Thus the solution

to the initial-value problem is $y(x) = \sum_{n=0}^{\infty} c_n x^n = \sum_{n=0}^{\infty} \frac{x^{2n}}{2^n n!} = \sum_{n=0}^{\infty} \frac{(x^2/2)^n}{n!} = e^{x^2/2}$.

11. Let $y(x) = \sum_{n=0}^{\infty} c_n x^n$. Then $y''(x) = \sum_{n=0}^{\infty} n(n-1) c_n x^{n-2} = \sum_{n=-1}^{\infty} (n+3)(n+2) c_{n+3} x^{n+1} =$

$2 c_2 + \sum_{n=0}^{\infty} (n+3)(n+2) c_{n+3} x^{n+1}$ and the differential equation becomes

$2 c_2 + \sum_{n=0}^{\infty} [(n+3)(n+2) c_{n+3} + (n+1) c_n] x^{n+1} = 0$. Then $c_2 = 0$ and the recursion relation is

$c_{n+3} = -\dfrac{(n+1) c_n}{(n+3)(n+2)}$, $n = 0, 1, 2, \ldots$. But $c_0 = y(0) = 0 = c_2$ and by the recursion relation

$c_{3n} = c_{3n+2} = 0$ for $n = 0, 1, 2, \ldots$. Also $c_1 = y'(0) = 1$ so

$c_4 = -\dfrac{2}{4 \cdot 3}$, $c_7 = -\dfrac{5 c_4}{7 \cdot 6} = (-1)^2 \dfrac{2 \cdot 5}{7 \cdot 6 \cdot 4 \cdot 3} = (-1)^2 \dfrac{2^2 5^2}{7!}$, $\ldots$, $c_{3n+1} = (-1)^n \dfrac{2^2 5^2 \cdots (3n-1)^2}{(3n+1)!}$. Thus

the solution is $y(x) = \sum_{n=0}^{\infty} c_n x^n = x + \sum_{n=0}^{\infty} \left[(-1)^n \dfrac{2^2 5^2 \cdots (3n-1)^2 x^{3n+1}}{(3n+1)!} \right]$.

▀18 Review

CONCEPT CHECK

1. (a) $ay'' + by' + cy = 0$ where a, b, and c are constants.

(b) $ar^2 + br + c = 0$

(c) If the auxiliary equation has two distinct real roots r_1 and r_2, the solution is $y = c_1 e^{r_1 x} + c_2 e^{r_2 x}$. If the roots are real and equal, the solution is $y = c_1 e^{rx} + c_2 x e^{rx}$ where r is the common root. If the roots are complex, we can write $r_1 = \alpha + i\beta$ and $r_2 = \alpha - i\beta$, and the solution is $y = e^{\alpha x}(c_1 \cos \beta x + c_2 \sin \beta x)$.

2. (a) An initial-value problem consists of finding a solution y of a second-order differential equation that also satisfies given conditions $y(x_0) = y_0$ and $y'(x_0) = y_1$, where y_0 and y_1 are constants.

(b) A boundary-value problem consists of finding a solution y of a second-order differential equation that also satisfies given boundary conditions $y(x_0) = y_0$ and $y(x_1) = y_1$.

3. (a) $ay'' + by' + cy = G(x)$ where a, b, and c are constants and G is a continuous function.

(b) The complementary equation is the related homogeneous equation $ay'' + by' + cy = 0$. If we find the general solution y_c of the complementary equation and y_p is any particular solution of the original differential equation, then the general solution of the original differential equation is $y(x) = y_p(x) + y_c(x)$.

(c) See Examples 1–5 and the associated discussion in Section 18.2 [ET 17.2].

(d) See the discussion on pages 1170-1171 [ET 1136-1137].

4. Second-order linear differential equations can be used to describe the motion of a vibrating spring or to analyze an electric circuit; see the discussion in Section 18.3 [ET 17.3].

5. See Example 1 and the preceding discussion in Section 18.4 [ET 17.4].

TRUE-FALSE QUIZ

1. True. See Theorem 18.1.3 [ET 17.1.3].

3. True. $\cosh x$ and $\sinh x$ are linearly independent solutions of this linear homogeneous equation.

EXERCISES

1. The auxiliary equation is $r^2 - 2r - 15 = 0 \implies (r - 5)(r + 3) = 0 \implies r = 5, r = -3$. Then the general solution is $y = c_1 e^{5x} + c_2 e^{-3x}$.

3. The auxiliary equation is $r^2 + 3 = 0 \implies r = \pm\sqrt{3}\,i$. Then the general solution is $y = c_1 \cos\left(\sqrt{3}\,x\right) + c_2 \sin\left(\sqrt{3}\,x\right)$.

5. $y_c(x) = e^{-x}(c_1 + c_2 x)$ and try $y_p(x) = A\cos 3x + B\sin 3x$. Then
$-9A\cos 3x - 9B\sin 3x - 6A\sin 3x + 6B\cos 3x + A\cos 3x + B\sin 3x = \sin 3x \implies$
$6B - 8A = 0$ and $-6A - 8B = 1 \implies A = -\frac{3}{50}$, $B = -\frac{2}{25}$ and the general solution is
$y(x) = e^{-x}(c_1 + c_2 x) - \frac{3}{50}\cos 3x - \frac{2}{25}\sin 3x$.

7. $y_c(x) = c_1 \cos\left(\frac{3}{2}x\right) + c_2 \sin\left(\frac{3}{2}x\right)$ and try $y_p(x) = Ax^2 + Bx + C$. Then $8A + 9Ax^2 + 9Bx + 9C = 2x^2 - 3$
$\implies A = \frac{2}{9}$, $B = 0$, $C = -\frac{43}{81}$, and the general solution is $y(x) = c_1 \cos\left(\frac{3}{2}x\right) + c_2 \sin\left(\frac{3}{2}x\right) + \frac{2}{9}x^2 - \frac{43}{81}$.

9. $y_c(x) = c_1 e^x + c_2 e^{2x}$ so try $y_p(x) = Axe^{2x}$. Then $(4Ax + 4A - 6Ax - 3A)e^{2x} + 2Axe^{2x} = e^{2x} \implies A = 1$. Thus the general solution is $y(x) = c_1 e^x + c_2 e^{2x} + xe^{2x}$.

11. The auxiliary equation is $r^2 + 6r = 0$ and the general solution is $y(x) = c_1 + c_2 e^{-6x} = k_1 + k_2 e^{-6(x-1)}$. But $3 = y(1) = k_1 + k_2$ and $12 = y'(1) = -6k^2$. Thus $k_2 = -2$, $k_1 = 5$ and the solution is $y(x) = 5 - 2e^{-6(x-1)}$.

13. The auxiliary equation is $r^2 - 5r + 4 = 0$ and the general solution is $y(x) = c_1 e^x + c_2 e^{4x}$. But
$0 = y(0) = c_1 + c_2$ and $1 = y'(0) = c_1 + 4c_2$, so the solution is $y(x) = \frac{1}{3}\left(e^{4x} - e^x\right)$.

15. Let $y(x) = \sum_{n=0}^{\infty} c_n x^n$. Then $y''(x) = \sum_{n=0}^{\infty} n(n-1)c_n x^{n-2} = \sum_{n=0}^{\infty} (n+2)(n+1)c_{n+2} x^n$ and the
differential equation becomes $\sum_{n=0}^{\infty} \left[(n+2)(n+1)c_{n+2} + (n+1)c_n\right] x^n = 0$. Thus the recursion relation is
$c_{n+2} = -c_n/(n+2)$ for $n = 0, 1, 2, \ldots$. But $c_0 = y(0) = 0$, so $c_{2n} = 0$ for $n = 0, 1, 2, \ldots$. Also
$c_1 = y'(0) = 1$, so $c_3 = -\dfrac{1}{3}$, $c_5 = \dfrac{(-1)^2}{3 \cdot 5}$, $c_7 = \dfrac{(-1)^3}{3 \cdot 5 \cdot 7} = \dfrac{(-1)^3 \, 2^3 3!}{7!}$, $\ldots$, $c_{2n+1} = \dfrac{(-1)^n \, 2^n \, n!}{(2n+1)!}$ for

$n = 0, 1, 2, \ldots$. Thus the solution to the initial-value problem is $y(x) = \displaystyle\sum_{n=0}^{\infty} c_n x^n = \sum_{n=0}^{\infty} \frac{(-1)^n \, 2^n \, n!}{(2n+1)!} x^{2n+1}$.

17. Here the initial-value problem is $2Q'' + 40Q' + 400Q = 12$, $Q(0) = 0.01$, $Q'(0) = 0!$. Then
$Q_c(t) = e^{-10t}(c_1 \cos 10t + c_2 \sin 10t)$ and we try $Q_p(t) = A!$. Thus the general solution is
$Q(t) = e^{-10t}(c_1 \cos 10t + c_2 \sin 10t) + \frac{3}{100}!$. But $0.01 = Q(0) = c_1 + 0.03$ and $0 = Q'(0) = -10c_1 + 10c_2$,
so $c_1 = -0.02 = c_2!$. Hence the charge is given by $Q(t) = -0.02e^{-10t}(\cos 10t + \sin 10t) + 0.03$.

19. (a) Since we are assuming that the earth is a solid sphere of uniform density, we can calculate the density ρ as

follows: $\rho = \dfrac{\text{mass of earth}}{\text{volume of earth}} = \dfrac{M}{\frac{4}{3}\pi R^3}$. If V_r is the volume of the portion of the earth which lies within a

distance r of the center, then $V_r = \frac{4}{3}\pi r^3$ and $M_r = \rho V_r = \dfrac{Mr^3}{R^3}$. Thus $F_r = -\dfrac{GM_r m}{r^2} = -\dfrac{GMm}{R^3} r$.

(b) The particle is acted upon by a varying gravitational force during its motion. By Newton's Second Law of

Motion, $m\dfrac{d^2 y}{dt^2} = F_y = -\dfrac{GMm}{R^3} y$, so $y''(t) = -k^2 y(t)$ where $k^2 = \dfrac{GM}{R^3}$. At the surface,

$-mg = F_R = -\dfrac{GMm}{R^2}$, so $g = \dfrac{GM}{R^2}$. Therefore $k^2 = \dfrac{g}{R}$.

(c) The differential equation $y'' + k^2 y = 0$ has auxiliary equation $r^2 + k^2 = 0$. (This is the r of Section 18.1
[ET 17.1], not the r measuring distance from the earth's center.) The roots of the auxiliary equation are $\pm ik$, so
by (11) in Section 18.1 [ET 17.1], the general solution of our differential equation for t is
$y(t) = c_1 \cos kt + c_2 \sin kt$. It follows that $y'(t) = -c_1 k \sin kt + c_2 k \cos kt$. Now $y(0) = R$ and $y'(0) = 0$,
so $c_1 = R$ and $c_2 k = 0$. Thus $y(t) = R \cos kt$ and $y'(t) = -kR \sin kt$. This is simple harmonic motion (see
Section 18.3 [ET 17.3]) with amplitude R, frequency k, and phase angle 0. The period is $T = 2\pi/k$.
$R \approx 3960$ mi $= 3960 \cdot 5280$ ft and $g = 32$ ft/s^2, so $k = \sqrt{g/R} \approx 1.24 \times 10^{-3}\,\text{s}^{-1}$ and
$T = 2\pi/k \approx 5079$ s ≈ 85 min.

(d) $y(t) = 0 \quad \Leftrightarrow \quad \cos kt = 0 \quad \Leftrightarrow \quad kt = \frac{\pi}{2} + \pi n$ for some integer $n \quad \Rightarrow$
$y'(t) = -kR \sin\left(\frac{\pi}{2} + \pi n\right) = \pm kR$. Thus the particle passes through the center of the earth with speed
$kR \approx 4.899$ mi/s $\approx 17,600$ mi/h.

Appendixes

G Complex Numbers

1. $(3+2i)+(7-3i)=(3+7)+(2-3)\,i=10-i$

3. $(3-i)\,(4+i)=12+3i-4i-(-1)=13-i$

5. $\overline{12+7i}=12-7i$

7. $\dfrac{2+3i}{1-5i}=\dfrac{2+3i}{1-5i}\cdot\dfrac{1+5i}{1+5i}=\dfrac{2+10i+3i+15\,(-1)}{1-25\,(-1)}=\dfrac{-13+13i}{26}=-\tfrac{1}{2}+\tfrac{1}{2}i$

9. $\dfrac{1}{1+i}=\dfrac{1}{1+i}\cdot\dfrac{1-i}{1-i}=\dfrac{1-i}{1-(-1)}=\dfrac{1-i}{2}=\dfrac{1}{2}-\dfrac{1}{2}i$

11. $i^3=i^2\cdot i=(-1)\,i=-i$

13. $\sqrt{-25}=\sqrt{25}\,i=5i$

15. $\overline{3+4i}=3-4i,\ |3+4i|=\sqrt{3^2+4^2}=\sqrt{25}=5$

17. $\overline{-4i}=\overline{0-4i}=0+4i=4i,\ |-4i|=\sqrt{0^2+(-4)^2}=4$

19. $4x^2+9=0\ \Leftrightarrow\ 4x^2=-9\ \Leftrightarrow\ x^2=-\tfrac{9}{4}\ \Leftrightarrow\ x=\pm\sqrt{-\tfrac{9}{4}}=\pm\sqrt{\tfrac{9}{4}}\,i=\pm\tfrac{3}{2}i.$

21. By the quadratic formula, $x^2-8x+17=0\ \Leftrightarrow$

$$x=\frac{8\pm\sqrt{(-8)^2-4\,(1)\,(17)}}{2\,(1)}=\frac{8\pm\sqrt{-4}}{2}=\frac{8\pm 2i}{2}=4\pm i.$$

23. By the quadratic formula, $z^2+z+2=0\ \Leftrightarrow\ z=\dfrac{-1\pm\sqrt{1^2-4\,(1)\,(2)}}{2\,(1)}=\dfrac{-1\pm\sqrt{-7}}{2}=-\dfrac{1}{2}\pm\dfrac{\sqrt{7}}{2}i.$

25. For $z=-3+3i,\ r=\sqrt{(-3)^2+3^2}=3\sqrt{2}$ and $\tan\theta=\tfrac{3}{-3}=-1\ \Rightarrow\ \theta=\tfrac{3}{4}\pi$ (since z lies in the second quadrant). Therefore, $-3+3i=3\sqrt{2}\left(\cos\tfrac{3\pi}{4}+i\sin\tfrac{3\pi}{4}\right).$

27. For $z=3+4i,\ r=\sqrt{3^2+4^2}=5$ and $\tan\theta=\tfrac{4}{3}\ \Rightarrow\ \theta=\tan^{-1}\tfrac{4}{3}$ (since z lies in the second quadrant). Therefore, $3+4i=5\left[\cos\left(\tan^{-1}\tfrac{4}{3}\right)+i\sin\left(\tan^{-1}\tfrac{4}{3}\right)\right].$

29. For $z = \sqrt{3} + i$, $r = \sqrt{\left(\sqrt{3}\right)^2 + 1^2} = 2$ and $\tan\theta = \frac{1}{\sqrt{3}}$ $\Rightarrow$ $\theta = \frac{\pi}{6}$ $\Rightarrow$ $z = 2\left(\cos\frac{\pi}{6} + i\sin\frac{\pi}{6}\right)$. For

$w = 1 + \sqrt{3}i$, $r = 2$ and $\tan\theta = \sqrt{3}$ $\Rightarrow$ $\theta = \frac{\pi}{3}$ $\Rightarrow$ $w = 2\left(\cos\frac{\pi}{3} + i\sin\frac{\pi}{3}\right)$.

Therefore, $zw = 2 \cdot 2\left[\cos\left(\frac{\pi}{6} + \frac{\pi}{3}\right) + i\sin\left(\frac{\pi}{6} + \frac{\pi}{3}\right)\right] = 4\left(\cos\frac{\pi}{2} + i\sin\frac{\pi}{2}\right)$,

$z/w = \frac{2}{2}\left[\cos\left(\frac{\pi}{6} - \frac{\pi}{3}\right) + i\sin\left(\frac{\pi}{6} - \frac{\pi}{3}\right)\right] = \cos\left(-\frac{\pi}{6}\right) + i\sin\left(-\frac{\pi}{6}\right)$, and $1 = 1 + 0i = 1\left(\cos 0 + i\sin 0\right)$ $\Rightarrow$

$1/z = \frac{1}{2}\left[\cos\left(0 - \frac{\pi}{6}\right) + i\sin\left(0 - \frac{\pi}{6}\right)\right] = \frac{1}{2}\left[\cos\left(-\frac{\pi}{6}\right) + i\sin\left(-\frac{\pi}{6}\right)\right]$. For $1/z$, we could also use the formula

that precedes Example 5 to obtain $1/z = \frac{1}{8}\left(\cos\frac{\pi}{6} - i\sin\frac{\pi}{6}\right)$.

31. For $z = 2\sqrt{3} - 2i$, $r = \sqrt{\left(2\sqrt{3}\right)^2 + (-2)^2} = 4$ and $\tan\theta = \frac{-2}{2\sqrt{3}} = -\frac{1}{\sqrt{3}}$ $\Rightarrow$

$\theta = -\frac{\pi}{6}$ $\Rightarrow$ $z = 4\left[\cos\left(-\frac{\pi}{6}\right) + i\sin\left(-\frac{\pi}{6}\right)\right]$. For $w = -1 + i$, $r = \sqrt{2}$,

$\tan\theta = \frac{1}{-1} = -1$ $\Rightarrow$ $\theta = \frac{3\pi}{4}$ $\Rightarrow$ $z = \sqrt{2}\left(\cos\frac{3\pi}{4} + i\sin\frac{3\pi}{4}\right)$. Therefore,

$zw = 4\sqrt{2}\left[\cos\left(-\frac{\pi}{6} + \frac{3\pi}{4}\right) + i\sin\left(-\frac{\pi}{6} + \frac{3\pi}{4}\right)\right] = 4\sqrt{2}\left(\cos\frac{7\pi}{12} + i\sin\frac{7\pi}{12}\right)$,

$z/w = \frac{4}{\sqrt{2}}\left[\cos\left(-\frac{\pi}{6} - \frac{3\pi}{4}\right) + i\sin\left(-\frac{\pi}{6} - \frac{3\pi}{4}\right)\right] = \frac{4}{\sqrt{2}}\left[\cos\left(-\frac{11\pi}{12}\right) + i\sin\left(-\frac{11\pi}{12}\right)\right]$

$= 2\sqrt{2}\left(\cos\frac{13\pi}{12} + i\sin\frac{13\pi}{12}\right)$, and

$1/z = \frac{1}{4}\left[\cos\left(-\frac{\pi}{6}\right) - i\sin\left(-\frac{\pi}{6}\right)\right] = \frac{1}{4}\left(\cos\frac{\pi}{6} + i\sin\frac{\pi}{6}\right)$.

33. For $z = 1 + i$, $r = \sqrt{2}$ and $\tan\theta = \frac{1}{1} = 1$ $\Rightarrow$ $\theta = \frac{\pi}{4}$ $\Rightarrow$ $z = \sqrt{2}\left(\cos\frac{\pi}{4} + i\sin\frac{\pi}{4}\right)$. So by

De Moivre's Theorem,

$$(1 + i)^{20} = \left[\sqrt{2}\left(\cos\frac{\pi}{4} + i\sin\frac{\pi}{4}\right)\right]^{20} = \left(2^{1/2}\right)^{20}\left(\cos\frac{20 \cdot \pi}{4} + i\sin\frac{20 \cdot \pi}{4}\right)$$

$$= 2^{10}\left(\cos 5\pi + i\sin 5\pi\right) = 2^{10}\left[-1 + i\left(0\right)\right] = -2^{10} = -1024$$

35. For $z = 2\sqrt{3} + 2i$, $r = 4$ and $\tan\theta = \frac{2}{2\sqrt{3}} = \frac{1}{\sqrt{3}}$ $\Rightarrow$ $\theta = \frac{\pi}{6}$ $\Rightarrow$ $z = 4\left(\cos\frac{\pi}{6} + i\sin\frac{\pi}{6}\right)$. So by

De Moivre's Theorem,

$$\left(2\sqrt{3} + 2i\right)^5 = \left[4\left(\cos\frac{\pi}{6} + i\sin\frac{\pi}{6}\right)\right]^5 = 4^5\left(\cos\frac{5\pi}{6} + i\sin\frac{5\pi}{6}\right) = 1024\left[-\frac{\sqrt{3}}{2} + \frac{1}{2}i\right] = -512\sqrt{3} + 512i$$

37. $1 = 1 + 0i = 1\left(\cos 0 + i\sin 0\right)$. Using Equation 3 with $r = 1$, $n = 8$, and $\theta = 0$, we have

$$w_k = 1^{1/8}\left[\cos\left(\frac{0 + 2k\pi}{8}\right) + i\sin\left(\frac{0 + 2k\pi}{8}\right)\right] = \cos\frac{k\pi}{4} + i\sin\frac{k\pi}{4}, \text{ where } k = 0, 1, 2, \ldots, 7.$$

$w_0 = 1\left(\cos 0 + i\sin 0\right) = 1$, $w_1 = 1\left(\cos\frac{\pi}{4} + i\sin\frac{\pi}{4}\right) = \frac{1}{\sqrt{2}} + \frac{1}{\sqrt{2}}i$,

$w_2 = 1\left(\cos\frac{\pi}{2} + i\sin\frac{\pi}{2}\right) = i$, $w_3 = 1\left(\cos\frac{3\pi}{4} + i\sin\frac{3\pi}{4}\right) = -\frac{1}{\sqrt{2}} + \frac{1}{\sqrt{2}}i$,

$w_4 = 1\left(\cos\pi + i\sin\pi\right) = -1$, $w_5 = 1\left(\cos\frac{5\pi}{4} + i\sin\frac{5\pi}{4}\right) = -\frac{1}{\sqrt{2}} - \frac{1}{\sqrt{2}}i$,

$w_6 = 1\left(\cos\frac{3\pi}{2} + i\sin\frac{3\pi}{2}\right) = -i$, $w_7 = 1\left(\cos\frac{7\pi}{4} + i\sin\frac{7\pi}{4}\right) = \frac{1}{\sqrt{2}} - \frac{1}{\sqrt{2}}i$

39. $i = 0 + i = 1\left(\cos\frac{\pi}{2} + i\sin\frac{\pi}{2}\right)$. Using Equation 3 with $r = 1$, $n = 3$, and $\theta = \frac{\pi}{2}$, we have

$$w_k = 1^{1/3}\left[\cos\left(\frac{\frac{\pi}{2} + 2k\pi}{3}\right) + i\sin\left(\frac{\frac{\pi}{2} + 2k\pi}{3}\right)\right], \text{ where } k = 0, 1, 2.$$

$$w_0 = \left(\cos\frac{\pi}{6} + i\sin\frac{\pi}{6}\right) = \frac{\sqrt{3}}{2} + \frac{1}{2}i$$

$$w_1 = \left(\cos\frac{5\pi}{6} + i\sin\frac{5\pi}{6}\right) = -\frac{\sqrt{3}}{2} + \frac{1}{2}i$$

$$w_2 = \left(\cos\frac{9\pi}{6} + i\sin\frac{9\pi}{6}\right) = -i$$

41. Using Euler's formula (6) with $y = \frac{\pi}{2}$, we have $e^{i\pi/2} = \cos\frac{\pi}{2} + i\sin\frac{\pi}{2} = 0 + 1i = i$.

43. Using Euler's formula with $y = \frac{3\pi}{4}$, we have $e^{i3\pi/4} = \cos\frac{3\pi}{4} + i\sin\frac{3\pi}{4} = -\frac{1}{\sqrt{2}} + \frac{1}{\sqrt{2}}i$.

45. Using Equation 7 with $x = 2$ and $y = \pi$, we have $e^{2+i\pi} = e^2 e^{i\pi} = e^2\left(\cos\pi + i\sin\pi\right) = e^2\left(-1 + 0\right) = -e^2$.

47. $F(x) = e^{rx} = e^{(a+bi)x} = e^{ax+bxi} = e^{ax}\left(\cos bx + i\sin bx\right) = e^{ax}\cos bx + i\left(e^{ax}\sin bx\right) \Rightarrow$

$\quad F'(x) = \left(e^{ax}\cos bx\right)' + i\left(e^{ax}\sin bx\right)' = \left(ae^{ax}\cos bx - be^{ax}\sin bx\right) + i\left(ae^{ax}\sin bx + be^{ax}\cos bx\right)$

$\qquad = a\left[e^{ax}\left(\cos bx + i\sin bx\right)\right] + b\left[e^{ax}\left(-\sin bx + i\cos bx\right)\right] = ae^{rx} + b\left[e^{ax}\left(i^2\sin bx + i\cos bx\right)\right]$

$\qquad = ae^{rx} + bi\left[e^{ax}\left(\cos bx + i\sin bx\right)\right] = ae^{rx} + bie^{rx} = (a + bi)e^{rx} = re^{rx}$